MicroRNAs: Key Regulators of Oncogenesis

Sadegh Babashah
Editor

MicroRNAs: Key Regulators of Oncogenesis

Springer

Editor
Sadegh Babashah
Department of Molecular Genetics
Faculty of Biological Sciences
Tarbiat Modares University
Tehran, Iran

ISBN 978-3-319-03724-0 ISBN 978-3-319-03725-7 (eBook)
DOI 10.1007/978-3-319-03725-7
Springer Cham Heidelberg New York Dordrecht London

Library of Congress Control Number: 2014933208

Printed on acid-free paper

Springer is part of Springer Science+Business Media (www.springer.com)

This book is honorably dedicated to Prof. Majid Sadeghizadeh, head of Department of Molecular Genetics, Faculty of Biological Sciences, Tarbiat Modares University. Prof. Sadeghizadeh founded the PhD program for molecular genetics at Tarbiat Modares University and is known to be one of the pioneers in the field of molecular biology in Iran.

Preface

The past decade witnessed significant efforts and progresses in the area of microRNA (miRNA) research in the field of cancer. Indeed, recent miRNA studies have been a major leap in the understanding of the molecular pathogenesis of cancer. It is crystal clear that miRNAs as either onco- or tumor suppressor genes can alter biological processes fundamental to tumor initiation and progression. The connections between miRNAs and oncogenesis are widespread enough to hold miRNAs as potential therapeutic targets and novel biomarkers. In this regard, identifying the expression signatures of miRNAs provides exciting opportunities for the diagnosis, prognosis, and therapy of cancer.

This edited book *MicroRNAs: Key Regulators of Oncogenesis* aims to offer a broad framework to understand the state-of-the-art knowledge of miRNA function and illustrate features of specific miRNAs in the regulation of cancer. It has been written for graduate students, postdoctoral fellows, and scientists in cancer research, and it is also well suited for clinical oncologists and other researchers interested in this field. The contents of this book was scrupulously designed and explicitly written. Ranging from the fundamental concepts to clinical applications, this book is composed of seventeen chapters organized in two parts. The first part is devoted to delving deep into the importance of miRNAs in cancer biology. In Chap. 1, I describe the dual function of miRNAs as either oncogenes or tumor suppressors in cancer and elucidate the link between aberrant miRNA expression and cancer development and progression. Since cancer is associated with accumulation of epigenetic and genetic alterations, Chap. 2 focuses on the relationship between epigenetics and miRNA. In line with this, Chap. 3 deals exclusively with the effects of dietary agents on miRNAs and their targets in the context of cancer biology. Chapter 4 elaborates on the interplay between miRNAs and oncogenes/tumor suppressors in tumor metabolism introducing another layer of complexity to the regulatory network of metabolic pathways in cancer cells. In Chaps. 5 and 6, authors clearly describe the crucial roles of miRNAs in different types of solid tumors and hematological malignancies and also discuss the feasibility of using miRNAs as potential biomarkers. The importance of miRNAs in the pathogenesis of oncogenic viruses and the link between miRNAs and liver inflammation during hepatocarcinogenesis is represented in Chaps. 7 and 8, respectively. Chapter 9 deciphers the mechanisms through which miRNAs modulate the activity of regulatory pathways in tumorigenesis, in particular miRNAs targeting potential pathways for therapeutic intervention. Chapter 10 reveals the regulatory mechanisms of miRNAs in apoptotic and autophagic networks, with the merit of finding application as potential drug targets for future cancer therapy. Chapter 11 describes the part played by miRNAs in drug resistance and drug sensitivity. In Chap. 12, the potential role of cancer stem cell-related miRNAs during tumor development and progression is clarified. The second part of the book highlights the clinical implications of miRNAs in cancer. In this part, Chap. 13 introduces miRNAs as potential biomarkers for diagnosis, prognosis and therapeutic intervention of cancer and the capacity of integrating miRNA data into clinical trials is discussed. Chapter 14 offers a precise description to the function of miRNAs in breast cancer and discusses their possible translation into molecular diagnostics. Chapter 15 particularizes the impact of miRNAs on drug resistance from a clinical point of view, thereby highlighting their capability to be exploited as predictors or modifiers of resistance towards chemo- and radio-therapeutics. In Chap. 16, authors meticulously address a variety of nanocarriers exploited for miRNA delivery in crafting therapeutic platforms for cancer and provide an in-depth analysis of different attributes of these nanovehicles. Finally, Chap. 17 outlines the advantages of a pulmonary drug delivery system and the strategies for miRNA-based treatment of lung cancer.

This book would not have come to fruition without the continuous support and administrative assistance of Melania Ruiz, along with additional administrative help by Marleen Moor from Springer International Publishing Switzerland. I also want to thank Dr. Seyed Javad Mowla, Dr. Masoud Soleimani and Dr. William Cho for their valuable comments during the editing process of the book. Ultimately, I wish to express my profound gratitude to all the authors for their time and efforts in bringing this project to completion. I am truly honored to have the opportunity to work with such a prestigious team.

Sadegh Babashah, Ph.D.

Department of Molecular Genetics
Faculty of Biological Sciences
Tarbiat Modares University, Tehran, Iran
January, 2014

Contents

Contributors

Aamir Ahmad Department of Pathology, Karmanos Cancer Institute, Wayne State University, Detroit, MI, USA

Shadan Ali Department of Oncology, Karmanos Cancer Institute, Wayne State University, Detroit, MI, USA

Sumit Arora Department of Urology, San Francisco Veterans Affairs Medical Center and University of California, San Francisco, CA, USA

Asfar S. Azmi Department of Pathology, Karmanos Cancer Institute, Wayne State University, Detroit, MI, USA

Sadegh Babashah Department of Molecular Genetics, Faculty of Biological Sciences, Tarbiat Modares University, Tehran, Iran

Sahar Babashah Department of Pharmacology and Toxicology, Pharmaceutical Sciences Branch and Pharmaceutical Sciences Research Center, Islamic Azad University, Tehran, Iran

Babak Bakhshinejad Department of Molecular Genetics, Faculty of Biological Sciences, Tarbiat Modares University, Tehran, Iran

Sanjeev Banerjee Department of Pathology, Karmanos Cancer Institute, Wayne State University, Detroit, MI, USA

Bin Bao Department of Pathology, Karmanos Cancer Institute, Wayne State University, Detroit, MI, USA

Yang Cao School of Life Science, University of Science and Technology of China, Hefei, China

William C. Cho Department of Clinical Oncology, Queen Elizabeth Hospital, Hong Kong, China

Sabine Colnot Oncogenesis of Digestive Epithelia, Inserm, U1016, Institut Cochin, Paris, France

Cnrs, UMR8104, Paris, France

Université Paris Descartes, Sorbonne Paris Cité, Paris, France

Rajvir Dahiya Department of Urology, San Francisco Veterans Affairs Medical Center and University of California, San Francisco, CA, USA

Kaiyuan Deng Department of General Surgery and Center of Translational Medicine, Nanjing Medical University Affiliated Wuxi Second Hospital, Wuxi, Jiangsu, China

Gianpiero Di Leva Department of Molecular Immunology, Virology and Medical Genetics, Ohio State University, Columbus, OH, USA

Manuela Ferracin Department of Morphology, Surgery and Experimental Medicine, Laboratory for Technologies of Advanced Therapies, University of Ferrara, Ferrara, Italy

Yu Fujita Division of Molecular and Cellular Medicine, National Cancer Center Research Institute, Tsukiji, Chuo-ku, Tokyo, Japan

Division of Respiratory Diseases, Department of Internal Medicine, Jikei University School of Medicine, Nishi-shinbashi, Minato-ku, Tokyo, Japan

Ping Gao School of Life Science, University of Science and Technology of China, Hefei, China

Division of Hematology, Department of Medicine, The Johns Hopkins University School of Medicine, Baltimore, MD, USA

Michela Garofalo Department of Molecular Immunology, Virology and Medical Genetics, Ohio State University, Columbus, OH, USA

Angélique Gougelet Oncogenesis of Digestive Epithelia, Inserm, U1016, Institut Cochin, Paris, France

Cnrs, UMR8104, Paris, France

Université Paris Descartes, Sorbonne Paris Cité, Paris, France

Xiaoqiang Guo Department of General Surgery and Center of Translational Medicine, Nanjing Medical University Affiliated Wuxi Second Hospital, Wuxi, Jiangsu, China

Joerg Haier Comprehensive Cancer Center Muenster (CCCM), University Hospital Muenster, Muenster, Germany

Xiaoping He School of Life Science, University of Science and Technology of China, Hefei, China

Richard Hummel Department of General and Visceral Surgery, University Hospital Muenster, Muenster, Germany

Hisashi Iizasa Division of Stem Cell Biology, Institute for Genetic Medicine, Hokkaido University, Sapporo, Japan

Mohammad Amin Javidi Department of Molecular Genetics, Faculty of Biological Sciences, Tarbiat Modares University, Tehran, Iran

Teru Kanda Division of Virology, Aichi Cancer Center Research Institute, Nagoya, Japan

Dejuan Kong Department of Pathology, Karmanos Cancer Institute, Wayne State University, Detroit, MI, USA

Kazuyoshi Kuwano Division of Respiratory Diseases, Department of Internal Medicine, Jikei University School of Medicine, Nishi-shinbashi, Minato-ku, Tokyo, Japan

Charles H. Lawrie Oncology Research Area, Biodonostia Research Institute, San Sebastián, Spain

Ikerbasque, Basque Foundation for Science, Bilbao, Spain

Nuffield Department of Clinical Laboratory Sciences, University of Oxford, Oxford, UK

Yiwei Li Department of Pathology, Karmanos Cancer Institute, Wayne State University, Detroit, MI, USA

Yong Li College of Life Science, Ningxia University, Yinchuan, Ningxia, China

Raquel T. Lima Cancer Drug Resistance Group, Instituto de Patologia e Imunologia Molecular da Universidade do Porto (IPATIMUP), Porto, Portugal

Centro de Química Medicinal da Universidade do Porto (CEQUIMED-UP), Faculdade de Farmácia, Universidade do Porto, Porto, Portugal

Kirsten Lindner Department of General and Visceral Surgery, University Hospital Muenster, Muenster, Germany

Bo Liu State Key Laboratory of Biotherapy, West China Hospital, Sichuan University, Chengdu, China

Xiaoming Liu Institute of Stem Cell Research, General Hospital of Ningxia Medical University, Yinchuan, Ningxia, China

College of Life Science, Ningxia University, Yinchuan, Ningxia, China

Laura Lupini Department of Morphology, Surgery and Experimental Medicine, Laboratory for Technologies of Advanced Therapies, University of Ferrara, Ferrara, Italy

Shahana Majid Department of Urology, San Francisco Veterans Affairs Medical Center and University of California, San Francisco, CA, USA

Yoshiki Murakami Department of Hepatology, Graduate School of Medicine, Osaka City University, Osaka, Japan

Takahiro Ochiya Division of Molecular and Cellular Medicine, National Cancer Center Research Institute, Tsukiji, Chuo-ku, Tokyo, Japan

Sharanjot Saini Department of Urology, San Francisco Veterans Affairs Medical Center and University of California, San Francisco, CA, USA

Fazlul H. Sarkar Department of Pathology, Karmanos Cancer Institute, Wayne State University, Detroit, MI, USA

Department of Oncology, Karmanos Cancer Institute, Wayne State University, Detroit, MI, USA

Hugo Seca Cancer Drug Resistance Group, Instituto de Patologia e Imunologia Molecular da Universidade do Porto (IPATIMUP), Porto, Portugal

Laboratório de Microbiologia, Departamento de Ciências Biológicas, Faculdade de Farmácia, Universidade do Porto, Porto, Portugal

Linchong Sun School of Life Science, University of Science and Technology of China, Hefei, China

Muriel Thirion Division of Molecular and Cellular Medicine, National Cancer Center Research Institute, Tsukiji, Chuo-ku, Tokyo, Japan

M. Helena Vasconcelos Cancer Drug Resistance Group, Instituto de Patologia e Imunologia Molecular da Universidade do Porto (IPATIMUP), Porto, Portugal

Laboratório de Microbiologia, Departamento de Ciências Biológicas, Faculdade de Farmácia, Universidade do Porto, Porto, Portugal

Jun Wei Institute of Stem Cell Research, General Hospital of Ningxia Medical University, Yinchuan, Ningxia, China

Yu-quan Wei State Key Laboratory of Biotherapy, West China Hospital, Sichuan University, Chengdu, China

Jürgen Wittmann Division of Molecular Immunology, Department of Internal Medicine III, Nikolaus-Fiebiger Center of Molecular Medicine, University of Erlangen-Nürnberg, Erlangen, Germany

Jiazeng Xia Department of General Surgery and Center of Translational Medicine, Nanjing Medical University Affiliated Wuxi Second Hospital, Wuxi, Jiangsu, China

Huafeng Zhang School of Life Science, University of Science and Technology of China, Hefei, China

Abbreviations

A

α-SMA	Smooth muscle alpha-actin
AA	Arachidonic acid
AAT	Androgen ablation therapy
AAV	Adeno-associated virus
ABCA1	ATP-binding cassette transporter sub-family A member 1
ABCB1	ATP-binding cassette, sub-family B (MDR/TAP), member 1
ABCC	Multidrug resistance-associated protein
ABCG2	ATP-binding cassette sub-family G member 2
AD	Adenocarcinoma
ADAM17	A Disintegrin and Metalloprotease-17
AdenoCA	Adenocarcinoma
ADM	Adriamycin
ADMA	Asymmetric ω-N^{G}, N^{G}-dimethylarginine
AFP	Alpha-fetoprotein
Ago	Argonaute
AI	Androgen independence
AID	Activation-induced cytidine deaminase
ALCL	Anaplastic large cell lymphoma
ALD	Alcoholic liver disease
ALDH1A3	Aldehyde dehydrogenase 1 family, member A3
ALK	Activin receptor-like kinase
ALL	Acute lymphoblastic leukemia
Ambra1	activating molecule in Beclin-1-regulated autophagy protein 1
AML	Acute myeloid leukemia
AMPK	AMP-activated protein kinase
AnD	Androgen-dependent
AnI	Androgen-independent
ANP32A	Acidic nuclear phosphoprotein 32 family member A
AP-1	Activator protein-1

Apaf-1	Apoptotic peptidase activating factor 1
APC	Adenomatous Polyposis Coli
APL	Acute promyelocytic leukemia
APO	Apolipoprotein
AR	Androgen receptor
ARHs	ADP-ribosylhydrolases
ARTs	ADP-ribosyltransferases
ASO	Antisense oligonucleotide
Atg	Autophagy-related gene
ATM	Ataxia telangiectasia mutated
ATO	Arsenic trioxide
ATRA	All-trans-retinoic acid
5-aza-dC	5-aza-2′-deoxycytidine

B

B-CLL	B-cell chronic lymphocytic leukemia
BAc	Barrett's adenocarcinoma
BAK1	Bcl-2 homologous antagonist/killer
BBC3	BCL2 binding component 3
BC	Breast cancer
BCL-2	B-cell leukemia/lymphoma-2BE, Barrett's esophagus
BCL6	B-cell lymphoma 6
BCLAF1	Bcl-2-associated transcription factor 1
BDNF	Brain-derived neurotrophic factor
Bif-1	BAX-interacting factor-1
BIM	BCL2-like 11
BL	Burkitt lymphoma
BM	Barrett's mucosa
BMF	Bcl-2-modifying factor
BMP7	Bone morphogenetic protein 7
BNIP2	BCL2/adenovirus E1B 19 kDa protein-interacting protein 2
BR-DIM	BioResponse 3,3′-diindolylmethane
BT-ICs	Breast tumor-initiating cells

C

c-FLIP	Cellular FLICE-like inhibitory protein
CAB39	Calcium binding protein 39
CAFs	Cancer associated fibroblasts
cALCL	Cutaneous anaplastic large cell lymphoma
CBP	CREB-binding protein
cccDNA	Covalently closed circular DNA
CCND1	Cyclin D1
CCNG1	Cyclin G1
CCNJ	Cyclin J

CDC25A Cell division cycle 25 homolog A
CDC42 Cell division cycle 42
CDF Diflourinated-curcumin
CDK6 Cyclin-dependent protein kinase 6
CDKN1A Cyclin-Dependent Kinase Inhibitor 1A
CDKN1B Cyclin-Dependent Kinase Inhibitor 1B ($p27^{Kip1}$)
CDS2 CDP-diacylglycerol synthase (phosphatidate cytidylyltransferase) 2
CDX2 Caudal-related homeobox 2
CEBPA CCAAT/enhancer binding protein alpha
CFUs Colony forming units
CGI CpG islands
CHC Chronic hepatitis c
ChIP Chromatin immunoprecipitation
CIN Cervical intraepithelial neoplasia
CLL chronic lymphocytic leukemia
CML Chronic myeloid leukemia
CMPs Common myeloid progenitors
CNS Central Nervous System
COL2A1 Collagen, type II, alpha 1
COPD Chronic obstructive pulmonary disease
COPZ2 Coatomer protein complex, subunit zeta 2
COX-2 cyclooxygenase-2
CPT -1 Carnitine palmitoyl transferase-1
CPT1A Carnitine palmitoyltransferase 1A
CRC Colorectal cancer
CROT Carnitine O-octanoyltransferase
CSCs Cancer stem cells
CSF-1 Colony stimulating factor-1
CSLCs Cancer stem-like cells
CSS Cancer-specific survival
CTGF Connective tissue growth factor
CTX Cyclophosphamide

D

DEX Dexamethasone
DFRS Distant relapse-free survival
DHA Docosahexaenoic acid
DHFR Dihydrofolate reductase
DIM Dimeric product, 3,3′-diindolylmethane
DKC1 Dyskeratosis congenita gene 1
DLBCL Diffuse Large B-cell Lymphoma
DNMT DNA methyltransferase
DNMT3L DNA methylatransferase 3-like
DNR Daunorubicin

DOT1L	DOT1-like protein
DOX	Doxorubicin
DPIs	Dry powder inhalers
DR4	Death receptor-4
DTX	Docetaxel
DUBs	Deubiquitinating enzymes
DUSP5	Dual-specificity phosphatase 5

E

E2F	E2 transcription factor family
EAC	Esophageal adenocarcinoma
EAG1	Ether-à-go-go 1
EBV	Epstein-Barr virus
EC	Esophageal cancer
ECM	Extracellular matrix
eEF1A2	Eukaryotic translation elongation factor 1A2
EGCG	Epigallocatechingallate
EGF	Epidermal growth factor
EGFR	Epidermal growth factor receptor
EMT	Epithelial to mesenchymal transition
EOC	Epithelial ovarian cancer
EPA	Eicosapentaenoic acid
EPR	Enhanced permeability and retention
ER	Estrogen receptor
ERK5	Extracellular signal regulated kinase 5
ERRFI-1	ErbB receptor inhibitor-1
ESCC	Esophageal squamous cell carcinoma
EVI1	Ecotropic Virus Integration 1
EVL	Ena/Vasp-like
EZH2	Enhancer of zeste homolog 2

F

FAD	Flavin adenine dinucleotides
FAK	Focal Adhesion Kinase
FDA	Food and Drug Administration
FFPE	Formalin-fixed paraffin embedded
FGFR1	Fibroblast growth factor receptor 1
FIP200	family interacting protein of 200 kDa
FL	Follicular Lymphoma
FNDC3B	Fibronectin type III domain containing 3B
FOXC2	Forkhead box C2
FOXF2	Forkhead Box F2
FOXO3A	Forkhead box O3
FOXO4	Forkhead box O4

FOXP1	Forkhead Box P1
Fra-1	Fos-related antigen 1
FSCN1	Fascin Homolog 1
5-FU	5-fluorouracil
Fzd3	Frizzled3

G

GAX	Growth arrest homeobox
GBM	Glioblastoma multiforme
GCN5	General control non-derepressible 5
GEMIN4	Component of gems 4
GLS1	Glutaminase 1
GLUT	Glucose transporter
GNAT	General N-acetyl-transferase
GSTP1	Glutathione S-transferase P1

H

HADHB	Hydroxyacyl-CoA dehydrogenase-3-ketoacyl-CoA thiolase-enoyl-CoA hydratase (trifunctional protein) β-subunit
HATs	Histone acetyl-transferases
HB-EGF	Heparin-binding EGF-like growth factor
HBP1	HMG box-containing protein 1
HBV	Hepatitis B virus
HBx	HBV X protein
HCC	Hepatocellular carcinoma
HCPT	Hydroxycamptothecin
HCV	Hepatitis C virus
HDACs	Histone deacetylases
HDL	High density lipoprotein
HDMs	Histone de-methylases
HER2	Human epidermal growth factor receptor 2
hESCs	Human embryonic stem cells
HGAL	Human Germinal-center Associated Lymphoma
HGS	Hepatocyte growth factor-regulated tyrosine kinase substrate
HIF-1α	Hypoxia-induced factor-1α
HIPK2	Homeodomain-interacting protein kinase-2
HK2	Hexokiase 2
HKMTs	Histone lysine methyltransferases
HLA	Human leukocyte antigen
HMEC	Human mammary epithelial cells
HMGA2	High mobility group AT-hook 2
hMSH2	Human mutS homolog 2
hMSH6	Human mutS homolog 6
HMTs	Histone methyl-transferases

HNSCC	Head and neck squamous cell carcinoma
HOXD10	Homeobox D10
HPV	Human papillomavirus
HSCs	Hematopoietic stem cells
HSCs	Hepatic stellate cells
HSPG2	Heparin Sulfate Proteoglycan 2
HSV	Herpes simplex virus

I

I3C	Indole-3-carbinol
IFITM1	Interferon-induced trans-membrane 1
IFN-α	Interferon-α
IGF1R	Insulin like growth factor 1 receptor
IL-1β	Interleukin-1β
Ilk	Integrin-linked kinase
IMP-1	Insulin-like growth factor 2 mRNA binding protein 1
IRAK1	Interleukin-1 receptor associated kinase 1
IRF	Interferon regulatory factor
ITGA5	Integrin α5

J

JHDM1	JmjC domain-containing histone demethylase 1
JMJD1A	Jumonji domain containing 1A

K

KEAP1	Kelch-like ECH-associated protein 1
KO	Knockout
KRAS	Kirsten rat sarcoma viral oncogene homolog
KS	Kaposi's sarcoma
KSHV	Kaposi's sarcoma-associated herpesvirus

L

L-OHP	Oxaliplatin
LA	Linoleic acid
LATS2	Large tumor suppressor 2
LC	Lung cancer
LCC	Large cell carcinoma
LE	Lovastatin plus ezetimibe
Lef-1	Lymphocyte enhancer transcription factor-1
LIN28A	*Lin-28* Homologue A
LKO	Liver-specific knockout
LMO2	LIM-only protein 2
LNA	Locked nucleic acid

lncRNA	long non-coding RNA
LPS	Lipopolysaccharide
LRRFIP1	Leucine rich repeat (in FLII) interacting protein 1
LSD1	Lysine-specific histone demethylase 1

M

M-RI P	Myosin phosphatase-Rho interacting protein
mAbs	Monoclonal antibodies
MAGE-A	Melanoma Antigen Family A
MALT	Mucosa-associated lymphoid tissue
MAPK	Mitogen-activated protein kinase
MAPT	Microtubule-associated protein tau
MARCKS	Myristoylated alanine-rich protein kinase c substrate
MAZ	Myc-associated zinc finger protein
MB	Medulloblastoma
MBDs	Methyl-CpG binding domains
MCL1	Myeloid cell leukemia 1
MCM	Minichromosome maintenance
mCRC	Metastatic colorectal cancer
MDIs	Metered-dose inhalers
MDS	Myelodysplastic syndrome
MDV	Marek's disease virus
MEF2C	Myocyte enhancer factor 2C
MET	Mesenchymal to epithelial cell transition
MIF	Macrophage inhibitory factor
MiR-SNPs	MicroRNA polymorphisms
MiRNA	MicroRNA
MLL	Mixed lineage leukemia
MM	Malignant melanoma
MM	Multiple myeloma
MMA	ω-N^{G}-monomethylarginine
MMC	Mitomycin C
MMP	Matrix metalloproteinase
2′-MOE	2′-O-methoxyethyl
MRI	Magnetic resonance imaging
MRP1	Multidrug resistance-associated protein 1
MSK1	Mitogen- and stress-activated protein kinase 1
MTDH	Metadherin
MTPN	Myotrophin
MTX	Methotrexate
MX	Mitoxantrone
MYB	Myeloblastosis
MYBL2	v-Myb myeloblastosis viral oncogene homolog-like2

N

NAD	Nicotinamide adenine dinucleotides
NAFLD	Non-alcoholic fatty liver disease
NASH	Non-alcoholic steatohepatitis
nc RNA	Non-coding RNA
NESTs	Normal esophageal squamous tissues
NF-κB	Nuclear factor-kappa B
NFIA	Nuclear factor I-A
NGS	Next-generation sequencing
NiS	Nickel sulfide
NLK	Nemo-like kinase
NK	Natural killer
NMZL	Nodal marginal zone lymphoma
NNMT	Nicotinamide N-methyltransferase
NPC	Nasopharyngeal carcinoma
NPC	Neural precursor cells
NR	Non-responder
NRAS	Neuroblastoma RAS viral oncogene homolog
NSCLC	Non-small cell lung cancer
NT	Non tumorous tissue
NZB strain	New Zealand Black strain

O

OC	Ovarian cancer
OCT4	Octamer-binding transcription factor 4
OncomiRs	Oncogenic microRNAs
OPN	Osteopontin
ORP	Oxysterol-binding-protein-related protein
ORR	Object response rate
OS	Overall survival
OSCC	Oral squamous cell carcinoma

P

p-Akt	Phosphorylated Akt
p53AIP1	p53-regulated apoptosis-inducing protein 1
PAA	Poly(amidoamine)
PAE	Poly (amino-co-ester)
PARGs	Poly-ADP-ribose-glycohydrolases
PARP	Poly ADP-ribose polymerase
4-PBA	4-phenylbutyric acid
PBMCs	Peripheral blood mononuclear cells
PC	Pancreatic cancer
PCa	Prostate cancer

PCR	Polymerase chain reaction
PDCD4	Programmed cell death protein 4
PDGFR	Platelet-derived growth factor receptor
PDK1	3-phosphoinositide-dependent protein kinase-1
PDMAEMA	Poly (2-N,N-dimethylaminoethylmethacrylate)
PDPK1	3′-phosphoinositidedependent protein kinase-1
PEG	Polyethylene glycol
PEI	Polyethyleneimine
PEITC	Phenethylisothiocyanate
PEL	Primary effusion lymphoma
PFS	Progression-free survival
PGI	Phosphogulcose isomerase
PgRNA	Pregenomic RNA
PHLPP	PH domain leucine-rich repeat protein phosphatase 2
PI3K	Phosphatidylinositol 3-kinase
PIK3R1	Phosphatidylinositol 3-kinase regulatory subunit alpha
PIK3R2	Phosphatidylinositol 3-kinase regulatory subunit Beta
piRNAs	PIWI-interacting RNAs
PKB	Protein kinase B
PKCε	P kinase C epsilon
PLA	Poly lactic acid
PLGA	Poly lactic-co-glycolic acid
PLK1	Serine/threonine-protein kinase
PLL	Poly-L-lysine
PLX4720	Specific inhibitor of B-RAF^{V600E}
PML	Promyelocytic leukemia
PP2A	Protein phosphatase 2
PPAR	Peroxisome proliferator-activated receptor
PPP2R1B	Protein Phosphatase 2, Regulatory Subunit A, Beta isoform
PPP2R2A	Protein Phosphatase 2, Regulatory Subunit B, Alpha isoform
PPP2R2A	PP2A regulatory subunit B α isoform
PR	Progesterone receptor
PRC	Polycomb Repressive Complex
PRDM1	PR domain zinc finger protein 1
PRKAA1	AMP-activated protein kinase catalytic subunit alpha-1
PRMTs	Protein arginine methyltransferases
Pre-miRNA	Precursor microRNA
Pri-miRNA	Primary microRNA
PSA	Prostate-specific antigen
PTEN	Phosphatase and tensin homolog
PTLD	Post-transplantation lymphoproliferative disorder
PTPN12	protein tyrosine phosphatase, non-receptor type 12
PTPN22	Protein tyrosine phosphatase, non-receptor type 22
PTX	Paclitaxel
PU-PEI	Polyurethane-short branch Polyethyleneimine
PUFAs	Polyunsaturated fatty acids

Q

Q-RT-PCR	Quantitative reverse transcriptase polymerase chain reaction

R

RA	Retinoic acid
RAB5A	Ras-related protein Rab-5A
RAP1A	Ras-related protein Rap-1A
RAR-α	Retinoic acid receptor-α
RASSF1A	Ras association domain family member 1A
Rb	Retinoblastoma
RBL2	Retinoblastoma-Like 2
RCC	Renal clear cell carcinoma
RdRp	RNA-dependent RNA polymerase
RDX	Radixin
RES	Reticuloendothelial system
RhoB	Ras homolog gene family member B
RIP-ChIP	Ribonucleoprotein chromatin immunoprecipitation
RISC	RNA-induced silencing complex
RKIP	the Raf kinase inhibitor protein
RNAi	RNA interference
ROS	Reactive oxygen species
RREB1	Ras-responsive element-binding protein
RRM2	Ribonucleotide reductase M2
RSSs	RNA-silencing suppressors
RSV	Respiratory syncytial virus
RTKN2	Rabbit polyclonal anti–Rhotekin 2
RVG	Rabies virus glycoprotein

S

S-TRAIL	Tumor necrosis factor related apoptosis inducing ligand
SAHA	Suberoylanilide hydroxamic acid
SAM	S-adenosylmethionine
SCC	Squamous cell carcinoma
SCLC	Small cell lung cancer
SDMA	Symmetric ω-N^{G}, N'^{G}-dimethylarginine
SECs	Sinusoidal endothelial cells
SFPQ	Splicing factor proline and glutamate-rich
SHIP	the SH2-domain-containing inositol 5-phosphatase
Shp	Small heterodimer partner
SIN	Self-inactivating
Sir2	Silent information regulator 2
siRNAs	Small interfering RNAs
SIRT1	Sirtuin 1

SIRT1	Silent mating type information regulation 2 homolog 1
SLC4A4	Na/bicarbonate cotransporter 1
SLN	Solid lipid nanoparticle
SMARCA4	SWI/SNF-related, matrix-associated, actin-dependent regulator chromatin, subfamily A, member 4
SMARCC1	SWI/SNF Related Matrix Associated actin dependent regulator of chromatin subfamily C member 1
snoRNAs	small nucleolar RNAs
SNP	Single nucleotide polymorphism
SOCS3	Suppressor of cytokine signaling 3
SOCS5	Suppressor of cytokine signaling 5
SOX4	SRY-related high-mobility group box 4
SPRED1	Sprouty-related protein 1
SPRY2	Sprouty 2
SRC	Sarcoma viral oncogene homolog
SREBP	Sterol regulatory element binding transcription factor
SRSF2	Serine/arginine-rich splicing factor 2
STARD13	StAR-related lipid transfer domain containing 13
STAT	Signal Transducer and activator of transcription
STMN1	Stathmin 1/oncoprotein 18
STS	Staurosporine
SuFu	suppressor of fused
SVR	Sustained virological response
SzS	Sézary syndrome

T

TAM	Tamoxifen
TCA	Tricarboxylic acid
TCF	T-cell factor
TCL1	T-cell leukemia 1
TCRα	T cell receptor alpha
TDX	Tomudex
TG	Triglyceride
TGF	Transforming growth factor
TGIF2	Transforming growth factor β-induced factor homeobox 2
THBS1	Thrombospondin1
TICs	Tumor initiating cells
TIMP3	Tissue inhibitor of metalloproteinase 3
TKIs	Tyrosine kinase inhibitors
TMZ	Temozolomide
TNC	Tenascin C
TNFα	Tumor necrosis factor α
TNFAIP3	Tumor necrosis factor-alpha-induced protein 3
TOP2B	DNA topoisomerase 2-beta

TP53INP1	Tumor protein p53 inducible nuclear protein 1
TPM1	Tropomyosin 1
TRAF6	tumor necrosis factor receptor associated factor 6
TRAIL	Tumor necrosis factor-related apoptosis-inducing ligand
TSA	Trichostain A
TSP-1	Thrombospondin-1
TWF1	Actin-binding protein twinfilin 1
TYMS	Thymidylate synthase

U

UCC	Urothelial carcinoma of the bladder
uPA	Urokinase-type plasminogen activator
3′ UTR	3′-Untranslated region
5′ UTR	5′-Untranslated region
UVRAG	UV irradiation resistance-associated gene

V

VCR	Vincristine
VDR	Vitamin D receptor
VE-cadherin	Vascular endothelial cadherin
VEGF	Vascular endothelial growth factor
VEGFR	Vascular endothelial growth factor receptor
VHL	Von Hippel-Lindau tumor-suppressor gene
VLP	Virus-like particle

W

WIF1	Wnt inhibitory factor-1

Y

YAP1	Yes-associated protein 1

Z

ZAP-70	Zeta-chain(TCR)-associated protein kinase 70 kDa
ZBTB10	Zinc finger and BTB domain containing 10
ZEB	Zinc finger E-box binding homeobox

Part I
MicroRNAs: Biology and Implications in Cancer

Chapter 1
MicroRNAs and Cancer: An Overview

Sadegh Babashah

Contents

Abstract MicroRNAs (miRNAs) constitute an evolutionarily conserved class of small, noncoding RNA molecules that regulate gene expression by targeting specific mRNAs for degradation and/or translational repression. MiRNAs have been widely investigated due to their potential role in regulating a variety of cellular processes, including proliferation, differentiation, and apoptosis. Many miRNAs are implicated in various human cancers. Functional analysis of cancer-related miRNAs has proposed that they might act as either oncogenes or tumor suppressors. In fact, the link between aberrant miRNA expression and cancer development and progression can be observed either through the loss of tumor suppressor miRNAs or the over-expression of oncogenic miRNAs. This chapter aims to provide a succinct framework to gain insight into miRNA function in cancer.

S. Babashah (✉)
Department of Molecular Genetics, Faculty of Biological Sciences, Tarbiat Modares University, Tehran 1411713116, Iran
e-mail: sadegh.babashah@gmail.com

S. Babashah (ed.), *MicroRNAs: Key Regulators of Oncogenesis*,
DOI 10.1007/978-3-319-03725-7_1,

Keywords MicroRNA • Cancer • Oncogene • Tumor suppressor • Metastasis • Angiogenesis

1 MicroRNAs: Biogenesis, Processing and Mode of Action

MicroRNAs (miRNAs or miRs) are a class of non-coding small RNAs of ~22 nucleotides that regulate gene expression by targeting specific mRNAs bearing partially complementary target sequences for degradation and/or translational repression (Liu et al. 2008; Babashah and Soleimani 2011). The first discovery of a small non-coding RNA dates back to 1993, when Victor Ambros and collaborators identified lin-4 in *Caenorhabditis elegans* (Lee et al. 1993). Lin-4 was believed to be a unique species until year 2000 when another small non-coding RNA, let-7, was reported in C. elegans (Reinhart et al. 2000) and in a variety of other organisms (Pasquinelli et al. 2000). Since then, hundreds of small non-coding RNA sequences (now known to be miRNAs) have been identified in a wide range of organisms from nematodes to vertebrates, plants and human. Currently, the official miRNA database miRBase lists 1,872 human miRNA gene loci, generating 2,578 mature miRNA sequences (http://www.mirbase.org, Release 20.0, June 2013). Precise attribution of miRNA effects on gene expression can be complicated by the fact that often each miRNA may control several hundred target genes directly or indirectly, whereas a single protein coding gene target could be regulated by more than one miRNA. In fact, miRNAs are predicted to target up to one-third of human transcripts (Zhong et al. 2012; Friedman et al. 2009).

The biogenesis of miRNAs begins in the nucleus with the synthesis of a relatively long double-stranded RNA molecule, known as primary (pri)-miRNA, by RNA polymerase II or III. The resultant pri-miRNA transcript is often more longer than 1 kb in length and includes a stable stem-loop hairpin structure that contains the sequence for the mature miRNA. The hairpin structure is excised in the nucleus from pri-miRNA as a ~70-nucleotide long precursor (pre)-miRNA by the nuclear RNase III endonuclease Drosha and DGCR8 (the "microprocessor complex") (Lee et al. 2003; Denli et al. 2004; Gregory et al. 2004). DGCR8 is essential as a molecular anchor for Drosha's activity on pri-mRNAs, as it recognizes the pri-miRNA at double-stranded RNA – single-stranded RNA junction and directs Drosha to cleave approximately 11 nucleotides from the base of the stem to free the hairpin from the primary transcript (Han et al. 2006). Members of the microprocessor complex have additional cellular functions, as Drosha is also involved in the processing of ribosomal RNA (Wu et al. 2000) and DGCR8 also acts as a heme-binding protein (Faller et al. 2007). The resultant pre-miRNA contains a 5′ phosphate and a distinctive 3′ two-nucleotide overhang which is signal to transport into the cytoplasm by a protein complex consisting of Exportin-5 and Ran-GTPase (Yi et al. 2003; Lund et al. 2004; Bohnsack et al. 2004) (Fig. 1.1). In cytoplasm, further processing facilitated by the second RNase III endonuclease Dicer, cuts off

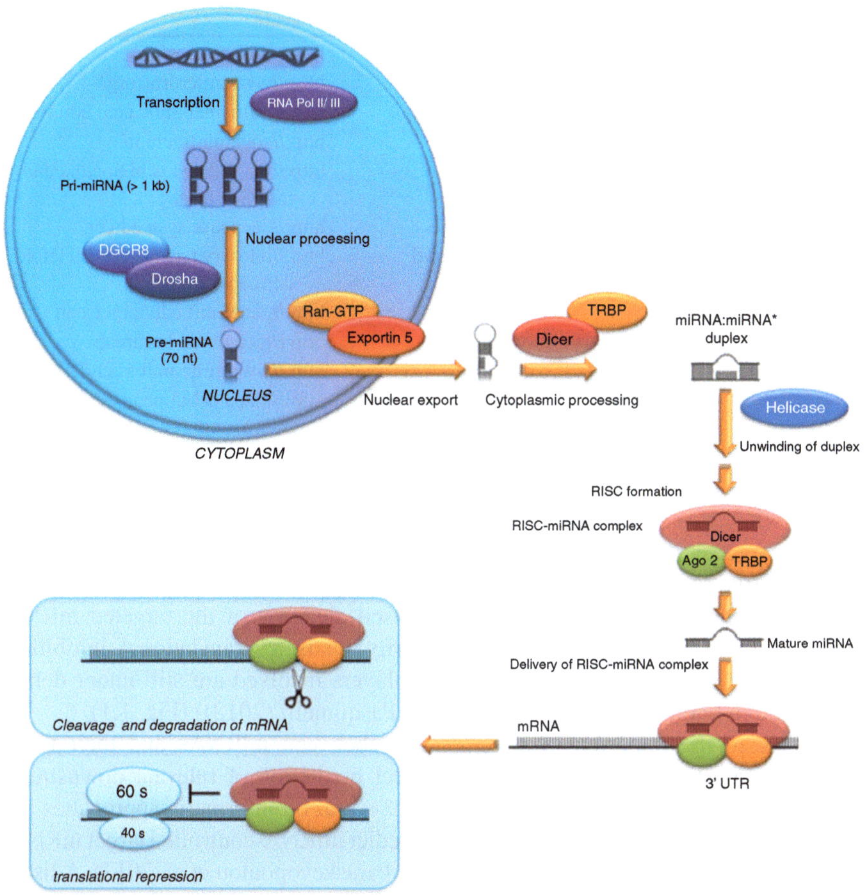

Fig. 1.1 Schematic representation of biogenesis, processing and function of microRNA. The biogenesis of miRNAs begins in the nucleus and is completed in the cytoplasm. For more details, see the text. *Pri-miRNA* primary miRNA, *Pre-miRNA* precursor miRNA, *Drosha* RNase III endonuclease, *DGCR8* DiGeorge syndrome critical region 8, *Dicer* RNase III endonuclease, *RISC* RNA-induced silencing complex

the terminal loop and generates an imperfect double-stranded RNA with about 17-26-nucleotide in length. This duplex molecule contains the mature miRNA (often designated miR) and its complementary miRNA*. The duplex binds to one of four proteins of the Argonaute (Ago) family, which are part of the RNA-induced silencing complex (RISC). After unwinding the double-stranded RNA and discarding and degrading the passenger strand (miRNA*), the mature miRNA is loaded onto the RISC and interacts with the complementary sequences that are mostly located in the 3′ untranslated region (3′ UTR) of the targeted mRNAs (Cullen 2004; Liu et al. 2008; Ikeda et al. 2007). Subsequent mechanisms by which miRNAs regulate gene expression depend on the degree of complementarity between

Table 1.1 MicroRNA databases

Name	Website
miRBase	http://mirbase.org/
miRanda	http://www.microrna.org/
miRNA map	http://mirnamap.mbc.nctu.edu.tw/
coGemiR	http://cogemir.tigem.it/
miRGen	http://www.diana.pcbi.upenn.edu/miRGen.html
deepBase	http://deepbase.sysu.edu.cn/
RNAhybrid	http://bibiserv.techfak.uni-bielefeld.de/rnahybrid
TargetScan	http://genes.mit.edu/targetscan
PicTar	http://pictar.mdc-berlin.de
EIMMo	http://www.mirz.unibas.ch/ElMMo3/
DIANA-microT	http://diana.pcbi.upenn.edu/cgi-bin/micro_t.cgi

mRNA target sites and the nucleotide sequence from position 2–8 at the 5′ end of miRNAs (the seed region). The rare occasion of perfect (or near perfect) Watson-Crick complementarity leads to Ago-catalysed cleavage of the targeted mRNA. More commonly, imperfect complementarity leads to translational inhibition, although the precise mechanisms and the players involved are still under debate (reviewed in Fabian and Sonenberg (2012); Pasquinelli (2012)) (Fig. 1.1).

Owing to the imperfect complementarity between miRNAs and their target mRNAs almost observed in mammals, direct prediction of relevant downstream targets of a miRNA is particularly difficult. Several bioinformatic approaches and various algorithms have been developed to predict miRNA-controlled target mRNAs *in silico* (Lewis et al. 2003; Krek et al. 2005; Paraskevopoulou et al. 2013). A list of computational tools for miRNA target prediction is summarized in Table 1.1. However, as the bioinformatic approach focuses on identifying conserved targets in the 3′-UTR of an mRNA, many non-conserved targets are missed. In addition, there are several lines of evidence indicating that miRNAs can also regulate gene expression through binding to "seedless" 3′-UTR miRNA recognition elements (Lal et al. 2009) or to sites located within the coding regions of transcript (Lee et al. 2009). Therefore, the efficacy of such a bioinformatic approach needs to be validated by *in vitro* or *in vivo* experiments.

MiRNAs are involved in the control of a variety of biological processes, including cellular proliferation, tissue differentiation, organ development, maintenance of stem cell potency and apoptosis (Babashah and Soleimani 2011; Cheng et al. 2005; Chen et al. 2004; Ambros 2004). Given this wide variety of functions, it is not surprising that miRNAs are affected in many diseases such as cancer. In fact, dysregulation of miRNAs has been widely observed in different types and stages of cancer and mounting evidence points to their important roles in the development of a variety of human cancers (Bandyopadhyay et al. 2010; Esquela-Kerscher and Slack 2006; Lu et al. 2005; Volinia et al. 2006).

2 The Oncogenic and Tumor Suppressive Roles of MicroRNAs in Cancer

Aberrant expression of miRNAs has been frequently noted in almost all types of cancer (Croce 2009; Farazi et al. 2011). Functional analysis of these aberrantly expressed miRNAs indicates that they might function as either oncogenes or tumor suppressors. The oncogenic miRNAs, called as "oncomiRs", are up-regulated in cancer and usually promote tumor development by inhibiting tumor suppressor genes and/or genes that control cell differentiation or apoptosis. On the contrary, there are many down-regulated miRNAs which may be considered as tumor suppressors in cancer. These miRNAs are called as "TSmiRs" and may function by inhibiting oncogenes and/or genes that inhibit cell differentiation or apoptosis (Bandyopadhyay et al. 2010; Esquela-Kerscher and Slack 2006; Lu et al. 2005; Babashah and Soleimani 2011). Deregulation of miRNA expression frequently results from genetic mutations and/or epigenetic alterations, represented by deletions, amplifications, point mutations and aberrant DNA methylation events. Indeed, about half of the cancer-related miRNA genes are located at fragile sites of the genome as well as in minimal regions with loss of heterozygosity, minimal regions of amplification or common breakpoint regions (Calin et al. 2002, 2004b).

The first evidence for the involvement of miRNAs in tumorigenesis was reported by Calin et al. (2002) in describing a chromosome region containing the miR-15a/miR-16-1 cluster, which is frequently lost or down-regulated in B-cell chronic lymphocytic leukemia (B-CLL). Down-regulation of the miR-15a/miR-16-1 cluster in CLL and several solid tumors raised the question whether they might function as tumor suppressors (Calin et al. 2002). Cimmino et al. (2005) demonstrated that both miR-15a and miR-16-1 promote the normal apoptotic response by directly targeting the anti-apoptotic gene BCL-2, indicating the possible tumor suppressive role of these two miRNAs in tumorigenesis.

A common tumor suppressive role for the let-7 family of miRNAs has been described in different types of human tissues, particularly in lung. It has been shown that let-7 is able to negatively regulate the expression of various oncogenes such as RAS and MYC as well as other cell cycle progression genes (Johnson et al. 2005; Bhat-Nakshatri et al. 2009). Reduced expression of Let-7 has been observed in different types of cancers, including lung, breast and prostate cancers. It has been shown that down-regulation of let-7 correlates with increased lymph node metastasis and proliferation capacity, suggesting a potential tumor suppressive role for this family of miRNAs in cancer progression (Lynam-Lennon et al. 2009; Liu et al. 2012). Although it has been demonstrated that induction of let-7 reduces tumor growth in a murine model of lung cancer (Esquela-Kerscher et al. 2008; Kumar et al. 2008), the regulation of individual let-7 targets on tumorigenesis needs to be further investigated in more *in vivo* models of human cancers.

The miR-17-92 cluster (containing seven homologous miRNAs: miR-17-3p, miR-17-5p, miR-18a, miR-20a, miR-19a, miR-19b-1 and miR-92a-1; with genomic positions on chromosomes X, 7 and 13) is the first and well-studied miRNA cluster with oncogenic activity. He et al. (2005) investigated the potential oncogenic role of the miR-17-92 cluster. They demonstrated that over-expression of the miR-17-92 cluster in the hematopoietic system acted with c-myc expression to accelerate tumor development and progression in a transgenic mouse model of B-cell lymphoma. Importantly, tumors resulting from combined c-Myc and miR-17-92 expression were able to evade from normal apoptotic responses that were otherwise prevalent in tumors lacking the cluster. O'Donnell et al. (2005) found that c-Myc activates expression of a set of six miRNAs on human chromosome 13 that was tied to the development of human lymphoma. They also found that expression of E2F1 was negatively regulated by two miRNAs in this cluster, miR-17-5p and miR-20a. These findings reveal a mechanism through which the c-Myc simultaneously promotes E2F1 transcription and represses following translation, indicating a tightly controlled proliferative signal. Woods et al. (2007) proposed a model in which the miR-17-92 cluster promotes cell proliferation by shifting the E2F transcriptional balance away from the pro-apoptotic E2F1 and toward the proliferative E2F3 transcriptional network. The miR-17-92 cluster might also inhibit apoptosis by negatively regulating the tumor suppressor PTEN and the pro-apoptotic protein Bim (Xiao et al. 2008; Mendell 2008). Bim is induced by Myc in B-cells and is able to antagonize anti-apoptotic proteins such as Bcl-2. Therefore, down-regulation of Bim by the miR-17-92 cluster may contribute to the ability of these miRNAs to exacerbate disease progression in a mouse model of B-cell leukemia (Egle et al. 2004).

As stated above, miRNAs can function either as oncogenes or tumor suppressors. However, it has been demonstrated that a miRNA can exploit both functions according to the cellular context of their target genes. For instance, there is a body of evidence pointing to the tumor suppressive activity of the miR-17-92 cluster, which contrasts with the hypothesized oncogenic role observed in other cancers (Yu et al. 2008). This implies that the tissue- and developmental-stage-specific expression decisively controls appropriate function of a miRNA.

3 MicroRNAs and Tumor Metastasis

Tumor invasion and metastasis are major characteristics of aggressive phenotypes observed in human cancers (Steeg 2003). During the "invasion-metastasis cascade", cancer cells (a) are detached and migrate out of the primary tumor site; (b) invade the basement membrane to enter the circulatory system (intravasation); (c) are translocated through the vasculature; (d) exit circulatory vessels at the metastatic site (extravasation); (e) survive within the foreign microenvironment; and finally (f) re-initiate their proliferative machinery to establish macroscopic secondary tumors (colonization) (Fig. 1.2) (Harquail et al. 2012; Fidler 2003). Despite the clinical significance of metastasis for determining disease outcome in human cancers, our current understanding on how cancer cells actually migrate out of primary tumors, adapt to distant tissues and organs, and form a secondary tumor are still not completely understood (Gupta and Massague 2006).

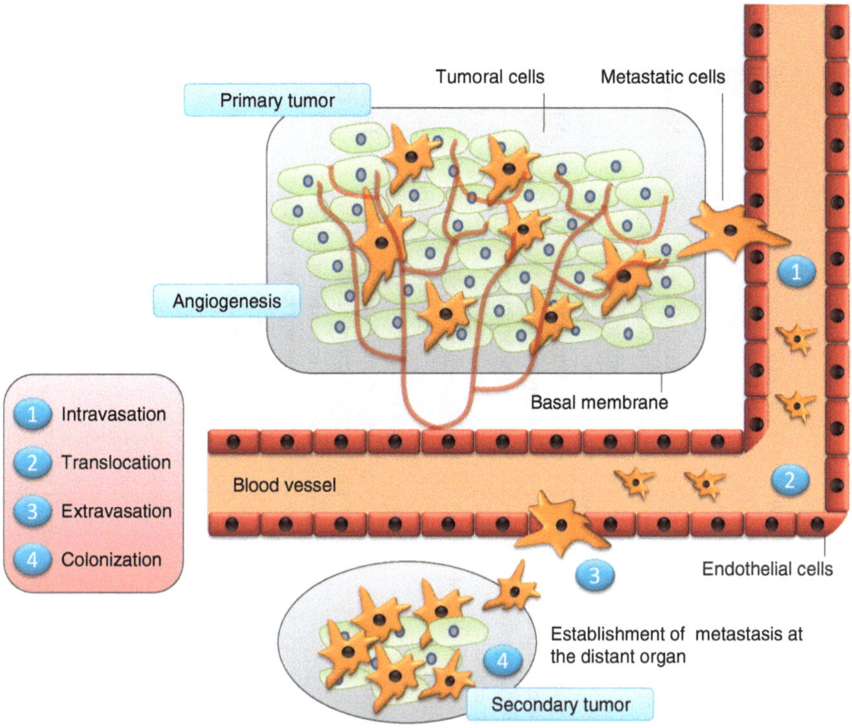

Fig. 1.2 Schematic representation of multistep metastatic process by which primary tumor cells are detached from the primary tumor site, consequently adapt into distant tissues and organs, and form a secondary tumor

MiRNAs have recently been more widely investigated due to their potential role as critical regulators of tumor metastasis in cancer development. The link between altered expression levels of miRNAs and cancer development and metastasis can be observed either through the loss of tumor suppressor miRNAs or the over-expression of oncogenic miRNAs in different cancer cells. Some miRNAs involved in metastasis are summarized in Table 1.2, most of which will be discussed in more detail in the sections below.

3.1 *Pro-metastatic miRNAs*

Multiple lines of evidence highlight the contribution of certain miRNAs to promoting tumor metastasis. MiR-10b is the first miRNA identified to positively regulate the metastatic potential of human cancer cell. Ma et al (2007) showed that miR-10b over-expression endowed otherwise non-metastatic breast cancer cells with the capacity to acquire invasive and metastatic behavior. MiR-10b is able to induce migration and invasion capacities in breast cancer cells through direct targeting of homeobox

Table 1.2 Some microRNAs involved in the regulation of the multistep metastatic process

MicroRNA	Molecular mechanism	Function	Pathogenesis and clinical significance	Reference(s)
miR-10b	Directly regulated by the transcription factor Twist, a metastasis-promoting gene Directly targets HOXD-10 and Syndecan-1	Pro-metastatic: promotes cell motility and invasiveness	Highly expressed in metastatic breast tumors Initiates tumor invasion *in vivo* Its expression in primary breast carcinomas correlates with clinical progression	Ibrahim et al. (2012) and Ma et al. (2007)
miR-21	Targets tumor suppressor genes including Pdcd4, Maspin, and TPM1	Pro-metastatic: promotes tumor growth, cell motility, invasiveness, and intravasation	Its high expression correlates with: Advanced stages of tumor progression Poor clinical outcome in breast and pancreatic tumors	Nicoloso et al. (2009), Asangani et al. (2008), and Zhu et al. (2008)
miR-31	Targets a cohort of pro-metastasis genes, including ITGA5, RhoA, MMP16, Fzd3, RDX, M-RIP	Anti-metastatic: exerted its suppressive effects on distinct steps of the metastatic process including cell motility, invasiveness, and post-intravasation events: extravasation or initial survival at a distant site, and metastatic colonization	Its expression correlates inversely with metastasis in human breast tumors Down-regulated in metastatic primary breast tumors Its expression suppresses metastasis *in vivo* and may correlate with more favorable outcome in clinical breast tumors.	Valastyan et al. (2009)
miR-34b/c	Targets oncogenic genes including c-myc, CDK6, E2F3	Anti-metastatic: Inhibits cell motility and invasiveness	Hypermethylated in metastatic human breast, lung and colon tumors Its epigenetic silencing contributes to metastasis formation and have important clinical and therapeutic outcome.	Lujambio et al. (2008)
miR-125b	Targets the tumor suppressor gene STARD13 Significantly up-regulates the expression of EMT markers (i.e. vimentin and α-SMA but not E-cadherin) leading to a high metastasis potentiality	Pro-metastatic: promotes cell motility and invasiveness	Its high expression induces luminal-like breast cancer cells with EMT properties and is associated with an aggressive phenotype and poor clinical outcome.	Tang et al. (2012)
miR-126	Targets?	Anti-metastatic: Inhibits cell proliferation and tumor growth	Loss of its expression is strongly associated with metastatic relapse in primary breast tumors, and is associated with poor distal metastasis-free survival	Tavazoie et al. (2008)

miR-148a	Targets oncogenic gene TGIF2	Anti-metastatic: Inhibits cell motility and invasiveness	Hypermethylated in metastatic human breast, lung and colon tumors Its epigenetic silencing contributes to metastasis formation and have important clinical and therapeutic outcome.	Lujambio et al. (2008)
miR-193b	Targets tumor invasion inducer uPA	Anti-metastatic: Inhibits cell motility and invasiveness	Loss of miR-193b is closely associated with clinical metastasis suggesting a potential role in prognostic stratification of breast cancer patients	Li et al. (2009)
miR-194	Targets cytoskeleton protein Talin-2	Anti-metastatic: Inhibits cell motility and invasiveness	Contributes to the anti-tumor activity of trastuzumab on HER2-over-expressing breast cancer cells	Le et al. (2012)
miR-206	Targets?	Anti-metastatic: Inhibits cell motility and invasiveness	Down-regulated in metastatic primary breast tumors	Tavazoie et al. (2008)
miR-335	Targets the progenitor cell transcription factor SOX-4 and extracellular matrix component TNC	Anti-metastatic: Inhibits cell motility and invasiveness	Loss of its expression is strongly associated with metastatic relapse in primary breast tumors, and is associated with poor distal metastasis-free survival	Tavazoie et al. (2008)
miR-373 and miR-520c	Both miRNAs cooperate in metastasis by suppressing CD44	Pro-metastatic: Promotes cell motility and invasiveness	Up-regulated in metastatic primary breast tumors miR-373 is not strong enough biomarker Loss of CD44 is associated with induction of breast cancer metastasis to the lung and led to poor clinical outcome	Huang et al. (2008) and Lopez et al. (2005)
miR-224	Targets the RKIP tumor suppressor, a repressor of stromal genes required for tumor metastasis	Pro-metastatic: Promotes cell motility and invasiveness	Plays an important role in metastasis of human breast cancer cells to the bone.	Huang et al. (2012)

Abbreviations: *α-SMA* smooth muscle alpha-actin, *CDK6* cyclin dependent kinase 6, *E2F-3* E2 transcription factor family-3, *EMT* epithelial to mesenchymal transition, *Fzd3* Frizzled3, *HOXD-10* homeobox D10, *ITGA5* Integrin α5, *M-RI P* myosin phosphatase-Rho interacting protein, *MMP16* matrix metallopeptidase 16, *PDCD4* programmed cell death 4, *RDX* Radixin, *RKIP* the Raf kinase inhibitor protein, *SOX4* SRY-related HMG-box, *STARD13* StAR-related lipid transfer domain containing 13, *TGIF2* TGFβ-induced factor homeobox 2, *TPM1* Tropomyosin 1, *TNC* Tenascin C, *uPA* urokinase-type plasminogen activator

D10 (HOXD10), a receptor of genes involved in cell migration and extracellular matrix remodeling. Notably, systemic treatment of breast tumor-bearing mice with miR-10b antagomirs decreased the metastatic tumor burden, providing promising evidence that antagomirs can be efficiently delivered to rapidly growing tumor cells *in vivo*, preventing metastasis (Ma et al. 2010). To identify miRNAs that have the capacity to promote metastasis, Huang et al. (2008) set up a genetic screen involving over-expression of approximately 450 miRNAs in non-metastatic, human breast tumor cell line. They found that miR-373 and miR-520c (both belonging to a miRNA family that shares similar seed sequence) can induce tumor cell migration and invasion *in vitro* and *in vivo*, and that the migratory phenotype of certain cancer cell lines depends on endogenous miR-373 expression. They proposed that suppression of cell migration by an anti-miR-373 oligonucleotide may be a potential strategy for developing efficient therapies against tumor metastasis. After that, two independent studies indicated that apart from the oncogenic role of miR-21 in tumorigenesis, this miRNA also plays a critical role in invasion and metastasis of human breast and colorectal carcinoma cells (Asangani et al. 2008; Zhu et al. 2008). These studies suggest that suppression of miR-21 might offer another promising therapeutic approach against advanced cancers (Table 1.2).

3.2 Anti-metastatic miRNAs

Multiple lines of evidence highlight the contribution of certain miRNAs to suppressing tumor metastasis. MiR-31 expression levels correlate inversely with metastasis in human breast cancer patients. By deploying gain- and loss-of-function strategies, Valastyan et al. (2009) demonstrated that miR-31 is capable of suppressing the metastatic potential of human breast tumor cells. They also successfully showed that miR-31 is involved during the multiple step metastatic process *in vivo*, including local invasion, extravasation or initial survival at a distant site, and metastatic colonization. MiR-126 and miR-335 have been identified as human breast cancer metastasis suppressor miRNAs that exert their unique effects on distinct steps of the invasion-metastasis cascade. By performing array-based miRNA profiling, Tavazoie et al. (2008) revealed that the expression of both miRNAs is lost in the majority of primary breast tumors with metastatic relapse, and the loss of expression of either miRNA is associated with poor distal metastasis-free survival. Importantly, *in vivo* experiments showed that miR-126 restoration reduced overall tumor growth and proliferation (at both primary site and distant organs), whereas miR-335 caused a significant reduction in cell motility and invasive capacity. The strong association of the loss of miR-335 and miR-126 expression with clinical metastatic relapse suggests the potential for the use of these miRNAs in prognostic assessment of breast cancer patients in addition to conventional clinical and pathological staging markers. Moreover, another study identified that miR-193b significantly inhibited the growth and dissemination of xenograft breast tumors in an immunodeficient mouse model. This study showed that the loss of miR-193b confers the metastatic colonization ability to the cells. As the loss of miR-193b expression is strongly correlated with metastasis, the use of this miRNA in addition to conventional clinical and pathological staging markers could be an attractive option for the prognostic stratification of patients with breast cancer (Li et al. 2009) (Table 1.2).

3.3 MiRNAs and Epithelial to Mesenchymal Transition

Epithelial to mesenchymal transition (EMT), in which polarized epithelial cells are converted into motile cells, plays an important role in tumor invasion and metastasis (Thiery 2002; Yang and Weinberg 2008; Togawa et al. 2011). The effect of miR-125b on metastatic activities of breast cancer cells was studied by Tang et al. They reported that miR-125b significantly up-regulates the expression of two EMT markers (i.e. vimentin and α-SMA expression) but another EMT marker (E-Cadherin) shows no significant change. Elevating vimentin and α-SMA expression results in a high metastasis potentiality and some mesenchymal cell characteristics in breast cancer cells (Tang et al. 2012). A large body of evidence indicates that the miR-200 family inhibits EMT and cancer cell migration by enhancing E-cadherin expression through direct targeting of the EMT-promoting transcription factors Zeb1 and Zeb2 (Korpal et al. 2008; Gregory et al. 2008; Park et al. 2008; Burk et al. 2008; Bracken et al. 2008). However, a study reported that over-expressing miR-200 in Murine breast cancer cell line 4TO7 enhances the ability of these cells to metastasize to lung and liver. This study reported that miR-200 expression leads to promote a mesenchymal to epithelial cell transition (MET) by suppressing Zeb2 expression. This finding contrasts with the EMT hypothesis of cancer metastasis that implies that the induction of epithelial characteristics would inhibit the formation of metastasis. This apparent contradiction could be explained on the basis that for some tumors, a reversion of the mesenchymal phenotype of malignant cells may facilitate tumor colonization at metastatic sites. This suggests that the epithelial nature of a tumor does not predict metastatic outcome. Moreover, these results imply that the cellular context of miRNA expression decisively controls the function of a miRNA (Dykxhoorn et al. 2009).

4 MicroRNAs and Tumor Angiogenesis

Angiogenesis is characterized by growth of new blood vessels from pre-existing vasculature in response to physiological or pathophysiological stimuli. This process, which involves proliferation, migration, and maturation of endothelial cells, plays an important role during tumor growth and metastasis (Urbich et al. 2008; Chung et al. 2010).

Evidence for the significance of miRNAs as regulators of angiogenesis comes from observations that Dicer is a critical component for embryonic angiogenesis. It has been shown that blood vessel formation/maintenance in Dicer-deficient mice embryos and their yolk sacs was severely compromised, suggesting a possible role for Dicer in angiogenesis through its function in the processing of miRNAs (Yang et al. 2005). Consistent with this observation, another studies showed that genetic silencing of Dicer in endothelial cells leads to down-regulation of several key positive regulators of the angiogenic phenotype and impairs tube formation activity *in vitro* and *in vivo* (Suarez et al. 2007; Kuehbacher et al. 2007). Mounting studies suggest that a number of angiogenesis-related miRNAs affect cancerous phenotype of malignant cells. MiRNAs can modulate angiogenesis by targeting positive or negative regulators in angiogenic signaling pathways (Hong et al. 2013; Landskroner-Eiger et al. 2013). Some miRNAs involved in tumor angiogenesis are summarized in Table 1.3, most of which will be discussed in more detail in the sections below.

4.1 Pro-angiogenic miRNAs

Up-regulation of pro-angiogenic growth factor receptors (such as platelet-derived growth factor receptor, "PDGFR" and vascular endothelial growth factor receptor, "VEGFR") on endothelial cells is a common feature of angiogenesis (Batchelor et al. 2007; Shih and Holland 2006). Wurdinger et al. (2008) showed that glioma- or growth factor-mediated induction of miR-296 in endothelial cells leads to increased levels of pro-angiogenic growth factor receptors VEGFR2 and PDGFR-β. Possible role of miR-296 in promoting angiogenesis in tumor was further supported when inhibition of miR-296 with antagomirs reduced angiogenesis in tumor xenografts *in vivo*.

Some other miRNAs, such as miR-378 and miR-17-92 cluster, have been also implicated in tumor angiogenesis. MiR-378 functions as an oncogene by enhancing tumor cell survival, blood vessel expansion, and tumor growth by targeting two tumor suppressors, SuFu (suppressor of fused) and Fus-1 (Lee et al. 2007a). The miR-17-92 cluster not only augments angiogenesis in endothelial cells during normal development (Suarez et al. 2008), but also its upregulation in cancer cells can serve to promote angiogenesis during tumor growth in a xenograft model (Dews et al. 2006). Importantly, this angiogenic effect is exerted through down-regulation of anti-angiogenic thrombospondin-1 (TSP-1) and related proteins, such as connective tissue growth factor (CTGF) (Dews et al. 2006).

One study showed that many miRNAs derived from tumor cells are packaged into microvesicles and then directly delivered to their microenvironment. These tumor-secreted microvesicles are then capable of interacting with proximal endothelial cells to transport miRNAs in endothelial cells. Among these miRNAs, it was shown that tumor-secreted miR-9 promotes endothelial cell migration and tumor angiogenesis by activating JAK-STAT pathway, one of the major oncogenic signaling pathways activated in a variety of human malignancies. Importantly, administration of miR-9 antagomiRs (anti-miR-9) or JAK inhibitors impaired microvesicles-induced cell migration *in vitro* and decreased tumor burden *in vivo*. Taken together, these observations support a novel intercellular communication in which tumor-secreted miRNAs function as pro-angiogenic mediators during tumorigenesis (Zhuang et al. 2012) (Table 1.3).

Table 1.3 Some microRNAs involved in the regulation of tumor angiogenesis

MicroRNA	Function	Validated target(s)	Reference(s)
miR-9	Pro-angiogenic: promotes endothelial cell migration and tumor angiogenesis by activating the JAK-STAT pathway	SOCS5	Zhuang et al. (2012)
miR-34a	Anti-angiogenic: its down-regulation promotes tumor growth and tumor angiogenesis in head and neck squamous cell carcinoma	E2F3 (directly), Survivin (indirectly)	Kumar et al. (2012)
miR-17-92	Pro-angiogenic: promotes angiogenesis within tumor cells and in dicer-depleted endothelial cells	TSP-1, CTGF	Suarez et al. (2008) and Dews et al. (2006)
miR-98	Anti-angiogenic: plays a regulatory role in tumor angiogenesis and invasion in a highly aggressive breast cancer model *in vitro* and *in vitro*	ALK4, MMP11	Siragam et al. (2012)
miR-125b	Anti-angiogenic: inhibits tube formation of endothelial cells *in vitro*	VE-cadherin	Muramatsu et al. (2013)
miR-126	Pro-angiogenic: required for vascular integrity and angiogenesis *in vivo*. It plays a role in regulating the adhesion of leukocytes to the endothelium.	SPRED-1, PIK3R2, VCAM1	Fish et al. (2008), Harris et al. (2008), and Wang et al. (2008a)
miR-130a	Pro-angiogenic: its expression antagonized the inhibitory effects of GAX or HOXA5 on endothelial cell tube formation *in vitro*.	HOXA5, GAX	Chen and Gorski (2008)
miR-221/222	Anti-angiogenic: inhibit normal erythropoiesis and erythroleukemic cell growth and prevent endothelial cell migration, proliferation and angiogenesis *in vitro*.	c-KIT	Poliseno et al. (2006), Felli et al. (2005), and Urbich et al. (2008)
miR-296	Pro-angiogenic: its induction in endothelial cells results in increased levels of pro-angiogenic growth factor receptors VEGFR2 and PDGFR-β. Its inhibition reduces angiogenesis in tumor xenografts *in vivo*.	HGS	Wurdinger et al. (2008)
miR-378	Pro-angiogenic: promotes tumorigenesis and angiogenesis *in vitro*	SuFu, Fus-1	Lee et al. (2007a)

Abbreviations: *SOCS5* suppressor of cytokine signaling 5, *TSP-1* Thrombospondin-1, *CTGF* connective tissue growth factor, *GAX* growth arrest homeobox, *HGS* hepatocyte growth factor-regulated tyrosine kinase substrate, *SuFu* suppressor of fused, *VE-cadherin* vascular endothelial cadherin, *ALK4* activin receptor-like kinase-4, *MMP11* matrix metalloproteinase-11

4.2 Anti-angiogenic miRNAs

The miR-34 family of miRNAs (miR-34a, b and c) as direct, conserved p53 target genes presumably induces apoptosis, cell cycle arrest and senescence (Bommer et al. 2007; Chang et al. 2007). MiR-34a functions as a tumor suppressor that is frequently down-regulated in various tumor types. Kumar et al. (2012) demonstrated that miR-34a expression is significantly down-regulated in head and neck squamous cell carcinoma tumors and cell lines. Ectopic expression of miR-34a reduced head and neck tumor cell proliferation, colony formation and migration and also significantly inhibited tumor growth and tumor angiogenesis in a SCID mouse xenograft model. This *in vivo* tumor growth study revealed that miR-34a inhibits tumor angiogenesis by down-regulating VEGF, a key angiogenic factor.

Siragam et al. (2012) defined a regulatory role for miR-98 in tumor angiogenesis and invasion using a highly aggressive breast cancer model *in vitro*. They showed that miR-98 inhibits tumor angiogenesis and invasion by repressing activin receptor-like kinase-4 (ALK4) and matrix metalloproteinase-11 (MMP11) expression.

Another study showed that transient induction of miR-125b inhibits *in vitro* tube formation of endothelial cells through suppression of vascular endothelial (VE)-cadherin. Importantly, induction of miR-125b induced non-functional blood vessels, resulting in inhibition of tumor growth. It seems that prolonged over-expression of miR-125b could be an option in cancer therapy by causing collapse of the lumen of endothelial cells (Muramatsu et al. 2013) (Table 1.3).

5 MicroRNA Profiling by High-Throughput Technologies

Considering the fact that current cancer detection tests have their own limitations, the use of miRNAs as promising biomarkers for diagnosis and prognosis of cancer has aroused intense research interests. Additionally, distinctive pattern of miRNA expression also serves as markers of important histopathologic features such as tumor stage, proliferative capacity and vascular invasion (Lynam-Lennon et al. 2009).

Many expression profiling studies of miRNA genes have been performed on different types of cancer. However, the results of analyses of the same type of cancer by different groups are not always consistent. The disparity in these results might attribute to the different platforms for miRNA profiling in each case and the use of different sample storage methods (Calin and Croce 2009).

Currently, the most widely used methods for miRNA profiling are based on sequencing, microarray, and real-time quantitative PCR. Microarray platforms have been used for miRNA profiling, but suffer from background and cross-hybridization problems and are generally restricted to identifying the relative abundance of previously discovered miRNAs (Calin et al. 2004a; Chen et al. 2009). Sequencing-based applications for identifying and profiling miRNAs have been hindered by laborious cloning techniques and the expense of capillary DNA sequencing (Pfeffer et al. 2005; Cummins et al. 2006). High-throughput sequencing-based approaches to generate miRNA profiles, hugely enabled by next-generation technologies, provide several advantages over probe-based methodologies, including the ability to discover novel miRNAs and the potential to detect variations in the mature miRNA length and miRNA editing (Morozova and Marra 2008). Next-generation sequencing technologies are able to identify low abundance miRNAs or those exhibiting modest expression differences among samples, which may not be detected by hybridization-based methods. Real-time quantitative PCR, another highly sensitive technique for miRNA quantification, is capable of distinguishing mature and precursor miRNA, and produces fewer false-positives and reduced bias when compared with microarray or sequencing approaches (Chen et al. 2009; Fuller et al. 2009; Petriv et al. 2010). Real-time PCR may be used to validate the expression of miRNAs discovered during high throughput arrays and study the expression of individual miRNAs. This method provides several important advantages for miRNA profiling studies including low cost, superior detection of low-abundance species and high throughput (Schmittgen et al. 2008). The emergence of novel high-throughput technologies will allow more sensitive and efficient miRNA detection in patient samples, and identification of novel miRNAs. However, standardization of these novel methods is necessary to overcome the variability observed when different miRNA-expression detection platforms are used.

6 Potential Use of MicroRNAs in Cancer Therapy

Dysregulation of miRNA has been widely observed in different types of human cancers (Table 1.4), and there is mounting evidence demonstrating their important roles during cancer development and progression. Uncovering the possible mechanisms underlying the importance of miRNAs in the pathogenesis of human cancers may lead to the development of miRNA-based therapeutic strategies or diagnostic/prognostic biomarkers.

Since cancer cells often have a distinctive expression pattern of oncogenic and tumor suppressive miRNAs (Babashah et al. 2012; Babashah and Soleimani 2011; Calin and Croce 2006), approaches that manipulate miRNA expression levels, either alone or in combination with currently used therapies, may prove to be therapeutically beneficial. Sequence-specific knockdown of oncogenic miRNAs by chemically engineered oligonucleotides termed "antagomirs" or locked nucleic acid (LNA)-modified oligonucleotides is a plausible therapeutic approach for inhibiting expression levels of oncogenic miRNAs in cancer (Orom et al. 2006; Krutzfeldt et al. 2005). In contrast, elevating the expression level of tumor suppressive miRNAs that could be achieved by viral or liposomal delivery of mimic miRNAs represents a potential therapeutic strategy against cancer (Calin and Croce 2006; Meng et al. 2006). However, many concerns need to be addressed before consideration of conducting miRNA-based therapy including dosage, safety, specificity, stability, efficacy, and problems of delivery to the target (Chen et al. 2010; Cho 2010b; Tong and Nemunaitis 2008; Wu et al. 2007).

Table 1.4 Some microRNAs aberrantly expressed in different cancers

Cancer type	Up-regulated	Down-regulated	References
Lung cancer	miR-17-92 cluster, miR-21, miR-106a, miR-155, miR-191, miR-205, miR-210	let- 7 family, miR-1, miR-7, miR-15a/16, miR-29 family, miR-34 family, miR-124a, miR-143, miR-145	Johnson et al. (2005), Lowery et al. (2008), Yanaihara et al. (2006), Hayashita et al. (2005), Takamizawa et al. (2004), Webster et al. (2009), Bandi et al. (2009), Fabbri et al. (2007), and Nasser et al. (2008)
Breast cancer	miR-10b, miR-21, miR-22, miR-23, miR-27a, miR-29b-2, miR-96, miR-155, miR-181, miR-182, miR-191, miR-210, miR-221, miR-222, miR-328, miR373, miR-520c	let-7, miR-7, miR-9-1, miR-9-3, miR-10b, miR-17/miR-20, miR-27b, miR-31, miR-34 family, miR-125a/b, miR-143, miR-145, miR-146, miR-155, miR-200 family, miR-205, miR-206, miR-335	Cho (2010a), Kondo et al. (2008), Gregory et al. (2008), Bhaumik et al. (2008), Scott et al. (2007), Valastyan et al. (2009), Lehmann et al. (2008), Miller et al. (2008), Camps et al. (2008), Kong et al. (2008), Mertens-Talcott et al. (2007), Zhu et al. (2007), Ma et al. (2007), Huang et al. (2008), Yu et al. (2007), Webster et al. (2009), and Lowery et al. (2008)
Thyroid cancer	miR-146b, miR-221, miR-222, miR-181b, miR-155, miR-197, miR-224, miR-346	miR-30d, miR-125b, miR-26a, miR-30a-5p	Nikiforova et al. (2008), Visone et al. (2007), and Lowery et al. (2008)
Colorectal cancer	miR-18, miR-224, miR-10a, miR-17-92 cluster, miR-21, miR-24-1, miR-29b-2, miR-31, miR-96, miR-135b, miR-183	let-7, miR-30-3p, miR-34a/b/c, miR-124a, miR-127, miR-129, miR-133b, miR-143, miR-145, miR-328, miR-342	Lowery et al. (2008), Barbarotto et al. (2008), Saito et al. (2006), Grady et al. (2008), Tazawa et al. (2007), Slaby et al. (2007), and Asangani et al. (2008)
Ovarian cancer	miR-18a, miR-93, miR-200a/b/c, miR-141, miR-214, miR-429	let7, miR-34b/c, miR-125a/b, miR-140, miR-145, miR-199a, miR-200 family	Iorio et al. (2007), Nam et al. (2008), Hu et al. (2009), Yang et al. (2008), and Corney et al. (2007)
Glioblastoma	miR-21, miR-221, miR-222	miR-7, miR-181a/b/c, miR-125a/b	Spizzo et al. (2009), Webster et al. (2009), Chan et al. (2005), and Gillies and Lorimer (2007)
Bladder	miR-17, miR-23a, b, miR-26b, miR-103-1, miR-185, miR-203, miR-205, miR-221, miR-223	miR-29c, miR-26a, miR-30c, miR-30e-5p, miR-145, miR-30a-3p, miR-133a/b, miR-195, miR-125b, miR-199a	Ichimi et al. (2009) and Gottardo et al. (2007)

Pancreas	miR-221, miR-376a, miR301, miR-21, miR-24-2, miR-100, miR-103, miR-107, miR-125b-1, miR-155, miR-181, miR-106, miR-363, miR-301, miR-376a, miR-212, miR-34a	miR-375, let 7, miR-200, miR-200b	Lee et al. (2007b), Cho (2010a), Lowery et al. (2008), Barbarotto et al. (2008), and Dutta et al. (2007)
Prostate cancer	let-7d, miR-21, miR-106, miR-181, miR-195, miR-203, miR-221, miR-222, miR-363	miR-15a/16-1 cluster, miR-23a/b, miR-101, miR-125a/b, miR-127, miR-128a, miR-143, miR-145, miR-200, miR-330, miR-331, miR-449a	Gartel and Kandel (2008), Lowery et al. (2008), Galardi et al. (2007), Noonan et al. (2009), Bonci et al. (2008), Saito et al. (2006), and Varambally et al. (2008)
CLL	miR-21, miR-23b, miR-24-1, miR 106b, miR-146, miR-155, miR-195, miR-221, miR-222	let7a, miR-15a, miR16-1, miR-29b/c, miR-30d, miR-34a, miR-45, miR-92, miR-143, miR-145, miR-150, miR-181a/b, miR-223	Cimmino et al. (2005), Calin et al. (2002), (2005), Zhao et al. (2010), Marton et al. (2008), Pekarsky et al. (2006), Stamatopoulos et al. (2009), and Akao et al. (2007)
Lymphoma	miR-17-92 cluster, miR-155	miR-143, miR-145	Akao et al. (2007), Kluiver et al. (2005), and He et al. (2005)
Melanoma	miR-221, miR-222	Let-7, miR-34a	Yan et al. (2009), Muller and Bosserhoff (2008), and Felicetti et al. (2008)
Hepatocellular carcinoma	miR-17-92 cluster, miR-18, miR-21, miR-33, miR-130b, miR-135a, miR-143, miR-221, miR-224, miR-301	miR-1, miR-26a, miR-101, miR-122a, miR-125a, miR-139, miR-150, miR-195, miR-199a/b, miR-200a/b, miR-214, miR-223	Jiang et al. (2008), Lowery et al. (2008), Wang et al. (2008b), Zhang et al. (2009), Connolly et al. (2008), Su et al. (2009), Datta et al. (2008), and Gramantieri et al. (2007)
Head and neck squamous cell carcinoma	miR-21	Let-7d, miR-138, miR-205	Chang et al. (2008), Liu et al. (2009), and Childs et al. (2009)

7 Conclusions and Perspectives

- As miRNAs can regulate various target genes, precise attribution of their functions on gene expression is very complicated. However, the critical involvement of miRNAs in many aspect of cancer biology is irrefutable.
- Although miRNAs are postulated to function as either oncogenes or tumor suppressors in human cancers, further studies establishing such roles for miRNAs using *in vivo* experimental models are needed to elucidate precise mechanisms of miRNAs functions in cancer.
- MiRNA expression profiling of human cancers has identified diagnostic and prognostic signatures. Additionally, miRNA signatures could be used for cancer classification and prediction of therapeutic efficacy.
- The association of miRNA dysregulation with oncogenesis demonstrates the feasibility of manipulating miRNA levels as a potential strategy for therapeutic purposes.
- Given the potential involvement of candidate miRNAs in the pathogenesis of human cancers, it seems that pharmacological modulation of miRNA expression will have a brilliant future and become a promising option in cancer therapy.

Acknowledgments I would like to thank all authors responsible for the insights that I attempted to summarize. I apologize to the colleagues whose work was not cited because of space considerations.

References

Akao Y, Nakagawa Y, Kitade Y, Kinoshita T, Naoe T (2007) Downregulation of microRNAs-143 and −145 in B-cell malignancies. Cancer Sci 98(12):1914–1920

Ambros V (2004) The functions of animal microRNAs. Nature 431(7006):350–355

Asangani IA, Rasheed SA, Nikolova DA, Leupold JH, Colburn NH, Post S et al (2008) MicroRNA-21 (miR-21) post-transcriptionally downregulates tumor suppressor Pdcd4 and stimulates invasion, intravasation and metastasis in colorectal cancer. Oncogene 27(15):2128–2136

Babashah S, Soleimani M (2011) The oncogenic and tumour suppressive roles of microRNAs in cancer and apoptosis. Eur J Cancer 47(8):1127–1137

Babashah S, Sadeghizadeh M, Tavirani MR, Farivar S, Soleimani M (2012) Aberrant microRNA expression and its implications in the pathogenesis of leukemias. Cell Oncol (Dordr) 35(5):317–334

Bandi N, Zbinden S, Gugger M, Arnold M, Kocher V, Hasan L et al (2009) miR-15a and miR-16 are implicated in cell cycle regulation in a Rb-dependent manner and are frequently deleted or down-regulated in non-small cell lung cancer. Cancer Res 69(13):5553–5559

Bandyopadhyay S, Mitra R, Maulik U, Zhang MQ (2010) Development of the human cancer microRNA network. Silence 1(1):6

Barbarotto E, Schmittgen TD, Calin GA (2008) MicroRNAs and cancer: profile, profile, profile. Int J Cancer 122(5):969–977

Batchelor TT, Sorensen AG, di Tomaso E, Zhang WT, Duda DG, Cohen KS et al (2007) AZD2171, a pan-VEGF receptor tyrosine kinase inhibitor, normalizes tumor vasculature and alleviates edema in glioblastoma patients. Cancer Cell 11(1):83–95

Bhat-Nakshatri P, Wang G, Collins NR, Thomson MJ, Geistlinger TR, Carroll JS et al (2009) Estradiol-regulated microRNAs control estradiol response in breast cancer cells. Nucleic Acids Res 37(14):4850–4861

Bhaumik D, Scott GK, Schokrpur S, Patil CK, Campisi J, Benz CC (2008) Expression of microRNA-146 suppresses NF-kappaB activity with reduction of metastatic potential in breast cancer cells. Oncogene 27(42):5643–5647

Bohnsack MT, Czaplinski K, Gorlich D (2004) Exportin 5 is a RanGTP-dependent dsRNA-binding protein that mediates nuclear export of pre-miRNAs. RNA 10(2):185–191

Bommer GT, Gerin I, Feng Y, Kaczorowski AJ, Kuick R, Love RE et al (2007) p53-mediated activation of miRNA34 candidate tumor-suppressor genes. Curr Biol 17(15):1298–1307

Bonci D, Coppola V, Musumeci M, Addario A, Giuffrida R, Memeo L et al (2008) The miR-15a-miR-16-1 cluster controls prostate cancer by targeting multiple oncogenic activities. Nat Med 14(11):1271–1277

Bracken CP, Gregory PA, Kolesnikoff N, Bert AG, Wang J, Shannon MF et al (2008) A double-negative feedback loop between ZEB1-SIP1 and the microRNA-200 family regulates epithelial-mesenchymal transition. Cancer Res 68(19):7846–7854

Burk U, Schubert J, Wellner U, Schmalhofer O, Vincan E, Spaderna S et al (2008) A reciprocal repression between ZEB1 and members of the miR-200 family promotes EMT and invasion in cancer cells. EMBO Rep 9(6):582–589

Calin GA, Croce CM (2006) MicroRNA signatures in human cancers. Nat Rev Cancer 6(11):857–866

Calin GA, Croce CM (2009) Chronic lymphocytic leukemia: interplay between noncoding RNAs and protein-coding genes. Blood 114(23):4761–4770

Calin GA, Dumitru CD, Shimizu M, Bichi R, Zupo S, Noch E et al (2002) Frequent deletions and down-regulation of micro- RNA genes miR15 and miR16 at 13q14 in chronic lymphocytic leukemia. Proc Natl Acad Sci U S A 99(24):15524–15529

Calin GA, Liu CG, Sevignani C, Ferracin M, Felli N, Dumitru CD et al (2004a) MicroRNA profiling reveals distinct signatures in B cell chronic lymphocytic leukemias. Proc Natl Acad Sci U S A 101(32):11755–11760

Calin GA, Sevignani C, Dumitru CD, Hyslop T, Noch E, Yendamuri S et al (2004b) Human microRNA genes are frequently located at fragile sites and genomic regions involved in cancers. Proc Natl Acad Sci U S A 101(9):2999–3004

Calin GA, Ferracin M, Cimmino A, Di Leva G, Shimizu M, Wojcik SE et al (2005) A microRNA signature associated with prognosis and progression in chronic lymphocytic leukemia. N Engl J Med 353(17):1793–1801

Camps C, Buffa FM, Colella S, Moore J, Sotiriou C, Sheldon H et al (2008) hsa-miR-210 Is induced by hypoxia and is an independent prognostic factor in breast cancer. Clin Cancer Res 14(5):1340–1348

Chan JA, Krichevsky AM, Kosik KS (2005) MicroRNA-21 is an antiapoptotic factor in human glioblastoma cells. Cancer Res 65(14):6029–6033

Chang TC, Wentzel EA, Kent OA, Ramachandran K, Mullendore M, Lee KH et al (2007) Transactivation of miR-34a by p53 broadly influences gene expression and promotes apoptosis. Mol Cell 26(5):745–752

Chang SS, Jiang WW, Smith I, Poeta LM, Begum S, Glazer C et al (2008) MicroRNA alterations in head and neck squamous cell carcinoma. Int J Cancer 123(12):2791–2797

Chen Y, Gorski DH (2008) Regulation of angiogenesis through a microRNA (miR-130a) that down-regulates antiangiogenic homeobox genes GAX and HOXA5. Blood 111(3):1217–1226

Chen CZ, Li L, Lodish HF, Bartel DP (2004) MicroRNAs modulate hematopoietic lineage differentiation. Science 303(5654):83–86

Chen Y, Gelfond JA, McManus LM, Shireman PK (2009) Reproducibility of quantitative RT-PCR array in miRNA expression profiling and comparison with microarray analysis. BMC Genomics 10:407

Chen J, Odenike O, Rowley JD (2010) Leukaemogenesis: more than mutant genes. Nat Rev Cancer 10(1):23–36

Cheng AM, Byrom MW, Shelton J, Ford LP (2005) Antisense inhibition of human miRNAs and indications for an involvement of miRNA in cell growth and apoptosis. Nucleic Acids Res 33(4):1290–1297

Childs G, Fazzari M, Kung G, Kawachi N, Brandwein-Gensler M, McLemore M et al (2009) Low-level expression of microRNAs let-7d and miR-205 are prognostic markers of head and neck squamous cell carcinoma. Am J Pathol 174(3):736–745

Cho WC (2010a) MicroRNAs in cancer – from research to therapy. Biochim Biophys Acta 1805(2):209–217

Cho WC (2010b) MicroRNAs: potential biomarkers for cancer diagnosis, prognosis and targets for therapy. Int J Biochem Cell Biol 42(8):1273–1281

Chung AS, Lee J, Ferrara N (2010) Targeting the tumour vasculature: insights from physiological angiogenesis. Nat Rev Cancer 10(7):505–514

Cimmino A, Calin GA, Fabbri M, Iorio MV, Ferracin M, Shimizu M et al (2005) miR-15 and miR-16 induce apoptosis by targeting BCL2. Proc Natl Acad Sci U S A 102(39):13944–13949

Connolly E, Melegari M, Landgraf P, Tchaikovskaya T, Tennant BC, Slagle BL et al (2008) Elevated expression of the miR-17-92 polycistron and miR-21 in hepadnavirus-associated hepatocellular carcinoma contributes to the malignant phenotype. Am J Pathol 173(3):856–864

Corney DC, Flesken-Nikitin A, Godwin AK, Wang W, Nikitin AY (2007) MicroRNA-34b and MicroRNA-34c are targets of p53 and cooperate in control of cell proliferation and adhesion-independent growth. Cancer Res 67(18):8433–8438

Croce CM (2009) Causes and consequences of microRNA dysregulation in cancer. Nat Rev Genet 10(10):704–714

Cullen BR (2004) Derivation and function of small interfering RNAs and microRNAs. Virus Res 102(1):3–9

Cummins JM, He Y, Leary RJ, Pagliarini R, Diaz LA Jr, Sjoblom T et al (2006) The colorectal microRNAome. Proc Natl Acad Sci U S A 103(10):3687–3692

Datta J, Kutay H, Nasser MW, Nuovo GJ, Wang B, Majumder S et al (2008) Methylation mediated silencing of MicroRNA-1 gene and its role in hepatocellular carcinogenesis. Cancer Res 68(13):5049–5058

Denli AM, Tops BB, Plasterk RH, Ketting RF, Hannon GJ (2004) Processing of primary microRNAs by the Microprocessor complex. Nature 432(7014):231–235

Dews M, Homayouni A, Yu D, Murphy D, Sevignani C, Wentzel E et al (2006) Augmentation of tumor angiogenesis by a Myc-activated microRNA cluster. Nat Genet 38(9):1060–1065

Dutta KK, Zhong Y, Liu YT, Yamada T, Akatsuka S, Hu Q et al (2007) Association of microRNA-34a overexpression with proliferation is cell type-dependent. Cancer Sci 98(12):1845–1852

Dykxhoorn DM, Wu Y, Xie H, Yu F, Lal A, Petrocca F et al (2009) miR-200 enhances mouse breast cancer cell colonization to form distant metastases. PLoS One 4(9):e7181

Egle A, Harris AW, Bouillet P, Cory S (2004) Bim is a suppressor of Myc-induced mouse B cell leukemia. Proc Natl Acad Sci U S A 101(16):6164–6169

Esquela-Kerscher A, Slack FJ (2006) Oncomirs – microRNAs with a role in cancer. Nat Rev Cancer 6(4):259–269

Esquela-Kerscher A, Trang P, Wiggins JF, Patrawala L, Cheng A, Ford L et al (2008) The let-7 microRNA reduces tumor growth in mouse models of lung cancer. Cell Cycle 7(6):759–764

Fabbri M, Garzon R, Cimmino A, Liu Z, Zanesi N, Callegari E et al (2007) MicroRNA-29 family reverts aberrant methylation in lung cancer by targeting DNA methyltransferases 3A and 3B. Proc Natl Acad Sci U S A 104(40):15805–15810

Fabian MR, Sonenberg N (2012) The mechanics of miRNA-mediated gene silencing: a look under the hood of miRISC. Nat Struct Mol Biol 19(6):586–593

Faller M, Matsunaga M, Yin S, Loo JA, Guo F (2007) Heme is involved in microRNA processing. Nat Struct Mol Biol 14(1):23–29

Farazi TA, Horlings HM, Ten Hoeve JJ, Mihailovic A, Halfwerk H, Morozov P et al (2011) MicroRNA sequence and expression analysis in breast tumors by deep sequencing. Cancer Res 71(13):4443–4453

Felicetti F, Errico MC, Bottero L, Segnalini P, Stoppacciaro A, Biffoni M et al (2008) The promyelocytic leukemia zinc finger-microRNA-221/-222 pathway controls melanoma progression through multiple oncogenic mechanisms. Cancer Res 68(8):2745–2754
Felli N, Fontana L, Pelosi E, Botta R, Bonci D, Facchiano F et al (2005) MicroRNAs 221 and 222 inhibit normal erythropoiesis and erythroleukemic cell growth via kit receptor down-modulation. Proc Natl Acad Sci U S A 102(50):18081–18086
Fidler IJ (2003) The pathogenesis of cancer metastasis: the 'seed and soil' hypothesis revisited. Nat Rev Cancer 3(6):453–458
Fish JE, Santoro MM, Morton SU, Yu S, Yeh RF, Wythe JD et al (2008) miR-126 regulates angiogenic signaling and vascular integrity. Dev Cell 15(2):272–284
Friedman RC, Farh KK, Burge CB, Bartel DP (2009) Most mammalian mRNAs are conserved targets of microRNAs. Genome Res 19(1):92–105
Fuller CW, Middendorf LR, Benner SA, Church GM, Harris T, Huang X et al (2009) The challenges of sequencing by synthesis. Nat Biotechnol 27(11):1013–1023
Galardi S, Mercatelli N, Giorda E, Massalini S, Frajese GV, Ciafre SA et al (2007) miR-221 and miR-222 expression affects the proliferation potential of human prostate carcinoma cell lines by targeting p27Kip1. J Biol Chem 282(32):23716–23724
Gartel AL, Kandel ES (2008) miRNAs: little known mediators of oncogenesis. Semin Cancer Biol 18(2):103–110
Gillies JK, Lorimer IA (2007) Regulation of p27Kip1 by miRNA 221/222 in glioblastoma. Cell Cycle 6(16):2005–2009
Gottardo F, Liu CG, Ferracin M, Calin GA, Fassan M, Bassi P et al (2007) Micro-RNA profiling in kidney and bladder cancers. Urol Oncol 25(5):387–392
Grady WM, Parkin RK, Mitchell PS, Lee JH, Kim YH, Tsuchiya KD et al (2008) Epigenetic silencing of the intronic microRNA hsa-miR-342 and its host gene EVL in colorectal cancer. Oncogene 27(27):3880–3888
Gramantieri L, Ferracin M, Fornari F, Veronese A, Sabbioni S, Liu CG et al (2007) Cyclin G1 is a target of miR-122a, a microRNA frequently down-regulated in human hepatocellular carcinoma. Cancer Res 67(13):6092–6099
Gregory RI, Yan KP, Amuthan G, Chendrimada T, Doratotaj B, Cooch N et al (2004) The Microprocessor complex mediates the genesis of microRNAs. Nature 432(7014):235–240
Gregory PA, Bert AG, Paterson EL, Barry SC, Tsykin A, Farshid G et al (2008) The miR-200 family and miR-205 regulate epithelial to mesenchymal transition by targeting ZEB1 and SIP1. Nat Cell Biol 10(5):593–601
Gupta GP, Massague J (2006) Cancer metastasis: building a framework. Cell 127(4):679–695
Han J, Lee Y, Yeom KH, Nam JW, Heo I, Rhee JK et al (2006) Molecular basis for the recognition of primary microRNAs by the Drosha-DGCR8 complex. Cell 125(5):887–901
Harquail J, Benzina S, Robichaud GA (2012) MicroRNAs and breast cancer malignancy: an overview of miRNA-regulated cancer processes leading to metastasis. Cancer Biomark 11(6):269–280
Harris TA, Yamakuchi M, Ferlito M, Mendell JT, Lowenstein CJ (2008) MicroRNA-126 regulates endothelial expression of vascular cell adhesion molecule 1. Proc Natl Acad Sci U S A 105(5):1516–1521
Hayashita Y, Osada H, Tatematsu Y, Yamada H, Yanagisawa K, Tomida S et al (2005) A polycistronic microRNA cluster, miR-17-92, is overexpressed in human lung cancers and enhances cell proliferation. Cancer Res 65(21):9628–9632
He L, Thomson JM, Hemann MT, Hernando-Monge E, Mu D, Goodson S et al (2005) A microRNA polycistron as a potential human oncogene. Nature 435(7043):828–833
Hong L, Han Y, Zhou Y, Nita A (2013) Angiogenesis-related microRNAs in colon cancer. Expert Opin Biol Ther 13(1):77–84
Hu X, Macdonald DM, Huettner PC, Feng Z, El Naqa IM, Schwarz JK et al (2009) A miR-200 microRNA cluster as prognostic marker in advanced ovarian cancer. Gynecol Oncol 114(3):457–464
Huang Q, Gumireddy K, Schrier M, le Sage C, Nagel R, Nair S et al (2008) The microRNAs miR-373 and miR-520c promote tumour invasion and metastasis. Nat Cell Biol 10(2):202–210

Huang L, Dai T, Lin X, Zhao X, Chen X, Wang C et al (2012) MicroRNA-224 targets RKIP to control cell invasion and expression of metastasis genes in human breast cancer cells. Biochem Biophys Res Commun 425(2):127–133

Ibrahim SA, Yip GW, Stock C, Pan JW, Neubauer C, Poeter M et al (2012) Targeting of syndecan-1 by microRNA miR-10b promotes breast cancer cell motility and invasiveness via a Rho-GTPase- and E-cadherin-dependent mechanism. Int J Cancer 131(6):E884–E896

Ichimi T, Enokida H, Okuno Y, Kunimoto R, Chiyomaru T, Kawamoto K et al (2009) Identification of novel microRNA targets based on microRNA signatures in bladder cancer. Int J Cancer 125(2):345–352

Ikeda S, Kong SW, Lu J, Bisping E, Zhang H, Allen PD et al (2007) Altered microRNA expression in human heart disease. Physiol Genomics 31(3):367–373

Iorio MV, Visone R, Di Leva G, Donati V, Petrocca F, Casalini P et al (2007) MicroRNA signatures in human ovarian cancer. Cancer Res 67(18):8699–8707

Jiang J, Gusev Y, Aderca I, Mettler TA, Nagorney DM, Brackett DJ et al (2008) Association of microRNA expression in hepatocellular carcinomas with hepatitis infection, cirrhosis, and patient survival. Clin Cancer Res 14(2):419–427

Johnson SM, Grosshans H, Shingara J, Byrom M, Jarvis R, Cheng A et al (2005) RAS is regulated by the let-7 microRNA family. Cell 120(5):635–647

Kluiver J, Poppema S, de Jong D, Blokzijl T, Harms G, Jacobs S et al (2005) BIC and miR-155 are highly expressed in Hodgkin, primary mediastinal and diffuse large B cell lymphomas. J Pathol 207(2):243–249

Kondo N, Toyama T, Sugiura H, Fujii Y, Yamashita H (2008) miR-206 expression is down-regulated in estrogen receptor alpha-positive human breast cancer. Cancer Res 68(13):5004–5008

Kong W, Yang H, He L, Zhao JJ, Coppola D, Dalton WS et al (2008) MicroRNA-155 is regulated by the transforming growth factor beta/Smad pathway and contributes to epithelial cell plasticity by targeting RhoA. Mol Cell Biol 28(22):6773–6784

Korpal M, Lee ES, Hu G, Kang Y (2008) The miR-200 family inhibits epithelial-mesenchymal transition and cancer cell migration by direct targeting of E-cadherin transcriptional repressors ZEB1 and ZEB2. J Biol Chem 283(22):14910–14914

Krek A, Grun D, Poy MN, Wolf R, Rosenberg L, Epstein EJ et al (2005) Combinatorial microRNA target predictions. Nat Genet 37(5):495–500

Krutzfeldt J, Rajewsky N, Braich R, Rajeev KG, Tuschl T, Manoharan M et al (2005) Silencing of microRNAs in vivo with 'antagomirs'. Nature 438(7068):685–689

Kuehbacher A, Urbich C, Zeiher AM, Dimmeler S (2007) Role of Dicer and Drosha for endothelial microRNA expression and angiogenesis. Circ Res 101(1):59–68

Kumar MS, Erkeland SJ, Pester RE, Chen CY, Ebert MS, Sharp PA et al (2008) Suppression of non-small cell lung tumor development by the let-7 microRNA family. Proc Natl Acad Sci U S A 105(10):3903–3908

Kumar B, Yadav A, Lang J, Teknos TN, Kumar P (2012) Dysregulation of microRNA-34a expression in head and neck squamous cell carcinoma promotes tumor growth and tumor angiogenesis. PLoS One 7(5):e37601

Lal A, Navarro F, Maher CA, Maliszewski LE, Yan N, O'Day E et al (2009) miR-24 inhibits cell proliferation by targeting E2F2, MYC, and other cell-cycle genes via binding to "seedless" 3′UTR microRNA recognition elements. Mol Cell 35(5):610–625

Landskroner-Eiger S, Moneke I, Sessa WC (2013) miRNAs as modulators of angiogenesis. Cold Spring Harb Perspect Med 3(2):a006643

Le XF, Almeida MI, Mao W, Spizzo R, Rossi S, Nicoloso MS et al (2012) Modulation of microRNA-194 and cell migration by HER2-targeting trastuzumab in breast cancer. PLoS One 7(7):e41170

Lee RC, Feinbaum RL, Ambros V (1993) The C. elegans heterochronic gene lin-4 encodes small RNAs with antisense complementarity to lin-14. Cell 75(5):843–854

Lee Y, Ahn C, Han J, Choi H, Kim J, Yim J et al (2003) The nuclear RNase III Drosha initiates microRNA processing. Nature 425(6956):415–419

Lee DY, Deng Z, Wang CH, Yang BB (2007a) MicroRNA-378 promotes cell survival, tumor growth, and angiogenesis by targeting SuFu and Fus-1 expression. Proc Natl Acad Sci U S A 104(51):20350–20355

Lee EJ, Gusev Y, Jiang J, Nuovo GJ, Lerner MR, Frankel WL et al (2007b) Expression profiling identifies microRNA signature in pancreatic cancer. Int J Cancer 120(5):1046–1054

Lee I, Ajay SS, Yook JI, Kim HS, Hong SH, Kim NH et al (2009) New class of microRNA targets containing simultaneous 5′-UTR and 3′-UTR interaction sites. Genome Res 19(7):1175–1183

Lehmann U, Hasemeier B, Christgen M, Muller M, Romermann D, Langer F et al (2008) Epigenetic inactivation of microRNA gene hsa-mir-9-1 in human breast cancer. J Pathol 214(1):17–24

Lewis BP, Shih IH, Jones-Rhoades MW, Bartel DP, Burge CB (2003) Prediction of mammalian microRNA targets. Cell 115(7):787–798

Li XF, Yan PJ, Shao ZM (2009) Downregulation of miR-193b contributes to enhance urokinase-type plasminogen activator (uPA) expression and tumor progression and invasion in human breast cancer. Oncogene 28(44):3937–3948

Liu X, Fortin K, Mourelatos Z (2008) MicroRNAs: biogenesis and molecular functions. Brain Pathol 18(1):113–121

Liu X, Jiang L, Wang A, Yu J, Shi F, Zhou X (2009) MicroRNA-138 suppresses invasion and promotes apoptosis in head and neck squamous cell carcinoma cell lines. Cancer Lett 286(2):217–222

Liu Q, Lv GD, Qin X, Gen YH, Zheng ST, Liu T et al (2012) Role of microRNA let-7 and effect to HMGA2 in esophageal squamous cell carcinoma. Mol Biol Rep 39(2):1239–1246

Lopez JI, Camenisch TD, Stevens MV, Sands BJ, McDonald J, Schroeder JA (2005) CD44 attenuates metastatic invasion during breast cancer progression. Cancer Res 65(15):6755–6763

Lowery AJ, Miller N, McNeill RE, Kerin MJ (2008) MicroRNAs as prognostic indicators and therapeutic targets: potential effect on breast cancer management. Clin Cancer Res 14(2):360–365

Lu J, Getz G, Miska EA, Alvarez-Saavedra E, Lamb J, Peck D et al (2005) MicroRNA expression profiles classify human cancers. Nature 435(7043):834–838

Lujambio A, Calin GA, Villanueva A, Ropero S, Sanchez-Cespedes M, Blanco D et al (2008) A microRNA DNA methylation signature for human cancer metastasis. Proc Natl Acad Sci U S A 105(36):13556–13561

Lund E, Guttinger S, Calado A, Dahlberg JE, Kutay U (2004) Nuclear export of microRNA precursors. Science 303(5654):95–98

Lynam-Lennon N, Maher SG, Reynolds JV (2009) The roles of microRNA in cancer and apoptosis. Biol Rev Camb Philos Soc 84(1):55–71

Ma L, Teruya-Feldstein J, Weinberg RA (2007) Tumour invasion and metastasis initiated by microRNA-10b in breast cancer. Nature 449(7163):682–688

Ma L, Reinhardt F, Pan E, Soutschek J, Bhat B, Marcusson EG et al (2010) Therapeutic silencing of miR-10b inhibits metastasis in a mouse mammary tumor model. Nat Biotechnol 28(4):341–347

Marton S, Garcia MR, Robello C, Persson H, Trajtenberg F, Pritsch O et al (2008) Small RNAs analysis in CLL reveals a deregulation of miRNA expression and novel miRNA candidates of putative relevance in CLL pathogenesis. Leukemia 22(2):330–338

Mendell JT (2008) miRiad roles for the miR-17-92 cluster in development and disease. Cell 133(2):217–222

Meng F, Henson R, Lang M, Wehbe H, Maheshwari S, Mendell JT et al (2006) Involvement of human micro-RNA in growth and response to chemotherapy in human cholangiocarcinoma cell lines. Gastroenterology 130(7):2113–2129

Mertens-Talcott SU, Chintharlapalli S, Li X, Safe S (2007) The oncogenic microRNA-27a targets genes that regulate specificity protein transcription factors and the G2-M checkpoint in MDA-MB-231 breast cancer cells. Cancer Res 67(22):11001–11011

Miller TE, Ghoshal K, Ramaswamy B, Roy S, Datta J, Shapiro CL et al (2008) MicroRNA-221/222 confers tamoxifen resistance in breast cancer by targeting p27Kip1. J Biol Chem 283(44):29897–29903

Morozova O, Marra MA (2008) Applications of next-generation sequencing technologies in functional genomics. Genomics 92(5):255–264

Muller DW, Bosserhoff AK (2008) Integrin beta 3 expression is regulated by let-7a miRNA in malignant melanoma. Oncogene 27(52):6698–6706

Muramatsu F, Kidoya H, Naito H, Sakimoto S, Takakura N (2013) microRNA-125b inhibits tube formation of blood vessels through translational suppression of VE-cadherin. Oncogene 32(4):414–421

Nam EJ, Yoon H, Kim SW, Kim H, Kim YT, Kim JH et al (2008) MicroRNA expression profiles in serous ovarian carcinoma. Clin Cancer Res 14(9):2690–2695

Nasser MW, Datta J, Nuovo G, Kutay H, Motiwala T, Majumder S et al (2008) Down-regulation of micro-RNA-1 (miR-1) in lung cancer. Suppression of tumorigenic property of lung cancer cells and their sensitization to doxorubicin-induced apoptosis by miR-1. J Biol Chem 283(48):33394–33405

Nicoloso MS, Spizzo R, Shimizu M, Rossi S, Calin GA (2009) MicroRNAs–the micro steering wheel of tumour metastases. Nat Rev Cancer 9(4):293–302

Nikiforova MN, Tseng GC, Steward D, Diorio D, Nikiforov YE (2008) MicroRNA expression profiling of thyroid tumors: biological significance and diagnostic utility. J Clin Endocrinol Metab 93(5):1600–1608

Noonan EJ, Place RF, Pookot D, Basak S, Whitson JM, Hirata H et al (2009) miR-449a targets HDAC-1 and induces growth arrest in prostate cancer. Oncogene 28(14):1714–1724

O'Donnell KA, Wentzel EA, Zeller KI, Dang CV, Mendell JT (2005) c-Myc-regulated microRNAs modulate E2F1 expression. Nature 435(7043):839–843

Orom UA, Kauppinen S, Lund AH (2006) LNA-modified oligonucleotides mediate specific inhibition of microRNA function. Gene 372:137–141

Paraskevopoulou MD, Georgakilas G, Kostoulas N, Vlachos IS, Vergoulis T, Reczko M et al (2013) DIANA-microT web server v5.0: service integration into miRNA functional analysis workflows. Nucleic Acids Res 41(Web Server issue):W169–W173

Park SM, Gaur AB, Lengyel E, Peter ME (2008) The miR-200 family determines the epithelial phenotype of cancer cells by targeting the E-cadherin repressors ZEB1 and ZEB2. Genes Dev 22(7):894–907

Pasquinelli AE (2012) MicroRNAs and their targets: recognition, regulation and an emerging reciprocal relationship. Nat Rev Genet 13(4):271–282

Pasquinelli AE, Reinhart BJ, Slack F, Martindale MQ, Kuroda MI, Maller B et al (2000) Conservation of the sequence and temporal expression of let-7 heterochronic regulatory RNA. Nature 408(6808):86–89

Pekarsky Y, Santanam U, Cimmino A, Palamarchuk A, Efanov A, Maximov V et al (2006) Tcl1 expression in chronic lymphocytic leukemia is regulated by miR-29 and miR-181. Cancer Res 66(24):11590–11593

Petriv OI, Kuchenbauer F, Delaney AD, Lecault V, White A, Kent D et al (2010) Comprehensive microRNA expression profiling of the hematopoietic hierarchy. Proc Natl Acad Sci U S A 107(35):15443–15448

Pfeffer S, Sewer A, Lagos-Quintana M, Sheridan R, Sander C, Grasser FA et al (2005) Identification of microRNAs of the herpesvirus family. Nat Methods 2(4):269–276

Poliseno L, Tuccoli A, Mariani L, Evangelista M, Citti L, Woods K et al (2006) MicroRNAs modulate the angiogenic properties of HUVECs. Blood 108(9):3068–3071

Reinhart BJ, Slack FJ, Basson M, Pasquinelli AE, Bettinger JC, Rougvie AE et al (2000) The 21-nucleotide let-7 RNA regulates developmental timing in Caenorhabditis elegans. Nature 403(6772):901–906

Saito Y, Liang G, Egger G, Friedman JM, Chuang JC, Coetzee GA et al (2006) Specific activation of microRNA-127 with downregulation of the proto-oncogene BCL6 by chromatin-modifying drugs in human cancer cells. Cancer Cell 9(6):435–443

Schmittgen TD, Lee EJ, Jiang J, Sarkar A, Yang L, Elton TS et al (2008) Real-time PCR quantification of precursor and mature microRNA. Methods 44(1):31–38

Scott GK, Goga A, Bhaumik D, Berger CE, Sullivan CS, Benz CC (2007) Coordinate suppression of ERBB2 and ERBB3 by enforced expression of micro-RNA miR-125a or miR-125b. J Biol Chem 282(2):1479–1486

Shih AH, Holland EC (2006) Platelet-derived growth factor (PDGF) and glial tumorigenesis. Cancer Lett 232(2):139–147
Siragam V, Rutnam ZJ, Yang W, Fang L, Luo L, Yang X et al (2012) MicroRNA miR-98 inhibits tumor angiogenesis and invasion by targeting activin receptor-like kinase-4 and matrix metalloproteinase-11. Oncotarget 3(11):1370–1385
Slaby O, Svoboda M, Fabian P, Smerdova T, Knoflickova D, Bednarikova M et al (2007) Altered expression of miR-21, miR-31, miR-143 and miR-145 is related to clinicopathologic features of colorectal cancer. Oncology 72(5–6):397–402
Spizzo R, Nicoloso MS, Croce CM, Calin GA (2009) SnapShot: microRNAs in cancer. Cell 137(3):586–586 e581
Stamatopoulos B, Meuleman N, Haibe-Kains B, Saussoy P, Van Den Neste E, Michaux L et al (2009) microRNA-29c and microRNA-223 down-regulation has in vivo significance in chronic lymphocytic leukemia and improves disease risk stratification. Blood 113(21):5237–5245
Steeg PS (2003) Metastasis suppressors alter the signal transduction of cancer cells. Nat Rev Cancer 3(1):55–63
Su H, Yang JR, Xu T, Huang J, Xu L, Yuan Y et al (2009) MicroRNA-101, down-regulated in hepatocellular carcinoma, promotes apoptosis and suppresses tumorigenicity. Cancer Res 69(3):1135–1142
Suarez Y, Fernandez-Hernando C, Pober JS, Sessa WC (2007) Dicer dependent microRNAs regulate gene expression and functions in human endothelial cells. Circ Res 100(8):1164–1173
Suarez Y, Fernandez-Hernando C, Yu J, Gerber SA, Harrison KD, Pober JS et al (2008) Dicer-dependent endothelial microRNAs are necessary for postnatal angiogenesis. Proc Natl Acad Sci U S A 105(37):14082–14087
Takamizawa J, Konishi H, Yanagisawa K, Tomida S, Osada H, Endoh H et al (2004) Reduced expression of the let-7 microRNAs in human lung cancers in association with shortened postoperative survival. Cancer Res 64(11):3753–3756
Tang F, Zhang R, He Y, Zou M, Guo L, Xi T (2012) MicroRNA-125b induces metastasis by targeting STARD13 in MCF-7 and MDA-MB-231 breast cancer cells. PLoS One 7(5):e35435
Tavazoie SF, Alarcon C, Oskarsson T, Padua D, Wang Q, Bos PD et al (2008) Endogenous human microRNAs that suppress breast cancer metastasis. Nature 451(7175):147–152
Tazawa H, Tsuchiya N, Izumiya M, Nakagama H (2007) Tumor-suppressive miR-34a induces senescence-like growth arrest through modulation of the E2F pathway in human colon cancer cells. Proc Natl Acad Sci U S A 104(39):15472–15477
Thiery JP (2002) Epithelial-mesenchymal transitions in tumour progression. Nat Rev Cancer 2(6):442–454
Togawa H, Nakanishi K, Mukaiyama H, Hama T, Shima Y, Sako M et al (2011) Epithelial-to-mesenchymal transition in cyst lining epithelial cells in an orthologous PCK rat model of autosomal-recessive polycystic kidney disease. Am J Physiol Renal Physiol 300(2):F511–F520
Tong AW, Nemunaitis J (2008) Modulation of miRNA activity in human cancer: a new paradigm for cancer gene therapy? Cancer Gene Ther 15(6):341–355
Urbich C, Kuehbacher A, Dimmeler S (2008) Role of microRNAs in vascular diseases, inflammation, and angiogenesis. Cardiovasc Res 79(4):581–588
Valastyan S, Reinhardt F, Benaich N, Calogrias D, Szasz AM, Wang ZC et al (2009) A pleiotropically acting microRNA, miR-31, inhibits breast cancer metastasis. Cell 137(6):1032–1046
Varambally S, Cao Q, Mani RS, Shankar S, Wang X, Ateeq B et al (2008) Genomic loss of microRNA-101 leads to overexpression of histone methyltransferase EZH2 in cancer. Science 322(5908):1695–1699
Visone R, Pallante P, Vecchione A, Cirombella R, Ferracin M, Ferraro A et al (2007) Specific microRNAs are downregulated in human thyroid anaplastic carcinomas. Oncogene 26(54):7590–7595
Volinia S, Calin GA, Liu CG, Ambs S, Cimmino A, Petrocca F et al (2006) A microRNA expression signature of human solid tumors defines cancer gene targets. Proc Natl Acad Sci U S A 103(7):2257–2261

Wang S, Aurora AB, Johnson BA, Qi X, McAnally J, Hill JA et al (2008a) The endothelial-specific microRNA miR-126 governs vascular integrity and angiogenesis. Dev Cell 15(2):261–271

Wang Y, Lee AT, Ma JZ, Wang J, Ren J, Yang Y et al (2008b) Profiling microRNA expression in hepatocellular carcinoma reveals microRNA-224 up-regulation and apoptosis inhibitor-5 as a microRNA-224-specific target. J Biol Chem 283(19):13205–13215

Webster RJ, Giles KM, Price KJ, Zhang PM, Mattick JS, Leedman PJ (2009) Regulation of epidermal growth factor receptor signaling in human cancer cells by microRNA-7. J Biol Chem 284(9):5731–5741

Woods K, Thomson JM, Hammond SM (2007) Direct regulation of an oncogenic micro-RNA cluster by E2F transcription factors. J Biol Chem 282(4):2130–2134

Wu H, Xu H, Miraglia LJ, Crooke ST (2000) Human RNase III is a 160-kDa protein involved in preribosomal RNA processing. J Biol Chem 275(47):36957–36965

Wu W, Sun M, Zou GM, Chen J (2007) MicroRNA and cancer: current status and prospective. Int J Cancer 120(5):953–960

Wurdinger T, Tannous BA, Saydam O, Skog J, Grau S, Soutschek J et al (2008) miR-296 regulates growth factor receptor overexpression in angiogenic endothelial cells. Cancer Cell 14(5):382–393

Xiao C, Srinivasan L, Calado DP, Patterson HC, Zhang B, Wang J et al (2008) Lymphoproliferative disease and autoimmunity in mice with increased miR-17-92 expression in lymphocytes. Nat Immunol 9(4):405–414

Yan D, Zhou X, Chen X, Hu DN, Dong XD, Wang J et al (2009) MicroRNA-34a inhibits uveal melanoma cell proliferation and migration through downregulation of c-Met. Invest Ophthalmol Vis Sci 50(4):1559–1565

Yanaihara N, Caplen N, Bowman E, Seike M, Kumamoto K, Yi M et al (2006) Unique microRNA molecular profiles in lung cancer diagnosis and prognosis. Cancer Cell 9(3):189–198

Yang J, Weinberg RA (2008) Epithelial-mesenchymal transition: at the crossroads of development and tumor metastasis. Dev Cell 14(6):818–829

Yang WJ, Yang DD, Na S, Sandusky GE, Zhang Q, Zhao G (2005) Dicer is required for embryonic angiogenesis during mouse development. J Biol Chem 280(10):9330–9335

Yang H, Kong W, He L, Zhao JJ, O'Donnell JD, Wang J et al (2008) MicroRNA expression profiling in human ovarian cancer: miR-214 induces cell survival and cisplatin resistance by targeting PTEN. Cancer Res 68(2):425–433

Yi R, Qin Y, Macara IG, Cullen BR (2003) Exportin-5 mediates the nuclear export of pre-microRNAs and short hairpin RNAs. Genes Dev 17(24):3011–3016

Yu F, Yao H, Zhu P, Zhang X, Pan Q, Gong C et al (2007) let-7 regulates self renewal and tumorigenicity of breast cancer cells. Cell 131(6):1109–1123

Yu Z, Wang C, Wang M, Li Z, Casimiro MC, Liu M et al (2008) A cyclin D1/microRNA 17/20 regulatory feedback loop in control of breast cancer cell proliferation. J Cell Biol 182(3):509–517

Zhang X, Liu S, Hu T, He Y, Sun S (2009) Up-regulated microRNA-143 transcribed by nuclear factor kappa B enhances hepatocarcinoma metastasis by repressing fibronectin expression. Hepatology 50(2):490–499

Zhao H, Wang D, Du W, Gu D, Yang R (2010) MicroRNA and leukemia: tiny molecule, great function. Crit Rev Oncol Hematol 74(3):149–155

Zhong X, Coukos G, Zhang L (2012) miRNAs in human cancer. Methods Mol Biol 822:295–306

Zhu S, Si ML, Wu H, Mo YY (2007) MicroRNA-21 targets the tumor suppressor gene tropomyosin 1 (TPM1). J Biol Chem 282(19):14328–14336

Zhu S, Wu H, Wu F, Nie D, Sheng S, Mo YY (2008) MicroRNA-21 targets tumor suppressor genes in invasion and metastasis. Cell Res 18(3):350–359

Zhuang G, Wu X, Jiang Z, Kasman I, Yao J, Guan Y et al (2012) Tumour-secreted miR-9 promotes endothelial cell migration and angiogenesis by activating the JAK-STAT pathway. EMBO J 31(17):3513–3523

Chapter 2
Epigenetics, MicroRNAs and Human Cancer

Jiazeng Xia, Xiaoqiang Guo, and Kaiyuan Deng

Contents

Abstract Classical genetics alone cannot explain how cancer occurs pretty well, and the proposal of concept of epigenetics provides a partial explanation about the cause of cancers. DNA methylation and histone modifications are the best-known epigenetic marks. MicroRNAs (miRNAs), a class of endogenous, single-stranded, non-coding small RNA with 18–22 nucleotides in length, play a critical role in initiation, progression, metastasis and invasion of cancers. It is widely recognized that deregulation of miRNAs is a hallmark of cancer. The expression of miRNAs can be regulated by several mechanisms, including epigenetic changes. Furthermore, it has been discovered that a subgroup of miRNAs, which are known as epi-miRNAs,

J. Xia (✉) • X. Guo • K. Deng
Department of General Surgery and Center of Translational Medicine, Nanjing Medical University Affiliated Wuxi Second Hospital, 214002 Wuxi, Jiangsu, China
e-mail: jiazengxia@yahoo.com

S. Babashah (ed.), *MicroRNAs: Key Regulators of Oncogenesis*,
DOI 10.1007/978-3-319-03725-7_2, © Springer International Publishing Switzerland 2014

can regulate the expression of effectors of the epigenetic mechanisms by directly or indirectly targeting these epigenetic-modifying enzymes and molecules. This chapter will focus on how epigenetic changes regulate the miRNAs expression as well as how epi-miRNAs affect the epigenome, and how to translate these findings into clinical application.

Keywords Epigenetics • DNA methylation • Histone modifications • MicroRNA • Human cancer

1 Brief Introduction of Epigenetics and Common Epigenetic Mechanisms

The human genome contains two types of genetic information: one is genetic information in the traditional sense; the other is the epigenetic information. In 1942, the term "epigenetics" was proposed for the first time by Waddington in a paper entitled "The epigenotype Endeavour", and it was defined as "the causal interactions between genes and their products which bring the phenotype into being" (Waddington 2012). "Epigenetics" is derived from "epigenesis" and can be split into two parts, which are "epi" and "genetics". "Epi" is a Greek prefix that means "above" or beyond, hence the word indicated that epigenetic events were needed to be studied above or beyond genetics. However, the word "epigenetics" was rarely mentioned in the following three decades. Even if it was mentioned, the meaning of it was not the same as the definition proposed by Waddington. In 1980s, some scholars began to use the word "epigenetics" in the same meaning as Waddington's definition. By the early 1990s, epigenetics entered into the fast development period. With a deeper knowledge of the phenomenon of epigenetics, the meaning of "epigenetics" has been constantly evolved. Nowadays, the meaning of "epigenetics" is different from the definition given by Waddington. At present, the widely accepted definition of "epigenetic" is that there are mitotically and potential meiotically heritable alternations in gene expression without any concomitant changes in original DNA sequence (Taby and Issa 2010). In other words, the phenotype changes while genotype does not change. The epigenetics mainly includes three characteristics: heritability, reversibility and no underlying DNA sequence alteration. There are several epigenetic modifications that can influence DNA, RNA or protein expression, the most common of which are DNA methylation and histone modifications.

1.1 DNA Methylation

DNA methylation is the epigenetic change which is studied earliest and most completely. It occurs when a methyl group from active methylene compounds, such as S-adenosylmethionine (SAM), is transferred into C5 position of cytosine ring in CpG dinucleotide under the catalysis of DNA methyltransferases (DNMTs). There are two kinds of distribution of CpG dinucleotides in human genome: genome-wide dispersed distribution and local concentration distribution. Approximately 80 % of CpG dinucleotides are genome-wide dispersed distribution and are usually located in repetitive DNA sequences, such as LINE and Alu sequences, and the other CpG dinucleotides are mainly concentrated in CpG islands (CGI). CGI, a 1–2 kb genomic region with CpG dinucleotides cluster in mammals, is mainly located in promoter region of approximately 50 % of human genes, but sometimes it can also be found in the first exon or 5′ untranslated region of genes (Dunn 2003). In a healthy cell, the genome-wide scattered distribution of CpG dinucleotides are always heavily methylated, whereas the CpG dinucleotides in CGI are protected from methylation. Aberrant DNA methylation of CGI has been shown to play an important role in gene silencing, genomic imprinting, inactivation of X chromosome in women and carcinogenesis (Bird 2002). In addition, DNA methylation of CGI in promoter region can regulate expression of genes through repression of transcription factors, such as E2F, binding to corresponding locus of DNA sequences and recruitment of methyl-CpG binding domain proteins (MBDs), including MBD1-4 and MeCP2, which can recruit enzymatic machinery to establish silent chromatin (Campanero et al. 2000; Bogdanovic and Veenstra 2009).

The addition of a methyl group to C5 position of cytosine is catalyzed by a family of enzymes, DNMTs. The active DNMTs mainly include DNMT1, DNMT3a and DNMT3b. DNMT1, as a maintenance methyltransferase, plays a critical role in maintaining the methylation patterns via replication of methylation patterns during S phase of mitosis (Leonhardt et al. 1992). Nevertheless, DNMT3a and DNMT3b act as *de novo* methyltransferases that methylate the unmethylated genomic regions (Okano et al. 1999). Hence, DNMT1 can unite with DNMT3a/3b to establish and maintain the DNA methylation patterns. DNMT2 is also the member of DNMT family, but its catalytic activity is very weak. DNA methyltransferase 3-like (DNMT3L) is a member of DNMT3 family, and it has been found to enhance the catalytic activity of DNMT3a/3b through binding to their catalytic domains (Gowher et al. 2005). DNA methylation is a reversible process, and DNMT inhibitors can be used to reduce the level of methylation. Among these inhibitors, 5-aza-2′-deoxycytidine (5-aza-dC) may be the most commonly used one. The 5-aza-dC is a cytidine analog, and inhibits the function of DNMTs by covalently bonded with them.

1.2 Histone Modifications

In eukaryotic cells, the basic unit of chromatin is the nucleosome, which consists of 147 bp of DNA wrapped around two copies of each of the core histones H2A, H2B, H3 and H4. Moreover, H1, known as a linker histone, plays a role in linking nucleosome to each other. In the early 1960s, Allfery first reported the acetylation and methylation of histones and supposed that the two histone modifications might play role in regulation of transcription (Allfrey et al. 1964). Through studies for almost half century, we have known that there are many other post-translational histone modifications. Moreover, in addition to the effect on transcription, histone modifications have also been found to be associated with various other fundamental biological processes, such as DNA repair, DNA replication and chromosome condensation (Kouzarides 2007). C-terminal globular domain of histones is involved in interactions between histones and organization of the DNA wrapped around the histones (Peterson and Laniel 2004), whereas the N-terminal tails can undergo a variety of post-translational modifications, including acetylation, methylation, phosphorylation, sumoylation, ubiquitination, ADP ribosylation, etc. By these chemical modifications, histone proteins can store epigenetic information, which is defined as "histone code", and induce the switch versus heterochromatin or euchromatin (Iorio et al. 2010). So far, the most studied histone modifications should be methylation and acetylation of special residues on histones H3 and H4. Moreover, several enzymes have been found to play the role in activation or repression of transcription through transferring or removing methyl group and acetyl group to histones, including histone methyl-transferases (HMTs), histone de-methylases (HDMs), histone acetyl-transferases (HATs) and histone de-acetylases (HDACs).

1.2.1 Histone Acetylation

Since the first report by Allfery et al. (1964), histone acetylation has been paid more and more attention. It has been found that the lysine's positive charge can be neutralized by combined with acetyl groups with negative charge, which may decrease the affinity between Histone octamer and DNA (Hong et al. 1993). Hence, histone acetylation can lead to relaxation of nucleosome structure, allowing that the transcription factors and cofactors bind to DNA binding sites, which active the gene transcription. As we known, histone acetylation that occurs at lysine residues in N-terminal tails is a dynamic process, and it is reversible. Histone acetylation is regulated by two enzymatic families, HATs and HDACs, which play the opposite effect. HATs can add the acetyl group to ε-amino group of lysine residues with the help of acetyl-CoA cofactor, and induces activation of transcription. According to the acting position and function, HATs can be divided into two types: the cytoplasmic type-B HATs which acetylate free histone and the nuclear type-A HATs which acetylate nucleosome histones and play a major role in transcriptional regulation. In addition to acetylate histones, HATs have also been found to acetylate non-histone proteins, such as p53 (Glozak et al. 2005). Compared with type-B HATs, type-A HATs are more diversified. Type-A HATs can be divided into three major families based on the sequence homology and structure: general control non-derepressible 5 (GCN5)-related N-acetyl-transferase (GNAT), MYST family and p300/CREB-binding protein (CBP) (Hodawadekar and Marmorstein 2007). These enzymes have been found to exist in multisubunit complexes, in which the specific functions of catalytic subunits are regulated by the other non-catalytic subunits (Lee and Workman 2007). Different HAT complexes that are made up of unique sub-strates generally have different residences and distinctive features, whereas several complexes which share many same subunits can also have specific functions. For example, SAGA complex and NuA3 complex have overlapped substrates, but the former gives priority to modify H3K9 and the latter preferentially modifies H3K14 (John et al. 2000).

Nevertheless, HDACs are a family of enzymes that reverse the acetylation of histone proteins through removing the acetyl group from lysine residues, which can induce euchromosome into heterochromosome, resulting in repression of transcription. Based on the sequence homology as well as structural and functional differences, HDACs can be divided into four classes: Classes I, II, III and IV (Brandl et al. 2009). Class I HDACs that include HDAC 1–3 and 8 are homologous to yeast scRpd3 and are widely expressed, whereas Class II HDACs that include HDAC 4–7 and HDAC 9–10 are most closely related to yeast scHda1. Moreover, according to domain organization, Class II HDACs can be further subdivided into Class IIa HDACs which include HDAC 4, 5, 7 and 9 and Class IIb HDACs which include HDAC 6 and 10, and they are expressed in a cell-special manner. Class IV HDACs only have one member, HDAC11, and its function is poorly understood. Unlike the above mentioned three classes that require a zinc metal ion to play a catalytic role, the Class III HDACs, including sirtuin 1–7, that are homologous to yeast scSir2 have deacetylase activity depending on a specific cofactor, NAD (nicotinamide adenine dinucleotide)$^+$. Similar to HATs, HDACs can target histone proteins as well as non-histone proteins, and play catalytic activity by existing in multisubunit complexes, such as Sin3, NCoR/SMR, NuRD and CoREST (Hayakawa and Nakayama 2011).

1.2.2 Histone Methylation

Histone methylation is another common histone modification. Different from histone acetylation, histone methylation does not neutralize the positive charge of histones, and it can not only occur on lysine residues, but also on the arginine and histidine residues. Lysine residues can be mono-, di-, or tri-methylated, and arginine residues can be mono-methylated, symmetrically or asymmetrically di-methylated. Histidine residues have also been found to be mono-methylated, whereas this type of methylation is seldom studied. The histone methylation which occurs on the lysine residues may be the most widely studied one, and the methylated sites of lysine mainly include H3 lysine 4 (H3K4), H3K9, H3K27, H3K36, H3K79, and H4K20 (Greer and Shi 2012). The addition of a methyl group from SAM to ε-amino group of lysine residues is catalyzed by a family of enzymes, histone lysine methyltransferases (HKMTs). Since the first HKMT, SUV39H1, which can specifically methylate H3K9 was reported in human (Rea et al. 2000), a number of HKMTs have been found. HKMTs can be divided into two types: SET domain-containing HKMTs and non-SET domain HKMTs. Almost all of the HKMTs contain the SET domain which harbours the methyltransferase activity, whereas DOT1-like protein (DOT1L) as the unique H3K79 methyltransferase does not contain the SET domain. The reason for the structural difference of DOT1L is not sufficiently clear, but DOT1L-mediated H3K79 methylation has been found to be involved in regulation of gene expression, regulation of cell cycle and DNA damage response, etc. (Nguyen and Zhang 2011).

Calmodulin methyltransferase is another non-SET domain HKMT, and it has been reported to trimethylate Lys-115 (Magnani et al. 2010). HKMTs can specifically target lysine residues and modify the certain lysines to certain degree of methylation. For example, human HKMT SET7/9 targets H3K4 and can exclusively mono-methylate the target (Xiao et al. 2003). Arginine methylation is also an important type of histone methylation, and the methylated sites of arginine mainly include H3 arginine 2 (H3R2), H3R8, H3R17, H3R26 and H4R3 (Greer and Shi 2012). Arginine methylation has been found to play a critical role in many cellular processes, such as transcriptional regulation, DNA repair, RNA process, cellular proliferation and signal transduction (Wolf 2009). The addition of a methyl group from SAM to arginine's guanidinyl group is catalyzed by a family of enzymes, protein arginine methyltransferases (PRMTs). So far, 11 human PRMTs, PRMT1-11, have been identified. PRMTs can be divided into type I, type II or type III. Type I and type II PRMTs can mono- and di-methylate the arginine's ω-guanidino nitrogen atoms in human, whereas the type III PRMTs enzymes mono-methylate the arginine's δ-guanidino nitrogen atom and are only found in yeast. In humans, both type I PRMTs, including PRMT1, PRMT3, PRMT4/CARM1, PRMT6 and PRMT8, and type II PRMTs, including PRMT5, PRMT7 and PRMT9, can catalyze the production of ω-N^G-monomethylarginine (MMA), but type I PRMTs further produce asymmetric ω-N^G, N^G-dimethylarginine (ADMA) and type II PRMTs further produce symmetric ω-N^G, N'^G-dimethylarginine (SDMA). Nevertheless, the type of the remaining three PRMTs, PRMT2, 10 and 11, is still not confirmed.

Histone methylation is not identified as a dynamic and reversible modification until the identification of the first HDM, lysine-specific histone demethylase 1 (LSD1), in 2004 (Shi et al. 2004). There are two families of HDMs that have been found: LSD family with two members, LSD1 and LSD2, and JMJC demethylases which contain JmjC domain. Due to the fact that LSD family needs a free electron pair at the methylated lysines to initiate demethylation, LSD1 and LSD2 are only able to demethylate the mono- or di-methyllysine. It has been found that LSD1/2 catalyzes demethylation through the amine oxidation reaction which is flavin adenine dinucleotide (FAD)-dependent. Since JmjC domain-containing histone demethylase 1 (JHDM1) that can specifically target H3K36 was discovered in 2006 (Tsukada et al. 2006), another family of HDMs, JMJC family, has been identified. The demethylation reaction catalyzed by JMJC family is an iron-dependent and α-ketoglutarate-dependent dioxygenase reaction. Unlike LSD family that needs protonated nitrogen for demethylation reaction and, hence, JMJC family can also remove methyl groups from tri-methylated substrates. The first identified JmjC domain proteins that can reverse the tri-methylation status of histone Lysines were JMJD2 family of proteins, which are able to catalyze demethylation of tri-methylated H3K9 and H3K36 (Whetstine et al. 2006). To date, the members of JMJC family have been found to target Lys4, 9, 27 and 36 on H3 and Lys20 on H4 (Iwase et al. 2007; Agger et al. 2007; Whetstine et al. 2006; Tsukada et al. 2006; Liu et al. 2010). Nevertheless, the histone arginine demethylases are still poorly understood.

1.2.3 Other Histone Modifications

There are also other post-translational modifications that can target histone proteins, including phosphorylation, ubiquitination and ADP ribosylation. Histone phosphorylation can target serine, threonine and tyrosine residues, and it is regulated by two enzymes, kinases and phosphatases, which have opposite effects. Kinases can result in phosphorylation of histones, whereas phosphatases are involved in dephosphorylation. Similar to histone acetylation, phosphorylation of specific sites can neutralize the positive charge of histone, which may influence the stability of chromatin structure. Histone phosphorylation can occur on four nucleosome core histones as well as histone H1, and it usually interplays with other post-translational modifications, called crosstalk, which is correlated with a number of cellular events, such as chromosome condensation, DNA repair, transcriptional regulation and cell cycle progression (Banerjee and Chakravarti 2011). It has been found that histones can also be modified by ubiquitin, a 76-amino-acid protein. Histone ubiquitination occurs on highly conserved lysine residues, and is catalyzed by three kinds of enzymes, ubiquitin-activating enzyme (E1), ubiquitin-conjugating enzyme (E2) and ubiquitin-protein ligase (E3) (Pickart 2001). Histones can be mono-ubiquitinated and poly-ubiquitinated, and all types of histone proteins have been found to be modified by ubiquitin. Among these ubiquitinated histones, mono-ubiquitinated H2A and H2B may be the most studied. Histone ubiquitination is also a dynamic and reversible modification, and it is mainly removed from histone lysines by a class of cysteine proteases, namely deubiquitinating enzymes (DUBs). ADP-ribosylation of histones is also a reversible post-translational modification, and it is associated with cellular process, such as DNA repair, cell cycle regulation, replication or transcription. ADP-ribosylation of histones occurs when an ADP-ribosyl group from NAD^+ is transferred to specific amino acid residues of histone tails or to histone-bound ADP-ribose under the catalysis of ADP-ribosyltransferases (ARTs), which lead to mono- ADP-ribosylation and poly-ADP-ribosylation, respectively. Histone ADP-ribosylation can be degradated by two families of enzymes: ADP-ribosylhydrolases (ARHs) and poly-ADP-ribose-glycohydrolases (PARGs). So far, only one RARG and three ARHs, ARH1-3, have been identified in human (Oka et al. 2006; Koch-Nolte et al. 2008; Mortusewicz et al. 2011).

2 The Role of Epigenetic Changes in Cancer

It is well known that tumorogenesis is a multistep process, which is involved in genetic and epigenetic alterations. Growing evidence has indicated that epigenetic modifications play critical roles in carcinogenesis. In cancers, aberrant methylation is classified into hypermethylation and hypomethylation. DNA hypomethylation is mostly genomic hypomethylation which often occurs in DNA-repetitive regions. Nevertheless, there is also gene-specific hypomethylation which usually occurs in promoter-associated CGI. Global hypomethylation results in genome instability, loss of imprinting and

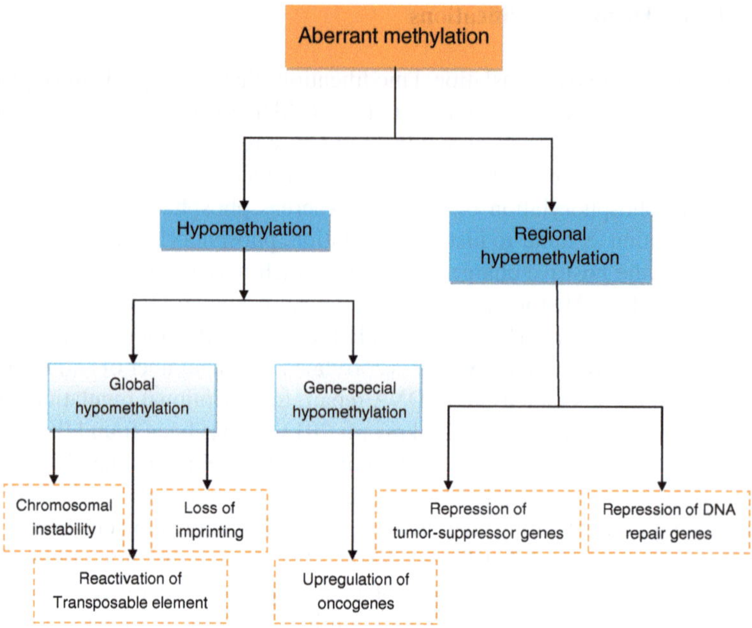

Fig. 2.1 Two types of aberrant methylation and their functional outcomes

reactivation of transposable elements (Dunn 2003; Eden et al. 2003), and gene-specific hypomethylation is associated with aberrant expression of oncogenes (Kwon et al. 2011; Tsai et al. 2010). However, the aberrant hypermethylation in CGI leads to transcription-silencing of tumor suppressor genes or DNA repair genes, such as tumor suppressor gene P16 and DNA repair gene MLH1 (Hossain et al. 2012), which contributes to tumor formation (Fig. 2.1). Moreover, based on the two-hit hypothesis, hypermethylation of tumor suppressor genes may act as the second hit after genes mutation, which is the first hit (Knudson 2001). In addition to aberrant DNA methylation, cancer cells also suffer from disruption of histone post-translational modifications. Understanding the critical roles of histone modifications in fundamental biological processes, such as transcription regulation, DNA repair, DNA replication and chromosome condensation, it is not surprising that aberrant histone modifications can be linked to carcinogenesis. Nevertheless, compared with DNA methylation, the information regarding histone modifications in cancers is limited, and only a fraction of modified residues in histone tails have been identified to be associated with carcinogenesis so far. In addition, a few histone modifiers, such as HDACs, HMTs, HDMs and HMT Polycomb group protein, EZH2, have been found to present abnormal expression in cancer, which may alter the levels and patterns of histone modifications and turn out to be deregulation of chromatin-based processes, finally resulting in development and progression of cancer. Furthermore, there is an epigenetic regulatory crosstalk between histone modifications and DNA methylation, which is associated with transcription regulation and abnormal silencing of genes in cancer (Vaissiere et al. 2008). With the development of research on critical role of microRNAs (miRNAs) in tumorogenesis, the complicated network between epigenetic regulation and gene expression has become more complex.

3 Epigenetic Regulation of MicroRNAs in Cancer

3.1 MiRNA

Noncoding RNAs also belong to category of epigenetics. Noncoding RNAs (ncRNAs) can be divided into small (under 200 nucleotides) or large ncRNAs. Moreover, small ncRNAs can be further subdivided into microRNAs (miRNAs), endogenous small interfering RNAs (endo-siRNAs), PIWI-interacting RNAs (piRNAs) and small nucleolar RNAs (snoRNAs). Owing to the critical role of miRNAs in the process of life, this class of small ncRNAs has been attracting more and more attention from researchers. MiRNAs is a class of endogenous, single-stranded, small ncRNAs with ~22 nucleotides (nts) in length, which remain highly conservative in phylogeny. Since the first report in 1993 and truly recognized in early 2000 (Lee et al. 1993; Reinhart et al. 2000), miRNAs has been one of the most popular and fastest growing research areas in molecular biology. So far, there have been more than 1,000 miRNAs identified in human, which regulate approximately 30 % of human protein-coding genes (Filipowicz et al. 2008). MiRNAs can modulate gene expression at post-transcriptional level through base pairing to mRNAs. In mammals, miRNAs mainly sequence-specifically bind to 3′-untranslated region (3′UTR) of target mRNAs. The perfect complementarity between miRNA and target mRNA can trigger RNA interference (RNAi), which results in mRNA cleavage. However, partial complementarity between them leads to translational inhibition of target mRNAs. Recently, some studies showed that miRNAs could bind to 5′-untranslated region (5′UTR) and protein-coding region of several mRNAs and, however, their functions require to be further studied (Lal et al. 2009; Lee et al. 2009).

Due to the characteristic that miRNAs can bind to 3′UTR of target mRNAs by partial complementarity, single miRNA may target a number of mRNAs and, in contrast, one mRNA may be the targets of many miRNAs. Therefore, abnormal expression of miRNAs may affect the normal expression of numerous genes and ultimately deregulate the control of biological processes, resulting in development and progression of cancer. Moreover, genome-wide studies have indicated that approximately 50 % of miRNAs are located at genomic regions of loss of heterozygosity or amplification, fragile sites of chromosomes or other cancer-associated regions (Calin et al. 2004b), which further confirms the critical role of miRNAs in carcinogenesis. Since deregulation of two miRNAs, miR-15 and miR-16, was first reported in chronic lymphocytic leukemia (CLL) (Calin et al. 2002), a number of genome-wide profiling studies have identified signatures with deregulated miRNAs in a variety of cancers. Further study on these deregulated miRNAs indicated that they have dual role in carcinogenesis as new oncogenes or tumor suppressor genes.

For instance, miR-21 is universal over-expressed in cancers, and it acts as oncogene through inhibiting the expression of targets genes which can regulate cellular proliferation, differentiation, and apoptosis as tumor suppressor proteins, such as phosphatase and tensin homolog (PTEN) and programmed cell death 4 (PDCD4) (Frankel et al. 2008; Meng et al. 2007). However, the miR-34 family, including miR-34a and miR-34b/c, targeted by p53 is universal downregulated in a variety of cancers, and the miR-34 members can play the role of tumor suppressor genes by regulating the expression of their targets. For example, the expression of miR-34a and miR-34c have been identified to be significantly down-regulated in breast cancers with lymph node metastasis, and reactivation of miR-34a/c can inhibit the metastasis and invasion of breast cancer through directly binding to 3′UTR of Fos-related antigen 1 (Fra-1) oncogene and repressing its expression (Yang et al. 2012). Interestingly, a few miRNAs can function as tumor suppressor genes in some cancers, while they can act as oncogenes in other cancers. For instance, miR-25 has been found to be down-regulated in human colon cancer, and it could inhibit growth and migration of colon cancer cells via repression of a direct target, Smad7 (Li et al. 2013). Nevertheless, another report found that miR-25 presented over-expression in esophageal squamous cell carcinoma (ESCC), and it might induce migration and invasion of ESCC through directly targeting 3′UTR of E-cadherin (CDH1) and consequently inhibiting the expression of CDH1 (Xu et al. 2012). It demonstrates that deregulation of some miRNAs may be tissue specific.

Increased data has demonstrated that miRNAs play critical roles in cellular processes associated with differentiation, proliferation, apoptosis, metastasis and invasion, and aberrant expression of which may be associated with development, progression and prognosis of cancers. Hence, it is necessary to investigate the mechanisms which lead to aberrant miRNAs expression in cancer, and a large number of studies have been conducted to explore the regulatory mechanisms for miRNAs.

3.2 Mechanisms of miRNA Deregulation in Human Cancer

Recent high-throughput studies have shown that the expression of miRNAs is deregulated in most cancer types. Some studies suggest that miRNA expression may be widely down-regulated in human tumors relative to normal tissues, whereas other studies demonstrate a tumor-specific mixed pattern of down-regulation and up-regulation of miRNA genes. Even in the same cancer type, some studies have also shown that miRNA expression signatures are associated not only with specific tumor subtypes but with clinical outcomes as well. However, the underlying mechanisms of miRNA deregulation in human cancer are still not thoroughly understood. Increasing evidence indicates that transcriptional deregulations, epigenetic alterations, mutations, DNA copy number abnormalities and defects in the miRNA biogenesis machinery might be the possible mechanisms, these mechanisms may each contribute, either alone or more likely together to miRNA deregulation in human cancer (Deng et al. 2008). Furthermore, expression of miRNAs can also be affected by other miRNAs, thus creating a complex level of reciprocal interaction and regulation (Iorio and Croce 2012) (Fig. 2.2).

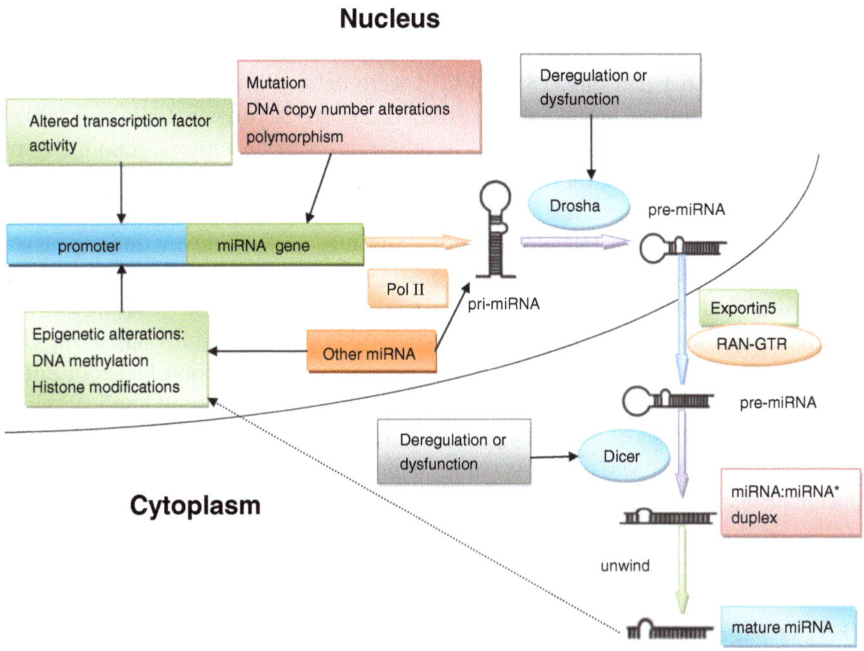

Fig. 2.2 Mechanisms of microRNA deregulation in human cancers

Transcriptional regulation is one of the important mechanisms in the control of miRNAs expression, which is based on the fact that most miRNA genes are derived from primary miRNA transcripts (pri-miRNAs) produced by Pol II and contain a 5′ cap and a poly (A) tail. For example, miR-34a family has been shown to be directly induced by the tumor suppressor p53, and p53 inactivation is thought to decrease miR-34 expression in human cancers (He et al. 2007b). C-Myc oncogene and transcriptional factor HIF are amplified and over-expressed in several types of human cancers, which might contribute to up-regulated miR-17-92 expression and miR-210 expression in cancer, respectively (Chang et al. 2008; Camps et al. 2008). Taken together, the increasing evidence shows that deregulation of miRNAs can be a result of increased or decreased transcription due to an altered transcription factor activity.

Unlike protein-coding genes, at present information on the mutation and polymorphism of miRNAs in cancer is just emerging, but the studies are increasing. For example, inherited mutations in the primary transcripts of miR-15a and miR-16-1 were found to be responsible for the reduced expression of the two miRNAs *in vitro* and *in vivo* in CLL (Calin et al. 2004a). In solid human tumors one sequence variation in a miRNA precursor and 15 variations in primary miRNAs were also identified, but no functional consequence was observed as the results of these aberrations (Diederichs and Haber 2006). Other studies suggested that polymorphism of miRNAs might be associated with hepatocellular carcinoma (HCC) or breast cancer. These data showed that genetic structural genetic alterations contribute to the deregulation of miRNAs as it does to other protein-coding genes.

In addition, DNA copy number alterations of miRNAs are frequently found in cancer and may account in part for the miRNA gene deregulation. The first evidence was that miR-16-1 and miR-15a at 13q14 were deleted in more than 50 % of the CLL patients, with concurrent reduced expression in ~65 % patients (Calin et al. 2002). A recent study has also shown that in advanced ovarian tumors genomic copy number loss may account for the down-regulation of approximately 15 % of miRNAs (Zhang et al. 2008). Moreover, some key proteins in the miRNAs biogenesis pathway may be dysfunction or deregulated in cancer, thus may further enhance tumorigenesis. For example, conditional deletion of Dicerl enhanced tumor development in a *K-Ras*-induced mouse model of lung cancer (Kumar et al. 2007). However, this mechanism might be tissue or cancer specific, because another study has shown that in ovarian cancer, Drosha and Dicer are not altered in expression levels of either mRNA or protein, therefore, functional assays might have important implications to examine the activities of these key proteins (Zhang et al. 2008). Furthermore, similar to protein-coding genes, miRNAs expression can also be modulated by epigenetic mechanisms, such as DNA methylation and histone acetylation.

3.3 Epigenetic Regulation of miRNAs in Cancers

The first evidence regarding deregulation of miRNAs due to epigenetic mechanisms was reported by Scott et al. (2006). Using miRNA array analysis, they found that the expression of 27 miRNAs in SKBr3 cell line altered rapidly by treatment of this cell line with an HDAC inhibitor, LAQ824, suggesting the relationship between epigenetic factors and miRNAs expression. Shortly thereafter, Saito et al. (2006) reported that 17 of 313 miRNAs in T24 bladder cancer cells were up-regulated after treatment of the cells with a DNMT inhibitor, 5-aza-dC, and a HDAC inhibitor, 4-phenylbutyric acid (4-PBA). Among these up-regulated miRNAs, miR-127, as a member of a miRNA cluster which also includes miR-136 and miR-431-433, is expressed in normal tissues, but it is down-regulated or silenced in T24 cells and many primary tumor tissues, such as bladder, prostate and colon cancers. MiR-127 is located in a CGI, and its re-expression is induced by DNA de-methylation, histone H3 acetylation and tri-methylation of H3K4 after treated with two epigenetic drugs. However, either 5-aza-dC or 4-PBA used alone had no function, which indicated that combined effects of histone modification and DNA methylation are necessary to regulate miR-127 expression. Moreover, the human oncogene B-cell lymphoma 6 (BCL6) may be a potential target of miR-127, and re-expression of miR-127 led to down-regulation of BCL6, indicating that miR-127 can function as a tumor suppressor gene. Thus, pharmacological unmasking by DNMT and/or HDAC inhibitors is often used to identify the epigenetically silenced miRNAs in various cancers. In addition, genetic unmasking, such as knockout of DNMTs, is also a common method. Moreover, with the development of technology, methylation status of miRNAs in cancer cells can be identified more intuitively by genome-wide DNA methylation analysis.

Recently, a large number of studies on epigenetic silencing of miRNAs have been reported in various cancers, and the relationship between aberrant DNA methylation and deregulated miRNAs may be most frequently and deeply studied. CGI hypermethylation is identified as an epigenetic aberration resulting in silencing of miRNAs which act as tumor suppressor genes. As previously mentioned, the miR-34 family is targeted by transcriptional factor p53, and this family has been found to epigenetically inactivate due to CGI hypermethylation in various cancers. MiR-34a is located on chromosomal locus 1p36 and, however, miR-34b and miR-34c as dicistronic cluster in one transcription unit, BC021736, locate on chromosome 11q23. Inactivation of both miR-34a and miR-34b/c due to abnormal promoter-associated CGI methylation has been identified in multiple cancer cell lines and primary tumors (Vogt et al. 2011). Re-expression of miR-34 family members can induce cell cycle arrest at G1 phase, apoptosis and senescence, and inhibit migration or invasion (Hermeking 2010). MiR-34 family members play the above-mentioned potential roles by down-regulation of candidate target genes as, for instance, Bcl2, E2F transcription factor-3 (E2F3), c-MYC, hepatocyte growth factor receptor (MET), cyclin E2 (CCNE2), CD44 and cyclin-dependent kinase 4/6 (CDK4/6) (Yamakuchi et al. 2008; Bommer et al. 2007; Lujambio et al. 2008; He et al. 2007a; Liu et al. 2011). Moreover, it has been found that DNA methylation of miR-34b/c is related to H. pylori infection in normal individuals, and the methylation levels of miR-34b/c in non-cancerous gastric mucosae of patients with multiple GC are higher than that of patients with single GC, suggesting that methylation of miR-34b/c was involved in epigenetic field defect (Suzuki et al. 2010). Similar to miR-34b/c, H. pylori infection induces promoter methylation of miR-124 family (miR-124-1, miR-124-2, and miR-124-3), and methylation level of miR-124 family is significantly higher in non-cancerous gastric mucosae of patients with GC than that in normal mucosae of healthy individuals, indicating that methylation of miR-124 family is also associated with epigenetic field defect (Ando et al. 2009). Epigenetic loss of miR-124 was first reported in colorectal line HCT-116 deficient in both DNMT3b and DNMT1, and its inactivation contributed to activation of CDK6, an oncogenic factor, and tumor suppressor gene retinoblastoma (Rb) phosphorylation (Lujambio et al. 2007). Epigenetic inactivation of miR-124 due to promoter hypermethylation has also been reported in many other malignancies, including acute lymphoblastic leukemia (ALL), hepatocellular cancer (HCC), clear cell renal cell carcinoma, cervical cancer and pancreatic cancer (Wang et al. 2013b; Wilting et al. 2010; Agirre et al. 2009; Gebauer et al. 2013; Furuta et al. 2010). Furthermore, miR-124 plays the role of tumor suppressor gene not only through directly targeting CDK6, but also through repression of other potential targets, such as SMYD3, VIM, IQGAP1 and Rac1 (Furuta et al. 2010; Wang et al. 2013b). Methylation of miR-181c has also been reported to be associated with formation of epigenetic field defect in non-cancerous tissues corresponded to GC samples, and it acts as tumor suppressor gene via repressing target genes, NOTCH4 and KRAS (Hashimoto et al. 2010).

Epithelial-mesenchymal transition (EMT) is considered as an important step in metastasis and invasion of cancer, and the miR-200 family members, including miR-200a, miR-200b, miR-200c, miR-141, and miR-429, and miR-205 can play critical role in regulation of EMT through down-regulation of two direct targets, zinc finger E-box binding homeobox 1 (ZEB1) and ZEB2, which can lead to repression of CDH1 expression (Gregory et al. 2008). The epigenetic silencing of miR-200c/141 cluster due to aberrant methylation was discovered in human breast and prostate cancer cell lines (Vrba et al. 2010). Inactivation of miR-200c by promoter-associated hypermethylation has also been identified in non-small cell lung cancer (NSCLC), which may be involved in formation of aggressive and chemoresistant phenotype (Ceppi et al. 2010). Thereafter, the whole miR-200 family was reported epigenetic repression by hypermethylation of CGI in colon, lung and breast cancer (Davalos et al. 2012). In invasive bladder cancer, coordinated inactivation of miR-200 family and miR-205 is correlated with promoter hypermethylation and repressive histone marks (Wiklund et al. 2011). In addition, silencing of miR-200b, miR-200c, and miR-205 by DNA methylation was identified in carcinogen-treated lung epithelial cells and contributed to EMT induction, which could be involved in initiation of lung cancer (Tellez et al. 2011). In a recent study, loss of miR-200 family expression due to both DNA methylation and histone modifications was discovered to have the capability to promote transition from to a non-stem to a stem-like phenotype, which might occur during progression of breast cancer (Lim et al. 2013).

In human, miR-9 family has three members, miR-9-1, miR-9-2 and miR-9-3, which locate on chromosomes 1, 5 and 15, respectively. Epigenetic repression of miR-9-1 due to aberrant promoter hypermethylation was first reported in breast cancer (Lehmann et al. 2008). Shortly thereafter, CGI hypermethylation of miR-9-1 was identified in pancreatic cancer using methylated CGI amplification microarrays (Omura et al. 2008). The promoter-associated CGI hypermethylation-mediated silencing of all three miR-9 family members was observed in metastatic cancer cell lines, indicating that epigenetic inactivation of miR-9 family can induce metastasis formation (Lujambio et al. 2008). Consistent with this finding, the methylation of both miR-9-1 and miR-9-3 plays a role in metastatic recurrence of clear cell renal cell carcinoma (Hildebrandt et al. 2010). Reactivation of miR-9 family members shows tumor suppressor features through down-regulation of some target genes, such as fibroblast growth factor receptor 1 (FGFR1), CDK6 and caudal-related homeobox 2 (CDX2) (Rodriguez-Otero et al. 2011; Rotkrua et al. 2011). However, a study showed that miR-9 is over-expressed in breast cancer, which leads to down-regulation of CDH1, and induces the formation of invasive phenotype (Wang et al. 2013a). These findings indicate that miR-9 may play a dual role. In a recent research, frequent hypermethylation and concomitant inactivation of miR-9-2, miR-9-3, miR-124 family, miR-129-2, miR-596 and miR-1247 was observed in human HCC, which can be reversed through DNMT1 knockdown or DNMT inhibition, and united hypermethylation of three or more miRNAs can be viewed as a new diagnostic and prognostic marker of HCC with high specificity (Anwar et al. 2013). Similar to miR-9 family, specific CGI hypermethylation-mediated silencing of miR-148a was also observed in metastatic cancer cell lines, which mediated the activation of metastatic gene transforming growth factor-beta-induced factor-2 (TGIF2), and reintroduction of miR-148a could repress motility, growth and metastasis of cancer cells (Lujambio et al. 2008). In addition, the presence of hypermethylation of miR-148 in promoter-associated CGI has also been observed in breast, lung, colon, head and neck cancers and melanomas, and is implicated in lymph node metastasis (Lujambio et al. 2008).

MiR-335 located at chromosome 7q32.2 was reported to be silenced by epigenetic promoter hypermethylation, and it can inhibit metastasis and migration of breast cancer by inhibition of its targets, SRY-related high-mobility group box 4 (SOX4) and extracellular matrix component tenascin C (Tavazoie et al. 2008; Png et al. 2011). Over-expression of SOX4 has also been found to be associated with epigenetic inactivation of miR-192-2 due to aberrant DNA methylation in endometrial cancer, gastric cancer and HCC (Chen et al. 2013; Huang et al. 2009; Shen et al. 2010). In addition, the high methylation level of miR-129-2 has been identified in ESCC and colorectal cancer (Chen et al. 2012; Bandres et al. 2009).

There are still many other miRNAs that are silenced by CGI hypermethylation in human cancers, including miR-137, which targets CDK6, LSD1 and cell division cycle 42 (CDC42), and is aberrantly methylated in multiple cancers, such as colon cancer, gastric cancer and lung cancer (Zhu et al. 2013; Balaguer et al. 2010; Chen et al. 2011); miR-375, which targets 3-phosphoinositide-dependent protein kinase-1 (PDK1) and insulin-like growth factor 1 receptor (IGF1R), and undergoes methylation-associated silencing in oesophageal cancer, cervical cancer and human melanoma (Mazar et al. 2011; Li et al. 2011; Kong et al. 2012; Wilting et al. 2013); and miR-203, which targets ATP-binding cassette, subfamily E, member 1 (ABCE1), CDK6 and ABL1, and is epigenetically silenced in Gastric B-cell lymphoma of mucosa-associated lymphoid tissue (MALT lymphoma), HCC, cervical cancer and hematological malignancies (Wilting et al. 2013; Furuta et al. 2010; Craig et al. 2011; Chim et al. 2011).

MiRNAs can be hosted in intronic regions of coding genes, and the connection between miRNAs and their host genes should be elucidated. Intronic miRNAs can co-express with the host genes, and they are subjected to the same regulation by epigenetic changes. One example is miR-342, which is encoded in an intron of its host gene Ena/Vasp-like (EVL) (Grady et al. 2008). The EVL promoter-associated CGI hypermethylation occurs in the early stages of colorectal cancer, which leads to silencing of both the protein and miR-342. In addition, methylation of EVL/miR-342 is significantly more common in non-cancerous colorectal mucosae of patients with colorectal cancer than that in normal mucosae of healthy individuals, indicating that methylation of EVL/miR-342 is involved in epigenetic field defect. MiR-126 is hosted in an intron of the EGFL7, and both are down-regulated in cancer cell lines as well as in primary prostate and bladder cancers (Saito et al. 2009a). Interestingly, both mature miRNA-126 and one of the EGFL7 transcripts, which owns a CGI promoter, are concomitantly up-regulated through epigenetic therapy using DNA methylation and histone deacetylation inhibitors, indicating that epigenetic changes of host genes can affect the expression of intronic miRNAs. MiR-152 as a tumor suppressor gene is frequently silenced in endometrial cancer cell lines and primary samples, and it is located within intron 1 of its host gene, coatomer protein complex, subunit zeta 2 (COPZ2), which is also inactivated in endometrial cancer (Tsuruta et al. 2011). The silencing of both miR-152 and its host gene is caused by DNA hypermethylation, which can be reversed by treatment with 5-aza-dCyd. Moreover, DNMT1, E2F3, MET, and Rictor have been identified as the potential targets of miR-152, illustrating how miR-152 plays its role in endometrial carcinogenesis. A recent study has shown that the expression of miR-335 and its host gene MEST can be repressed by DNA hypermethylation in HCC, which may be involved in distant metastasis (Dohi et al. 2013).

In contrast, some miRNAs, which act as oncogenes, are up-regulated in cancers due to DNA hypomethylation. For instance, the CGI of human let-7a-3 gene, which is located at 22q13.31, is always heavily methylated in normal human tissues, whereas it is hypomethylated in lung cancer tissues, leading to epigenetic activation of let-7a-3 (Brueckner et al. 2007). Over-expression of let-7a-3 in lung cancer cells contributes to formation of tumor phenotypes, and plays its oncogenic role through affecting the expression of several target genes, which are associated with cell proliferation and cell adhesion processes. The expression of miR-21, miR-205, and miR-203 are up-regulated in epithelial ovarian cancer (EOC) compared with the normal tissues, and their expression levels can significantly elevated by treatment of OVCAR3 cells with demethylation drug 5-aza-dC, indicating that their over-expression may be due to DNA hypomethylation (Iorio et al. 2007). Another study showed that expression of miR-200a and miR-200b is significantly up-modulated in pancreatic cancer compared with the normal tissues, which is due to CGI hypomethylation (Li et al. 2010a). In addition, miR-196b functions as an oncogene in primary GC, and its elevated expression is induced by abnormal CGI hypomethylation statue in promoter regions (Tsai et al. 2010). Furthermore, miR-191 is highly expressed in HCC due to aberrant hypomethylation, and it exerts the role in inducing EMT by directly targeting and repressing the expression of tissue inhibitor of metalloproteinase 3 (TIMP3) (He et al. 2011).

Silencing of tumor suppressor miRNAs and over-expression of oncogenic miRNAs due to DNA hypermethylation and hypomethylation, respectively, could play critical role in initiation and progression of various cancers. Interestingly, let-7a-3 has been described to present over-expression and aberrant hypomethylation in lung cancer cells (Brueckner et al. 2007), whereas it has also been found to be silenced by hypermethylation in EOC (Lu et al. 2007). In addition, miR-203, miR-200a and miR-200b are down-regulated in many types of cancer due to hypermethylation and, however, miR-203 is identified to be up-regulated in EOC, and miR-200a/b is up-modulated in pancreatic cancer due to CGI hypomethylation (Iorio et al. 2007; Li et al. 2010a). These data indicates that epigenetic regulation of some miRNAs may be tissue specific, which is likely to be associated with their expression of tissue specificity.

DNA methylation is not the only epigenetic change that affects the normal expression of miRNAs. Similar to protein-coding genes, epigenetic regulation of miRNAs is also closely correlated with histone modification. The first evidence regarding deregulation of miRNAs due to histone modification was reported by Scott et al. (2006). Using miRNA microarray analysis, they discovered that the expression of 27 miRNAs altered rapidly by treatment of SKBr3 breast cancer cell line with HDAC inhibitor. Another study showed that miR-127 was silenced in human T24 breast cancer cells, and DNA demethylation, histone H3 acetylation and tri-methylation of H3K4 by 5-Aza-dC and PBA treatment can lead to its re-activation (Saito et al. 2006).

Similarly, re-expression of miR-512-5p may be associated with DNA demethylation, histone H3 acetylation and di-methylation of H3K4 by epigenetic treatment with 5-aca-dC and PBA, and its activation leads to inhibition of Mcl-1 and induces apoptosis of gastric cancer cells (Saito et al. 2009b). Furthermore, a study showed that expression of miRNAs is positively associated with tri-methylation of H3K4 and negatively associated with tri-methylation of H3K27 in promoter regions of human miRNAs based on the data of miRNA expression microarrays and chromatin immunoprecipitation (ChIP)-on-chip (Ke et al. 2009). In ALL, the CGIs of 13 miRNAs exist two abnormal histone modifications, increased di-methylation of H3K9 and decreased tri-methylation of H3K4, which may be involved in silencing of these miRNAs (Roman-Gomez et al. 2009). Accumulated tri-methylation of H3K27 has been identified to inhibit the transcriptional expression of miR-22 in ALL, and its expression can be up-regulated after treatment with trichostain A (TSA), a HDAC inhibitor (Li et al. 2010b). In addition, analysis of CHIP revealed that microbial stimulus promotes NFkappaB p50-C/EBPbeta silencer complex binding to promoter region of let-7i, which leads to histone H3 deacetylation and silencing of miRNA let-7i (O'Hara et al. 2010). More recently, a study showed that accumulation of repressive histone marks, including tri-methylation of H3K9 and H3K27, is an important epigenetic change, which results in silencing of miR-125b1 in breast cancer (Soto-Reyes et al. 2012). Taking together, histone H3 acetylation and di-/tri-methylation of H3K4 can promote transcriptional activation of miRNAs, whereas tri-methylation of H3K27 and di/tri-methylation of H3K9 can lead to epigenetic inactivation of miRNAs. Moreover, aberrant histone modifications often coexist with DNA methylation in epigenetically deregulated miRNAs in cancer.

Interestingly, it has been founded that some oncoproteins can bind to promoter regions of miRNAs and recruit epigenetic effectors, leading to deregulation of miRNAs. For example, Fazi et al. (2007) reported that epigenetic silencing of miR-223 can be induced by AML1/ETO oncoprotein, the product of AML-associated t(8;21) translocation, through binding to pre-miR-223 region and recruiting epigenetic effectors, such as DNMTs, MeCP2 and HDAC. However, there is interaction between epigenetic machinery and miRNAs. MiRNAs can not only be regulated by epigenetic machinery, but also regulate the components of epigenetic machinery, which is associated with methylation or acetylation.

4 MicroRNAs Can Regulate the Epigenetic Effectors in Cancer

It has been discovered that a subgroup of miRNAs can regulate the expression of effectors of the epigenetic mechanisms by directly or indirectly targeting these epigenetic-modifying enzymes and molecules, such as DNMTs, HATs, HMTs, HDACs, Retinoblastoma-Like 2 (RBL2), enhancer of zeste homolog 2 (EZH2) and Polycomb Repressive Complex (PRC). These miRNAs, called "epi-miRNAs", have been widely investigated in the past few years because their deregulation may be

Table 2.1 Epi-miRNAs in human cancers

microRNA	Targets	Tissue type	Reference(s)
miR-29a	DNMT3a, DNMT3b	Lung cancer	Fabbri et al. (2007)
miR-29b	DNMT3a/3b, sp1	Lung cancer, AML	Fabbri et al. (2007) and Garzon et al. (2009)
miR-148a	DNMT3b, DNMT1	Cervical cancer, Gastric cancer, Cholangiocarcinoma	Duursma et al. (2008), Braconi et al. (2010), and Zhu et al. (2012)
miR-152	DNMT1	Cholangiocarcinoma NiS-transformed cells	Braconi et al. (2010) and Ji et al. (2013)
miR-301	DNMT1	Cholangiocarcinoma	Braconi et al. (2010)
miR-143	DNMT-3a	Colorectal cancer	Ng et al. (2009)
miR-342	DNMT1	Colorectal cancer	Wang et al. (2011)
miR-185	DNMT1	Hepatocellular carcinoma	Zhang et al. (2011)
miR-290 cluster	RBL-2	Mouse ES cells	Benetti et al. (2008) and Sinkkonen et al. (2008)
miR-34a	SIRT1	Colon cancer	Yamakuchi et al. (2008)
miR-449	SIRT1, HDAC1	Gastric cancer, Prostate cancer	Bou Kheir et al. (2011) and Noonan et al. (2009)
miR-200 family	SIRT1	Breast cancer	Eades et al. (2011)
miR-1	HDAC4	Skeletal muscle tissue	Chen et al. (2006)
miR-140	DNMT1	Hepatocarcinoma	Takata et al. (2013)
	HDAC4	Mouse cartilage tissue	Tuddenham et al. (2006)
miR-101	EZH2	Prostate cancer Bladder cancer	Friedman et al. (2009) and Varambally et al. (2008)
miR-K12-4-5p	RBL-2	KSHV	Lu et al. (2010)

Abbreviation: *DNMT* DNA methyltransferase, *RBL2* Retinoblastoma-Like 2, *SIRT1* sirtuin 1, *HDAC* histone de-acetylases, *EZH2* enhancer of zeste homolog 2, *AML* acute myeloid leukemia, *KSHV* Kaposi's sarcoma-associated herpesvirus

closely correlated with human carcinogenesis (Table 2.1). Recent studies have suggested that epi-miRNAs may occupy an important position in tumor mediation by modulating various cellular processes like cell proliferation, apoptosis, cellular movement and metastasis.

The existence of epi-miRNAs was first discovered in lung cancer, where miR-29 family has been proved to directly target DNMT3a and DNMT3b through some interesting complementarities to the 3′UTR of DNMT3 (Fabbri et al. 2007). In addition, Garzon et al. (2009) have demonstrated that miR-29b is also able to indirectly suppress DNMT1 in AML cells by directly targeting the transactivator Sp1. Studies have shown that expression of miR-29s is inversely correlated with the expression levels of both DNMT3a and DNMT3b in lung cancer and AML. The enforced expression of miR-29s can induce disruption of *de novo* DNA methylation and contribute to promoter-associated CGI demethylation of epigentically silenced tumor suppressor genes, such as WHOX, FHIT, and p151NK4B, due to promoter hypermethylation, which ultimately inhibits tumorigenesis by inducing reactivation of these tumor suppressor genes. These findings indicate a role of miR-29s in tumor suppression and provide a basic principle for the development of miRNA-based approaches for cancer therapy.

After that, miRNAs have drawn more attention than ever and emerged as a new category of tumor inhibitors or regulators of signal transduction. A few reports have revealed that miR-148a can regulate DNMT3b expression through an unusual binding site within the coding region instead of the 3′UTR of DNMT3b mRNA (Duursma et al. 2008). Studies also suggest that DNMT1 is directly modulated by miR-148a, along with miR-152 and miR-301, in cholangiocarcinoma (Braconi et al. 2010). The expression of these miRNAs is down-regulated and emerges tumor suppressive abilities in cancer cells. Zhu et al. found that silencing of miR-148a in gastric cancer was related to aberrant methylation in promoter region, which could contribute to activation of DNMT1, and this may in turn lead to inactivation of miR-148a through promoting DNA methylation (Zhu et al. 2012). Additionally, the expression of miR-152 in nickel sulfide (NiS)-transformed cells can directly suppress DNMT1 by targeting the 3′ untranslated regions of its transcript, which demonstrates a significant interaction between miR-152 and DNMT1 via a double-negative feedback mechanism involved in NiS-induced malignant transformation (Ji et al. 2013). Another miRNA with tumor suppressive abilities linked to DNMT3a is miR-143. Studies indicate a direct and specific crosstalk between miR-143 and DNMT3a 3′UTR, thus ectopic miR143 expression can reduce DNMT3a expression and repress cell proliferation in colorectal cancer (Ng et al. 2009). MiR-342 has also been found to inhibit colorectal cancer cell proliferation. MiR-342 presents decreased expression in colorectal cancer samples and cells, and its reactivation can directly inhibiting the expression of DNMT1, which reactivating several tumor suppressor genes, such as ADAM metallopeptidase domain 23 (ADAM23) and Ras association domain family member 1A (RASSF1A), through promoter demethylation (Wang et al. 2011). Furthermore, miR-185 can also modulate methylation levels of several gene promoters by targeting DNMT1 in human glioma (Zhang et al. 2011). More recently, miR-140 was found to be down-regulated in HCC due to deficiency of DDX20, and reactivation of its expression can increase metallothionein expression through directly targeting DNMT1, leading to decreased NF-κB activity resulting in inhibition of hepatocarcinogenesis (Takata et al. 2013).

In addition, a study was performed with Dicer-deficient mouse embryonic stem cells. In this study the level of miR-290 cluster was identified to be able to indirectly regulate the expression of DNMT3a and DNMT3b by directly silencing RBL2, a suppressor of DNMT3 genes (Benetti et al. 2008; Sinkkonen et al. 2008). The experiments have shown that the expression of miR-290 cluster is dramatically down-regulated in Dicer1-null embryonic stem cells with respect to wild-type controls, which results in decreased DNMT3 expression and DNA-methylation defects in these cells via loss of RBL2 silencing. Decreased DNMT expression leads to a remarkable hypomethylation of the genome, which impairs the embryonic stem cells differentiation program and promotes telomere elongation and telomere recombination. Furthermore, re-expression of the miR-290 cluster is able to rescue the down-regulation of DNMT3 expression and restore the normal methylation pattern via simultaneously silencing the expression of

RBL2. Altogether, these results demonstrate an important role of the miR-290 cluster in the regulatory of DNMT3 enzymes and global DNA methylation, which in turn mediate the appearance of telomeric phenotypes.

Over-expression of HDACs has been observed in a broad range of cancer types, which is a significant mechanism to promote proliferation and reduce apoptosis of cancer cells by repressing some important growth suppressive genes. HDACs play a key role in epigenetic modifications of cancers and are also under regulation of epi-miRNAs. It has been found that miR-34a can regulate the pathway that mediates cellular aging and limits longevity in various cancers including colon cancer, by interacting with the expression of mammalian sirtuin 1 (SIRT1) and p53 (Yamakuchi et al. 2008). SIRT1, a homologue of yeast gene silent information regulator 2 (Sir2), is a class III HDAC with an enzymatic activity dependent on NAD^+. P53 is an important tumor suppressor, which can affect cell proliferation, cell apoptosis, DNA repair and angiogenesis by modulating a variety of physical responses to many cancer-related stress signals. SIRT1 is frequently up-regulated in human cancers and inactivates apoptosis of cancer cells by inducing deacetylation of p53. Studies demonstrate that miR-34a directly inhibits SIRT1 expression by binding to the 3′ UTR region of SIRT1, which indirectly results in an increase of acetylated p53 and in turn prevent cell cycle arrest and induce apoptosis. Thereby, miR-34a functions as a tumor inhibitor, in part, through a SIRT1-p53 pathway.

A recent study using gastric cancer cells in a mouse model elucidated that SIRT1 was directly regulated by miR-449 (Bou Kheir et al. 2011). Researchers have found that miR-449 is part of the miR-34 family and may act as a tumor suppressor, which is down-regulated in gastric cancer and prostate cancer (Bou Kheir et al. 2011; Noonan et al. 2009). Restoration of miR-449 expression contributes to decreased expression of SIRT1. Moreover, re-introduction of miR-449 into cancer cell lines inhibits cell proliferation and induces cell cycle arrest, apoptosis and senescent-like phenotype by targeting various cell cycle regulators concomitant with the activation of p53 pathway. Interestingly, miR-449 was recently found to be epigenetically modulated and possessed a key position in a negative feedback loop in which E2F1 activated the transcription of miR-449 that in turn exerted tumor suppressive ability by directly targeting CDC25A and CDK6 (Yang et al. 2009).

A crosstalk between miR-200 family and SIRT1 has also been reported in recent studies. The miR-200 family is an important regulator of EMT, which is a common embryological process linked to various pathologies including cancer metastasis and tumorigenicity. The miR-200 family is down-regulated and acts as an inhibitor in renal, prostate, breast, bladder, pancreatic, and gastric cancers. Up-regulation of SIRT1 is observed in breast cancer tissues, which is associated with decreased expression of miR200a. Additionally, re-introduction of miR-200a or knockdown of SIRT1 inhibits transformation of normal mammary epithelial cells and prevents cancer metastasis, elucidating that miR-200a may be a potential tumor suppressor involved in breast cancer metastasis (Eades et al. 2011). Several other miRNAs have also been associated with the modulation of HDACs, such as miR-1, involved in myogenesis and related disease, and miR-140, reduced in various cancer types and suppressed cell proliferation in colon cancer cells, both of which directly target HDAC4 (Chen et al. 2006; Tuddenham et al. 2006).

EZH2 is the catalytic submit of the PRC2 and mediates heterochromatin formation by trimethylating histone H3K27, contributing to aberrant silencing of several tumor suppressor genes in cancer. As recently shown in both prostate and bladder cancers, the expression of miR-101 was down-regulated during cancer progression and inversely correlated with up-regulation of EZH2 (Friedman et al. 2009; Varambally et al. 2008). Further studies showed that transfection of miR-101 resulted in a stable EZH2 knockdown and decreased levels of trimethylation of H3K27, which significantly suppressed proliferation, migration, clonogenicity, and tumorigenicity of cancer cells, suggesting a role as tumor suppressor for miR-101.

A regulatory loop where miRNAs can also be used by virus to modulate the epigenetic mechanism of the host cell has been revealed recently. Researchers found the Kaposi's sarcoma-associated herpesvirus (KSHV) miR-K12-4-5p can directly target RBL2 3′ UTR and decrease RBL2 protein levels, which indirectly up-regulated mRNA levels of DNMT1, -3a, and -3b, thus regulating global epigenetic reprogramming (Lu et al. 2010).

Besides, a subtle and fascinating regulatory mechanism of methylation pattern has been reported: it seems that activation of epigenetic silencing by DNA methylation is dependent on the ratio of miRNA and its target RNA (Khraiwesh et al. 2010). In most physcomitrella patens mutants, the loss of DICER-LIKE 1b gene leads to a maturation of normal miRNAs, which in turn suppresses the disruption of target RNAs and causes accumulation of miRNAs. As a result, target RNA duplexes, hypermethylation of the genes encoding target RNA and ultimate gene silencing.

5 Clinical Utility of Epigenetically Silenced MicroRNAs in Cancer

With the help of microarray technology, we can detect thousands of miRNAs expression at the same time and establish miRNAs expression profiles of various cancers. From a clinical point of view, the miRNAs expression profile has potential to be a useful tool for diagnosis, prognosis judgment and prediction of treatment response. The discovery of circulating miRNAs in human blood serum and plasma has been frequently reported recently. Moreover, the expression levels of these circulating miRNAs are stable, reproducible and consistent among individuals of the same species, and they can be detected more easily. At present, circulating miRNAs profiles have been successfully evaluated in a number of solid cancers as novel early diagnostic markers. In addition to aberrant expression levels of miRNAs, the abnormal epigenetic marks of miRNAs can also be useful diagnostic and prognostic biomarkers. For instance, by analyzing bone marrow samples from 353 ALL patients, at least one of 13 methylated miRNA genes could be found in 65 % of all ALL cases, which had a significantly higher relapse and mortality rate, indicating that miRNA gene methylation can be an critical prognostic factor of ALL (Agirre et al. 2012).

Given that epigenetically inactivated miRNAs lead to initiation and progression of cancer, reactivation of these miRNAs may have great potential to treat cancer.

It has been widely proved that restoration of epigenetically silenced miRNA by epigenetic agents can result in down-regulation of oncogenic target genes and inhibition of development and progression of cancers. DNMT inhibitors can be used to reduce the level of methylation. Among these inhibitors, 5-aza-dC and 5-azacytidine have been approved by US Food and Drug Administration (FDA) to treat acute myeloid leukemia and myelodysplastic syndromes, respectively (Rodriguez-Paredes and Esteller 2011). In addition, current data reveal that miRNA mimics instead of miRNAs can effectively make up the loss of miRNAs expression and may be a potential therapeutic strategy. Moreover, HDAC inhibitors have also been approved by FAD to treat certain types of lymphomas. However, we cannot ignore that these epigenetic drugs are nonselective, and their side effects are still poorly understood.

6 Concluding Remarks

Cancer is associated with accumulation of epigenetic and genetic alterations. The epigenetic changes can result in silencing of tumor suppressor genes and up-regulation of oncogenes. Owing to that miRNAs can target a large number of mRNAs involved in control of various biological processes, they can also play the role of tumor suppressor genes or oncogenes (Babashah and Soleimani 2011; Babashah et al. 2012). The aberrant expression of miRNAs is the common pathogenesis of human cancers and, hence, it is of great significance in treatment of cancers to study the regulation mechanism of miRNAs. Currently, experimental evidence indicates that deregulation of miRNAs is closely associated with epigenetic changes, including abnormal DNA methylation and histone modifications. Moreover, with the development of the detection technology, it is possible for us to detect a wide range of epigenetic deregulation of miRNAs in a cancer, which can be used as new diagnostic and prognostic marker of the cancer. In addition, a group of miRNAs, which are known as epi-miRNAs, have been discovered to be able to regulate the expression of epigenetic effectors, such as DNMTs, HDACs and polycomb genes, which indirectly leads to epigenetic changes and in turn affects the expression of miRNAs. Hence, there is interaction between miRNAs and epigenetic machinery.

The relationship between epigenetics and miRNAs has been widely studied, but the understanding of it is still far from perfect. Further studies should be focused on the regulation system involving epigenetic machinery and miRNAs and how to translate research results into clinical treatment of cancers. In addition, epigenetic alteration is a reversible process, and treatment with epigenetic drugs, such as DNMT or HDAC inhibitors, can lead to re-expression of the silenced miRNAs and restore their normal function, which may achieve the purpose of cancer therapy. Hence, future studies also need to be conducted to develop the miRNA- and epi-miRNA-based treatment. It can be predicted that, with the continuous deepening research, the diagnosis and treatment of cancer will make great progress.

References

Agger K, Cloos PA, Christensen J, Pasini D, Rose S, Rappsilber J et al (2007) UTX and JMJD3 are histone H3K27 demethylases involved in HOX gene regulation and development. Nature 449(7163):731–734

Agirre X, Vilas-Zornoza A, Jimenez-Velasco A, Martin-Subero JI, Cordeu L, Garate L et al (2009) Epigenetic silencing of the tumor suppressor microRNA Hsa-miR-124a regulates CDK6 expression and confers a poor prognosis in acute lymphoblastic leukemia. Cancer Res 69(10):4443–4453

Agirre X, Martinez-Climent JA, Odero MD, Prosper F (2012) Epigenetic regulation of miRNA genes in acute leukemia. Leukemia 26(3):395–403

Allfrey VG, Faulkner R, Mirsky AE (1964) Acetylation and methylation of histones and their possible role in the regulation of RNA synthesis. Proc Natl Acad Sci U S A 51:786–794

Ando T, Yoshida T, Enomoto S, Asada K, Tatematsu M, Ichinose M et al (2009) DNA methylation of microRNA genes in gastric mucosae of gastric cancer patients: its possible involvement in the formation of epigenetic field defect. Int J Cancer 124(10):2367–2374

Anwar SL, Albat C, Krech T, Hasemeier B, Schipper E, Schweitzer N et al (2013) Concordant hypermethylation of intergenic microRNA genes in human hepatocellular carcinoma as new diagnostic and prognostic marker. Int J Cancer 133(3):660–670

Babashah S, Soleimani M (2011) The oncogenic and tumour suppressive roles of microRNAs in cancer and apoptosis. Eur J Cancer 47(8):1127–1137

Babashah S, Sadeghizadeh M, Tavirani MR, Farivar S, Soleimani M (2012) Aberrant microRNA expression and its implications in the pathogenesis of leukemias. Cell Oncol (Dordr) 35(5):317–334

Balaguer F, Link A, Lozano JJ, Cuatrecasas M, Nagasaka T, Boland CR et al (2010) Epigenetic silencing of miR-137 is an early event in colorectal carcinogenesis. Cancer Res 70(16):6609–6618

Bandres E, Agirre X, Bitarte N, Ramirez N, Zarate R, Roman-Gomez J et al (2009) Epigenetic regulation of microRNA expression in colorectal cancer. Int J Cancer 125(11):2737–2743

Banerjee T, Chakravarti D (2011) A peek into the complex realm of histone phosphorylation. Mol Cell Biol 31(24):4858–4873

Benetti R, Gonzalo S, Jaco I, Munoz P, Gonzalez S, Schoeftner S et al (2008) A mammalian microRNA cluster controls DNA methylation and telomere recombination via Rbl2-dependent regulation of DNA methyltransferases. Nat Struct Mol Biol 15(3):268–279

Bird A (2002) DNA methylation patterns and epigenetic memory. Genes Dev 16(1):6–21

Bogdanovic O, Veenstra GJ (2009) DNA methylation and methyl-CpG binding proteins: developmental requirements and function. Chromosoma 118(5):549–565

Bommer GT, Gerin I, Feng Y, Kaczorowski AJ, Kuick R, Love RE et al (2007) p53-mediated activation of miRNA34 candidate tumor-suppressor genes. Curr Biol 17(15):1298–1307

Bou Kheir T, Futoma-Kazmierczak E, Jacobsen A, Krogh A, Bardram L, Hother C et al (2011) miR-449 inhibits cell proliferation and is down-regulated in gastric cancer. Mol Cancer 10:29

Braconi C, Huang N, Patel T (2010) MicroRNA-dependent regulation of DNA methyltransferase-1 and tumor suppressor gene expression by interleukin-6 in human malignant cholangiocytes. Hepatology 51(3):881–890

Brandl A, Heinzel T, Kramer OH (2009) Histone deacetylases: salesmen and customers in the post-translational modification market. Biol Cell 101(4):193–205

Brueckner B, Stresemann C, Kuner R, Mund C, Musch T, Meister M et al (2007) The human let-7a-3 locus contains an epigenetically regulated microRNA gene with oncogenic function. Cancer Res 67(4):1419–1423

Calin GA, Dumitru CD, Shimizu M, Bichi R, Zupo S, Noch E et al (2002) Frequent deletions and down-regulation of micro- RNA genes miR15 and miR16 at 13q14 in chronic lymphocytic leukemia. Proc Natl Acad Sci U S A 99(24):15524–15529

Calin GA, Liu CG, Sevignani C, Ferracin M, Felli N, Dumitru CD et al (2004a) MicroRNA profiling reveals distinct signatures in B cell chronic lymphocytic leukemias. Proc Natl Acad Sci U S A 101(32):11755–11760

Calin GA, Sevignani C, Dumitru CD, Hyslop T, Noch E, Yendamuri S et al (2004b) Human microRNA genes are frequently located at fragile sites and genomic regions involved in cancers. Proc Natl Acad Sci U S A 101(9):2999–3004

Campanero MR, Armstrong MI, Flemington EK (2000) CpG methylation as a mechanism for the regulation of E2F activity. Proc Natl Acad Sci U S A 97(12):6481–6486

Camps C, Buffa FM, Colella S, Moore J, Sotiriou C, Sheldon H et al (2008) hsa-miR-210 Is induced by hypoxia and is an independent prognostic factor in breast cancer. Clin Cancer Res 14(5):1340–1348

Ceppi P, Mudduluru G, Kumarswamy R, Rapa I, Scagliotti GV, Papotti M et al (2010) Loss of miR-200c expression induces an aggressive, invasive, and chemoresistant phenotype in non-small cell lung cancer. Mol Cancer Res 8(9):1207–1216

Chang TC, Yu D, Lee YS, Wentzel EA, Arking DE, West KM et al (2008) Widespread microRNA repression by Myc contributes to tumorigenesis. Nat Genet 40(1):43–50

Chen JF, Mandel EM, Thomson JM, Wu Q, Callis TE, Hammond SM et al (2006) The role of microRNA-1 and microRNA-133 in skeletal muscle proliferation and differentiation. Nat Genet 38(2):228–233

Chen Q, Chen X, Zhang M, Fan Q, Luo S, Cao X (2011) miR-137 is frequently down-regulated in gastric cancer and is a negative regulator of Cdc42. Dig Dis Sci 56(7):2009–2016

Chen X, Hu H, Guan X, Xiong G, Wang Y, Wang K et al (2012) CpG island methylation status of miRNAs in esophageal squamous cell carcinoma. Int J Cancer 130(7):1607–1613

Chen X, Zhang L, Zhang T, Hao M, Zhang X, Zhang J et al (2013) Methylation-mediated repression of microRNA 129–2 enhances oncogenic SOX4 expression in HCC. Liver Int 33(3):476–486

Chim CS, Wong KY, Leung CY, Chung LP, Hui PK, Chan SY et al (2011) Epigenetic inactivation of the hsa-miR-203 in haematological malignancies. J Cell Mol Med 15(12):2760–2767

Craig VJ, Cogliatti SB, Rehrauer H, Wundisch T, Muller A (2011) Epigenetic silencing of microRNA-203 dysregulates ABL1 expression and drives Helicobacter-associated gastric lymphomagenesis. Cancer Res 71(10):3616–3624

Davalos V, Moutinho C, Villanueva A, Boque R, Silva P, Carneiro F et al (2012) Dynamic epigenetic regulation of the microRNA-200 family mediates epithelial and mesenchymal transitions in human tumorigenesis. Oncogene 31(16):2062–2074

Deng S, Calin GA, Croce CM, Coukos G, Zhang L (2008) Mechanisms of microRNA deregulation in human cancer. Cell Cycle 7(17):2643–2646

Diederichs S, Haber DA (2006) Sequence variations of microRNAs in human cancer: alterations in predicted secondary structure do not affect processing. Cancer Res 66(12):6097–6104

Dohi O, Yasui K, Gen Y, Takada H, Endo M, Tsuji K et al (2013) Epigenetic silencing of miR-335 and its host gene MEST in hepatocellular carcinoma. Int J Oncol 42(2):411–418

Dunn BK (2003) Hypomethylation: one side of a larger picture. Ann N Y Acad Sci 983:28–42

Duursma AM, Kedde M, Schrier M, le Sage C, Agami R (2008) miR-148 targets human DNMT3b protein coding region. RNA 14(5):872–877

Eades G, Yao Y, Yang M, Zhang Y, Chumsri S, Zhou Q (2011) miR-200a regulates SIRT1 expression and epithelial to mesenchymal transition (EMT)-like transformation in mammary epithelial cells. J Biol Chem 286(29):25992–26002

Eden A, Gaudet F, Waghmare A, Jaenisch R (2003) Chromosomal instability and tumors promoted by DNA hypomethylation. Science 300(5618):455

Fabbri M, Garzon R, Cimmino A, Liu Z, Zanesi N, Callegari E et al (2007) MicroRNA-29 family reverts aberrant methylation in lung cancer by targeting DNA methyltransferases 3A and 3B. Proc Natl Acad Sci U S A 104(40):15805–15810

Fazi F, Racanicchi S, Zardo G, Starnes LM, Mancini M, Travaglini L et al (2007) Epigenetic silencing of the myelopoiesis regulator microRNA-223 by the AML1/ETO oncoprotein. Cancer Cell 12(5):457–466

Filipowicz W, Bhattacharyya SN, Sonenberg N (2008) Mechanisms of post-transcriptional regulation by microRNAs: are the answers in sight? Nat Rev Genet 9(2):102–114

Frankel LB, Christoffersen NR, Jacobsen A, Lindow M, Krogh A, Lund AH (2008) Programmed cell death 4 (PDCD4) is an important functional target of the microRNA miR-21 in breast cancer cells. J Biol Chem 283(2):1026–1033

Friedman JM, Liang G, Liu CC, Wolff EM, Tsai YC, Ye W et al (2009) The putative tumor suppressor microRNA-101 modulates the cancer epigenome by repressing the polycomb group protein EZH2. Cancer Res 69(6):2623–2629

Furuta M, Kozaki KI, Tanaka S, Arii S, Imoto I, Inazawa J (2010) miR-124 and miR-203 are epigenetically silenced tumor-suppressive microRNAs in hepatocellular carcinoma. Carcinogenesis 31(5):766–776

Garzon R, Liu S, Fabbri M, Liu Z, Heaphy CE, Callegari E et al (2009) MicroRNA-29b induces global DNA hypomethylation and tumor suppressor gene reexpression in acute myeloid leukemia by targeting directly DNMT3A and 3B and indirectly DNMT1. Blood 113(25):6411–6418

Gebauer K, Peters I, Dubrowinskaja N, Hennenlotter J, Abbas M, Scherer R et al (2013) Hsa-mir-124-3 CpG island methylation is associated with advanced tumours and disease recurrence of patients with clear cell renal cell carcinoma. Br J Cancer 108(1):131–138

Glozak MA, Sengupta N, Zhang X, Seto E (2005) Acetylation and deacetylation of non-histone proteins. Gene 363:15–23

Gowher H, Liebert K, Hermann A, Xu G, Jeltsch A (2005) Mechanism of stimulation of catalytic activity of Dnmt3A and Dnmt3B DNA-(cytosine-C5)-methyltransferases by Dnmt3L. J Biol Chem 280(14):13341–13348

Grady WM, Parkin RK, Mitchell PS, Lee JH, Kim YH, Tsuchiya KD et al (2008) Epigenetic silencing of the intronic microRNA hsa-miR-342 and its host gene EVL in colorectal cancer. Oncogene 27(27):3880–3888

Greer EL, Shi Y (2012) Histone methylation: a dynamic mark in health, disease and inheritance. Nat Rev Genet 13(5):343–357

Gregory PA, Bert AG, Paterson EL, Barry SC, Tsykin A, Farshid G et al (2008) The miR-200 family and miR-205 regulate epithelial to mesenchymal transition by targeting ZEB1 and SIP1. Nat Cell Biol 10(5):593–601

Hashimoto Y, Akiyama Y, Otsubo T, Shimada S, Yuasa Y (2010) Involvement of epigenetically silenced microRNA-181c in gastric carcinogenesis. Carcinogenesis 31(5):777–784

Hayakawa T, Nakayama J (2011) Physiological roles of class I HDAC complex and histone demethylase. J Biomed Biotechnol 2011:129383

He L, He X, Lim LP, de Stanchina E, Xuan Z, Liang Y et al (2007a) A microRNA component of the p53 tumour suppressor network. Nature 447(7148):1130–1134

He L, He X, Lowe SW, Hannon GJ (2007b) microRNAs join the p53 network–another piece in the tumour-suppression puzzle. Nat Rev Cancer 7(11):819–822

He Y, Cui Y, Wang W, Gu J, Guo S, Ma K et al (2011) Hypomethylation of the hsa-miR-191 locus causes high expression of hsa-mir-191 and promotes the epithelial-to-mesenchymal transition in hepatocellular carcinoma. Neoplasia 13(9):841–853

Hermeking H (2010) The miR-34 family in cancer and apoptosis. Cell Death Differ 17(2):193–199

Hildebrandt MA, Gu J, Lin J, Ye Y, Tan W, Tamboli P et al (2010) Hsa-miR-9 methylation status is associated with cancer development and metastatic recurrence in patients with clear cell renal cell carcinoma. Oncogene 29(42):5724–5728

Hodawadekar SC, Marmorstein R (2007) Chemistry of acetyl transfer by histone modifying enzymes: structure, mechanism and implications for effector design. Oncogene 26(37):5528–5540

Hong L, Schroth GP, Matthews HR, Yau P, Bradbury EM (1993) Studies of the DNA binding properties of histone H4 amino terminus. Thermal denaturation studies reveal that acetylation markedly reduces the binding constant of the H4 "tail" to DNA. J Biol Chem 268(1):305–314

Hossain MB, Vahter M, Concha G, Broberg K (2012) Environmental arsenic exposure and DNA methylation of the tumor suppressor gene p16 and the DNA repair gene MLH1: effect of arsenic metabolism and genotype. Metallomics 4(11):1167–1175

Huang YW, Liu JC, Deatherage DE, Luo J, Mutch DG, Goodfellow PJ et al (2009) Epigenetic repression of microRNA-129-2 leads to overexpression of SOX4 oncogene in endometrial cancer. Cancer Res 69(23):9038–9046

Iorio MV, Croce CM (2012) Causes and consequences of microRNA dysregulation. Cancer J 18(3):215–222

Iorio MV, Visone R, Di Leva G, Donati V, Petrocca F, Casalini P et al (2007) MicroRNA signatures in human ovarian cancer. Cancer Res 67(18):8699–8707

Iorio MV, Piovan C, Croce CM (2010) Interplay between microRNAs and the epigenetic machinery: an intricate network. Biochim Biophys Acta 1799(10–12):694–701

Iwase S, Lan F, Bayliss P, de la Torre-Ubieta L, Huarte M, Qi HH, Whetstine JR et al (2007) The X-linked mental retardation gene SMCX/JARID1C defines a family of histone H3 lysine 4 demethylases. Cell 128(6):1077–1088

Ji W, Yang L, Yuan J, Yang L, Zhang M, Qi D et al (2013) MicroRNA-152 targets DNA methyltransferase 1 in NiS-transformed cells via a feedback mechanism. Carcinogenesis 34(2):446–453

John S, Howe L, Tafrov ST, Grant PA, Sternglanz R, Workman JL (2000) The something about silencing protein, Sas3, is the catalytic subunit of NuA3, a yTAF(II)30-containing HAT complex that interacts with the Spt16 subunit of the yeast CP (Cdc68/Pob3)-FACT complex. Genes Dev 14(10):1196–1208

Ke XS, Qu Y, Rostad K, Li WC, Lin B, Halvorsen OJ et al (2009) Genome-wide profiling of histone h3 lysine 4 and lysine 27 trimethylation reveals an epigenetic signature in prostate carcinogenesis. PLoS One 4(3):e4687

Khraiwesh B, Arif MA, Seumel GI, Ossowski S, Weigel D, Reski R et al (2010) Transcriptional control of gene expression by microRNAs. Cell 140(1):111–122

Knudson AG (2001) Two genetic hits (more or less) to cancer. Nat Rev Cancer 1(2):157–162

Koch-Nolte F, Kernstock S, Mueller-Dieckmann C, Weiss MS, Haag F (2008) Mammalian ADP-ribosyltransferases and ADP-ribosylhydrolases. Front Biosci 13:6716–6729

Kong KL, Kwong DL, Chan TH, Law SY, Chen L, Li Y et al (2012) MicroRNA-375 inhibits tumour growth and metastasis in oesophageal squamous cell carcinoma through repressing insulin-like growth factor 1 receptor. Gut 61(1):33–42

Kouzarides T (2007) Chromatin modifications and their function. Cell 128(4):693–705

Kumar MS, Lu J, Mercer KL, Golub TR, Jacks T (2007) Impaired microRNA processing enhances cellular transformation and tumorigenesis. Nat Genet 39(5):673–677

Kwon OH, Park JL, Kim M, Kim JH, Lee HC, Kim HJ et al (2011) Aberrant up-regulation of LAMB3 and LAMC2 by promoter demethylation in gastric cancer. Biochem Biophys Res Commun 406(4):539–545

Lal A, Navarro F, Maher CA, Maliszewski LE, Yan N, O'Day E et al (2009) miR-24 Inhibits cell proliferation by targeting E2F2, MYC, and other cell-cycle genes via binding to "seedless" 3′UTR microRNA recognition elements. Mol Cell 35(5):610–625

Lee KK, Workman JL (2007) Histone acetyltransferase complexes: one size doesn't fit all. Nat Rev Mol Cell Biol 8(4):284–295

Lee RC, Feinbaum RL, Ambros V (1993) The C. elegans heterochronic gene lin-4 encodes small RNAs with antisense complementarity to lin-14. Cell 75(5):843–854

Lee I, Ajay SS, Yook JI, Kim HS, Hong SH, Kim NH et al (2009) New class of microRNA targets containing simultaneous 5′-UTR and 3′-UTR interaction sites. Genome Res 19(7):1175–1183

Lehmann U, Hasemeier B, Christgen M, Muller M, Romermann D, Langer F et al (2008) Epigenetic inactivation of microRNA gene hsa-mir-9-1 in human breast cancer. J Pathol 214(1):17–24

Leonhardt H, Page AW, Weier HU, Bestor TH (1992) A targeting sequence directs DNA methyltransferase to sites of DNA replication in mammalian nuclei. Cell 71(5):865–873

Li A, Omura N, Hong SM, Vincent A, Walter K, Griffith M et al (2010a) Pancreatic cancers epigenetically silence SIP1 and hypomethylate and overexpress miR-200a/200b in association with elevated circulating miR-200a and miR-200b levels. Cancer Res 70(13):5226–5237

Li X, Liu J, Zhou R, Huang S, Huang S, Chen XM (2010b) Gene silencing of MIR22 in acute lymphoblastic leukaemia involves histone modifications independent of promoter DNA methylation. Br J Haematol 148(1):69–79

Li X, Lin R, Li J (2011) Epigenetic silencing of microRNA-375 regulates PDK1 expression in esophageal cancer. Dig Dis Sci 56(10):2849–2856

Li Q, Zou C, Zou C, Huang H, Jin J, Han Z et al (2013) MicroRNA-25 functions as a potential tumor suppressor in colon cancer by targeting Smad7. Cancer Lett 335(1):168–174

Lim Y, Wright JA, Attema JL, Gregory PA, Bert AG, Smith E et al (2013) Epigenetic modulation of the miR-200 family is associated with transition to a breast cancer stem cell-like state. J Cell Sci 126(Pt 10):2256–2266

Liu W, Tanasa B, Tyurina OV, Zhou TY, Gassmann R, Liu WT et al (2010) PHF8 mediates histone H4 lysine 20 demethylation events involved in cell cycle progression. Nature 466(7305):508–512
Liu C, Kelnar K, Liu B, Chen X, Calhoun-Davis T, Li H et al (2011) The microRNA miR-34a inhibits prostate cancer stem cells and metastasis by directly repressing CD44. Nat Med 17(2):211–215
Lu L, Katsaros D, de la Longrais IA, Sochirca O, Yu H (2007) Hypermethylation of let-7a-3 in epithelial ovarian cancer is associated with low insulin-like growth factor-II expression and favorable prognosis. Cancer Res 67(21):10117–10122
Lu F, Stedman W, Yousef M, Renne R, Lieberman PM (2010) Epigenetic regulation of Kaposi's sarcoma-associated herpesvirus latency by virus-encoded microRNAs that target Rta and the cellular Rbl2-DNMT pathway. J Virol 84(6):2697–2706
Lujambio A, Ropero S, Ballestar E, Fraga MF, Cerrato C, Setien F et al (2007) Genetic unmasking of an epigenetically silenced microRNA in human cancer cells. Cancer Res 67(4):1424–1429
Lujambio A, Calin GA, Villanueva A, Ropero S, Sanchez-Cespedes M, Blanco D et al (2008) A microRNA DNA methylation signature for human cancer metastasis. Proc Natl Acad Sci U S A 105(36):13556–13561
Magnani R, Dirk LM, Trievel RC, Houtz RL (2010) Calmodulin methyltransferase is an evolutionarily conserved enzyme that trimethylates Lys-115 in calmodulin. Nat Commun 1:43
Mazar J, DeBlasio D, Govindarajan SS, Zhang S, Perera RJ (2011) Epigenetic regulation of microRNA-375 and its role in melanoma development in humans. FEBS Lett 585(15):2467–2476
Meng F, Henson R, Wehbe-Janek H, Ghoshal K, Jacob ST, Patel T (2007) MicroRNA-21 regulates expression of the PTEN tumor suppressor gene in human hepatocellular cancer. Gastroenterology 133(2):647–658
Mortusewicz O, Fouquerel E, Ame JC, Leonhardt H, Schreiber V (2011) PARG is recruited to DNA damage sites through poly(ADP-ribose)- and PCNA-dependent mechanisms. Nucleic Acids Res 39(12):5045–5056
Ng EK, Tsang WP, Ng SS, Jin HC, Yu J, Li JJ et al (2009) MicroRNA-143 targets DNA methyltransferases 3A in colorectal cancer. Br J Cancer 101(4):699–706
Nguyen AT, Zhang Y (2011) The diverse functions of Dot1 and H3K79 methylation. Genes Dev 25(13):1345–1358
Noonan EJ, Place RF, Pookot D, Basak S, Whitson JM, Hirata H et al (2009) miR-449a targets HDAC-1 and induces growth arrest in prostate cancer. Oncogene 28(14):1714–1724
O'Hara SP, Splinter PL, Gajdos GB, Trussoni CE, Fernandez-Zapico ME, Chen XM et al (2010) NFkappaB p50-CCAAT/enhancer-binding protein beta (C/EBPbeta)-mediated transcriptional repression of microRNA let-7i following microbial infection. J Biol Chem 285(1):216–225
Oka S, Kato J, Moss J (2006) Identification and characterization of a mammalian 39-kDa poly(ADP-ribose) glycohydrolase. J Biol Chem 281(2):705–713
Okano M, Bell DW, Haber DA, Li E (1999) DNA methyltransferases Dnmt3a and Dnmt3b are essential for de novo methylation and mammalian development. Cell 99(3):247–257
Omura N, Li CP, Li A, Hong SM, Walter K, Jimeno A et al (2008) Genome-wide profiling of methylated promoters in pancreatic adenocarcinoma. Cancer Biol Ther 7(7):1146–1156
Peterson CL, Laniel MA (2004) Histones and histone modifications. Curr Biol 14(14):R546–R551
Pickart CM (2001) Mechanisms underlying ubiquitination. Annu Rev Biochem 70:503–533
Png KJ, Yoshida M, Zhang XH, Shu W, Lee H, Rimner A et al (2011) MicroRNA-335 inhibits tumor reinitiation and is silenced through genetic and epigenetic mechanisms in human breast cancer. Genes Dev 25(3):226–231
Rea S, Eisenhaber F, O'Carroll D, Strahl BD, Sun ZW, Schmid M et al (2000) Regulation of chromatin structure by site-specific histone H3 methyltransferases. Nature 406(6796):593–599
Reinhart BJ, Slack FJ, Basson M, Pasquinelli AE, Bettinger JC, Rougvie AE et al (2000) The 21-nucleotide let-7 RNA regulates developmental timing in Caenorhabditis elegans. Nature 403(6772):901–906
Rodriguez-Otero P, Roman-Gomez J, Vilas-Zornoza A, Jose-Eneriz ES, Martin-Palanco V, Rifon J et al (2011) Deregulation of FGFR1 and CDK6 oncogenic pathways in acute lymphoblastic leukaemia harbouring epigenetic modifications of the MIR9 family. Br J Haematol 155(1):73–83

Rodriguez-Paredes M, Esteller M (2011) Cancer epigenetics reaches mainstream oncology. Nat Med 17(3):330–339

Roman-Gomez J, Agirre X, Jimenez-Velasco A, Arqueros V, Vilas-Zornoza A, Rodriguez-Otero P et al (2009) Epigenetic regulation of microRNAs in acute lymphoblastic leukemia. J Clin Oncol 27(8):1316–1322

Rotkrua P, Akiyama Y, Hashimoto Y, Otsubo T, Yuasa Y (2011) MiR-9 downregulates CDX2 expression in gastric cancer cells. Int J Cancer 129(11):2611–2620

Saito Y, Liang G, Egger G, Friedman JM, Chuang JC, Coetzee GA et al (2006) Specific activation of microRNA-127 with downregulation of the proto-oncogene BCL6 by chromatin-modifying drugs in human cancer cells. Cancer Cell 9(6):435–443

Saito Y, Friedman JM, Chihara Y, Egger G, Chuang JC, Liang G (2009a) Epigenetic therapy upregulates the tumor suppressor microRNA-126 and its host gene EGFL7 in human cancer cells. Biochem Biophys Res Commun 379(3):726–731

Saito Y, Suzuki H, Tsugawa H, Nakagawa I, Matsuzaki J, Kanai Y et al (2009b) Chromatin remodeling at Alu repeats by epigenetic treatment activates silenced microRNA-512-5p with downregulation of Mcl-1 in human gastric cancer cells. Oncogene 28(30):2738–2744

Scott GK, Mattie MD, Berger CE, Benz SC, Benz CC (2006) Rapid alteration of microRNA levels by histone deacetylase inhibition. Cancer Res 66(3):1277–1281

Shen R, Pan S, Qi S, Lin X, Cheng S (2010) Epigenetic repression of microRNA-129-2 leads to overexpression of SOX4 in gastric cancer. Biochem Biophys Res Commun 394(4):1047–1052

Shi Y, Lan F, Matson C, Mulligan P, Whetstine JR, Cole PA et al (2004) Histone demethylation mediated by the nuclear amine oxidase homolog LSD1. Cell 119(7):941–953

Sinkkonen L, Hugenschmidt T, Berninger P, Gaidatzis D, Mohn F, Artus-Revel CG et al (2008) MicroRNAs control de novo DNA methylation through regulation of transcriptional repressors in mouse embryonic stem cells. Nat Struct Mol Biol 15(3):259–267

Soto-Reyes E, Gonzalez-Barrios R, Cisneros-Soberanis F, Herrera-Goepfert R, Perez V, Cantu D et al (2012) Disruption of CTCF at the miR-125b1 locus in gynecological cancers. BMC Cancer 12:40

Suzuki H, Yamamoto E, Nojima M, Kai M, Yamano HO, Yoshikawa K et al (2010) Methylation-associated silencing of microRNA-34b/c in gastric cancer and its involvement in an epigenetic field defect. Carcinogenesis 31(12):2066–2073

Taby R, Issa JP (2010) Cancer epigenetics. CA Cancer J Clin 60(6):376–392

Takata A, Otsuka M, Yoshikawa T, Kishikawa T, Hikiba Y, Obi S et al (2013) MicroRNA-140 acts as a liver tumor suppressor by controlling NF-kappaB activity by directly targeting DNA methyltransferase 1 (Dnmt1) expression. Hepatology 57(1):162–170

Tavazoie SF, Alarcon C, Oskarsson T, Padua D, Wang Q, Bos PD et al (2008) Endogenous human microRNAs that suppress breast cancer metastasis. Nature 451(7175):147–152

Tellez CS, Juri DE, Do K, Bernauer AM, Thomas CL, Damiani LA et al (2011) EMT and stem cell-like properties associated with miR-205 and miR-200 epigenetic silencing are early manifestations during carcinogen-induced transformation of human lung epithelial cells. Cancer Res 71(8):3087–3097

Tsai KW, Hu LY, Wu CW, Li SC, Lai CH, Kao HW et al (2010) Epigenetic regulation of miR-196b expression in gastric cancer. Genes Chromosomes Cancer 49(11):969–980

Tsukada Y, Fang J, Erdjument-Bromage H, Warren ME, Borchers CH, Tempst P et al (2006) Histone demethylation by a family of JmjC domain-containing proteins. Nature 439(7078):811–816

Tsuruta T, Kozaki K, Uesugi A, Furuta M, Hirasawa A, Imoto I et al (2011) miR-152 is a tumor suppressor microRNA that is silenced by DNA hypermethylation in endometrial cancer. Cancer Res 71(20):6450–6462

Tuddenham L, Wheeler G, Ntounia-Fousara S, Waters J, Hajihosseini MK, Clark I et al (2006) The cartilage specific microRNA-140 targets histone deacetylase 4 in mouse cells. FEBS Lett 580(17):4214–4217

Vaissiere T, Sawan C, Herceg Z (2008) Epigenetic interplay between histone modifications and DNA methylation in gene silencing. Mutat Res 659(1–2):40–48

Varambally S, Cao Q, Mani RS, Shankar S, Wang X, Ateeq B et al (2008) Genomic loss of microRNA-101 leads to overexpression of histone methyltransferase EZH2 in cancer. Science 322(5908):1695–1699

Vogt M, Munding J, Gruner M, Liffers ST, Verdoodt B, Hauk J et al (2011) Frequent concomitant inactivation of miR-34a and miR-34b/c by CpG methylation in colorectal, pancreatic, mammary, ovarian, urothelial, and renal cell carcinomas and soft tissue sarcomas. Virchows Arch 458(3):313–322

Vrba L, Jensen TJ, Garbe JC, Heimark RL, Cress AE, Dickinson S et al (2010) Role for DNA methylation in the regulation of miR-200c and miR-141 expression in normal and cancer cells. PLoS One 5(1):e8697

Waddington CH (2012) The epigenotype. 1942. Int J Epidemiol 41(1):10–13

Wang H, Wu J, Meng X, Ying X, Zuo Y, Liu R et al (2011) MicroRNA-342 inhibits colorectal cancer cell proliferation and invasion by directly targeting DNA methyltransferase 1. Carcinogenesis 32(7):1033–1042

Wang J, Zhao H, Tang D, Wu J, Yao G, Zhang Q (2013a) Overexpressions of microRNA-9 and microRNA-200c in human breast cancers are associated with lymph node metastasis. Cancer Biother Radiopharm 28(4):283–288

Wang P, Chen L, Zhang J, Chen H, Fan J, Wang K et al (2013b) Methylation-mediated silencing of the miR-124 genes facilitates pancreatic cancer progression and metastasis by targeting Rac1. Oncogene. doi:10.1038/onc.2012.598. [Epub ahead of print]

Whetstine JR, Nottke A, Lan F, Huarte M, Smolikov S, Chen Z et al (2006) Reversal of histone lysine trimethylation by the JMJD2 family of histone demethylases. Cell 125(3):467–481

Wiklund ED, Bramsen JB, Hulf T, Dyrskjot L, Ramanathan R, Hansen TB et al (2011) Coordinated epigenetic repression of the miR-200 family and miR-205 in invasive bladder cancer. Int J Cancer 128(6):1327–1334

Wilting SM, van Boerdonk RA, Henken FE, Meijer CJ, Diosdado B, Meijer GA et al (2010) Methylation-mediated silencing and tumour suppressive function of hsa-miR-124 in cervical cancer. Mol Cancer 9:167

Wilting SM, Verlaat W, Jaspers A, Makazaji NA, Agami R, Meijer CJ et al (2013) Methylation-mediated transcriptional repression of microRNAs during cervical carcinogenesis. Epigenetics 8(2):220–228

Wolf SS (2009) The protein arginine methyltransferase family: an update about function, new perspectives and the physiological role in humans. Cell Mol Life Sci 66(13):2109–2121

Xiao B, Jing C, Wilson JR, Walker PA, Vasisht N, Kelly G et al (2003) Structure and catalytic mechanism of the human histone methyltransferase SET7/9. Nature 421(6923):652–656

Xu X, Chen Z, Zhao X, Wang J, Ding D, Wang Z et al (2012) MicroRNA-25 promotes cell migration and invasion in esophageal squamous cell carcinoma. Biochem Biophys Res Commun 421(4):640–645

Yamakuchi M, Ferlito M, Lowenstein CJ (2008) miR-34a repression of SIRT1 regulates apoptosis. Proc Natl Acad Sci U S A 105(36):13421–13426

Yang X, Feng M, Jiang X, Wu Z, Li Z, Aau M et al (2009) miR-449a and miR-449b are direct transcriptional targets of E2F1 and negatively regulate pRb-E2F1 activity through a feedback loop by targeting CDK6 and CDC25A. Genes Dev 23(20):2388–2393

Yang S, Li Y, Gao J, Zhang T, Li S, Luo A et al (2012) MicroRNA-34 suppresses breast cancer invasion and metastasis by directly targeting Fra-1. Oncogene 32(36):4294–4303

Zhang L, Volinia S, Bonome T, Calin GA, Greshock J, Yang N et al (2008) Genomic and epigenetic alterations deregulate microRNA expression in human epithelial ovarian cancer. Proc Natl Acad Sci U S A 105(19):7004–7009

Zhang Z, Tang H, Wang Z, Zhang B, Liu W, Lu H et al (2011) MiR-185 targets the DNA methyltransferases 1 and regulates global DNA methylation in human glioma. Mol Cancer 10:124

Zhu A, Xia J, Zuo J, Jin S, Zhou H, Yao L et al (2012) MicroRNA-148a is silenced by hypermethylation and interacts with DNA methyltransferase 1 in gastric cancer. Med Oncol 29(4):2701–2709

Zhu X, Li Y, Shen H, Li H, Long L, Hui L et al (2013) miR-137 inhibits the proliferation of lung cancer cells by targeting Cdc42 and Cdk6. FEBS Lett 587(1):73–81

[illegible] Frequent concomitant inactivation of [illegible] and p16-INK4a by CpG methylation in colorectal, pancreatic, [illegible] primary ovarian, urothelial, and renal cell carcinomas and soft tissue sarcomas. Virchows Arch. [illegible]

[illegible]

Chapter 3
MicroRNAs, Diet and Cancer Chemoprevention

Shahana Majid, Sumit Arora, Rajvir Dahiya, and Sharanjot Saini

Contents

Abstract A growing body of clinical and epidemiological evidence suggests that diet is one of the most important modifiable determinants for risk of developing cancer and contributes to differences in cancer incidence among various populations. Experimental evidence supports the role of dietary components as chemopreventive agents in various cancers. Dietary factors potentially influence fundamental cellular

S. Majid • S. Arora • R. Dahiya • S. Saini (✉)
Department of Urology, San Francisco Veterans Affairs Medical Center
and University of California, San Francisco, CA, USA
e-mail: sainis@urology.ucsf.edu

S. Babashah (ed.), *MicroRNAs: Key Regulators of Oncogenesis*,
DOI 10.1007/978-3-319-03725-7_3, © Springer International Publishing Switzerland 2014

processes involved in carcinogenesis; including apoptosis, cell-cycle control, angiogenesis, inflammation and DNA repair. Since microRNAs (miRNA) have emerged as critical gene/protein regulators that control basic cellular processes, there is much interest in determining the effect of dietary agents on modulation of miRNAs and their cognate targets in cancer. It has been proposed that dietary modulation of miRNA expression may contribute to the cancer-protective effects of dietary components. During the last few years, a plethora of studies have examined the effect of dietary agents on miRNAs and their targets in the context of cancer biology which suggest that dietary factors play an important role in carcinogenesis via modulation of miRNAs. In this chapter, we will summarize findings from these studies and will discuss their potential utility for cancer chemoprevention and therapeutics.

Keywords MicroRNAs • Cancer • Diet • Chemoprevention • Therapy

1 Introduction

Cancer remains one of the leading causes of mortality worldwide. A minority of cancers are caused by germline mutations and a vast majority of malignancies (~90 %) are linked to somatic mutations and environmental factors (Anand et al. 2008). Environmental and dietary factors are believed to contribute to differences in cancer incidence among populations with different dietary habits. A survey of cancer incidence shows that the age-adjusted cancer incidence in the Western world is much higher (above 300 cases per 100,000 populations) as compared to that in Asian countries (100 cases per 100,000). Life-style risk factors such as high-fat diet, tobacco, alcohol, etc. are major contributors of carcinogenesis. A growing body of clinical, experimental, and epidemiological evidence supports the concept that diet is one of the most important modifiable risk determinants for cancer. Studies suggest that a large fraction of cancer deaths may be prevented by modifying dietary composition e.g. the content of fiber, fat, cereals, spices, etc. Experimental evidence suggests that dietary components act as chemopreventive agents in carcinogenesis. Several natural dietary chemopreventive agents have been identified, some with well characterized pleiotropic actions in cancer cells (Parasramka et al. 2012; Saini et al. 2010; Shah et al. 2012). Dietary components potentially influence fundamental cellular processes involved in carcinogenesis; including apoptosis, cell-cycle control, angiogenesis, inflammation and DNA repair. It has been proposed that dietary modulation of microRNA (miRNA) expression may contribute to the cancer-protective effects of dietary components. Recent data suggest that bioactive dietary components play a role directly or indirectly in the modulation of miRNA expression to regulate carcinogenesis and thereby possess chemopreventive potential (Izzotti et al. 2012; Parasramka et al. 2012; Saini et al. 2010; Shah et al. 2012).

miRNAs are small (~22 nucleotides) endogenously expressed non-coding RNAs that regulate gene expression and play crucial roles in basic cellular processes including the cell cycle, apoptosis, development, differentiation and metabolism. It has been estimated that miRNAs regulate ~30 % of the human genome (Bartel 2004). Given the crucial role of these small RNAs in fundamental cellular processes, the deregulated expression of miRNAs leads to various disease states, including cancer (Calin 2009; Garzon et al. 2009). Examination of tumor specific miRNA expression profiles has revealed widespread dysregulation of these molecules in diverse cancers. miRNAs can act either as oncogenes or tumor suppressor genes (Babashah and Soleimani 2011). Over-expressed miRNAs in cancers, such as miR-17-92, may function as oncogenes and promote cancer development by negatively regulating tumor suppressor genes and/or genes that control cell differentiation or apoptosis. Also, some miRNAs are down-regulated in cancer and act as bona fide tumor suppressor genes, such as let-7 and miR-34a. Cancer initiation and progression can involve miRNAs and their expression profiles can be used for the classification, diagnosis, and prognosis of human malignancies. In addition, a role for miRNAs has been established in the later steps of tumorigenesis, progression, and metastasis. It has been proposed that alterations in miRNA genes play a critical role in the pathophysiology of many, perhaps all, human cancers (Calin 2009; Garzon et al. 2009).

Evidence suggests that the chemopreventive potential of dietary agents stem from their ability to modulate the expression of tumor suppressors/oncogenes involved in signal transduction pathways (Ashendel 1995; Shah et al. 2012). Since miRNAs have emerged as critical regulators of genes/proteins, there is much interest in determining the effect of dietary agents on modulation of miRNA and their targets in cancer. A plethora of studies have examined the effect of dietary agents on miRNAs and their targets in the context of cancer biology in the last 5 years (Izzotti et al. 2012; Parasramka et al. 2012; Saini et al. 2010; Shah et al. 2012). These studies suggest that dietary factors play an important role in carcinogenesis via modulation of miRNAs and their cognate targets. A few studies explored the mechanisms by which dietary factors modulate miRNA expression and function that underlie their chemopreventive potential.

This chapter will focus on summarizing the findings from studies focused on the effects of several dietary components in terms of miRNA expression and explore their chemopreventive and therapeutic potential. Overall, these studies suggest that miRNAs are preferred targets for chemoprotective dietary agents and may be used as indicators of the efficacy of dietary intervention. Further, several of the diet-regulated miRNAs target tumor suppressors or oncogenes that mediate the initiation and progression of carcinogenesis. However, this is an area that is still in its infancy, and warrants thorough investigation before the findings can be translated for cancer chemoprevention and therapy.

2 MicroRNAs and Essential Nutritional Factors

2.1 *Vitamin A/Retinoic Acid*

Vitamin A/Retinol is an essential dietary factor that is involved in vision, immune function, cell growth and differentiation. All-trans-retinoic acid (ATRA) is the most biologically active metabolite of vitamin A that acts as a tumor suppressor in prostate, breast, lung, liver and pancreatic cancer models (Sun and Lotan 2002). Several studies suggest that retinoic acid exposure of acute promyelocytic leukemia (APL) leads to miRNA modulation. APL is caused by the reciprocal translocation between the retinoic acid receptor-α (RAR-α) on chromosome 17 with the promyelocytic leukemia gene (PML) on chromosome 15 resulting in a novel fusion protein (PML-RAR-α) (Melnick and Licht 1999). In the absence of an adequate dose of retinoic acid (RA), this fusion protein interferes with myeloid differentiation by transcriptional repression of retinoic acid-responsive genes. Pharmacological doses of ATRA reverse these repressive effects (Melnick and Licht 1999). Garzon et al. reported that retinoic acid exposure of human acute promyelocytic leukemia (AML) cells led to up-regulation of miR-15a, miR-15b, miR-16-1, let-7a-3, let-7c, let-7d, miR-223, miR-342, miR-107 and down-regulated miR-181b (Garzon et al. 2007). This study identified a putative nuclear factor-kappa B (NF-κB) binding site in the upstream genomic region of let-7a-3/let-7b cluster as essential for the retinoic acid-mediated transactivation of this cluster. Another study reported the up-regulation of miR-186, miR-215, miR-223 and down-regulation of miR-17-5p, miR-25, miR-193, miR-195, let-7a upon RA exposure of acute promyelocytic leukemia cells (Rossi et al. 2010). Analyses of the expression of granulocytic differentiation signature miRNAs in a cohort of APL patients treated with ATRA in addition to chemotherapy reported the down-regulation of miR-181b and up-regulation of miR-15b, -16, -107, -223, -342 and let-7c (Careccia et al. 2009). Another group identified miR-342 as one of the miRNAs up-regulated by ATRA during APL differentiation that is a direct transcriptional target of the critical hematopoietic transcription factors PU.1 and interferon regulatory factor (IRF)-1 and IRF-9 (De Marchis et al. 2009). IRF-1 maintains miR-342 at low levels, whereas the binding of PU.1 and IRF-9 in the promoter region following ATRA-mediated differentiation, up-regulates miR-342 expression.

Retinoic acid-induced differentiation of the neuroblastoma cell line led to increased expression of tumor suppressive miR-34a and decreased expression of its target E2F3 (Welch et al. 2007). In breast cancer, miR-21 is selectively induced by ATRA in $ER\alpha^+$ breast carcinoma cells. Induction of pro-oncogenic miR-21 counteracts the anti-proliferative action of ATRA but has the potentially beneficial effect of reducing cell motility of $ER\alpha^+$ cells (Terao et al. 2011). A recent study suggests that retinoic acid positively regulates expression of miR-210 and miR-23a/24-2 in breast cancer and is counteracted by estrogen (Saumet et al. 2012).

miR-10a, a key mediator of metastatic behavior in pancreatic cancer that regulates metastasis via suppression of HOX genes, HOXB1 and HOXB3 is a retinoic acid target. Retinoic acid receptor antagonists effectively repress miR-10a expression and completely block metastasis (Weiss et al. 2009). In human embryonal carcinoma NT2 cells, miR-23 plays a critical role in the RA-induced neuronal differentiation (Kawasaki and Taira 2003).

2.2 *Vitamin D*

Vitamin D and its metabolites, 25-hydroxyvitamin D3 (25D) and 1,25-dihydroxyvitamin D3 (1,25D), have pleiotropic effects that include regulation of calcium homeostasis, anti-inflammatory and cancer preventive and/or therapeutic properties

(Giangreco and Nonn 2013; Fleet et al. 2012; Garland et al. 2011). Epidemiological studies have shown that vitamin D deficiency is associated with increased risk and mortality of various cancers including that of prostate (Shui et al. 2012; Tretli et al. 2009), breast (Engel et al. 2010, 2011) and colon (Garland et al. 1989). Vitamin D primarily exerts its effects through its active metabolite 1,25D that regulates gene transcription via binding to the vitamin D receptor (VDR) (Fleet 2004; Shah et al. 2012). Recent studies suggest that vitamin D exerts it effects by modulating miRNAs and their target genes (Giangreco and Nonn 2013; Shah et al. 2012) as described below.

To investigate the cross talk between androgen- and vitamin D-mediated intracellular signaling pathways in prostate cancer, Wang et al. (2013) investigated the individual and combined effects of testosterone and 1,25D on global gene expression in LNCaP prostate cancer cells. A set of 15 miRNAs were differentially regulated by combination treatment. Among these miR-22, miR-29ab, miR-134, miR-1207-5p and miR-371-5p were up regulated, while miR-17 and miR-20a, members of the miR-17/92 cluster were down regulated. A number of genes implicated in cell cycle progression, lipid synthesis and accumulation and calcium homeostasis constitute the mRNA targets of these miRNAs. Recently, they reported that vitamin D_3 and androgen coordinately regulate the expression of miR-17/92 cluster leading to up-regulation of the mRNA encoding peroxisome proliferator-activated receptor alpha (PPARA) and its downstream targets, leading to increased lipogenesis (Wang et al. 2013). Ting et al. screened miRNA profiles in response to vitamin D in prostate cancer cell line LNCaP and found that a tumor suppressive miRNA, miR-98, is transcriptionally induced by 1,25D that directly targets cyclin J (CCNJ) (Ting et al. 2013). Another study suggests that miR-106b is up-regulated by 1,25D in prostate cells that contributes to p21 mediated cell-cycle arrest (Thorne et al. 2011). Giangreco et al. demonstrated that miRNAs could be key physiologic mediators of vitamin D3 activity in prevention and early treatment of prostate cancer. They showed that Vitamin D3 supplementation augmented tumor suppressive miRNAs, miR-100 and miR-125b, in patient prostate tissue samples. *In vitro*, miR-100 and miR-125b were up-regulated while their targets PLK1 and E2F3 were down-regulated by 1,25D in a VDR-dependent manner (Giangreco et al. 2013).

In ovarian cancer cells, miR-498 induced by 1,25D decreased the mRNA expression of the human telomerase reverse transcriptase leading to reduced growth (Kasiappan et al. 2012). 1,25D induced the expression of miR-627, which down-regulated histone demethylase JMJD1A (Jumonji domain containing 1A), and suppressed growth of xenograft tumors from HCT-116 colon cancer cells in nude mice (Padi et al. 2013). In another study, miR-22 was shown to be induced by 1,25D in a time-, dose- and VDR-dependent manner contributing to its antiproliferative, antimigratory and gene regulatory effects in colon cancer cells (Alvarez et al. 2012). In lung cancer cells, 1,25D up-regulated the transcription of hsa-let-7a-2 leading to anti-proliferative effects (Guan et al. 2013). In human myeloid leukemia cells, 1,25D markedly induced expression of miR-32, which targets the proapoptotic factor Bim, leading to inhibition of Bim and AraC-induced apoptosis (Gocek et al. 2011). Also, treatment of human myeloid leukemia cells with 1,25D led to down-regulation of miR-181a and miR-181b, resulting in up-regulation of $p27^{Kip1}$ and

$p21^{Cip1}$, that led to G_1 cell cycle arrest (Wang et al. 2009). In breast epithelial cells, there is a significant protective role for 25D against cellular stress mediated by altered miRNA expression, particularly miR-182 expression (Peng et al. 2010). Endogenous VDR levels are under regulatory control of miRNAs. miR-125 has been shown to post transcriptionally repress human VDR by direct binding to its 3′ UTR. Since miR-125b levels are known to be down-regulated in cancer, such a decrease may result in the up-regulation of VDR in cancer and augmentation of the antitumor effects of 1,25D (Mohri et al. 2009).

While studies show that vitamin D or its metabolites alter the levels of specific miRNAs in various cancers via VDRE-mediated or alternate mechanisms, some reports also suggest that vitamin D may globally augment miRNA expression to counteract the widespread suppression of miRNAs observed in cancer (Giangreco and Nonn 2013; Giangreco et al. 2013). It has been proposed that the mechanistic basis of such an effect of vitamin D could be VDR-dependent chromatin opening that increases pri-miRNA transcription globally or by alteration of miRNA processing machinery by vitamin D and its metabolites (Giangreco and Nonn 2013).

2.3 Folic Acid

Folic acid plays an important role in DNA synthesis, repair and methylation and acts as a cofactor in certain biological reactions (Weinstein et al. 2003). A few studies suggest that folate can modulate miRNA expression. Experimental rats fed a folate, methionine- and choline-deficient diet developed hepatocellular carcinoma (HCC) at 54 weeks of age in the absence of carcinogen treatment (Kutay et al. 2006). miRNA expression analysis of liver tumors showed that let-7a, miR-21, miR-23, miR-130, miR-190, and miR-17-92 family of genes were up-regulated whilst miR-122, an abundant liver-specific miRNA, was down-regulated as compared to control rats. Rats switched to the folate and methyl-adequate diet after 36 weeks on the deficient diet showed an increase in miR-122 expression associated with the inhibition of hepatocarcinogenesis (Kutay et al. 2006). In human lymphoblastoid cells, folate deficiency induced a pronounced global increase in miRNA expression. MiR-222 was significantly over-expressed under folate-deficient conditions *in vitro* and also *in vivo* in human peripheral blood from individuals with low folate intake (Marsit et al. 2006).

2.4 Vitamin E

Vitamin E is an essential vitamin with important antioxidant properties and also plays a role in cell signaling (Rimbach et al. 2010). Evidence suggests that vitamin E can alter miRNA expression as well. Experimental rats fed a vitamin E-depleted diet for 6 months resulted in reduced concentrations of miR-122a and miR-125b, which may play an important role in lipid metabolism, carcinogenesis, and inflammation (Gaedicke et al. 2008). In non-small cell lung cancer cells (NSCLC), delta-tocotrienol (a vitamin E variant) led to up-regulation of miR-34a expression and down-regulation of the Notch-1 pathway leading to inhibition of NSCLC cellular proliferation, invasion and induction of apoptosis (Ji et al. 2012).

2.5 Selenium

Selenium, an essential trace element, plays an important role as a component of the active site of several cellular enzymes, including those involved in detoxification (Burk 2002). Thus, selenium is involved in the regulation of various cellular functions including growth and apoptosis and selenium deficiency is associated with increased cancer risk (Kim and Milner 2001). Selenium supplementation has emerged as an attractive new approach to intervene in a range of human cancers, in particular prostate cancer (Sarveswaran et al. 2010). Sarveswaran et al. reported the modulation of miR-34b and miR-34c by selenium in prostate cancer cells (Sarveswaran et al. 2010). Incubation of LNCaP human prostate cancer cells (p53 +/+) with a natural form of selenium triggered rapid transcriptional activation of p53, and up-regulation of the expression of p53-target genes as well as induction of the miR-34 class of miRNAs (Sarveswaran et al. 2010). Since the majority of early-stage human prostate cancers bear a functional p53 gene, their results suggest that selenite may be useful not only for prevention but also for treatment of human prostate cancer.

2.6 Fatty Acids

Increasing evidence suggests that diets rich in n-3-polyunsaturated fatty acids (n-3 PUFAs) (docosahexaenoic acid, DHA and eicosapentaenoic acid, EPA) play a protective role in human cancers including colon, prostate and breast cancer. On the other hand, diets rich in n-6 PUFAs (linoleic acid, LA and arachidonic acid, AA) enhance carcinogenesis (Reddy et al. 1991; Shah et al. 2012; Whelan and McEntee 2004; Williams et al. 2011). Studies suggest that the protective effects of PUFAs may be partly mediated by affecting miRNA expression. Davidson et al. explored the effect of n-3 PUFA diet (fish oil) on azoxymethane- induced colon carcinogenesis in rats (Davidson et al. 2009). Carcinogen exposure led to reduced expression of five different tumor suppressive miRNAs (let-7d, miR-15b, miR-107, miR-191 and miR-324-5p), an effect that was reversed in n-3 PUFA fed animals pointing to a novel role of fish oil in protecting the colon from carcinogen-induced miRNA dysregulation (Davidson et al. 2009). Mandal et al. demonstrated a novel mechanism for the therapeutic function of fish oil diet that blocks miR-21, thereby increasing PTEN levels to prevent expression of CSF-1(colony stimulating factor-1) in breast cancer cells (Mandal et al. 2012). In another study, unsaturated fatty acids were shown to inhibit PTEN expression in hepatocytes by up-regulating microRNA-21 synthesis via an mTOR/NF-kappaB-dependent mechanism (Vinciguerra et al. 2009). Consistent with these data, microRNA-21 expression was increased in the livers of rats fed high-fat diets and in human liver biopsies of obese patients having diminished PTEN expression and steatosis (Vinciguerra et al. 2009). In glio-bastoma cells, PUFA treatment led to differential expression of several miRNAs, in particular, down-regulation of miR-143 and up-regulation of miR-20b expression (Farago et al. 2011).

Butyrate, a short-chain fatty acid that is the metabolic end product of unabsorbed dietary fiber or obtained directly from butter, is a putative chemoprotective agent (Karius et al. 2012; Miller 2004). Butyrate acts as a histone deacetylase inhibitor that has been shown to decrease proliferation and increase apoptosis in cancer cells (Miller 2004). Hu et al. (2011) reported that treatment of human colon cancer cells HCT116 with butyrate led to changes in the expression of 44 miRNAs, including significant reduction of miR-17~92, miR~18b- 106a, and miR-106b~25 clusters. MicroRNAs altered by butyrate are primarily over-expressed in cancer cells as compared to normal cells, including miR-106b, that was shown to directly target p21 (Hu et al. 2011). Butyrate was also shown to alter miRNA patterns regulating endodermal differentiation of human embryonic stem cells (hESC) (Tzur et al. 2008). Upon sodium butyrate-induced differentiation of hESC, two of the most up-regulated miRNAs were miR-24 and miR-10a, whose target genes have been shown to inhibit endodermal differentiation (Tzur et al. 2008).

3 MicroRNAs and Phytochemicals

3.1 Polyphenols

Polyphenols constitute a ubiquitous group of phytochemicals that possess important anti-cancer properties (Arts and Hollman 2005; Scalbert et al. 2005; Spencer et al. 2008). Some of the most common cancer chemopreventive polyphenols include curcumin and analogues, resveratrol, ellagitanin, catechins. Mechanistic studies suggest that polyphenolic compounds affect intracellular signaling network molecules involved in initiation, promotion and progression of cancer. Recent evidence suggest that polyphenolic compounds modulate miRNAs and their targets in several cancer models as summarized in the following sections.

3.1.1 Curcumin and Analogues

Curcumin (diferuloylmethane), a naturally occurring flavanoid compound derived from the rhizome of *Curcuma longa*, has been shown to have protective effect against various cancers (Aggarwal et al. 2003; Everett et al. 2007; Li et al. 2004, 2005, 2007a, b; Liang et al. 2004; Nakamura et al. 2002). *In vitro* and *in vivo* preclinical studies have shown that curcumin has antioxidant, anti-inflammatory, antiproliferative, and proapoptotic activities. Curcumin putatively down-regulates the expression of gene products such as NF-κB, suppresses growth, induces apoptosis and modulates various signal transduction pathways and the expression of many oncogenes (Nakamura et al. 2002; Sarkar et al. 2009). Recent studies suggest that curcumin modulates miRNA expression patterns as well.

Curcumin has been shown to be a potentially effective agent in pancreatic cancer. Several studies have examined the modulation of miRNA expression by curcumin and its analogues in pancreatic cancer. An initial study demonstrated that curcumin alters the miRNA expression profiles in human pancreatic cancer cells (Sun et al. 2008). The effects of curcumin (10 μmol/L) on miRNA expression profiles were studied in a pancreatic cancer cell line, BxPC3 which showed that 11 miRNAs were significantly up-regulated, whereas 18 were significantly down-regulated by curcumin compared with the control. Significantly, miR-22 was up-regulated accompanied by repression of its target genes *SP1 transcription factor (SP1)* and *estrogen receptor 1 (ESR1)*, whereas miR-199a* was down-regulated by curcumin (Sun et al. 2008). It has been reported that curcumin reduces the expression of WT1, a transcription factor that plays an important role in cellular proliferation and survival of various cancer cells, and is frequently expressed in pancreatic cancer (Glienke et al. 2009). It was shown that curcumin down-regulates the expression of WT1 partly by up-regulating the expression of miR-15a/16-1 in leukemic cells (Gao et al. 2012).

Diflourinated-curcumin (CDF), a synthetic analog of curcumin, shows greater bioavailability in animal tissues than curcumin (Padhye et al. 2009). Studies suggest that CDF either alone or in combination with curcumin, alters miRNA expression profiles. Bao et al. showed that CDF down-regulates miR-200 and miR-21 expression, inducing the up-regulation of its target gene, PTEN, in pancreatic cancer cells (Bao et al. 2011a). Another report by the same group demonstrated that CDF inhibits pancreatic cancer tumor growth and aggressiveness by reducing EZH2 expression and increasing the expression of a panel of tumor suppressive miRNAs (including let-7a, b, c, d, miR-26a, miR-101, miR-146a, and miR-200b,c) that are typically lost in pancreatic cancer (Bao et al. 2012b). Hypoxia-induced aggressiveness of pancreatic cancer cells is due to increased expression of VEGF, IL-6 and miR-21, which could be attenuated by treatment with CDF (Bao et al. 2012a). Another study assessed the effect of CDF and BR-DIM (BioResponse 3,3′-diindolylmethane; a natural derivative of curcumin) on pancreatic cancer cells. The treatment augmented expression of the miR-200 family and PTEN while reducing expression of MT1-MMP (Soubani et al. 2012). Ali et al. reported that in pancreatic cancer, loss of expression of let-7 and miR-143, and increased expression of miR-21 lead to increased expression of Ras and its GTPase activity, which could be attenuated by CDF treatment (Ali et al. 2012).

In a few studies with other cancers, curcumin up-regulated the expression of miR- 15a and miR-16 expression in breast cancer MCF-7 cells. miR-15a and miR-16 inhibited the expression of Bcl-2, thereby inducing apoptosis in these cancer cells (Yang et al. 2010). In colorectal cancer cells, curcumin inhibited the transcriptional regulation of miR-21 via AP-1, suppressed cell proliferation, tumor growth, invasion and *in vivo* metastasis, and stabilized the expression of the tumor suppressor

Pdcd4 (Mudduluru et al. 2011). In A549/DDP multidrug-resistant human lung adenocarcinoma cells, curcumin significantly down-regulated the expression of miR-186* (Zhang et al. 2010). We showed that curcumin directly induces a tumor suppressive miRNA, miR-203, in bladder cancer (Saini et al. 2011). miR-203 is frequently down-regulated in bladder cancer due to DNA hypermethylation of its promoter. Functional analyses showed that miR-203 has tumor suppressive properties in bladder cancer and miR-203 directly targets Akt2 and Src. Curcumin induces hypomethylation of the miR-203 promoter and up-regulates miR-203 expression, leading to down-regulation of miR-203 target genes Akt2 and Src that culminates in decreased proliferation and increased apoptosis of bladder cancer cells. This was the first report that showed a direct effect of curcumin on inducing epigenetic changes at a miRNA promoter with direct biological consequences (Saini et al. 2011).

Expression of miR-21 is greatly increased in chemotherapy-resistant colon cancer cells that are enriched in undifferentiated cancer stem/stem-like cells (CSCs/CSLCs). Anti-miR-21 treatment induced differentiation in chemotherapy-resistant colon cancer cells. CDF was highly effective in inhibiting growth and reducing CSCs/CSLCs in anti-miR-21-induced differentiating colon cancer cells (Yu et al. 2013a). Recently, Rao et al. reported that CDF could be a novel demethylating agent for restoring the expression of miR-34 family members (miR-34a and miR-34c) that are frequently down-regulated in colon cancer due to promoter hypermethylation (Roy et al. 2012). Gandhy et al. evaluated the effect of curcumin and its synthetic cyclohexanone and piperidine RL197 analog in colon cancer cell lines. They reported that curcumin and RL197 (the most active synthetic piperidine analog) induced reactive oxygen species (ROS), leading to inhibition of growth and induction of apoptosis. They found that curcumin-/RL197-induced repression of specificity protein (Sp) transcription factors that was ROS-dependent. This was due to induction of the Sp repressors ZBTB10 and ZBTB4 and down-regulation of miR-27a, miR-20a and miR-17-5p that regulate these repressors (Gandhy et al. 2012). EF24, a curcumin analog with improved anticancer activity, has been shown to up-regulate the expression of potential tumor suppressor miRNAs and down-regulate oncogenic miRNAs, including miR-21, in human prostate cancer cells and murine melanoma cells (Yang et al. 2013). Collectively, these studies suggest that modulation of miRNAs may underlie the chemopreventive and therapeutic properties of curcumin.

3.1.2 Resveratrol

Resveratrol (trans-3,4′,5-trihydroxystilbene) is a dietary polyphenolic, non-flavonoid antioxidant derived from the skin of fruits, especially grapes, berries and other plant sources. Resveratrol has cancer preventive properties, and it is currently in preclinical trials for human cancer prevention (Bishayee 2009). The chemopreventive properties of resveratrol stem from its ability to modulate canonical signal transduction pathways that control cell division, growth, apoptosis, inflammation, angiogenesis, metastasis and recent studies suggest that resveratrol may exhibit these protective effects by modulating miRNAs (Bishayee 2009; Shah et al. 2012).

Tili et al. (2010b) studied resveratrol-mediated miRNA regulation in colon cancer cells (SW480) and established that resveratrol treatment decreases the levels of several oncogenic miRNAs (miR-17, miR-21, miR-25, miR-92a, and miR-196a) targeting genes encoding Dicer1, tumor suppressor factors such as PDCD4 or PTEN, as well as key effectors of the TGFβ signaling pathway, while increasing the levels of *miR-663*, a tumor suppressor miRNA targeting *TGFβ1* transcripts. They also reported that resveratrol treatment of human monocytic cells induced miR-663 that decreases endogenous activator protein-1 (AP-1) activity and impairs its up-regulation by lipopolysaccharides (LPS), at least in part by directly targeting JunB and JunD transcripts. Further, they showed that resveratrol impairs the up-regulation of miR-155 by LPS in a miR-663-dependent manner (Tili et al. 2010a). In colon cancer cells, resveratrol in combination with quercetin repressed oncogenic microRNA-27a and induced zinc finger protein ZBTB10, that in turn led to repression of Specificity protein (Sp) transcription factors (Del Follo-Martinez et al. 2013).

Dhar et al. (2011) examined the differential miRNA expression in prostate cancer cells treated with resveratrol and found that 23 miRNAs were significantly down-regulated and 28 miRNAs were significantly up-regulated. The down-regulated miRNAs included miR-17-92 and miR-106ab clusters with well recognized oncogenic properties while the up-regulated miRNAs included miR-150, miR-149, and miR-1290. Also, PTEN, a predicted target of some of these miRNAs, was up-regulated. Another study showed that resveratrol inhibited prostate cancer growth and metastasis, in part, through inhibition of Akt/miR-21 signaling pathway (Sheth et al. 2012). In addition to reduced miR-21 and pAkt, resveratrol increased the expression of PDCD4 and maspin.

In lung cancer cells, resveratrol inhibited cell mobility through induction of mesenchymal-epithelial transition (MET) and induction of miR-520c expression that repressed FOXC2 (forkhead box C2) via miRNA-520h-|PP2A/C-|Akt → NF-κB → FOXC2 signal cascade. This study identified a new miRNA-520h-related signal cascade involved in resveratrol-mediated tumor suppression activity and indicated the clinical significance of miR-520h, PP2A/C and FOXC2 in lung cancer patients (Yu et al. 2013b). Hagiwara et al. demonstrated that resveratrol promotes expression and activity of Argonaute2 (Ago2), a central RNA interference (RNAi) component, thereby inhibiting breast cancer stem-like cell characteristics by increasing the expression of a number of tumor suppressive miRNAs, including miR-16, miR-141, miR-143, and miR-200c. They also found that pterostilbene, a natural dimethylated resveratrol analogue, is capable of mediating Ago2-dependent anti-cancer activity in a manner mechanistically similar to that of resveratrol (Hagiwara et al. 2012). In human non-small cell lung cancer cells (A549), resveratrol treatment significantly up-regulated miR-194, miR-299, miR-338, miR-582, miR-758 and down-regulated miR-92a. Furthermore, this study identified target genes related to apoptosis, cell cycle regulation, cell proliferation, and differentiation as predicted targets of these miRNAs (Bae et al. 2011). In breast cancer MCF7 cells, miR-663 and miR-744 inhibited expression of the proto-oncogene eukaryotic translation elongation factor 1A2 (eEF1A2) resulting in retardation of cellular proliferation. Resveratrol-treatment of MCF7 cells led to up-regulation of miR-663 and mir-744 accompanied by down-regulation of EEF1A2 mRNA (Vislovukh et al. 2013). miR-622 functions as a tumor suppressor by targeting K-Ras and enhances the anticarcinogenic effect of resveratrol (Han et al. 2012). These reports suggest that miRNA modulation form a significant mechanistic basis underlying the chemopreventive effects of resveratrol.

3.1.3 Ellagitanin

Ellagitannins are a diverse class of hydrolyzable tannins that were initially characterized for their anti-oxidant and free radical scavenging activity. These polyphenol compounds also possess anti-inflammatory, anti-tumor promoting, anti-proliferative, and apoptosis-inducing properties (Okuda et al. 1989; Parasramka et al. 2012).

Ellagitannin (BJA3121), 1,3-Di-O-galloyl-4,6-(s)-HHDP-b-D-glucopyranose, is a new natural polyphenol compound isolated from the plant Balanophora Japonica MAKINO. This ellagitannnin was reported to have antiproliferative effects in human HepG(2) liver cancer cells (Wen et al. 2009). miRNA profiling showed that BJA3121 can regulate the expression of 25 miRNAs, including 17 up-regulated and 8 downregulated miRNAs in HepG(2) cells including let-7 family members, miR-370, miR-373, and miR-526b. Prediction and functional analyses identified putative targets with roles in cell proliferation and differentiation (Parasramka et al. 2012; Wen et al. 2009).

3.1.4 Catechins

Catechins from tea extracts are considered to be potential chemotherapeutic drugs as well as cancer prevention agents. Chemopreventive effect of catechins have been described in preclinical models for all major sites of cancers, including prostate, breast, colon, lung, liver, and skin. Epigallocatechin gallate (EGCG) is a major type of green tea polyphenol that has been shown to alter cancer growth by targeting various cancer signaling pathways in a pleiotropic manner (Ahmad and Mukhtar 1999; Bailey and Mukhtar 2013; Mukhtar and Ahmad 1999a, b; Shah et al. 2012). Recent data suggests that EGCG can modulate miRNAs expression.

In human hepatocellular carcinoma HepG2 cells, EGCG treatment was found to down-regulate expression of 48 miRNAs and up-regulate 13 miRNAs, including miR-16. EGCG treatment led to down-regulation of miR-16 target, Bcl2, leading to induction of apoptosis (Tsang and Kwok 2010). In prostate cancer, EGCG functionally antagonizes the action of androgen resulting in inhibition of PCa growth. EGCG treatment led to significant down-regulation of androgen-regulated miR-21 and up-regulation of a tumor suppressor, miR-330 in prostate tumor xenografts (Siddiqui et al. 2011). In lung cancer cell lines, tea catechins up-regulated tumor suppressor miRNAs, let-7a-1 and let-7g, leading to repression of their targets, C-MYC and the regulatory protein of LIN-28 (Zhong et al. 2012). In human and mouse lung cancer cells, EGCG specifically up-regulated expression of miR-210, a major miRNA regulated by HIF-1α (Wang et al. 2011). Polyphenon-60 (green tea extract) treatment of MCF-7 breast cancer cells led to significant miRNA alterations, predominantly down-regulation of oncogenic miR-21 and miR-27 (Fix et al. 2010).

3.2 *Isoflavones*

Soy isoflavones such as genistein, daidzein and glycitein have been reported to have chemopreventive and therapeutic effects, via modulation of estrogen receptor binding in various tissues (Parasramka et al. 2012; Saini et al. 2010; Shah et al. 2012). Genistein, a well-studied isoflavone isolated from soybean, has been found to be a potent antitumor agent. A large body of evidence suggests that the inhibition of human cancer cell growth by genistein is mediated via the modulation of genes that are related to the control of cell cycle, apoptosis, angiogenesis, invasion, and metastasis (Banerjee et al. 2008; Khan et al. 2008; Saini et al. 2010). Genistein induces the p21WAF1/CIP, p16INK4a and other tumor suppressor genes in prostate cancer cells by epigenetic mechanisms involving active chromatin modifications (Kikuno et al. 2008; Majid et al. 2008, 2009a, b). It has been shown that genistein inhibits the activation of NF-kappaB and Akt signaling pathways, antagonizes estrogen- and androgen-mediated signaling pathways, possesses antioxidant properties, and is a potent inhibitor of angiogenesis and metastasis. Due to its pleiotropic anti-cancer activities, genistein is a promising agent for cancer chemoprevention (Banerjee et al. 2008; Khan et al. 2008) and is undergoing clinical trials in prostate, bladder, kidney and breast cancers (Taylor et al. 2009). Our group has been exploring the chemopreventive effects of genistein in various urological malignancies. Recent evidence suggest that miRNAs are molecular targets of this isoflavone as summarized below.

Genistein significantly inhibits human uveal melanoma cell growth in a time- and dose-related manner by regulating miR-27a and its target gene zinc finger and BTB domain containing 10 (ZBTB10) (Sun et al. 2009). Genistein suppressed prostate cancer growth through inhibition of oncogenic miR-151 (Chiyomaru et al. 2012) and up-regulation of miR-145 (Suh et al. 2011). In prostate cancer cells, isoflavone (70.5 % genistein, 26.3 % daidzein and 0.31 % glycetein) treatment led to increased levels of miR-29a and miR-1256 via demethylation of their promoters, causing decreased expression of target genes TRIM68 and PGK-1 culminating in inhibition of PCa cell growth and invasion (Li et al. 2012a). Bao et al. reported that in pancreatic cancer, activation of Notch-1 signaling contributes to the acquisition of EMT phenotype, which is in part mediated through the regulation of miR-200b and CSC self-renewal capacity, and these processes could be attenuated by genistein treatment (Bao et al. 2011b). We reported miRNA mediated alteration of minichromosome maintenance (MCM) genes by genistein in prostate cancer. MCM genes that are involved in DNA replication are commonly dysregulated in cancer cells. In prostate cancer cells treated with genistein, the expression of tumor suppressive miR-1296 was induced leading to down-regulation of MCM2 accompanied with cell cycle arrest in S-phase (Majid et al. 2010). Also, genistein could up-regulate the expression of ARH1 tumor suppressor gene in prostate cancer cells by down-regulating miR-221 and miR-222 (Chen et al. 2011). Genistein inhibited expression of oncogenic miR-21 and miR-23b-3p in RCC cell lines (Zaman et al. 2012a, b). Additionally, genistein inhibited Wnt-signaling by regulating onco-miR miR-1260b expression in renal cancer cells (Hirata et al. 2013). Genistein treatment of pancreatic cancer cells led to increased miR-146a expression and inhibited cancer cell invasion by down-regulation of EGFR, MTA-2, IRAK-1, and NF-kappaB (Li et al. 2010). Another study showed that genistein treatment of pancreatic cancer cells resulted in the up-regulation of let-7 and miR-200. This was associated with the down-regulation of EMT transcription factors, ZEB1, slug, and vimentin (Li et al. 2009). In ovarian cancer cells, genistein treatment led to differential expression of 53 miRNAs with the induction of ERalpha and ERbeta expression and decreased cellular growth (Parker et al. 2009). Genistein synergized with tumor suppressive miR-16 to induce apoptosis in a murine chronic lymphocytic leukemia (CLL) model suggesting that genistein can potentially modulate the biological effects of miRNAs (Salerno et al. 2009).

3.3 *Indoles*

Indole compounds, found in cruciferous vegetables, are potent anti-cancer agents. Studies with indole-3-carbinol (I3C) and its dimeric product, 3,3′-diindolylmethane (DIM) suggest that these compounds have the ability to deregulate multiple cellular signaling pathways, including PI3K/Akt/mTOR signaling, NF-κB signaling which may explain their ability to inhibit invasion, angiogenesis, reverse epithelial-to-mesenchymal transition (EMT) phenotype and drug resistance (Ahmad et al. 2013).

Kong et al. (2012) reported that treatment of prostate cancer cells with BR-DIM (formulated DIM: 3,3′-diindolylmethane by Bio Response, Boulder, CO, abbreviated as BR-DIM) up-regulated let-7 and down-regulated EZH2 expression, consistent with inhibition of self-renewal and clonogenic capacity. Moreover, BR-DIM intervention in a phase II clinical trial in patients prior to radical prostatectomy showed up-regulation of let-7 consistent with down-regulation of EZH2 expression in prostate cancer tissue specimens after BR-DIM intervention (Kong et al. 2012). In another prostate cancer study, formulated isoflavone and 3,3′-diindolylmethane (BR-DIM) down-regulated the expression of miR-92a, which is known to be associated with RANKL signaling, EMT and cancer progression. Their data suggests that isoflavone and BR-DIM could be useful for the prevention of PCa progression, by attenuating bone metastasis mechanisms (Li et al. 2012b).

Treatment of pancreatic cancer cells with DIM or isoflavone increased miR-146a expression, causing down-regulation of EGFR, MTA-2, IRAK-1, and NF-kappaB, resulting in an inhibition of pancreatic cancer cell invasion (Li et al. 2010). In another study, treatment of gemcitabine-resistant pancreatic cancer cells with either DIM or isoflavone resulted in up-regulation of miR-200 and let-7, leading to the down-regulation of ZEB1, slug, and vimentin that in turn, led to morphologic reversal of EMT phenotype (Li et al. 2009). Further, indole-3-carbinol could effectively enhance the sensitivity of pancreatic cancer cells to gemcitabine via down-regulation of miR-21 (Paik et al. 2013).

Melkamu et al. (2010) studied the modulation of miRNA expression by indole-3-carbinol in induced mouse lung tumors and found that the levels of several oncogenic miRNAs including miR-21, mir-31, miR-130a, miR-146b were reduced in indole-3-carbinol fed mice relative to the level in mice treated with the carcinogen alone. Altered miRNA expression profiles were observed in lung tissues of rats exposed to environmental cigarette smoke (Izzotti et al. 2009). Upon treatment with five dietary agents, including I3C, restoration of miRNAs targeting p53 function (miR-34b), TGF-β expression (miR-26a), ERBB2 activation (miR-125a), and angiogenesis (miR-10a) was observed (Izzotti et al. 2009; Parasramka et al. 2012). Treatment of MCF-7 breast cancer cells with DIM increased miR-21 expression and down-regulated Cdc25A, resulting in inhibition of breast cancer cell proliferation (Jin 2011). Collectively, these preliminary results in various cancer models suggest that miRNAs are promising molecular targets of chemopreventive dietary indoles.

3.4 Isothiocyanates

Isothiocyanates derived from cruciferous vegetables have been shown to modulate carcinogen metabolism in different tissues and also exert other chemoprotective effects (Higdon et al. 2007; Parasramka et al. 2012; Shah et al. 2012). Phenethyl isothiocyanate (PEITC) has been evaluated in various studies for its ability to modulate the expression of miRNAs. Studies have demonstrated that PEITC modulates miRNA expression in lung and liver of rodents exposed to cigarette smoke. Izzotti et al. evaluated the modulation of mRNA expression as related to exposure of neonatal mice to environmental cigarette smoke and to treatment with chemopreventive agents. PEITC protected the lung from cigarette smoke-induced alterations of miRNA expression but exhibited some adverse effects in liver. In the lung, PEITC counteracted the biological effect of cigarette smoke by modulating several miRNAs including let-7a, let-7c, miR-200, miR- 26a, miR-31, miR-125b, miR-135, and miR-382 in the lung, whereas in the liver, mixed alterations were observed (Izzotti et al. 2010b). In another study by the same group, PEITC was able to counteract miRNA alterations induced by cigarette smoke in mouse lung tissue (normal or affected by pneumonia) but not in lung cancer tissue (Izzotti et al. 2011, 2012). They also evaluated the effect of PEITC on miRNA alterations induced by cigarette smoke in rats and found that PEITC was very effective in restoring down-regulated miRNAs including let-7a, let-7c, miR-26a, miR-125b, miR-123, miR-99a, miR-146 and miR-222 (Izzotti et al. 2010a, 2012). In prostate cancer cells, PEITC down-regulated miR-141 leading to increased expression of small heterodimer partner

(Shp) that is a co-repressor to androgen receptor (AR) and inhibited AR transcriptional activity (Xiao et al. 2012). In a recent study, Slaby et al. identified a set of miRNAs (miR-155, miR-23b, miR-27b) regulated by isothiocyanates (sulforaphane, iberin) in colorectal cancer (Slaby et al. 2013).

Collectively, these studies support the concept that bioactive dietary components such as vitamins, fatty acids, trace elements, isoflavones, polyphenols, indoles and isothiocyanates play a chemopreventive role in the context of cancer biology by directly or indirectly modulating miRNA expression. Several of the downstream targets of these miRNAs are tumor suppressors or oncogenes that mediate the initiation and progression of carcinogenesis.

4 Potential for Cancer Therapeutics

Numerous alterations in multiple cellular signaling pathways often underlie various malignancies. As a result, a 'mono-modal therapy' or 'one gene-one target' approach has often failed as a therapeutic intervention for cancers (Parasramka et al. 2012). In fact, with the realization of the extreme molecular complexity of cancer, therapeutic strategies are being envisioned to disable multiple networks of tumor maintenance, rather than individual signaling pathways (Altieri et al. 2009). Recently, with the increasing understanding of miRNA regulation and functionality, miRNA based cancer therapeutic strategies are evolving, presently most of the strategies are based on the principle of gain- or loss- of-function (Parasramka et al. 2012). The potential of dietary agents to modulate miRNA expression patterns suggest that these agents will likely dovetail with ongoing research on the therapeutic potential of miRNAs. As detailed in the preceding sections, dietary factors have been reported to impact miRNAs involved in various stages of carcinogenesis, including early chemoprevention and also late-stage therapeutic effects. The advantage of dietary factors is that usually there are low associated toxicities. Currently, some of the dietary agents like genistein and curcumin are undergoing clinical trials in prostate, bladder, kidney and breast cancers (Taylor et al. 2009). Recent studies suggest that miRNAs in serum can serve as non-invasive biomarkers for cancer. In this regard, determining the change in miRNA levels in serum after exposure to dietary agents could be utilized as a diagnostic tool to monitor the effects of treatment over time.

Also there is great interest in combining conventional therapeutics with natural bioactive agents having chemoprotective properties. However, much work needs to be done to clarify the mechanisms by which dietary factors regulate miRNAs and other signaling pathways. In addition, issues such as *in vivo* bioavailability, selective targeting and the generation of bioactive metabolites need to be thoroughly examined. Currently, synthetic formulations of natural products with enhanced bioavailability, or encapsulation via nanoparticles and liposomes are being examined as alternative approaches (Parasramka et al. 2012).

5 Conclusions

In conclusion, experimental evidence suggests that dietary components such as essential nutritional factors, polyphenols, isoflavones, indoles and isothiocyanates can affect carcinogenesis through modulation of miRNA expression patterns and their target genes. Various dietary factors are now being investigated as modulators of miRNA profiles in cancer, and considering the growing awareness of the regulatory role of miRNA and their targets during different stages of cancer, there is much promise in the area from a chemopreventive and therapeutic standpoint. However, our current understanding in this area is still limited as the studies to date have been performed using cancer cell lines and are mostly highly descriptive. A vast majority of the studies have explored the global effect of dietary factors on miRNA expression and very few studies have validated the downstream targets. *In vivo* animal studies are required as these approaches tend to be more physiologically relevant and observations using cell lines are not always recapitulated *in vivo*. Also, *in vivo* approaches will take into account the pharmacokinetic and metabolic features of the dietary factors (Shah et al. 2012). In addition, more focused, detailed, mechanistic studies are required to move the field forward in the context of miRNAs involved in cancer chemoprevention (Ashendel 1995; Parasramka et al. 2012; Saini et al. 2011; Shah et al. 2012). It is necessary to understand the upstream mediators that are responsible for diet-mediated miRNA alterations. Some of the dietary agents such as curcumin, butyrate, flavonoids have been reported to alter epigenetic modifications which can modulate gene/miRNA transcription leading to changes in cell proliferation, differentiation, and cell survival (Shah et al. 2012). Overall, a better understanding of diet- regulated miRNA pathways will potentially aid in developing novel chemopreventive and therapeutic strategies for the management of cancer. Such studies are urgently needed given the fact that dietary components have been shown to play a crucial chemopreventive role in various cancers.

References

Aggarwal BB, Kumar A, Bharti AC (2003) Anticancer potential of curcumin: preclinical and clinical studies. Anticancer Res 23(1A):363–398

Ahmad N, Mukhtar H (1999) Green tea polyphenols and cancer: biologic mechanisms and practical implications. Nutr Rev 57(3):78–83

Ahmad A, Biersack B, Li Y, Kong D, Bao B, Schobert R et al (2013) Targeted regulation of PI3K/Akt/mTOR/NF-kappaB signaling by indole compounds and their derivatives: mechanistic details and biological implications for cancer therapy. Anti-cancer Agents Med Chem 13(7):1002–1013

Ali S, Ahmad A, Aboukameel A, Bao B, Padhye S, Philip PA et al (2012) Increased Ras GTPase activity is regulated by miRNAs that can be attenuated by CDF treatment in pancreatic cancer cells. Cancer Lett 319(2):173–181

Altieri DC, Languino LR, Lian JB, Stein JL, Leav I, van Wijnen AJ et al (2009) Prostate cancer regulatory networks. J Cell Biochem 107(5):845–852

Alvarez J, Wasse H, Tangpricha V (2012) Vitamin D supplementation in pre-dialysis chronic kidney disease: a systematic review. Dermato-Endocrinol 4(2):118–127

Anand P, Kunnumakkara AB, Sundaram C, Harikumar KB, Tharakan ST, Lai OS et al (2008) Cancer is a preventable disease that requires major lifestyle changes. Pharm Res 25(9):2097–2116

Arts IC, Hollman PC (2005) Polyphenols and disease risk in epidemiologic studies. Am J Clin Nutr 81(1 Suppl):317S–325S

Ashendel CL (1995) Diet, signal transduction and carcinogenesis. J Nutr 125(3 Suppl):686S–691S

Babashah S, Soleimani M (2011) The oncogenic and tumour suppressive roles of microRNAs in cancer and apoptosis. Eur J Cancer 47(8):1127–1137

Bae S, Lee EM, Cha HJ, Kim K, Yoon Y, Lee H et al (2011) Resveratrol alters microRNA expression profiles in A549 human non-small cell lung cancer cells. Mol Cells 32(3):243–249

Bailey HH, Mukhtar H (2013) Green tea polyphenols and cancer chemoprevention of genitourinary cancer. American Society of Clinical Oncology educational book/ASCO American Society of Clinical Oncology Meeting 2013:92–96

Banerjee S, Li Y, Wang Z, Sarkar FH (2008) Multi-targeted therapy of cancer by genistein. Cancer Lett 269(2):226–242

Bao B, Ali S, Kong D, Sarkar SH, Wang Z, Banerjee S et al (2011a) Anti-tumor activity of a novel compound-CDF is mediated by regulating miR-21, miR-200, and PTEN in pancreatic cancer. PLoS One 6(3):e17850

Bao B, Wang Z, Ali S, Kong D, Li Y, Ahmad A et al (2011b) Notch-1 induces epithelial-mesenchymal transition consistent with cancer stem cell phenotype in pancreatic cancer cells. Cancer Lett 307(1):26–36

Bao B, Ali S, Ahmad A, Azmi AS, Li Y, Banerjee S et al (2012a) Hypoxia-induced aggressiveness of pancreatic cancer cells is due to increased expression of VEGF, IL-6 and miR-21, which can be attenuated by CDF treatment. PLoS One 7(12):e50165

Bao B, Ali S, Banerjee S, Wang Z, Logna F, Azmi AS et al (2012b) Curcumin analogue CDF inhibits pancreatic tumor growth by switching on suppressor microRNAs and attenuating EZH2 expression. Cancer Res 72(1):335–345

Bartel DP (2004) MicroRNAs: genomics, biogenesis, mechanism, and function. Cell 116(2): 281–297

Bishayee A (2009) Cancer prevention and treatment with resveratrol: from rodent studies to clinical trials. Cancer Prev Res (Phila) 2(5):409–418

Burk RF (2002) Selenium, an antioxidant nutrient. Nutr Clin Care 5(2):75–79

Calin GA (2009) MicroRNAs and cancer: what we know and what we still have to learn. Genome Med 1(8):78

Careccia S, Mainardi S, Pelosi A, Gurtner A, Diverio D, Riccioni R et al (2009) A restricted signature of miRNAs distinguishes APL blasts from normal promyelocytes. Oncogene 28(45):4034–4040

Chen Y, Zaman MS, Deng G, Majid S, Saini S, Liu J et al (2011) MicroRNAs 221/222 and genistein-mediated regulation of ARHI tumor suppressor gene in prostate cancer. Cancer Prev Res (Phila) 4(1):76–86

Chiyomaru T, Yamamura S, Zaman MS, Majid S, Deng G, Shahryari V et al (2012) Genistein suppresses prostate cancer growth through inhibition of oncogenic microRNA-151. PLoS One 7(8):e43812

Davidson LA, Wang N, Shah MS, Lupton JR, Ivanov I, Chapkin RS (2009) n-3 polyunsaturated fatty acids modulate carcinogen-directed non-coding microRNA signatures in rat colon. Carcinogenesis 30(12):2077–2084

De Marchis ML, Ballarino M, Salvatori B, Puzzolo MC, Bozzoni I, Fatica A (2009) A new molecular network comprising PU.1, interferon regulatory factor proteins and miR-342 stimulates ATRA-mediated granulocytic differentiation of acute promyelocytic leukemia cells. Leukemia 23(5):856–862

Del Follo-Martinez A, Banerjee N, Li X, Safe S, Mertens-Talcott S (2013) Resveratrol and quercetin in combination have anticancer activity in colon cancer cells and repress oncogenic microRNA-27a. Nutr Cancer 65(3):494–504

Dhar S, Hicks C, Levenson AS (2011) Resveratrol and prostate cancer: promising role for microRNAs. Mol Nutr Food Res 55(8):1219–1229

Engel P, Fagherazzi G, Boutten A, Dupre T, Mesrine S, Boutron-Ruault MC et al (2010) Serum 25(OH) vitamin D and risk of breast cancer: a nested case-control study from the French E3N cohort. Cancer Epidemiol Biomarkers Prev 19(9):2341–2350

Engel P, Fagherazzi G, Mesrine S, Boutron-Ruault MC, Clavel-Chapelon F (2011) Joint effects of dietary vitamin D and sun exposure on breast cancer risk: results from the French E3N cohort. Cancer Epidemiol Biomarkers Prev 20(1):187–198

Everett PC, Meyers JA, Makkinje A, Rabbi M, Lerner A (2007) Preclinical assessment of curcumin as a potential therapy for B-CLL. Am J Hematol 82(1):23–30

Farago N, Feher LZ, Kitajka K, Das UN, Puskas LG (2011) MicroRNA profile of polyunsaturated fatty acid treated glioma cells reveal apoptosis-specific expression changes. Lipids Health Dis 10:173

Fix LN, Shah M, Efferth T, Farwell MA, Zhang B (2010) MicroRNA expression profile of MCF-7 human breast cancer cells and the effect of green tea polyphenon-60. Cancer Genomics Proteomics 7(5):261–277

Fleet JC (2004) Rapid, membrane-initiated actions of 1,25 dihydroxyvitamin D: what are they and what do they mean? J Nutr 134(12):3215–3218

Fleet JC, DeSmet M, Johnson R, Li Y (2012) Vitamin D and cancer: a review of molecular mechanisms. Biochem J 441(1):61–76

Gaedicke S, Zhang X, Schmelzer C, Lou Y, Doering F, Frank J et al (2008) Vitamin E dependent microRNA regulation in rat liver. FEBS Lett 582(23–24):3542–3546

Gandhy SU, Kim K, Larsen L, Rosengren RJ, Safe S (2012) Curcumin and synthetic analogs induce reactive oxygen species and decreases specificity protein (Sp) transcription factors by targeting microRNAs. BMC Cancer 12:564

Gao SM, Yang JJ, Chen CQ, Chen JJ, Ye LP, Wang LY et al (2012) Pure curcumin decreases the expression of WT1 by upregulation of miR-15a and miR-16-1 in leukemic cells. J Exp Clin Cancer Res 31:27

Garland CF, Comstock GW, Garland FC, Helsing KJ, Shaw EK, Gorham ED (1989) Serum 25-hydroxyvitamin D and colon cancer: eight-year prospective study. Lancet 2(8673): 1176–1178

Garland CF, French CB, Baggerly LL, Heaney RP (2011) Vitamin D supplement doses and serum 25-hydroxyvitamin D in the range associated with cancer prevention. Anticancer Res 31(2):607–611

Garzon R, Pichiorri F, Palumbo T, Visentini M, Aqeilan R, Cimmino A et al (2007) MicroRNA gene expression during retinoic acid-induced differentiation of human acute promyelocytic leukemia. Oncogene 26(28):4148–4157

Garzon R, Calin GA, Croce CM (2009) MicroRNAs in cancer. Annu Rev Med 60:167–179

Giangreco AA, Nonn L (2013) The sum of many small changes: microRNAs are specifically and potentially globally altered by vitamin D3 metabolites. J Steroid Biochem Mol Biol 136:86–93

Giangreco AA, Vaishnav A, Wagner D, Finelli A, Fleshner N, Van der Kwast T et al (2013) Tumor suppressor microRNAs, miR-100 and -125b, are regulated by 1,25-dihydroxyvitamin D in primary prostate cells and in patient tissue. Cancer Prev Res (Phila) 6(5):483–494

Glienke W, Maute L, Wicht J, Bergmann L (2009) Wilms' tumour gene 1 (WT1) as a target in curcumin treatment of pancreatic cancer cells. Eur J Cancer 45(5):874–880

Gocek E, Wang X, Liu X, Liu CG, Studzinski GP (2011) MicroRNA-32 upregulation by 1,25-dihydroxyvitamin D3 in human myeloid leukemia cells leads to Bim targeting and inhibition of AraC-induced apoptosis. Cancer Res 71(19):6230–6239

Guan H, Liu C, Chen Z, Wang L, Li C, Zhao J et al (2013) 1,25-Dihydroxyvitamin D3 up-regulates expression of hsa-let-7a-2 through the interaction of VDR/VDRE in human lung cancer A549 cells. Gene 522(2):142–146

Hagiwara K, Kosaka N, Yoshioka Y, Takahashi RU, Takeshita F, Ochiya T (2012) Stilbene derivatives promote Ago2-dependent tumour-suppressive microRNA activity. Sci Rep 2:314

Han Z, Yang Q, Liu B, Wu J, Li Y, Yang C et al (2012) MicroRNA-622 functions as a tumor suppressor by targeting K-Ras and enhancing the anticarcinogenic effect of resveratrol. Carcinogenesis 33(1):131–139

Higdon JV, Delage B, Williams DE, Dashwood RH (2007) Cruciferous vegetables and human cancer risk: epidemiologic evidence and mechanistic basis. Pharmacol Res 55(3):224–236

Hirata H, Ueno K, Nakajima K, Tabatabai ZL, Hinoda Y, Ishii N et al (2013) Genistein downregulates onco-miR-1260b and inhibits Wnt-signalling in renal cancer cells. Br J Cancer 108(10):2070–2078

Hu S, Dong TS, Dalal SR, Wu F, Bissonnette M, Kwon JH et al (2011) The microbe-derived short chain fatty acid butyrate targets miRNA-dependent p21 gene expression in human colon cancer. PLoS One 6(1):e16221

Izzotti A, Calin GA, Arrigo P, Steele VE, Croce CM, De Flora S (2009) Downregulation of microRNA expression in the lungs of rats exposed to cigarette smoke. FASEB J 23(3): 806–812

Izzotti A, Calin GA, Steele VE, Cartiglia C, Longobardi M, Croce CM et al (2010a) Chemoprevention of cigarette smoke-induced alterations of MicroRNA expression in rat lungs. Cancer Prev Res (Phila) 3(1):62–72

Izzotti A, Larghero P, Cartiglia C, Longobardi M, Pfeffer U, Steele VE et al (2010b) Modulation of microRNA expression by budesonide, phenethyl isothiocyanate and cigarette smoke in mouse liver and lung. Carcinogenesis 31(5):894–901

Izzotti A, Larghero P, Balansky R, Pfeffer U, Steele VE, De Flora S (2011) Interplay between histopathological alterations, cigarette smoke and chemopreventive agents in defining microRNA profiles in mouse lung. Mutation Res 717(1–2):17–24

Izzotti A, Cartiglia C, Steele VE, De Flora S (2012) MicroRNAs as targets for dietary and pharmacological inhibitors of mutagenesis and carcinogenesis. Mutation Res 751(2):287–303

Ji X, Wang Z, Geamanu A, Goja A, Sarkar FH, Gupta SV (2012) Delta-tocotrienol suppresses Notch-1 pathway by upregulating miR-34a in nonsmall cell lung cancer cells. Int J Cancer Journal international du cancer 131(11):2668–2677

Jin Y (2011) 3,3′-Diindolylmethane inhibits breast cancer cell growth via miR-21-mediated Cdc25A degradation. Mol Cell Biochem 358(1–2):345–354

Karius T, Schnekenburger M, Dicato M, Diederich M (2012) MicroRNAs in cancer management and their modulation by dietary agents. Biochem Pharmacol 83(12):1591–1601

Kasiappan R, Shen Z, Tse AK, Jinwal U, Tang J, Lungchukiet P et al (2012) 1,25-Dihydroxyvitamin D3 suppresses telomerase expression and human cancer growth through microRNA-498. J Biol Chem 287(49):41297–41309

Kawasaki H, Taira K (2003) Functional analysis of microRNAs during the retinoic acid-induced neuronal differentiation of human NT2 cells. Nucleic Acids Res Suppl 3:243–244

Khan N, Afaq F, Mukhtar H (2008) Cancer chemoprevention through dietary antioxidants: progress and promise. Antioxid Redox Signal 10(3):475–510

Kikuno N, Shiina H, Urakami S, Kawamoto K, Hirata H, Tanaka Y et al (2008) Genistein mediated histone acetylation and demethylation activates tumor suppressor genes in prostate cancer cells. Int J Cancer Journal international du cancer 123(3):552–560

Kim YS, Milner J (2001) Molecular targets for selenium in cancer prevention. Nutr Cancer 40(1):50–54

Kong D, Heath E, Chen W, Cher ML, Powell I, Heilbrun L et al (2012) Loss of let-7 up-regulates EZH2 in prostate cancer consistent with the acquisition of cancer stem cell signatures that are attenuated by BR-DIM. PLoS One 7(3):e33729

Kutay H, Bai S, Datta J, Motiwala T, Pogribny I, Frankel W et al (2006) Downregulation of miR-122 in the rodent and human hepatocellular carcinomas. J Cell Biochem 99(3):671–678

Li L, Aggarwal BB, Shishodia S, Abbruzzese J, Kurzrock R (2004) Nuclear factor-kappaB and IkappaB kinase are constitutively active in human pancreatic cells, and their down-regulation by curcumin (diferuloylmethane) is associated with the suppression of proliferation and the induction of apoptosis. Cancer 101(10):2351–2362

Li L, Braiteh FS, Kurzrock R (2005) Liposome-encapsulated curcumin: in vitro and in vivo effects on proliferation, apoptosis, signaling, and angiogenesis. Cancer 104(6):1322–1331

Li L, Ahmed B, Mehta K, Kurzrock R (2007a) Liposomal curcumin with and without oxaliplatin: effects on cell growth, apoptosis, and angiogenesis in colorectal cancer. Mol Cancer Ther 6(4):1276–1282

Li M, Zhang Z, Hill DL, Wang H, Zhang R (2007b) Curcumin, a dietary component, has anticancer, chemosensitization, and radiosensitization effects by down-regulating the MDM2 oncogene through the PI3K/mTOR/ETS2 pathway. Cancer Res 67(5):1988–1996

Li Y, VandenBoom TG 2nd, Kong D, Wang Z, Ali S, Philip PA et al (2009) Up-regulation of miR-200 and let-7 by natural agents leads to the reversal of epithelial-to-mesenchymal transition in gemcitabine-resistant pancreatic cancer cells. Cancer Res 69(16):6704–6712

Li Y, Vandenboom TG 2nd, Wang Z, Kong D, Ali S, Philip PA et al (2010) miR-146a suppresses invasion of pancreatic cancer cells. Cancer Res 70(4):1486–1495

Li Y, Kong D, Ahmad A, Bao B, Dyson G, Sarkar FH (2012a) Epigenetic deregulation of miR-29a and miR-1256 by isoflavone contributes to the inhibition of prostate cancer cell growth and invasion. Epigenetics 7(8):940–949

Li Y, Kong D, Ahmad A, Bao B, Sarkar FH (2012b) Targeting bone remodeling by isoflavone and 3,3′-diindolylmethane in the context of prostate cancer bone metastasis. PLoS One 7(3):e33011

Liang T, Chen MJ, Zhou KY, Tang XD, Wang XG (2004) Induction of apoptosis in human nasopharyngeal carcinoma cell line CNE-2Z by curcumin. Ai Zheng 23(12):1651–1654

Majid S, Kikuno N, Nelles J, Noonan E, Tanaka Y, Kawamoto K et al (2008) Genistein induces the p21WAF1/CIP1 and p16INK4a tumor suppressor genes in prostate cancer cells by epigenetic mechanisms involving active chromatin modification. Cancer Res 68(8):2736–2744

Majid S, Dar AA, Ahmad AE, Hirata H, Kawakami K, Shahryari V et al (2009a) BTG3 tumor suppressor gene promoter demethylation, histone modification and cell cycle arrest by genistein in renal cancer. Carcinogenesis 30(4):662–670

Majid S, Dar AA, Shahryari V, Hirata H, Ahmad A, Saini S et al (2009b) Genistein reverses hypermethylation and induces active histone modifications in tumor suppressor gene B-Cell translocation gene 3 in prostate cancer. Cancer 116(1):66–76

Majid S, Dar AA, Saini S, Chen Y, Shahryari V, Liu J et al (2010) Regulation of minichromosome maintenance gene family by microRNA-1296 and genistein in prostate cancer. Cancer Res 70(7):2809–2818

Mandal CC, Ghosh-Choudhury T, Dey N, Choudhury GG, Ghosh-Choudhury N (2012) miR-21 is targeted by omega-3 polyunsaturated fatty acid to regulate breast tumor CSF-1 expression. Carcinogenesis 33(10):1897–1908

Marsit CJ, Eddy K, Kelsey KT (2006) MicroRNA responses to cellular stress. Cancer Res 66(22):10843–10848

Melkamu T, Zhang X, Tan J, Zeng Y, Kassie F (2010) Alteration of microRNA expression in vinyl carbamate-induced mouse lung tumors and modulation by the chemopreventive agent indole-3-carbinol. Carcinogenesis 31(2):252–258

Melnick A, Licht JD (1999) Deconstructing a disease: RARalpha, its fusion partners, and their roles in the pathogenesis of acute promyelocytic leukemia. Blood 93(10):3167–3215

Miller SJ (2004) Cellular and physiological effects of short-chain fatty acids. Mini Rev Med Chem 4(8):839–845

Mohri T, Nakajima M, Takagi S, Komagata S, Yokoi T (2009) MicroRNA regulates human vitamin D receptor. Int J Cancer Journal international du cancer 125(6):1328–1333

Mudduluru G, George-William JN, Muppala S, Asangani IA, Kumarswamy R, Nelson LD et al (2011) Curcumin regulates miR-21 expression and inhibits invasion and metastasis in colorectal cancer. Biosci Rep 31(3):185–197

Mukhtar H, Ahmad N (1999a) Green tea in chemoprevention of cancer. Toxicol Sci 52 (2 Suppl):111–117

Mukhtar H, Ahmad N (1999b) Mechanism of cancer chemopreventive activity of green Tea. Proc Soc Exp Biol Med 220(4):234–238

Nakamura K, Yasunaga Y, Segawa T, Ko D, Moul JW, Srivastava S et al (2002) Curcumin down-regulates AR gene expression and activation in prostate cancer cell lines. Int J Oncol 21(4):825–830

Okuda T, Yoshida T, Hatano T (1989) Ellagitannins as active constituents of medicinal plants. Planta Med 55(2):117–122

Padhye S, Banerjee S, Chavan D, Pandye S, Swamy KV, Ali S et al (2009) Fluorocurcumins as cyclooxygenase-2 inhibitor: molecular docking, pharmacokinetics and tissue distribution in mice. Pharm Res 26(11):2438–2445

Padi SK, Zhang Q, Rustum YM, Morrison C, Guo B (2013) MicroRNA-627 mediates the epigenetic mechanisms of vitamin D to suppress proliferation of human colorectal cancer cells and growth of xenograft tumors in mice. Gastroenterology 145(2):437–446

Paik WH, Kim HR, Park JK, Song BJ, Lee SH, Hwang JH (2013) Chemosensitivity induced by down-regulation of microRNA-21 in gemcitabine-resistant pancreatic cancer cells by indole-3-carbinol. Anticancer Res 33(4):1473–1481

Parasramka MA, Ho E, Williams DE, Dashwood RH (2012) MicroRNAs, diet, and cancer: new mechanistic insights on the epigenetic actions of phytochemicals. Mol Carcinog 51(3): 213–230

Parker LP, Taylor DD, Kesterson J, Metzinger DS, Gercel-Taylor C (2009) Modulation of microRNA associated with ovarian cancer cells by genistein. Eur J Gynaecol Oncol 30(6):616–621

Peng X, Vaishnav A, Murillo G, Alimirah F, Torres KE, Mehta RG (2010) Protection against cellular stress by 25-hydroxyvitamin D3 in breast epithelial cells. J Cell Biochem 110(6):1324–1333

Reddy BS, Burill C, Rigotty J (1991) Effect of diets high in omega-3 and omega-6 fatty acids on initiation and postinitiation stages of colon carcinogenesis. Cancer Res 51(2):487–491

Rimbach G, Moehring J, Huebbe P, Lodge JK (2010) Gene-regulatory activity of alpha-tocopherol. Molecules 15(3):1746–1761

Rossi A, D'Urso OF, Gatto G, Poltronieri P, Ferracin M, Remondelli P et al (2010) Non-coding RNAs change their expression profile after Retinoid induced differentiation of the promyelocytic cell line NB4. BMC Res Notes 3:24

Roy S, Levi E, Majumdar AP, Sarkar FH (2012) Expression of miR-34 is lost in colon cancer which can be re-expressed by a novel agent CDF. J Hematol Oncol 5:58

Saini S, Majid S, Dahiya R (2010) Diet, microRNAs and prostate cancer. Pharm Res 27(6): 1014–1026

Saini S, Arora S, Majid S, Shahryari V, Chen Y, Deng G et al (2011) Curcumin modulates microRNA-203-mediated regulation of the Src-Akt axis in bladder cancer. Cancer Prev Res (Phila) 4(10):1698–1709

Salerno E, Scaglione BJ, Coffman FD, Brown BD, Baccarini A, Fernandes H et al (2009) Correcting miR-15a/16 genetic defect in New Zealand Black mouse model of CLL enhances drug sensitivity. Mol Cancer Ther 8(9):2684–2692

Sarkar FH, Li Y, Wang Z, Kong D (2009) Cellular signaling perturbation by natural products. Cell Signal 21(11):1541–1547

Sarveswaran S, Liroff J, Zhou Z, Nikitin AY, Ghosh J (2010) Selenite triggers rapid transcriptional activation of p53, and p53-mediated apoptosis in prostate cancer cells: implication for the treatment of early-stage prostate cancer. Int J Oncol 36(6):1419–1428

Saumet A, Vetter G, Bouttier M, Antoine E, Roubert C, Orsetti B et al (2012) Estrogen and retinoic acid antagonistically regulate several microRNA genes to control aerobic glycolysis in breast cancer cells. Mol Biosyst 8(12):3242–3253

Scalbert A, Manach C, Morand C, Remesy C, Jimenez L (2005) Dietary polyphenols and the prevention of diseases. Crit Rev Food Sci Nutr 45(4):287–306

Shah MS, Davidson LA, Chapkin RS (2012) Mechanistic insights into the role of microRNAs in cancer: influence of nutrient crosstalk. Front Genet 3:305

Sheth S, Jajoo S, Kaur T, Mukherjea D, Sheehan K, Rybak LP et al (2012) Resveratrol reduces prostate cancer growth and metastasis by inhibiting the Akt/MicroRNA-21 pathway. PLoS One 7(12):e51655

Shui IM, Mucci LA, Kraft P, Tamimi RM, Lindstrom S, Penney KL et al (2012) Vitamin D-related genetic variation, plasma vitamin D, and risk of lethal prostate cancer: a prospective nested case–control study. J Natl Cancer Inst 104(9):690–699

Siddiqui IA, Asim M, Hafeez BB, Adhami VM, Tarapore RS, Mukhtar H (2011) Green tea polyphenol EGCG blunts androgen receptor function in prostate cancer. FASEB J 25(4): 1198–1207

Slaby O, Sachlova M, Brezkova V, Hezova R, Kovarikova A, Bischofova S et al (2013) Identification of microRNAs regulated by isothiocyanates and association of polymorphisms inside their target sites with risk of sporadic colorectal cancer. Nutr Cancer 65(2):247–254

Soubani O, Ali AS, Logna F, Ali S, Philip PA, Sarkar FH (2012) Re-expression of miR-200 by novel approaches regulates the expression of PTEN and MT1-MMP in pancreatic cancer. Carcinogenesis 33(8):1563–1571

Spencer JP, Abd El Mohsen MM, Minihane AM, Mathers JC (2008) Biomarkers of the intake of dietary polyphenols: strengths, limitations and application in nutrition research. Br J Nutr 99(1):12–22

Suh SO, Chen Y, Zaman MS, Hirata H, Yamamura S, Shahryari V et al (2011) MicroRNA-145 is regulated by DNA methylation and p53 gene mutation in prostate cancer. Carcinogenesis 32(5):772–778

Sun SY, Lotan R (2002) Retinoids and their receptors in cancer development and chemoprevention. Crit Rev Oncol/Hematol 41(1):41–55

Sun M, Estrov Z, Ji Y, Coombes KR, Harris DH, Kurzrock R (2008) Curcumin (diferuloylmethane) alters the expression profiles of microRNAs in human pancreatic cancer cells. Mol Cancer Ther 7(3):464–473

Sun Q, Cong R, Yan H, Gu H, Zeng Y, Liu N et al (2009) Genistein inhibits growth of human uveal melanoma cells and affects microRNA-27a and target gene expression. Oncol Rep 22(3): 563–567

Taylor CK, Levy RM, Elliott JC, Burnett BP (2009) The effect of genistein aglycone on cancer and cancer risk: a review of in vitro, preclinical, and clinical studies. Nutr Rev 67(7):398–415

Terao M, Fratelli M, Kurosaki M, Zanetti A, Guarnaccia V, Paroni G et al (2011) Induction of miR-21 by retinoic acid in estrogen receptor-positive breast carcinoma cells: biological correlates and molecular targets. J Biol Chem 286(5):4027–4042

Thorne JL, Maguire O, Doig CL, Battaglia S, Fehr L, Sucheston LE et al (2011) Epigenetic control of a VDR-governed feed-forward loop that regulates p21(waf1/cip1) expression and function in non-malignant prostate cells. Nucleic Acids Res 39(6):2045–2056

Tili E, Michaille JJ, Adair B, Alder H, Limagne E, Taccioli C et al (2010a) Resveratrol decreases the levels of miR-155 by upregulating miR-663, a microRNA targeting JunB and JunD. Carcinogenesis 31(9):1561–1566

Tili E, Michaille JJ, Alder H, Volinia S, Delmas D, Latruffe N et al (2010b) Resveratrol modulates the levels of microRNAs targeting genes encoding tumor-suppressors and effectors of TGFβ signaling pathway in SW480 cells. Biochem Pharmacol 80(12):2057–2065

Ting HJ, Messing J, Yasmin-Karim S, Lee YF (2013) Identification of microRNA-98 as a therapeutic target inhibiting prostate cancer growth and a biomarker induced by vitamin D. J Biol Chem 288(1):1–9

Tretli S, Hernes E, Berg JP, Hestvik UE, Robsahm TE (2009) Association between serum 25(OH) D and death from prostate cancer. Br J Cancer 100(3):450–454

Tsang WP, Kwok TT (2010) Epigallocatechin gallate up-regulation of miR-16 and induction of apoptosis in human cancer cells. J Nutr Biochem 21(2):140–146

Tzur G, Levy A, Meiri E, Barad O, Spector Y, Bentwich Z et al (2008) MicroRNA expression patterns and function in endodermal differentiation of human embryonic stem cells. PLoS One 3(11):e3726

Vinciguerra M, Sgroi A, Veyrat-Durebex C, Rubbia-Brandt L, Buhler LH, Foti M (2009) Unsaturated fatty acids inhibit the expression of tumor suppressor phosphatase and tensin homolog (PTEN) via microRNA-21 up-regulation in hepatocytes. Hepatology 49(4):1176–1184

Vislovukh A, Kratassiouk G, Porto E, Gralievska N, Beldiman C, Pinna G et al (2013) Proto-oncogenic isoform A2 of eukaryotic translation elongation factor eEF1 is a target of miR-663 and miR-744. Br J Cancer 108(11):2304–2311

Wang X, Gocek E, Liu CG, Studzinski GP (2009) MicroRNAs181 regulate the expression of p27Kip1 in human myeloid leukemia cells induced to differentiate by 1,25-dihydroxyvitamin D3. Cell Cycle 8(5):736–741

Wang H, Bian S, Yang CS (2011) Green tea polyphenol EGCG suppresses lung cancer cell growth through upregulating miR-210 expression caused by stabilizing HIF-1alpha. Carcinogenesis 32(12):1881–1889

Wang WL, Welsh J, Tenniswood M (2013) 1,25-Dihydroxyvitamin D3 modulates lipid metabolism in prostate cancer cells through miRNA mediated regulation of PPARA. J Steroid Biochem Mol Biol 136:247–251

Weinstein SJ, Hartman TJ, Stolzenberg-Solomon R, Pietinen P, Barrett MJ, Taylor PR et al (2003) Null association between prostate cancer and serum folate, vitamin B(6), vitamin B(12), and homocysteine. Cancer Epidemiol Biomarkers Prev 12(11 Pt 1):1271–1272

Weiss FU, Marques IJ, Woltering JM, Vlecken DH, Aghdassi A, Partecke LI et al (2009) Retinoic acid receptor antagonists inhibit miR-10a expression and block metastatic behavior of pancreatic cancer. Gastroenterology 137(6):2136–2145.e1-7

Welch C, Chen Y, Stallings RL (2007) MicroRNA-34a functions as a potential tumor suppressor by inducing apoptosis in neuroblastoma cells. Oncogene 26(34):5017–5022

Wen XY, Wu SY, Li ZQ, Liu ZQ, Zhang JJ, Wang GF et al (2009) Ellagitannin (BJA3121), an anti-proliferative natural polyphenol compound, can regulate the expression of MiRNAs in HepG2 cancer cells. Phytother Res 23(6):778–784

Whelan J, McEntee MF (2004) Dietary (n-6) PUFA and intestinal tumorigenesis. J Nutr 134 (12 Suppl):3421S–3426S

Williams CD, Whitley BM, Hoyo C, Grant DJ, Iraggi JD, Newman KA et al (2011) A high ratio of dietary n-6/n-3 polyunsaturated fatty acids is associated with increased risk of prostate cancer. Nutr Res 31(1):1–8

Xiao J, Gong AY, Eischeid AN, Chen D, Deng C, Young CY et al (2012) miR-141 modulates androgen receptor transcriptional activity in human prostate cancer cells through targeting the small heterodimer partner protein. Prostate 72(14):1514–1522

Yang J, Cao Y, Sun J, Zhang Y (2010) Curcumin reduces the expression of Bcl-2 by upregulating miR-15a and miR-16 in MCF-7 cells. Med Oncol 27(4):1114–1118

Yang CH, Yue J, Sims M, Pfeffer LM (2013) The curcumin analog EF24 targets NF-kappaB and miRNA-21, and has potent anticancer activity in vitro and in vivo. PLoS One 8(8):e71130

Yu Y, Sarkar FH, Majumdar AP (2013a) Down-regulation of miR-21 induces differentiation of chemoresistant colon cancer cells and enhances susceptibility to therapeutic regimens. Transl Oncol 6(2):180–186

Yu YH, Chen HA, Chen PS, Cheng YJ, Hsu WH, Chang YW et al (2013b) MiR-520h-mediated FOXC2 regulation is critical for inhibition of lung cancer progression by resveratrol. Oncogene 32(4):431–443

Zaman MS, Shahryari V, Deng G, Thamminana S, Saini S, Majid S et al (2012a) Up-regulation of microRNA-21 correlates with lower kidney cancer survival. PLoS One 7(2):e31060

Zaman MS, Thamminana S, Shahryari V, Chiyomaru T, Deng G, Saini S et al (2012b) Inhibition of PTEN gene expression by oncogenic miR-23b-3p in renal cancer. PLoS One 7(11):e50203

Zhang J, Zhang T, Ti X, Shi J, Wu C, Ren X et al (2010) Curcumin promotes apoptosis in A549/DDP multidrug-resistant human lung adenocarcinoma cells through an miRNA signaling pathway. Biochem Biophys Res Commun 399(1):1–6

Zhong Z, Dong Z, Yang L, Chen X, Gong Z (2012) Inhibition of proliferation of human lung cancer cells by green tea catechins is mediated by upregulation of let-7. Exp Ther Med 4(2):267–272

Chapter 4
MicroRNAs and Energy Metabolism in Cancer Cells

Linchong Sun, Xiaoping He, Yang Cao, Ping Gao, and Huafeng Zhang

Contents

Abstract In 1927, Otto Warburg demonstrated that cancer cells use glucose for growth and division in a manner that is different from normal cells, a phenomenon known nowadays as Warburg Effect. To date, overwhelming evidence indicates that aberration in metabolism plays important roles in cancer progression. More recently, for more than a decade biologists are fascinated by the functions of small RNAs known as microRNAs (miRNAs), which play vital roles in many important biological processes, such as cell proliferation, differentiation, EMT/MET transition, cell signaling, response to infection, induction of pluripotent stem cells and cell metabolism. As discussed in other excellent chapters of this

---

L. Sun • X. He • Y. Cao • H. Zhang (✉)
School of Life Science, University of Science and Technology of China,
Hefei 230027, China
e-mail: hzhang222@ustc.edu.cn

P. Gao (✉)
School of Life Science, University of Science and Technology of China,
Hefei 230027, China

Division of Hematology, Department of Medicine, The Johns Hopkins University
School of Medicine, Baltimore, MD 21205, USA
e-mail: pgao2@ustc.edu.cn

S. Babashah (ed.), *MicroRNAs: Key Regulators of Oncogenesis*,
DOI 10.1007/978-3-319-03725-7_4, © Springer International Publishing Switzerland 2014

book, the roles of miRNAs in cancer development have been extensively studied. Here in this chapter, we will discuss the significance of miRNAs in regulating cancer cell metabolism. Specifically, we will focus on the roles of miRNAs in mediating metabolism of three major energy substrates including glucose, lipid and glutamine metabolism in cancer development.

Keywords MicroRNA • Glucose metabolism • Lipid metabolism • Glutamine metabolism • Cancer metabolism

1 Introduction

The theory put forward by Otto Warburg many years ago is that the faster growth rate of cancer cells than that of their normal counterparts is largely because cancer cells tend to drive glycolysis followed by lactate fermentation in the cytosol as energy source rather than oxidation of pyruvate fed into mitochondria, even in the presence of oxygen. This claim is now known as Warburg Effect (Wang et al. 2009; Warburg 1956). After more than eight decades, discovery and characterization of specific metabolic changes in cancer cells continue to progress rapidly and reprogramming energy metabolism is considered an emerging hallmark of cancers (Hanahan and Weinberg 2011). During cancer development, in order to meet the excessive demand of cellular growth, cancer cells may adapt a serial of strategic alterations regarding energy metabolism. For instance, glucose, as a ubiquitous energy source in most organisms, can supply carbon, oxygen, hydrogen for anabolic processes, and its uptake is frequently enhanced in cancer cells. Glutamine, a non-essential amino acid, serves as carbon source for energy production, conduces to biosynthetic reaction by supplying carbon and nitrogen, regulates redox homeostasis through synthesis of glutathione (Dang 2010). Increased glutaminolysis is emerged as a novel feature of the metabolic profile of cancer cells, along with increased aerobic glycolysis. Another energy resource, fatty acid, which is seldom appreciated before but under the spotlight now, is required for energy storage, membrane synthesis and generation of signaling molecules (Currie et al. 2013).

Today, mutations in oncogenes and tumor suppressors are known to be answer for the malignant transformation. Furthermore, Warburg Effect is generally accepted as the result of these mutations, rather than a cause. Several famous oncogenes and tumor suppressors are conversantly participated in modifying glycolysis, glutaminolysis, lipid metabolism and various biosynthetic pathways. For instance, PI3K pathway/hypoxia/c-Myc promotes glucose uptake (Cairns et al. 2011). AMP-activated protein kinase (AMPK) regulates glucose switch through mTOR pathway (Cairns et al. 2011). Oncogenic Kras enhances glucose import at gene expression and protein levels as well as regulates glutamine metabolism in tissue-specific PDAC model (Ying et al. 2012; Son et al. 2013). Both c-Myc and p53 can manipulate glutaminolysis through glutaminase1 or glutaminase2 (Gao et al. 2009; Hu et al. 2010; Suzuki et al. 2010). Hypoxia can mediate lipogenesis by IDH1 through reductive glutamine metabolism (Metallo et al. 2011).

In the meantime, nearly two decades ago, microRNAs (miRNAs), which are approximately 22 nucleotides long, were first discovered in *Caenorhabditis elegans* (Lee et al. 1993; Reinhart et al. 2000). They are encoded by miRNA genes which are transcribed by RNA polymerase II or III, then forming the primary miRNA in the nucleus, the primary miRNA is processed by Drosha and DGCR8, resulting in pre-miRNA, then it is exported from nucleus into cytoplasm by exportin 5 and reprocess by Dicer. Dicer is a RNAse III enzyme that processes the pre-miRNA to a 22 nucleotides miRNA. Presumably, miRNAs regulate a third of human genes through base pairing with regions within the 3′ untranslated region of the target mRNA (Cannell et al. 2008). There is abundant evidence on the multitudinous roles of miRNAs in biological processes, covering proliferation, differentiation, aging, apoptosis, metastasis, signal transduction and so on (Babashah and Soleimani 2011).

Evidence is emerging that miRNAs are involved in multiple cancer development process including cancer metabolism. Whether we can cut off the specific energy supply for cancer cell growth by restoring normal aerobic metabolism, and how miRNAs involved in this alteration, are among the fiercely investigated topics for cancer researchers. Here, we will focus on the special roles of miRNAs played on tumor metabolism, including glucose, glutamine and lipid metabolism.

2 MicroRNAs Regulate Glucose Metabolism

2.1 MiRNAs and Glucose Homeostasis

As summarized in Table 4.1, miRNAs are shown to regulate insulin secretion, biosynthesis and sensitivity to maintain the glucose homeostasis. Lin28/let-7 axis is demonstrated as the central regulator of mammalian glucose metabolism in mouse models (Zhu et al. 2011). Let-7 is the well-known tumor suppressor, whose biogenesis blocked by the RNA-binding protein Lin28. In mice, lin28 and let-7 play the opposite role in glucose metabolism through insulin-PI3K-mTOR pathway. Global knockdown of the let-7 family with an antimiR was sufficient to prevent and treat impaired glucose tolerance in mice with diet-induced obesity (Frost and Olson 2011). Let-7 and let-7-targets such as c-Myc and Kras all have significant effects on cancer metabolism, hence, the regulation of glucose homeostasis by lin28/let-7 axis may have profound implications in cancer metabolism (Zhu et al. 2010).

The expression of miR-143 is up-regulated in the liver of genetic and dietary mouse model of obesity (Jordan et al. 2011). Highly expressed miR-143 inhibits insulin-stimulated AKT activation and impairs glucose metabolism by directly targeting oxysterol-binding-protein-related protein (ORP). In the regulation of glucose homeostasis, the pancreatic specific miRNA miR-375 (Kloosterman et al. 2007) has a profound effect. miR-375 as well as miR-124 and let-7b regulate insulin exocytosis in the pancreatic beta cell by suppressing myotrophin (MTPN) which helps secretary granules to exocytotic sites (Gauthier and Wollheim 2006). Recent

Table 4.1 MicroRNA regulations in glucose and glutamine metabolism

	MiRNAs	Targets	Function	References
Glucose homeostasis	Let-7	INSR; IRS2; IGF1R	Impairs glucose tolerance	Zhu et al. (2011), Frost and Olson (2011)
	miR-143	ORP8	Insulin resistance	Jordan et al. (2011)
	miR-375; miR-124; Let-7b	MTPN	Insulin secretion	Kloosterman et al. (2007), Gauthier and Wollheim (2006)
Glycolysis	miR-195-5p	GLUT3	Suppresses proliferation and promotes apoptosis	Fei et al. (2012)
	miR-233; miR-133	GLUT4	Increases glucose uptake	Lu et al. (2010), Horie et al. (2009)
	miR-143	HK2	Inhibits glycolysis	Fang et al. (2012), Jiang et al. (2012)
	miR-155	SOCS1 and C/EBPβ	Promotes glycolysis	Jiang et al. (2012)
	miR-200	ZEB1/ZEB2	Promotes invasion and metastasis	Davalos et al. (2012), Ahmad et al. (2011)
	miR-15a/16-1	ALDO A	Suppresses glycolysis	Calin et al. (2008)
	miR-122	ALDO A; PTP1B	Suppresses glycolysis and suppresses insulin signaling	Chen et al. (2012), Yang et al. (2012)
Induced or suppressed by glucose stress	miR-32-5p	PTEN	Promotes cell cycle	Zhu et al. (2013)
	miR-30a-5p	BETA2/NeuroD	Impairs glucose tolerance and beta cell function	Kim et al. (2013)
	miR-29c	SPRY1	Induces cell apoptosis, increased extracellular matrix protein accumulation	Long et al. (2011)
	miR-451	CAB39	Promotes cell growth, inhibits migration and survival	Godlewski et al. (2010)
	miR-466h-5p	Unknown	Promotes apoptosis	Druz et al. (2012)
Amino acid catabolism	miR-23a/b	GLS	Increases glutamine uptake and catabolism	Gao et al. (2009)
	miR-23b*	POX/PRODH	Proline catabolism	Liu et al. (2012)
	miR-277	Unknown	Metabolic switch in modulating amino acid catabolism	Stark et al. (2003)

study showed that miR-375 directly targets 3′-phosphoinositidedependent protein kinase-1(PDPK1), and significantly impairs the activity of PDPK1 (El Ouaamari et al. 2008), a master kinase related to human disease such as diabetes and cancer.

2.2 *MiRNAs and Glycolysis*

It is well known that the most fundamental metabolism alteration in tumor cells is increased levels of glucose uptake and glycolysis (also known as Warburg Effect). Glucose can't pass across the hydrophobic cell membrane freely. They need a transport, glucose transporter proteins, also called solute carrier family 2 (GLUT). The level of glucose uptake can be affected by the activity of GLUT in cell membranes. GLUT3 is inhibited by miR-195-5p (Fei et al. 2012), which directly targets GLUT3 3′-untranslated region in human bladder cancer T24 cell lines. Via this molecular mechanism, miR-195-5p suppresses proliferation and promotes apoptosis of cancer cells *in vitro*. GLUT4, another transporter of GLUT family, can be up-regulated by miR-223 (Lu et al. 2010) and miR-133 (Horie et al. 2009) in rat cardiomyocytes.

MiRNAs regulate glycolysis by directly targeting key metabolic enzymes. Hexokiase 2 (HK2), the first enzyme of glycolysis that phosphorylates glucose to produce glucose 6-phosphate, was identified as a target of miR-143 (Fang et al. 2012). MiR-143 was shown to down-regulate HK2 protein level and inhibit glucose metabolism in lung cancer. This phenomenon was also confirmed in breast cancer. Further it was found that pro-inflammatory cytokines promote glycolysis mediated by miR-155, which up-regulates HK2 in two cascades of miR-155-SOCS1-STAT3-HK2 and miR-155-C/EBPβ-miR-143-HK2 (Jiang et al. 2012). Phosphogulcose isomerase (PGI) is a housekeeping cytosolic enzyme that brings the interconversion of glucose-6-phoshpate and fructose-6-phosphate, playing an important role in glycolysis and gluconeogenesis. A decrease of PGI expression leads to over-expression of miR-200 (Ahmad et al. 2011), which is associated with reversal of mesenchymal-epithelial transition (EMT) phenotype in human breast cancer cells (Davalos et al. 2012). MiR-15a/16-1 cluster, the first tumor suppressor miRNAs described in human malignancies (Calin et al. 2002), are reported to target the enzyme Aldo A which catalyzes a reversible aldol reaction in glycolysis (Calin et al. 2008). MiR-122 is the most abundantly expressed miRNA in the liver, and may be involved in lipid and cholesterol metabolism. While PTP1B is a newly identified direct target of miR-122, it is a negative regulator of insulin signaling cascade (Yang et al. 2012). And miR-122 is also predicted to target aldo A (Chen et al. 2012). Thus, these miRNAs are reported to have roles in cancer glycolysis.

2.3 *MiRNAs and Glucose Stress*

The change of miRNA expression is also a cause of abnormal glucose condition and subsequent cellular functions. In mesenchymal stem cells, hyperglycemic condition lead to a decrease of miR-32-5p, which promotes cell cycle by targeting tensin homologs deleted on chromosome 10 (PTEN) (Zhu et al. 2013). MiR-30a-5p over-expression induce beta cell dysfunction *in vitro*, which can be induced by high glucose at the concentration of 33.3 mmol/l (Kim et al. 2013). Another glucose induced miRNA, miR-29c, is known as a tumor suppressor which induces cell apoptosis, increases extracellular matrix protein accumulation and activates Rho kinase in mouse diabetic nephropathy model (Long et al. 2011). In glioma cells, a single miRNA was shown to adapt the cells to different glucose conditions. Glucose induced miR-451 promotes cell growth. While decreased miR-451 expression in low glucose slows proliferation, it enhances migration and survival (Godlewski et al. 2010). Deprivation of glucose leads to a change of miRNAs such as miR-466h-5p (Druz et al. 2012). Cultivation of mouse cell line with glucose-free medium causes the oxidative stress. The inhibition of histone deacetylases (HDACs) activity by this oxidative stress increases acetylation in miR-466h-5p promoter region, which lead to the activation of this miRNA.

3 MicroRNAs Modulate Lipid Metabolism

Alteration and its impact of lipid metabolism in cancer cells have long been ignored compared to the intensive interest in two other star nutrients, glucose and glutamine. This sounds a bit strange considering the pivotal roles of fatty acid in energy storage, cell proliferation and generation of membrane signal. However, last few years have witnessed a rapid development of the field and now a growing body of research is focused on lipid metabolism. Herein, we will summarize the role of several miRNAs that are involved in regulation of lipid homeostasis (Fig. 4.1).

One of the most notable miRNAs involved in lipid metabolism is liver-specific miR-122, which accounts for approximately 70 % of total miRNAs in the liver. In addition to its known roles in regulating HCV/HBV (hepatitis C/B virus) replication and expression, hepatocyte differentiation and glucose metabolism (Fukuhara et al. 2012; Qiu et al. 2010; Hu et al. 2012), miR-122 is the first miRNA to be linked recently to fat and cholesterol metabolism control. Two studies demonstrated the fundament roles of miR-122 in regulating lipid metabolism by 2′-O-methoxyethyl (2′-MOE) phosphorothioate-modified antisense oligonucleotide (ASO) *in vivo* (Esau et al. 2006; Fabani and Gait 2007). More recently, two independent groups discovered the tumor suppressor role of miR-122 in mouse liver using genetic deletion technology (Hsu et al. 2012; Tsai et al. 2012). Hsu et al. utilized mice which is germline knockout (KO) or liver-specific knockout (LKO) of miR-122 to describe the characteristics of fatty acid metabolism *in vivo*. They found that in LKO mice, the hepatic lipid level, especially triglyceride (TG) accumulation, was increased by up-regulation of Agpat1 and Mogat1 which are responsible for TG biosynthesis. But the secretion of TG into serum is reduced, thus resulting in hepatic microsteatosis in early adult life of LKO mice. Moreover, many genes involved in development, cellular proliferation, death and cancer, such

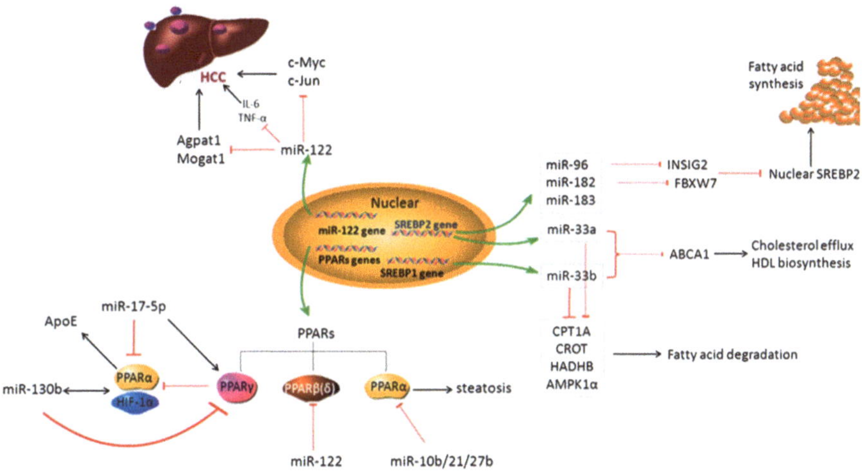

Fig. 4.1 MicroRNA-mediated regulation of lipid metabolism. Many miRNAs are involved in regulation of the expression of target coding genes essential for fat acid synthesis and degradation. Focus of this figure is on the vital roles of miR-122, miR-33a/b and miRNAs related with PPARs in lipid metabolism

as, c-Myc, c-Jun, CCND1, Igf2, Epcam, Rhoa, are aberrantly expressed in addition to the variation of lipid metabolism genes. In KO mice, increased monocytes and neutrophils are recruited to livers, leading to inflammation, the major factor attributed to malignant transformation of hepatocellular carcinoma. Production of pro-tumorigenic cytokines including IL-6 and TNF-α are also increased in LKO and KO mice which possess higher HCC incidence along with aging (Hsu et al. 2012). Tsai et al. also described the similar tumor suppressor role of miR-122 using genetic deletion in mice. They demonstrated that serum HDL and VLDL were significantly reduced in the miR-$122^{-/-}$ mice, which, on the contrary, causes liver steatosis. In addition, miR-122 can control hepatic fibrogenesis through one target named Krüppel-like factor 6 (KLF6) that is expressed mainly in hepatic carcinoma, whereas its expression is much lower in the normal hepatocytes of WT livers (Tsai et al. 2012). As a whole, these studies represented endeavors to decipher and manipulate the vital roles of miR-122 in regulating liver homeostasis using miR-122 antisense oligonucleotides or using an loss-of-function model *in vivo* (Wen and Friedman 2012).

SREBP, abbreviation of sterol regulatory element binding transcription factor, includes two protein named SREBP1 and SREBP2 whose vital function is played in sterol biosynthesis and cholesterol homeostasis by regulating transcription of sterol-regulated genes. Known miRNAs that are related with SREBP include miR-33a/b. MiR-33a is encoded in intron 16 of SREBP2 gene on chromosome 22, whereas miR-33b is encoded in intron 17 of SREBP1 gene on chromosome 17, respectively

(Najafi-Shoushtari et al. 2010). The two mature miRNAs only differ in two nucleotides, but can regulate plentiful overlapping target genes, such as ATP-binding cassette transporter sub-family A member 1 (ABCA1), which is a cholesterol transporter and can mediate cholesterol efflux from within the cell to lipid-free apolipoprotein A1 (APOA-I). When ABCA1 is decreased because of elevated miR-33, then high density lipoprotein (HDL) cholesterol levels formation is destroyed, so the incidence of cardiometabolic diseases increased (Najafi-Shoushtari et al. 2010; Horie et al. 2010; Rayner et al. 2010). In mouse macrophages, miR-33 also targets ATP-binding cassette, sub-family G (white), member 1 (ABCG1), reducing cholesterol efflux to nascent HDL (Rayner et al. 2010; Marquart et al. 2010). Meantime, miR-33a and miR-33b also contribute to the regulation of several proteins involved in fatty acid β-oxidation, modulating the expression of carnitine O-octanoyltransferase (CROT), carnitine palmitoyltransferase 1A (CPT1A), and hydroxyacyl-CoA dehydrogenase-3-ketoacyl-CoA thiolase-enoyl-CoA hydratase (trifunctional protein) β-subunit (HADHB). CROT, CPT1A, and HADHB regulate the transport and degradation of fatty acids in the mitochondria (Rayner et al. 2011; Davalos et al. 2011).

Nevertheless, Tae-Il Jeon and colleagues claimed that additional miRNAs may be more robustly activated by SREBPs than miR-33 to regulate intracellular cholesterol, due to the fact that intracellular cholesterol levels are tightly controlled (Jeon et al. 2013). They performed a genome-wide analysis searching for miRNAs that are differentially expressed in livers of mice through controlled dietary, normal diet supplemented with excess cholesterol versus another diet with a combination of lovastatin plus ezetimibe (LE), drugs inhibit both endogenous cholesterol synthesis and dietary absorption of cholesterol. Intriguing finding is that miR-96/182/183 cluster comes to light. They are expressed from a unique primary transcript and the promoter for this locus is a direct target regulated by SREBP2 in mice. In this paper, miR-96 and miR-182 negatively regulate the expression of INSIG-2 and FBXW7 by targeting their 3′UTR, respectively. INSIG-2 and FBXW7 are two proteins that weaken the maturation of SREBP2, especially nuclear SREBP2 which plays the role of a transcription factor. Further analyses revealed that, when mice are fed with diet with cholesterol, normal, LE ingredient, miR-182 presents the gradient increase tendency because of the different density of cholesterol. Inversely, FBXW7 will decrease and result in accumulation of nuclear SREBP2 in LE group. This phenomenon is also conserved in human cells. So it demonstrates the coordinate and reciprocal regulation of nuclear SREBP2 with miR-182 and FBXW7. Also, these miRNAs can modulate synthesis of lipid through regulate nuclear SREBP.

Peroxisome-proliferator-activated receptors (PPARs) are nuclear hormone receptors that exert a transcriptional activity regulating genes involved in cell proliferation, cell differentiation, energy homeostasis, diabetes, obesity, atherosclerosis, and other basic metabolically processes (Peyrou et al. 2012). Three major isoforms are known: PPAR-alpha, PPAR-beta/delta, PPAR-gamma. At least five different miRNAs have been reported to regulate PPARs expression in liver cells directly or indirectly. Such as miR-10b, miR-21/miR-122 can directly regulate PPARα/PPARβ(δ) respectively. MiR-27b/miR-132 modulate PPARα/PPARγ indirectly

(Zheng et al. 2010; Kida et al. 2011; Gatfield et al. 2009; Mann et al. 2010). The performance of relationship between these miRNAs and PPARs is embodied in steatosis, cholesterol and lipid metabolism. Recently, Alessio Papi et al. found that, when mammospheres derived from breast cancer stem cell are exposed with breast tumor fibroblasts supernatant, enhanced autocrine tumor necrosis factor-a (TNFα) lead to functional interplay between peroxisome proliferator activated receptor-α and hypoxia inducible factor-1α (PPARα/HIF1α), two nuclear transcriptional factors. But high PPARγ expression will antagonize the PPARα/HIF1α interplay. Further, they demonstrated that siHIF1α and siPPARα can decrease miR-130b expression, and pre-miR-130b in turn facilitates PPARα expression and decrease PPARγ. Another miRNA, miR17-5p, down-regulates PPARα, whereas it increases PPARγ expression. Through this regulatory network, they suggest that the antagonist interplay between PPARα and PPARγ is mediated by miR130b and miR17-5p. Next, they pinpoint that apolipoprotein E (ApoE) is over-expressed in mammospheres and PPARα over-expression can induce ApoE expression at mRNA and protein levels. Hence, PPARα/HIF1α interplay can regulate lipid homeostasis through control of ApoE expression (Zissel et al. 2013).

4 MicroRNAs Regulate Glutamine Metabolism

Since most cancer cells rely on aerobic glycolysis for proliferation and metabolism, fewer metabolites produced from glycolysis enter the tricarboxylic acid (TCA) cycle for energy supply. Thus, other alternative metabolites such as glutamine will play such a vital role as feeding the TCA cycle and redox homeostasis, not only as a nitrogen donor (Table 4.1). The c-Myc transcription factor, which is well known to regulate cell proliferation and glucose metabolism, has been validated to stimulate glutamine catabolism by repressing miR-23a/b that target glutaminase 1 (GLS1), resulting in an increased uptake and catabolism of glutamine (Gao et al. 2009). GLS1 catalyzes the conversion of glutamine to glutamate, the latter is further catabolized to α-ketoglutarate, which serves as a TCA substrate. Moreover, glutamate can also be converted to proline in metabolic process. C-Myc not only increases GLS, but also induces proline biosynthesis from glutamine. In 2012, Liu et al. found that c-Myc robustly suppresses the expression of POX/PRODH protein, the first enzyme in proline catabolism, primarily through increasing miR-23b*, which is processed from the same transcript as miR-23b (Liu et al. 2012).

In recent years, energy metabolism has been reported to play a major role in somatic reprogramming. TDH-mediated threonine catabolism could stimulate reprogramming process in mouse embryonic fibroblast (Han et al. 2013). In Drosophila, bioinformatics screen reveals that miR-277 acts as a metabolic switch in modulating amino acid catabolism. Hence, the regulation of amino acid metabolism by miRNAs is emerging to exert significant impact in cancer biology and beyond (Stark et al. 2003).

5 Summary and Future Perspectives

As we have discussed in this chapter, a plethora of studies now focus on the diverse roles of miRNAs played on tumor metabolism including glucose, glutamine and lipid metabolism with an aim to understand and combat cancers. Given that cancer cells adapt specific strategies for energy metabolism, one might expect to explore the regulatory mechanisms of cancer specific metabolism by miRNAs for cancer therapy. Indeed, restoration of miR-122 constitutes a novel approach that could be beneficial for therapy of chronic liver diseases and HCCs that express low levels of miR-122. Moreover, high miR-122 expression abolishes hepatic insulin resistance, resulting in lower incidence of diabetes. Nevertheless, the actual roles and regulatory networks of miRNAs in the real biological process of development and diseases are far more complicated than previously thought. In cancer metabolism, substantial cross-talks exist among the functions of miRNAs, oncogenes, nutrients enzymes and metabolites, dissection of which will greatly enhance our understanding of the complex process underlying human malignancies as well as provide insight for cancer therapy.

Acknowledgments Our work is supported in part by National Basic Key Research Program of China (2014CB910600, 2011CBA01103 and 2012CB910104), National Nature Science Foundation of China (31171385 and 31071257). H.Z is supported by Chinese Government "1000 Youth Talent Program".

References

Ahmad A, Aboukameel A, Kong D, Wang Z, Sethi S, Chen W et al (2011) Phosphoglucose isomerase/autocrine motility factor mediates epithelial-mesenchymal transition regulated by miR-200 in breast cancer cells. Cancer Res 71(9):3400–3409

Babashah S, Soleimani M (2011) The oncogenic and tumour suppressive roles of microRNAs in cancer and apoptosis. Eur J Cancer 47(8):1127–1137

Cairns RA, Harris IS, Mak TW (2011) Regulation of cancer cell metabolism. Nat Rev Cancer 11(2):85–95

Calin GA, Dumitru CD, Shimizu M, Bichi R, Zupo S, Noch E et al (2002) Frequent deletions and down-regulation of micro- RNA genes miR15 and miR16 at 13q14 in chronic lymphocytic leukemia. Proc Natl Acad Sci U S A 99(24):15524–15529

Calin GA, Cimmino A, Fabbri M, Ferracin M, Wojcik SE, Shimizu M et al (2008) MiR-15a and miR-16-1 cluster functions in human leukemia. Proc Natl Acad Sci U S A 105(13):5166–5171

Cannell IG, Kong YW, Bushell M (2008) How do microRNAs regulate gene expression? Biochem Soc Trans 36(6):1224

Chen B, Li H, Zeng X, Yang P, Liu X, Zhao X et al (2012) Roles of microRNA on cancer cell metabolism. J Transl Med 10:228

Currie E, Schulze A, Zechner R, Walther TC, Farese RV (2013) Cellular fatty acid metabolism and cancer. Cell Metab 18(2):153–161

Dang CV (2010) Glutaminolysis: supplying carbon or nitrogen, or both for cancer cells? Cell Cycle 9(19):3884–3886

Davalos A, Goedeke L, Smibert P, Ramirez CM, Warrier NP, Andreo U et al (2011) miR-33a/b contribute to the regulation of fatty acid metabolism and insulin signaling. Proc Natl Acad Sci U S A 108(22):9232–9237

Davalos V, Moutinho C, Villanueva A, Boque R, Silva P, Carneiro F et al (2012) Dynamic epigenetic regulation of the microRNA-200 family mediates epithelial and mesenchymal transitions in human tumorigenesis. Oncogene 31(16):2062–2074

Druz A, Betenbaugh M, Shiloach J (2012) Glucose depletion activates mmu-miR-466h-5p expression through oxidative stress and inhibition of histone deacetylation. Nucleic Acids Res 40(15): 7291–7302

El Ouaamari A, Baroukh N, Martens GA, Lebrun P, Pipeleers D, van Obberghen E (2008) miR-375 targets 3′-phosphoinositide-dependent protein kinase-1 and regulates glucose-induced biological responses in pancreatic cells. Diabetes 57(10):2708–2717

Esau C, Davis S, Murray SF, Yu XX, Pandey SK, Pear M et al (2006) miR-122 regulation of lipid metabolism revealed by in vivo antisense targeting. Cell Metab 3(2):87–98

Fabani MM, Gait MJ (2007) miR-122 targeting with LNA/2′-O-methyl oligonucleotide mixmers, peptide nucleic acids (PNA), and PNA-peptide conjugates. RNA 14(2):336–346

Fang R, Xiao T, Fang Z, Sun Y, Li F, Gao Y et al (2012) MicroRNA-143 (miR-143) regulates cancer glycolysis via targeting hexokinase 2 gene. J Biol Chem 287(27):23227–23235

Fei X, Qi M, Wu B, Song Y, Wang Y, Li T (2012) MicroRNA-195-5p suppresses glucose uptake and proliferation of human bladder cancer T24 cells by regulating GLUT3 expression. FEBS Lett 586(4):392–397

Frost RJA, Olson EN (2011) Control of glucose homeostasis and insulin sensitivity by the Let-7 family of microRNAs. Proc Natl Acad Sci 108(52):21075–21080

Fukuhara T, Kambara H, Shiokawa M, Ono C, Katoh H, Morita E et al (2012) Expression of microRNA miR-122 facilitates an efficient replication in nonhepatic cells upon infection with hepatitis C virus. J Virol 86(15):7918–7933

Gao P, Tchernyshyov I, Chang T-C, Lee Y-S, Kita K, Ochi T et al (2009) c-Myc suppression of miR-23a/b enhances mitochondrial glutaminase expression and glutamine metabolism. Nature 458(7239):762–765

Gatfield D, Le Martelot G, Vejnar CE, Gerlach D, Schaad O, Fleury-Olela F et al (2009) Integration of microRNA miR-122 in hepatic circadian gene expression. Genes Dev 23(11):1313–1326

Gauthier BR, Wollheim CB (2006) MicroRNAs: 'ribo-regulators' of glucose homeostasis. Nat Med 12(1):36–38

Godlewski J, Nowicki MO, Bronisz A, Nuovo G, Palatini J, De Lay M et al (2010) MicroRNA-451 regulates LKB1/AMPK signaling and allows adaptation to metabolic stress in glioma cells. Mol Cell 37(5):620–632

Han C, Gu H, Wang J, Lu W, Mei Y, Wu M (2013) Regulation of L-threonine dehydrogenase in somatic cell reprogramming. Stem Cells 31(5):953–965

Hanahan D, Weinberg RA (2011) Hallmarks of cancer: the next generation. Cell 144(5):646–674

Horie T, Ono K, Nishi H, Iwanaga Y, Nagao K, Kinoshita M et al (2009) MicroRNA-133 regulates the expression of GLUT4 by targeting KLF15 and is involved in metabolic control in cardiac myocytes. Biochem Biophys Res Commun 389(2):315–320

Horie T, Ono K, Horiguchi M, Nishi H, Nakamura T, Nagao K et al (2010) MicroRNA-33 encoded by an intron of sterol regulatory element-binding protein 2 (Srebp2) regulates HDL in vivo. Proc Natl Acad Sci 107(40):17321–17326

Hsu S-h, Wang B, Kota J, Yu J, Costinean S, Kutay H et al (2012) Essential metabolic, anti-inflammatory, and anti-tumorigenic functions of miR-122 in liver. J Clin Invest 122(8):2871–2883

Hu W, Zhang C, Wu R, Sun Y, Levine A, Feng Z (2010) Glutaminase 2, a novel p53 target gene regulating energy metabolism and antioxidant function. Proc Natl Acad Sci 107(16):7455–7460

Hu J, Xu Y, Hao J, Wang S, Li C, Meng S (2012) MiR-122 in hepatic function and liver diseases. Protein Cell 3(5):364–371

Jeon T-I, Esquejo RM, Roqueta-Rivera M, Phelan PE, Moon Y-A, Govindarajan SS et al (2013) An SREBP-responsive microRNA operon contributes to a regulatory loop for intracellular lipid homeostasis. Cell Metab 18(1):51–61

Jiang S, Zhang L-F, Zhang H-W, Hu S, Lu M-H, Liang S et al (2012) A novel miR-155/miR-143 cascade controls glycolysis by regulating hexokinase 2 in breast cancer cells. EMBO J 31(8):1985–1998

Jordan SD, Krüger M, Willmes DM, Redemann N, Wunderlich FT, Brönneke HS et al (2011) Obesity-induced overexpression of miRNA-143 inhibits insulin-stimulated AKT activation and impairs glucose metabolism. Nat Cell Biol 13(4):434–446

Kida K, Nakajima M, Mohri T, Oda Y, Takagi S, Fukami T et al (2011) PPARα is regulated by miR-21 and miR-27b in human liver. Pharm Res 28(10):2467–2476

Kim JW, You YH, Jung S, Suh-Kim H, Lee IK, Cho JH et al (2013) miRNA-30a-5p-mediated silencing of Beta2/NeuroD expression is an important initial event of glucotoxicity-induced beta cell dysfunction in rodent models. Diabetologia 56(4):847–855

Kloosterman WP, Lagendijk AK, Ketting RF, Moulton JD, Plasterk RH (2007) Targeted inhibition of miRNA maturation with morpholinos reveals a role for miR-375 in pancreatic islet development. PLoS Biol 5(8):e203

Lee RC, Feinbaum RL, Ambros V (1993) The C. elegans heterochronic gene lin-4 encodes small RNAs with antisense complementarity to lin-14. Cell 75(5):843–854

Liu W, Le A, Hancock C, Lane AN, Dang CV, Fan TW et al (2012) Reprogramming of proline and glutamine metabolism contributes to the proliferative and metabolic responses regulated by oncogenic transcription factor c-MYC. Proc Natl Acad Sci U S A 109(23):8983–8988

Long J, Wang Y, Wang W, Chang BH, Danesh FR (2011) MicroRNA-29c is a signature microRNA under high glucose conditions that targets Sprouty homolog 1, and its in vivo knockdown prevents progression of diabetic nephropathy. J Biol Chem 286(13):11837–11848

Lu H, Buchan RJ, Cook SA (2010) MicroRNA-223 regulates Glut4 expression and cardiomyocyte glucose metabolism. Cardiovasc Res 86(3):410–420

Mann J, Chu DCK, Maxwell A, Oakley F, Zhu NL, Tsukamoto H et al (2010) MeCP2 controls an epigenetic pathway that promotes myofibroblast transdifferentiation and fibrosis. Gastroenterology 138(2):705–714.e704

Marquart TJ, Allen RM, Ory DS, Baldan A (2010) miR-33 links SREBP-2 induction to repression of sterol transporters. Proc Natl Acad Sci U S A 107(27):12228–12232

Metallo CM, Gameiro PA, Bell EL, Mattaini KR, Yang J, Hiller K et al (2011) Reductive glutamine metabolism by IDH1 mediates lipogenesis under hypoxia. Nature 481(7381):380–384

Najafi-Shoushtari SH, Kristo F, Li Y, Shioda T, Cohen DE, Gerszten RE et al (2010) MicroRNA-33 and the SREBP host genes cooperate to control cholesterol homeostasis. Science 328(5985): 1566–1569

Peyrou M, Ramadori P, Bourgoin L, Foti M (2012) PPARs in liver diseases and cancer: epigenetic regulation by microRNAs. PPAR Res 2012:1–16

Qiu L, Fan H, Jin W, Zhao B, Wang Y, Ju Y et al (2010) miR-122-induced down-regulation of HO-1 negatively affects miR-122-mediated suppression of HBV. Biochem Biophys Res Commun 398(4):771–777

Rayner KJ, Suarez Y, Davalos A, Parathath S, Fitzgerald ML, Tamehiro N et al (2010) MiR-33 contributes to the regulation of cholesterol homeostasis. Science 328(5985):1570–1573

Rayner KJ, Sheedy FJ, Esau CC, Hussain FN, Temel RE, Parathath S et al (2011) Antagonism of miR-33 in mice promotes reverse cholesterol transport and regression of atherosclerosis. J Clin Invest 121(7):2921–2931

Reinhart BJ, Slack FJ, Basson M, Pasquinelli AE, Bettinger JC, Rougvie AE et al (2000) The 21-nucleotide let-7 RNA regulates developmental timing in Caenorhabditis elegans. Nature 403(6772):901–906

Son J, Lyssiotis CA, Ying H, Wang X, Hua S, Ligorio M et al (2013) Glutamine supports pancreatic cancer growth through a KRAS-regulated metabolic pathway. Nature 496(7443):101–105

Stark A, Brennecke J, Russell RB, Cohen SM (2003) Identification of drosophila microRNA targets. PLoS Biol 1(3):e60

Suzuki S, Tanaka T, Poyurovsky MV, Nagano H, Mayama T, Ohkubo S et al (2010) Phosphate-activated glutaminase (GLS2), a p53-inducible regulator of glutamine metabolism and reactive oxygen species. Proc Natl Acad Sci 107(16):7461–7466

Tsai WC, Hsu SD, Hsu CS, Lai TC, Chen SJ, Shen R et al (2012) MicroRNA-122 plays a critical role in liver homeostasis and hepatocarcinogenesis. J Clin Invest 122(8):2884–2897

Wang L, Tang H, Thayanithy V, Subramanian S, Oberg AL, Cunningham JM et al (2009) Gene networks and microRNAs implicated in aggressive prostate cancer. Cancer Res 69(24): 9490–9497

Warburg O (1956) On the origin of cancer cells. Science 123(3191):309–314

Wen J, Friedman JR (2012) miR-122 regulates hepatic lipid metabolism and tumor suppression. J Clin Invest 122(8):2773–2776

Yang YM, Seo SY, Kim TH, Kim SG (2012) Decrease of microRNA-122 causes hepatic insulin resistance by inducing protein tyrosine phosphatase 1B, which is reversed by licorice flavonoid. Hepatology 56(6):2209–2220

Ying H, Kimmelman AC, Lyssiotis CA, Hua S, Chu GC, Fletcher-Sananikone E et al (2012) Oncogenic Kras maintains pancreatic tumors through regulation of anabolic glucose metabolism. Cell 149(3):656–670

Zheng L, Lv G-c, Sheng J, Yang Y-d (2010) Effect of miRNA-10b in regulating cellular steatosis level by targeting PPAR-Î± expression, a novel mechanism for the pathogenesis of NAFLD. J Gastroenterol Hepatol 25(1):156–163

Zhu H, Shah S, Shyh-Chang N, Shinoda G, Einhorn WS, Viswanathan SR et al (2010) Lin28a transgenic mice manifest size and puberty phenotypes identified in human genetic association studies. Nat Genet 42(7):626–630

Zhu H, Shyh-Chang N, Segrè AV, Shinoda G, Shah SP, Einhorn WS et al (2011) The Lin28/let-7 axis regulates glucose metabolism. Cell 147(1):81–94

Zhu G, Chai J, Ma L, Duan H, Zhang H (2013) Downregulated microRNA-32 expression induced by high glucose inhibits cell cycle progression via PTEN upregulation and Akt inactivation in bone marrow-derived mesenchymal stem cells. Biochem Biophys Res Commun 433(4):526–531

Zissel G, Papi A, Storci G, Guarnieri T, De Carolis S, Bertoni S et al (2013) Peroxisome proliferator activated receptor-α/hypoxia inducible factor-1α interplay sustains carbonic anhydrase IX and apoliprotein E expression in breast cancer stem cells. PLoS One 8(1):e54968

Chapter 5
MicroRNAs in Solid Tumors

Gianpiero Di Leva and Michela Garofalo

Contents

Abstract MicroRNAs (miRNAs) are a class of non-coding RNAs that hybridize to mRNAs inducing either translation repression or mRNA cleavage. MiRNAs regulate a variety of biological processes in the cell, including development, cell differentiation, proliferation, apoptosis and their abnormal expression levels are closely associated with pathogenesis of cancers. In this scenario, several high-throughput technologies studies have revealed miRNA roles in classifying tumors and predicting patient outcome with high accuracy. Because of their ability to concurrently target multiple pathways miRNA-based anticancer therapies are being developed, either alone or in combination with current targeted therapies, with the goal to improve the response and increase cure rates. Here, we highlight recent advances on the use of miRNAs as a potential approach for diagnosis and prognosis of solid cancer.

G. Di Leva • M. Garofalo (✉)
Department of Molecular Immunology, Virology and Medical Genetics,
Ohio State University, Biological Research Tower 992, 460 W 12th Avenue,
Columbus, OH 43210, USA
e-mail: michela.garofalo@osumc.edu

S. Babashah (ed.), *MicroRNAs: Key Regulators of Oncogenesis*,
DOI 10.1007/978-3-319-03725-7_5,

Keywords MicroRNA • Solid tumors • Oncogene • Tumor suppressor

1 Introduction

The mechanism of action of MicroRNAs (miRNAs) has revolutionized the concept of gene expression regulation, because we now know that mRNA levels in a cell do not strictly correlate with protein expression (Bartel 2004). Functions of miRNAs are essential to control almost every biological process including cell cycle regulation, cell growth, cell differentiation, apoptosis and stress response (Mendell and Olson 2012). We can argue that miRNAs evolved primarily to play the role of a reinforcer, in that their activities overlap with transcriptional patterns to sharpen developmental transitions and establish cellular identities. MiRNAs also control fluctuations in gene expression and more faithfully signal outcomes in the context of certain regulatory networks. In this scenario, miRNA function is essential to keep a steady state of cellular machinery and maintain the cell type identity chosen during differentiation. MiRNAs are predicted to regulate a total of ~60 % of human genes generating an inhibitory network where multiple miRNAs can regulate single mRNA molecules and a single miRNA can act on a number of mRNA targets (Bartel 2004). It is important to note that a miRNA target gene has an expression threshold below which the gene is efficiently repressed and above which it can overwhelm the available miRNA (Mukherji et al. 2011). The miRNA and target concentration, the strength and number of miRNA-binding sites in the target determine this threshold level (Mukherji et al. 2011). Clearly, different tissues or a particular tissue under different conditions exhibit different miRNA/target expression profiles and thus have different thresholds for a given target gene.

As more miRNAs are unraveled and their roles dissected, it has become evident that the involvement of these molecules in cancer is much more extensive than initially thought (Di Leva and Croce 2010; Kasinski and Slack 2011). After the seminal study from our laboratory showing that the miR-15/16 cluster is frequently deleted in chronic lymphocytic leukemia (CLL), myriad studies were published defining a role for miRNAs in the diagnosis, prognosis and therapy of cancer (Calin et al. 2002). The most striking theme in the study of miRNAs and cancer is the significant alteration of miRNA expression in malignant cells compared with their normal counterparts (Calin and Croce 2006). All analyzed tumors have exhibited a specific miRNA signature, termed the miRNome, that characterizes the malignant state and defines some of their clinicopathological features (e.g., grade, stage, sex, age, aggressiveness, vascular invasion, proliferation index) (Calin and Croce 2006). Since miRNAs have a very restricted tissue-specific expression, the apparent miRNA modulation in cancer tissues may represent the manifestation of a different cell population in the tumor than in normal tissues. However, experimental approaches in human cancer cells and animal models have shown the functional link between the tumor appearance and progression and the miRNA modulation. For example, transgenic expression of miR-155 (Costinean et al. 2006) or miR-21

(Medina et al. 2010) so well loss of miR-15/16 are sufficient to initiate lymphomagenesis in mice (Klein et al. 2010). The expression data documenting dysregulation of these miRNAs in various tumor types and the genomic data linking these miRNA genes to recurrent amplifications or deletions in cancer cells have strongly supported that specific miRNAs act as oncogenes and tumor suppressors (Lu et al. 2005; Volinia et al. 2006; Calin et al. 2004; Babashah and Soleimani 2011). Consequently, dysfunctional miRNAs in tumor cells are attractive candidates for next-generation therapeutics. Two different strategies have been defined: (1) inhibition of oncogenic miRNAs by using miRNA antagonists, such as anti-miRs, locked nucleic acids (LNAs), or antagomiRs; (2) miRNA replacement involving the reintroduction of a tumor suppressor miRNA mimetic to restore its loss of function (Garofalo and Croce 2011). In this review, we synthesize our current understanding of the role of miRNA diagnostic, prognostic and therapeutics in solid cancers and the manner in which dysregulation of miRNA levels can contribute to all aspect of tumorigenesis.

2 MicroRNAs and Hepatocellular Carcinoma

Hepatocellular carcinoma (HCC) is the fifth most frequently diagnosed cancer worldwide and the second most frequent cause of cancer death in men. The overall 5-year survival rate of liver cancer patients remains low, ranging from 0 to 14 % (Siegel et al. 2011). This rate illustrates the importance of uncovering the cellular and molecular mechanisms involved in this aggressive HCC to develop more effective treatment options and improve the prognosis of HCC patients. Cancer can arise from a combination of epigenetic and genetic abnormalities that result in abnormal gene expression and function (Dumitrescu 2009; Piperi et al. 2008; Jones and Baylin 2007). However, like other cancers, the molecular mechanisms underlying the development and progression of HCC are still far from being understood. Recent studies have indicated that miRNAs directly contribute to HCC by targeting many critical regulatory genes. In an attempt to use miRNAs to create a molecular classification of HCC, Murakami et al. analyzed miRNA expression profiles in 24 HCC samples and 22 adjacent non tumorous tissue (NT) (Murakami et al. 2006). They identified 30 miRNA genes that were differentially expressed in the HCC and corresponding NT specimens. The expression levels of four miRNAs (miR-92, miR-20, miR-18 and precursor miR- 18) were significantly higher in poorly differentiated HCC samples, with moderate expression in moderately differentiated HCC and low expression in well-differentiated HCC. In addition, the expression levels of miR 92, miR-20, and miR-18 were inversely correlated with the degree of HCC differentiation, suggesting that these miRNAs contribute to both tumorigenesis and loss of tumor differentiation. In another study, Sato et al. developed a mathematical model to assess the risk of HCC recurrence after liver resection, based on miRNA expression profiling (Sato et al. 2011). They found that tumor miRNA profiles could predict early recurrence, whereas the profiles of the normal tissues were predictive of late recurrence, suggesting that the miRNA profile represents the

Table 5.1 MicroRNAs with clinical significance in Hepatocellular carcinoma

MicroRNA	Expression in HCC	Clinical outcome
miR-122	Down	Poor prognosis
miR-124	Down	More aggressive and/or poor prognosis
miR-139	Down	Poor prognosis
miR-145	Down	Short disease survival
miR-199b	Down	Poor overall survival and progression free survival
miR-22	Down	Poor prognosis
miR-26	Down	Short overall survival
miR-10b	Up	Poor prognosis
miR-125b	Up	Good survival
miR-155	Up	Poor recurrence free survival and overall survival
miR-21	Up	Poor prognosis
miR-221	Up	Poor prognosis
miR-222	Up	Poor prognosis

HCC hepatocellular carcinoma

malignant potential of primary tumors, associated with the presence of hepatic dissemination. MiRNA expression profiles have also been informative of the metastatic state of HCC. In this regard, in contrast with the results obtained by Wong et al., who did not find differences in miRNA expression pattern between primary HCCs and venous metastases (Wong et al. 2012), Budhu et al. identified a 20-miRNA tumor signature associated with HCC venous metastasis (Budhu et al. 2008). This signature predicted survival and recurrence of HCC in patients with multinodular or solitary tumors, including those with early-stage disease (Budhu et al. 2008). Overall the genome-wide analyses have demonstrated that miRNAs have a strong diagnostic and prognostic value in HCC (Table 5.1).

As previously stated, many *in vitro* and preclinical studies have either reintroduced onco/suppressive miRNAs or inhibited oncogenic miRNAs in cancer cells, showing that these treatments often result in impairment of cell proliferation and invasion or in an increased apoptosis. Regarding HCC, several reports have shown that either exogenous expression of tumor suppressor miRNAs or inhibition of oncomiRs resulted in impaired growth or invasive ability of HCC cell lines *in vitro* or in xenografts. Kota and colleagues demonstrated that HCC cells exhibit reduced expression of miR-26a (Kota et al. 2009). Expression of this miRNA in liver cancer cells *in vitro* induces cell-cycle arrest associated with direct targeting of cyclins D2 and E2. Importantly, systemic administration of miR-26a in a mouse model of HCC using adeno-associated virus resulted in inhibition of cancer cells proliferation, induction of tumor-specific apoptosis, and dramatic protection from disease progression without toxicity (Kota et al. 2009).

One of the most expressed miRNAs in liver cancer is miR-221, which accelerates hepatocyte proliferation by targeting important tumor suppressor genes such as PTEN, TIMP3, p27, p57 (Garofalo et al. 2009; le Sage et al. 2007; Fornari et al. 2008). The importance of miR-221 in liver cancer has recently been addressed by Callegari and coworkers (Callegari et al. 2012). These authors showed that, in a transgenic mouse model over-expressing miR-221, the emergence of spontaneous nodular liver lesions in approximately 50 % of male mice and a strong acceleration of tumor development in 100 % of mice treated with diethylnitrosamine. Of note, *in vivo* delivery of anti-miR-221 oligonucleotides led to a significant reduction of the number and size of tumor nodules. In another study, Park and colleagues used cholesterol-modified isoform of anti-miR-221 (chol-anti-miR-221) improving pharmacokinetics and liver tissue distribution compared with unmodified oligonucleotide. Chol-anti-miR-221 significantly reduced miR-221 levels in liver within a week of intravenous administration and in situ hybridization studies confirmed accumulation of the oligonucleotide in tumor cells *in vivo* (Park et al. 2011).

MiRNAs can influence the sensitivity of tumors, including HCC, to therapy. Up-regulation of miR-130a directly inhibited expression of tumor suppressor gene RUNX3, which resulted in activation of Wnt/β-catenin signaling and increased resistance to cisplatin in Huh7 cells (Xu et al. 2012b). Instead, MiR-193a-3p has been demonstrated to dictate the 5-FU resistance of HCC cells via repression of serine/arginine-rich splicing factor 2 (SRSF2) expression, which up-regulates the proapoptotic splicing form of caspase 2 (CASP2L) and sensitizes HCC cells to 5-FU (Ma et al. 2012).

3 MicroRNAs and Cervical Cancer

More than 400,000 new cases of cervical cancer that lead to approximately 270,000 deaths are diagnosed every year worldwide and, in underdeveloped countries, it is still one of the leading causes of death in female population (Parkin and Bray 2006). Cervical cancer etiology is strongly linked to human papilloma virus (HPV) infection and, although high efficacy of the screening has decreased the rates of cervical cancer diagnoses, the exact pathway leading from the infection to tumorigenesis has not been revealed so far (Jabbar et al. 2009; Yugawa and Kiyono 2009).

Emerging roles of miRNAs in cervical carcinoma has been postulated by an enlarging number of studies (Table 5.2) (Lui et al. 2007; Wang et al. 2008; Martinez et al. 2008; Lee et al. 2008). Among the altered miRNAs reported, all cervical carcinoma cell lines and tumor tissues examined contained no detectable miR-143 and miR-145. A recent study indicates that down-regulation of the miR-143/145 cluster requires the Ras-responsive element-binding protein (RREB1) to repress its promoter, but K-Ras and RREB1 are themselves targets of miR-143/145, arguing for a feed-forward mechanism that potentiates Ras signaling (Kent et al. 2010). However, miR-143/145 cluster repression does not seem to be specific for this type of cancer, as significant decrease in its expression was observed also in other tumors, e.g. colorectal, breast, and lymphoid cancers (Michael et al. 2003; Iorio et al. 2005; Akao et al. 2007a, b). As in many other solid tumors, the up-regulation of miR-21 was also observed in cervical cancer and linked to inhibition of several tumor suppressor genes involved in regulation of cells proliferation and apoptosis (Volinia et al. 2006; Iorio et al. 2005; Chan et al. 2005; Roldo et al. 2006; Frankel et al. 2008;

Table 5.2 MicroRNA profiling studies in cervical cancer

Authors	Samples	Techniques	Results
Lui et al. (2007)	6 cervical cancer cell lines and 5 normal cervical samples	Cloning-based technique, Northern blot	miR-21, let-7b, let-7c, miR-23b, miR-196b, miR-143
Wang et al. (2008)	HPV161, HPV181, HPV negative cervical cancer cell lines, primary cancer cell lines, cervical cancer vs. age matched normal cervical samples	Microarray platform, Northern blot	miR-15a, miR-223, miR-146a, miR-143, miR-145, miR-218, miR-424
Martinez et al. (2008)	HPV161, HPV181, HPV negative cervical cancer cell lines and cervical cancer vs. normal cervical samples	Microarray platform, Northern blot, quantitative RT-PCR	miR-182, miR-183, miR-210, miR-143, miR-145, miR-126, miR-195, miR-218, miR-368, miR-497
Lee et al. (2008)	10 cancer tissue samples vs. 10 normal cervical controls	Quantitative RT-PCR	miR-199a, miR-199b, miR-199s, miR-9, miR-145, miR-133a, miR-133b, miR-214, miR-127, miR-149, miR-203

Up- or down-regulated miRNAs are reported in *red* and *green*, respectively

Qi et al. 2009; Meng et al. 2007). For example, Yao et al. found that inhibition of miR-21 in HeLa cells resulted in diminished proliferation and increased expression of Pdcd4, a well-known tumor suppressor (Yao et al. 2009).

High-risk HPVs, such as HPV16, HPV18, and HPV31, have been detected in up to 99.7 % of cervical squamous cell carcinomas and 94–100 % of cervical adeno- and adenosquamous carcinomas (Bosch and de Sanjose 2003; Walboomers et al. 1999). The HPV oncoproteins, E6 and E7 contribute to cervical carcinogenesis by inactivating the cellular tumor suppressor proteins p53 and pRb, respectively (Scheffner et al. 1990; Dyson et al. 1989; Boyer et al. 1996). Because the miR-34a gene is a direct transcriptional target of p53, and its expression can be transactivated by the binding of p53 to a consensus p53 binding site in the miR-34a promoter region (He et al. 2007; Chang et al. 2007; Raver-Shapira et al. 2007), E6-mediated degradation of p53 leads to the reduction of miR-34a in raft cultures, cervical intraepithelial neoplasia (CIN), and cervical cancer tissues. Viral E6 also regulates the expression of miR-218 and miR-23b (Martinez et al. 2008; Au Yeung et al. 2011). Although how E6 reduces miR-218 expression remains unknown, down-regulation of miR-23b expression has been linked to p53. Au Yeung and colleagues (2011) found that the gene encoding miR-23b on chromosome 9 contains a promoter region with a p53 binding site, and in the presence of p53, the gene is activated to express miR-23b as a cluster of miR-23b/27b/24-1, providing another example of E6/p53 mediated down-regulation of miRNAs. High levels of miR-15a and miR-16-1 have been identified in cervical cancer relative to normal cervical tissues (Wang et al. 2008). Increased expression of the miR-15a/16-1 cluster in cervical cancer tissues is associated exclusively with the expression of the viral E7 oncoproteins in HPV16 or HPV18 infected cells. In fact, viral E7-mediated degradation of the tumor suppressor pRB frees E2F from the pRB–E2F complex to interact with the promoter regions of c-Myb and c-Myc, which in turn activate DLEU2 transcription and promote miR-15a/16-1 cluster expression. High-risk HPV E7 also represses the expression of miR-203 in human keratinocytes controlling the shift of keratinocytes in differentiating epithelia from a proliferative state to a nonproliferative state (Melar-New and Laimins 2010). Expression of E7 viral protein blocks miR-203 dependent differentiation of human foreskin keratinocytes with a corresponding increase in the expression of ΔNp63, a member of the p53 family that is highly expressed in proliferative undifferentiated basal keratinocytes, but poorly expressed in differentiated nonproliferative cells. The authors also showed that miR-203 expression in HPV-positive cervical cancer cells promotes viral genome amplification in the short term, but a high level of miR-203 expression interferes with viral genome amplification in the long term (Melar-New and Laimins 2010). Although the mechanism by which miR-203 is involved in viral DNA replication remains to be investigated, this is the first evidence that cellular miRNAs may also play an important role in the regulation of viral gene expression and DNA replication. In this regard, reduced miR-125b expression was found in cervical pre-cancerous lesions during early HPV infection and ectopic miR-125b suppresses HPV16 DNA replication, presumably through the sequence homology of HPV16 L2 and miR-125b (Nuovo et al. 2010).

4 MicroRNAs and Breast Cancer

Based on the American Cancer Society data (http://www.cancer.org), breast cancer still remains the most frequent cancer among American women, with approximately 230,000 new cases diagnosed and about 40,000 deaths in 2012. It is therefore essential to better understand the underlying molecular mechanisms and to develop more effective treatments.

The first report describing the existence of a miRNA signature characterizing human breast cancer was published in 2005, suggesting the involvement of miRNAs in the pathogenesis of this human neoplasm (Iorio et al. 2005). The authors described a breast cancer-specific miRNA signature performing a genome-wide miRNA expression analysis on a limited set of normal and tumor breast tissues, resulting in the identification of a list of 29 miRNAs differentially expressed, and able to classify tumors and normal tissues with an accuracy of 100 %. Among the miRNAs differentially expressed, miR-10b, miR-125b and miR-145 were down-modulated, while miR-21 and miR-155 were up-modulated, suggesting that these miRNAs could exert a role as tumor suppressor genes or oncogenes. Similar to mRNA expression profiling, which have been used to classify breast tumors into four or five intrinsic subtypes (Prat and Perou 2011), miRNA expression profiles may also be applied to classify breast tumors, such as luminal A, luminal B, basal-like, HER2, and normal-like (Blenkiron et al. 2007). Individual miRNAs, such as let-7 family members, have been found to be associated with tumor subtype, ER status, and tumor grade. A recent analysis by Volinia et al. (2012) identified by analyzing recently published deep-sequencing dataset the miRNA signature related to the transition from ductal carcinoma in situ to invasive ductal carcinoma, a key event in breast cancer progression.

A nine-miRNA signature was identified that differentiates invasive from in situ carcinoma. Specifically, let-7d, miR-210, and -221 were down-regulated in the in situ and up-regulated in the invasive transition, thus featuring an expression reversal along the cancer progression path. Indeed, five miRNAs (miR-210,-21,-106b*,-197, and let-7i) were associated with overall survival and time to metastasis and miR-210 was the only one also involved in the invasive transition.

In another study, Volinia and Croce (2013) identified the survival signature of breast cancer through the analysis of a cohort of 466 patients with primary invasive ductal carcinoma, the most frequent type of breast cancer, by integrating mRNA, miRNA, and DNA methylation next-generation sequencing data from The Cancer Genome Atlas (TCGA). Thirty mRNAs and seven miRNAs were associated with overall survival across different clinical and molecular subclasses of 466 invasive ductal carcinomas. The prognostic RNAs included PIK3CA, one of the two most frequently mutated genes in invasive breast cancer, and miRNAs such as miR-328, miR-484, and miR-874. The integrated RNA signature was successfully validated on eight different breast cancer cohorts, comprising a total of 2,399 patients, by showing that a combination of mRNA/miRNA signature has a superior performance for risk stratification with respect to other RNA predictors, including the mRNA signatures used in the commercial MammaPrint and Oncotype DX assays. Except for a few miRNAs, the functions of most prognostic miRNAs in breast cancer initiation and progression remain to be elucidated. The involvement of miR-200 and let-7 in the biology of breast cancer is a relatively well characterized phenomenon. From breast tumor profiling, these two miRNA families showed a similar expression patterns across breast tumor subtypes. First, both miRNA families are highly repressed in the breast cancer tumor-initiating cells compared to non-tumorigenic cancer cells (Shimono et al. 2009; Iliopoulos et al. 2010; Ma et al. 2007). Second, they resulted to be highly expressed in breast cancer samples with either lymph node metastasis or higher proliferation index, suggesting that low levels of these miRNAs could be associated with poor prognosis (Iorio et al. 2005). Shimono et al. (2009) demonstrated that enforced expression of miR-200c disrupts breast tumor-initiating cells proliferation and inhibits tumorigenesis *in vivo*. MiR-200b also regulates breast tumor-initiating cells growth by directly targeting Suz12, a subunit of PRC2 complex which regulates the epithelial to mesenchymal transition (EMT) by repressing E-cadherin (Iliopoulos et al. 2010). Many other groups have reported that miR-200 family members (miR-141, miR-200a, b and c, and miR-429) regulate EMT by targeting the transcriptional repressors ZEB1 and ZEB2, thereby increasing E-cadherin expression (Burk et al. 2008; Park et al. 2008).

Let-7 is a known tumor suppressor that targets important oncogenes such as RAS, Myc and HMGA2 (Yu et al. 2007; Johnson et al. 2005; Lee and Dutta 2007). let-7 members expression is markedly reduced in breast tumor-initiating cells and increased with differentiation (Yu et al. 2007). Specifically, reduced let-7 levels are required for functional SKBR3 stemness, as enforced expression of let-7 suppresses self-renewal and tumor initiation, while reduced let-7 level, mediated by a let-7a antisense oligonucleotide, inhibits cell differentiation without affecting proliferation but suppressing the stem cell-like properties of breast cancer cells.

Research in the past 5 years implies multiple functions of miRNAs in breast cancer metastasis. As for primary tumors, some miRNAs are remarkably reduced or eliminated, whereas other miRNAs are aberrantly accumulated in metastatic breast cancer specimens relative to non-metastatic specimens. The most represented miRNAs are miR-10b (Ma et al. 2007), miR-335 (Tavazoie et al. 2008), miR-373, and miR-520 (Huang et al. 2008), miR-210 (Volinia et al. 2012). In addition, miR-200 and let-7 families are also involved in the regulation of metastasis, by controlling the stemness of breast tumor initiating cells (Shimono et al. 2009; Iliopoulos et al. 2010; Yu et al. 2007). The expression of miR-10b, found down-modulated in breast carcinoma in comparison to normal tissue, has subsequently been analyzed in correlation with clinical progression and presence of metastasis by Weinberg's group. They observed that miR-10b was down-modulated in all the breast carcinomas from metastasis-free patients, but, surprisingly, 50 % of metastasis-positive patients had elevated miR-10b levels in their primary tumors. Induced by transcription factor Twist, miR-10b inhibits the translation of mRNA encoding homeobox D10, releasing the expression of the pro-metastatic gene RHOC and thus leading to tumor cell invasion and metastasis (Ma et al. 2007). By profiling the MDA-MB-231 parental cell line and its derived bone/lung-metastatic cell lines (BoM1 and LM2), Tavazoie et al. (2008) identified eight miRNAs (miR-335, miR-199a*, miR-122a, miR-126, miR-206, miR-203, miR-489, and miR-127) down-regulated in derived metastatic cells. Three of them, miR-335, miR-126, and miR-206, showed a similar reduced expression in other patient tumor-derived malignant cell lines (CN34-LM1 and CN34-BoM1) compared to parental cancer cells. The restoration of these three miRNAs significantly inhibited dissemination and metastases, with miR-335 and miR-206 suppressing migration and morphology and miR-126 reducing tumor growth and proliferation. Expression levels of miR-335 and miR-126 are significantly associated with metastasis-free survival of breast cancer patients.

Approximately 40 % of breast cancer patients that are initially responsive to tamoxifen and other endocrine therapies relapse. The mechanisms for acquired endocrine resistance, despite continued ERα expression, are complex. To date, the role of miRNAs in endocrine-resistance has been examined by relatively few investigators. miRNA-221/222 are over-expressed in tamoxifen-, fulvestrant-, and tumor necrosis factor (TNF)- resistant MCF-7 cell line derivatives and in ERα negative tumors (Miller et al. 2008; Rao et al. 2011; Zhao et al. 2008; Di Leva et al. 2010). A screen of miRNAs involved in estrogen-resistance in MCF-7 cells identified up-regulation of miR-101 as promoting estrogen-independent growth of MCF-7 cells, without affecting ERα levels or activity (Sachdeva et al. 2011). The mechanism involves down-regulation of MAGI-2, a scaffold protein required for PTEN activity by miR-101, and consequent increased Akt activity. As summarized in Fig. 5.1, many miRNAs have been identified as biomarkers and/or characterized as indispensable regulators in both normal mammary, and breast cancer development, including cancer initiation, metastasis, and therapy resistance. These miRNAs hold the potential for development of novel miRNA therapeutics, including both antagomirs to inhibit oncomiRs and miRNA replacement therapy to restore levels of tumor suppressor-like miRNAs.

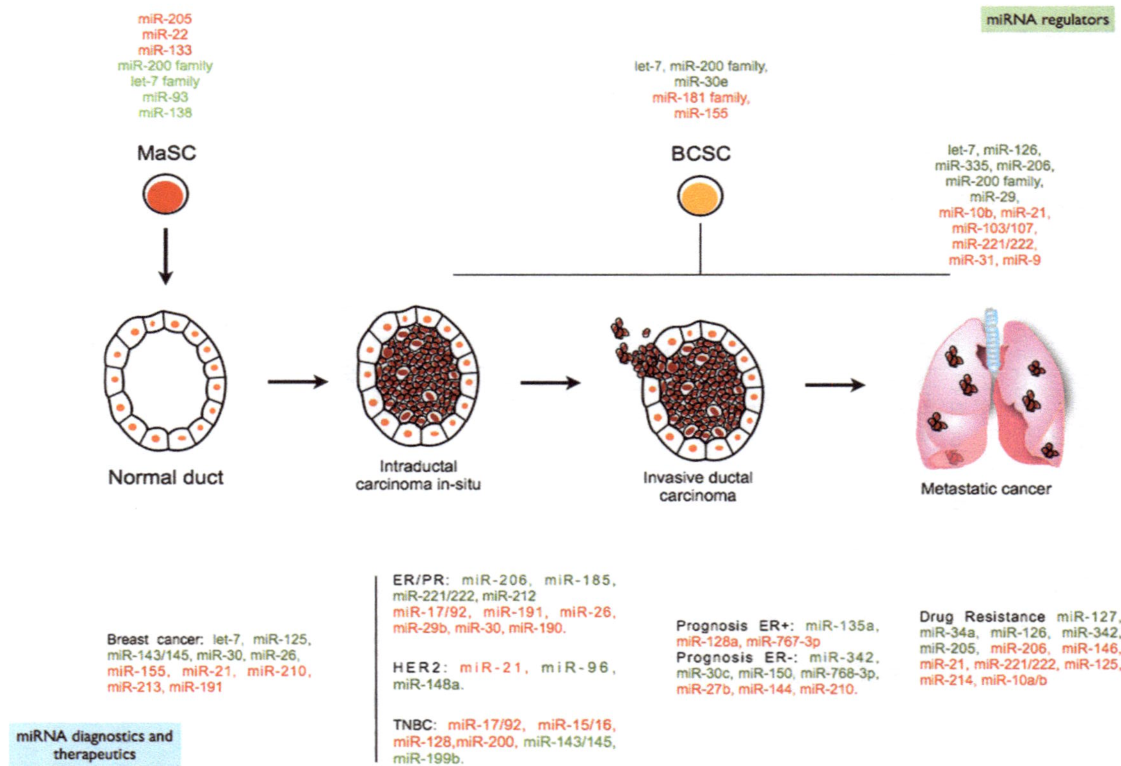

Fig. 5.1 Schematic representation of microRNA involvement in breast cancer progression. MiRNA signatures are reported for breast cancer tissues versus normal mammary gland, for breast cancer molecular subtypes, relative to prognosis in ER positive and negative breast tumors and associated to hormonal resistance phenotype. MiRNA expression is also reported for mammary stem cells and tumor-initiating cells. Up-regulated and down-regulated miRNAs were reported in *red* and *green* respectively. *MaSC* mammary stem cell, *BCSCs* breast cancer stem cells, *ER* estrogen receptor alpha, *PR* progesterone receptor, *TNBC* triple negative breast cancer

associated with tumor progression and poor prognosis. Studies have shown that miR-221 is capable of inducing NE differentiation in LNCaP cells in an androgen deprived environment, which may lead to Androgen Independence (AI) (Zheng et al. 2012). It was revealed that miR-221 stimulates the growth of LNCaP and LNCaP-AI cells, and it is consistent with findings that the ectopic introduction of miR-221 in low expressing LNCaP cells bolstered their growth potential by inducing a G1-S shift in cell cycle.

9 MicroRNAs and Brain Cancer

Gliomas are the most common primary brain tumors (over 50 % of primary brain tumors) and arise from the neuroepithelial support cells of the Central Nervous System (CNS), the glial cells. World Health Organization classifies gliomas by grades on the basis of their histological features: grade I (pilocytic astrocytoma), grade II (diffuse astrocytoma), grade III (anaplastic astro-cytoma), and grade IV (glioblastoma multiforme, GBM) (Louis et al. 2007). Glioblastoma multiforme is the most common and deadliest of all gliomas with approximately 10,000 new diagnosis every year in USA with a median survival of only 14 months despite aggressive surgery, radiation, and chemotherapies.

Microarray studies of glioma tissues have implicated a number of miRNAs involved in glioma formation and propagation. The majority of miRNAs are under-expressed in proliferating glioma cells with the important exception of miR-10b, miR-21, and miR-221. MiR-21 is strongly elevated in glioma and glioblastoma tumor samples and glioma cell lines when compared to non-neoplastic control samples, and its up-regulation is particularly prominent in grade IV gliomas. In a group of four patients with primary grade II gliomas that spontaneously progressed to grade IV, miR-21 levels gradually increased suggesting that miR-21 could have a role in glioma progression (Chan et al. 2005; Ciafrè et al. 2005; Conti et al. 2009; Zhi et al. 2010). Suppression of miR-21 in glioblastoma cells with antisense oligonucleotides decreased cell proliferation and it is associated with a marked increase in apoptosis and caspase activation (Chan et al. 2005). MiR-21 knockdown also leads to a considerable reduction of glioma volumes in mouse xenografts (Corsten et al. 2007). Between the targets of miR-21 that contribute to its tumor promoting and antiapoptotic activities in gliomas: PDCD4 mRNA is inversely correlates with miR-21 in a number of glioblastoma cell lines (T98G, A172, U87, and U251) and over-expression of miR-21 inhibits PDCD4-dependent apoptosis (Chen et al. 2008); RECK and TIMP3 repression both *in vitro* and *in vivo* leads to a reduction of glioma cell motility and invasion (Gabriely et al. 2008). MiR-21 represents also a promising target to improve the efficacy of chemotherapy in gliomas. Recent data demonstrated that increased levels of miR-21 play a key role in promoting human glioblastoma cells resistance against the antitumoral agent temozolomide (TMZ) (Ito et al. 2009). Indeed, ectopic over-expression of miR-21 significantly reduced TMZ-induced apoptosis in this cell line through a suppression of Bax/Bcl-2 ratio and caspase-3 activity (Shi et al. 2010). Additionally, suppression of miR-21 in glioblastoma cell lines enhances sensitivity of cancer cells to antineoplastic cytotoxic therapy with neural precursor cells (NPC) expressing a secretable variant of the cytotoxic agent tumor necrosis factor related apoptosis inducing ligand (S-TRAIL) (Corsten et al. 2007). Finally, down-regulation of miR-21 contributes to the antitumor effects of IFN-β on glioma cell and intracranial tumor xenografts and the activation of the transcription factor STAT3 may have a key role in the IFN-β mediated suppression of miR-21 (Ito et al. 2009). Conti et al. found both miR-21 and miR-221 up-regulated in glioma samples; however, they noted that whereas miR-21 was elevated in all gliomas, high levels of miR-221 were only found in high-grade gliomas (Conti et al. 2009). As well as for miR-21, knockdown of miR-221/222 through antisense oligonucleotides strongly reduced growth potential of glioma cells *in vitro* and *in vivo* (Zhang et al. 2009a, 2010a, b). Further evidence showed that knockdown of miR-221/222 induced a change of mitochondrial membrane potential and caspase-mediated apoptosis on glioblastoma cells. Moreover, miR-221/222 directly down-regulate the proapoptotic protein PUMA (mediator of p53-associated apoptosis), which leads in turn to decrease Bcl-2 and increase BAX (key regulators of mitochondrial apoptotic pathway). These results highlight the oncogenic role of miR-221/222 on glioblastoma (Zhang et al. 2010a).

MiRNAs have also been shown to function in gliomas as bona fide tumor suppressors. One example is represented by miR-128, a brain enriched miRNA, which is down-regulated in glioma cell lines and tissues when compared to normal brain samples (Zhang et al. 2009b). Increasing levels of miR-128 expression led to a reduced expression of the oncogene Bmi-1 with a concomitant over-expression of p21CIP1 and a decrease in phosphorylated Akt. Bmi-1 is a critical factor of normal stem cell maintenance and glioblastoma self-renewal and, consistently with these observation, miR-128 over-expression in human glioma neurospheres blocked glioma self-renewal reducing neurosphere number and size. In conclusion, mir-128 down-regulation is likely to enhance glioma tumorigenesis by promoting an undifferentiated phenotype and self renewing state through Bmi-1 increased expression. Another confirmed target of miR-128 is E2F3a, a transcription factor involved in cell cycle progression (Godlewski et al. 2008). Indeed, miR-128 and E2F3a levels are negatively correlated and miR-128 over-expression has similar inhibitory effects on proliferation of glioma cell lines as E2F3a knocking down. Ectopic over-expression of E2F3a partially reversed the effects of miR-128, suggesting that miR-128 could exert its antitumor effects at least partially by inhibiting E2F3a expression (Cui et al. 2010).

Medulloblastoma (MB) is the most common malignant brain tumor in children with an incidence of approximately 2 per 100,000 (Siegel et al. 2011). It is an embryonal tumor of the cerebellum that originates from aberrant growth of cerebellar granule cell precursors, which, during development, reside in the external granule cell layer. About 70 % of cases occur before the age of 16 years and approximately one-third disseminate in the CSF and up to 5 % spread systemically. Medulloblastoma treatment most often involves a combination of surgery and radiation therapy. Chemotherapy is usually reserved for children younger than 3 years of age or for recurrent tumors. Five-year survival rates for medulloblastoma have been estimated to range from 35 to 75 %. Therefore, there is an urgent need for a deep understanding of the molecular carcinogenesis of medulloblastoma, so that future effective and low-toxicity targeted therapies can be made available.

The first report of the involvement of miRNAs in medulloblastoma showed that miR-124 modulates cell-cycle regulation in medulloblastoma cells (Pierson et al. 2008). They showed that miR-124 expression is significantly decreased in medulloblastoma and that augmentation of miR-124 levels can slow tumor cell growth by targeting CDK6. Ferretti et al. (2009) profiled by using high-throughput screening the expression of miRNAs in 34 patients with medulloblastoma compared with normal adult and fetal cerebellar cells. They identified 78 miRNAs with altered expression in medulloblastoma, and the majority of these miRNAs were down-regulated in medulloblastoma compared to normal tissues, supporting a role for miRNAs as tumor suppressors. In particular, they detected an up-regulation of miR-21 and miR-17-92 cluster (miR-17-5p, miR-20a and miR-19a) and a down-regulation of miR-128a/b, let-7, miR-124a, miR-103, miR-134, miR-138, miR-149, miR-181b, miR-9 and miR-125a, most of them previously reported to be dysregulated in other brain tumor cell lines or nervous system cancers. MiR-9 and miR-125a, both down-regulated in MB, are involved in cell proliferation and, when transfected into MB cells, promote apoptosis and impair anchorage-independent growth by down-regulating the truncated isoform of the neurotropin receptor TrkC (t-TrkC). T-TrkC expression levels are higher in MB, inversely correlate with miR-9 and miR-125a levels and are responsible of enhanced cell proliferation and worse prognosis. Additionally, the authors identified specific miRNA expression patterns which distinguish medulloblastoma differing in histotypes (anaplastic, classic and desmoplastic), in molecular subtypes (ErbB2 or c-Myc over-expressing tumors) and disease-risk stratification. For example, let7g, miR-19a, miR-106b, and miR-191 were significantly up-regulated in anaplastic compared with desmoplastic medulloblastomas; let7g and miR-106b were differentially expressed in desmoplastic compared with classic medulloblastomas; and miR19a was up-regulated in anaplastic compared with classic medulloblastomas. Members of the miR-17/92 cluster were also identified by Uziel and coworkers (2009) as over-expressed in hedgehog-active medulloblastomas and was also associated with elevated c-Myc and n-Myc. Aberrant expression of miRNAs encoded by the miR-17/92 enhance the growth potential of medulloblastoma and that miRNA-mediated modulation of hedgehog signaling may be an important contributing factor to medulloblastoma pathogenesis.

10 MicroRNAs and Pancreatic Cancer

With a mortality rate of nearly 100 %, pancreatic cancer is one of the most lethal malignancies (Siegel et al. 2011) and it is the fourth most prominent cause for cancer-related deaths in both men and women in the United States. An extremely aggressive tumor growth rate and a high incidence of metastasis are characteristics of pancreatic cancer. At present, no curative therapies are available for patients with pancreatic cancer and the mainstay of treatment with the potential for cure is the complete, eradicative surgical resection of the primary carcinoma. However the 5-years survival rate is 6 %.

A large number of microarray data have provided a long list of differentially expressed miRNAs in pancreatic cancer. Schmittgen's team performed a hierarchical clustering of the 222 miRNA precursors on pancreatic adenocarcinoma samples, adjacent normal tissue, normal pancreas, chronic pancreatitis and several pancreatic cancer cell lines (Lee et al. 2007). They could evidence a significant difference between tumors and chronic pancreatitis, normal pancreas, pancreatic cellular lines, and 11 of the 15 adjacent normal samples. One hundred miRNA precursors were aberrantly expressed in pancreatic cancer, including miRNAs previously reported as differentially expressed in other human cancers (miR-155, miR-21, miR-221 and miR-222), and others not previously associated with cancer (miR-376a and miR-301). Another study performed by Carlo Croce's group compared the miRNA expression of 65 microdissected pancreatic adenocarcinomas, their adjacent non-cancerous pancreatic tissues and a set of chronic pancreatitis specimens (Bloomston et al. 2007). Thirty miRNAs were up-regulated and three were down-regulated in pancreatic cancers compared with normal pancreatic tissue. 12 out of these 30 up-regulated miRNAs are hypoxamiRs (miRNAs associated with hypoxia), which is in line with the low oxygen levels of the pancreatic tumors. Chronic pancreatitis and normal pancreas showed similar expression patterns, whereas the miRNA signature of normal tissue and pancreatic cancer were clearly distinct. In chronic pancreatitis only two of the 22 dysregulated miRNAs are hypoxia responsive. The authors also used miRNA microarray patterns to explore a relationship between miRNA expression and survival. They compared miRNA expression in node-positive patients who survived over 2 years and in patients who died of the disease during the same period of time. Six miRNAs were identified (miR-452, miR-105, miR-127, miR-518a-2, miR-187, and miR-30a-3p) whose expression differed in long-term survivors with node-positive disease. Two additional miRNAs related to survival were also identified, miR-196a-2 and miR-219. High expression of these miRNAs was found to predict poor survival. Surprisingly, these miRNAs that could discriminate between long- and short-term survivors were not among those specific to pancreatic cancer.

An important miRNA biomarker candidate for pancreatic cancer is represented by a significant up-regulation of miR-21 in pancreatic cancers (Bloomston et al. 2007; Dillhoff et al. 2008) but its expression did not correlate with tumor size, differentiation, nodal status, or tumor stage. The over-expression of miR-21 was predictive of a poorer outcome compared to the absence or faint/focal miR-21 expression in patients (Dillhoff et al. 2008). Therefore, in patients, over-expression of miR-21 indicates the limited survival and may be an important biological marker for prognosis. Profiling of four miRNAs such as miR-21, miR-210, miR-155, and miR-196a in the plasma of pancreatic cancer patients revealed that elevated expression of miR-155 and miR-196a was observed with the parallel progression of disease (Wang et al. 2009a). The assays for plasma levels analyses of these four miRNAs showed a sensitivity of 64 % and a specificity of 89 %. The study indicated that development of plasma miRNA profiling can be a sensitive and specific blood-based biomarker assay for pancreatic cancer. Further, elevated levels of miR-155, miR-203, miR-210 and miR-222 expression in pancreatic tumors were significantly associated with increased risk (6.2-fold) of death compared to patients with tumors having reduced expression of these miRNAs (Greither et al. 2010). The expression of let-7 is higher in normal acinar pancreatic cells but it is lost in poorly differentiated cancer cells (Torrisani et al. 2009). Enforced expression of the let-7 in pancreatic adenocarcinoma cell lines strongly inhibits the cell proliferation *in vitro* and *in vivo*, by repressing K-ras expression, and mitogen-activated protein kinase activation. Another study revealed that over-expression of miR-21 and -221 enhanced the malignant phenotype of the pancreatic cancer cells (Park et al. 2009). An inhibition of these miRNAs using antisense oligonucleotides revealed decreased proliferation and increased apoptosis of a pancreatic cancer cell line compared to control oligonucleotides. Specifically, knockdown of miR-21 and -221 expression with their corresponding antisense oligonucleotides in HS766T cells results in cell cycle arrest (G1 phase) which is accompanied by an increased expression of PTEN, RECK, and p27 respectively. Treatment of pancreatic cancer cells with the antisense oligonucleotides–gemcitabine combinations was synergistic. Similarly, increased expression of miR-10a was reported in metastatic pancreatic adenocarcinoma and it promotes the metastatic potential of pancreatic tumor cells by suppressing the expression of HOXB1 and HOXB3 genes (Weiss et al. 2009). Further, the retinoic acid receptor (RAR) antagonists effectively repress miR-10a expression that results in inhibition of pancreatic cancer cell metastasis. Therefore, these studies provided early steps for miRNA replacement therapy for pancreatic cancer.

11 Concluding Remarks

Discovery of the critical role of miRNAs in modulating gene expression has not only changed our concept of gene expression regulation, but has also offered a new opportunity for designing anticancer strategies and therapies. Profiling of human tumors based on miRNA expression has identified signatures associated with diagnosis, staging, progression, prognosis, and response to treatment. The more we understand about the etiological and biological nature of cancer, the better equipped we will be at designing effective preventive, diagnostic, and therapeutic tools to help reduce the burden of this disease. Practical issues that are being addressed include identifying effective therapeutic strategies, identifying the precise patient populations that may benefit from adjuvant chemotherapy and intensive screening for disease recurrence. Given the significant role of miRNAs in many aspects of cancer development such as proliferation, cell cycle control, invasion, EMT, and maintained tumor stem-cell phenotype, we remain hopeful that miRNA based therapeutics, diagnosis, and prognosis may emerge in the near future to benefit patients.

Acknowledgements Gianpiero Di Leva and Michela Garfolao researched data for the article and contributed to the writing, reviewing and editing of the manuscript. This work was supported in part by the Kimmel Cancer Award 2011 to Michela Garofalo.

References

Akao Y, Nakagawa Y, Kitade Y, Kinoshita T, Naoe T (2007a) Downregulation of microRNAs-143 and -145 in B-cell malignancies. Cancer Sci 98(12):1914–1920

Akao Y, Nakagawa Y, Naoe T (2007b) MicroRNA-143 and -145 in colon cancer. DNA Cell Biol 26(5):311–320

Akao Y, Nakagawa Y, Hirata I, Iio A, Itoh T, Kojima K et al (2010) Role of anti-oncomirs miR-143 and -145 in human colorectal tumors. Cancer Gene Ther 17(6):398–408

Aqeilan RI, Calin GA, Croce CM (2010) miR-15a and miR-16-1 in cancer: discovery, function and future perspectives. Cell Death Differ 17(2):215–220

Au Yeung CL, Tsang TY, Yau PL, Kwok TT (2011) Human papillomavirus type 16 E6 induces cervical cancer cell migration through the p53/microRNA-23b/urokinase-type plasminogen activator pathway. Oncogene 30(21):2401–2410

Babashah S, Soleimani M (2011) The oncogenic and tumour suppressive roles of microRNAs in cancer and apoptosis. Eur J Cancer 47(8):1127–1137

Bartel DP (2004) MicroRNAs: genomics, biogenesis, mechanism, and function. Cell 116(2):281–297

Blenkiron C, Goldstein LD, Thorne NP, Spiteri I, Chin SF, Dunning MJ et al (2007) MicroRNA expression profiling of human breast cancer identifies new markers of tumor subtype. Genome Biol 8(10):R214

Bloomston M, Frankel WL, Petrocca F, Volinia S, Alder H, Hagan JP et al (2007) MicroRNA expression patterns to differentiate pancreatic adenocarcinoma from normal pancreas and chronic pancreatitis. JAMA 297(17):1901–1908

Bo J, Yang G, Huo K, Jiang H, Zhang L, Liu D et al (2011) MicroRNA-203 suppresses bladder cancer development by repressing bcl-w expression. FEBS J 278(5):786–792

Bonci D, Coppola V, Musumeci M, Addario A, Giuffrida R, Memeo L et al (2008) The miR-15a-miR-16-1 cluster controls prostate cancer by targeting multiple oncogenic activities. Nat Med 14(11):1271–1277

Bosch FX, de Sanjose S (2003) Human papillomavirus and cervical cancer-burden and assessment of causality. J Natl Cancer Inst Monogr 31:3–13

Boyer SN, Wazer DE, Band V (1996) E7 protein of human papilloma virus-16 induces degradation of retinoblastoma protein through the ubiquitin-proteasome pathway. Cancer Res 56(20): 4620–4624

Budhu A, Jia HL, Forgues M, Liu CG, Goldstein D, Lam A et al (2008) Identification of metastasis-related microRNAs in hepatocellular carcinoma. Hepatology 47(3):897–907

Burk U, Schubert J, Wellner U, Schmalhofer O, Vincan E, Spaderna S et al (2008) A reciprocal repression between ZEB1 and members of the miR-200 family promotes EMT and invasion in cancer cells. EMBO Rep 9(6):582–589

Calin GA, Croce CM (2006) MicroRNA signatures in human cancers. Nat Rev Cancer 6(11): 857–866

Calin GA, Dumitru CD, Shimizu M, Bichi R, Zupo S, Noch E et al (2002) Frequent deletions and down-regulation of micro-RNA genes miR15 and miR16 at 13q14 in chronic lymphocytic leukemia. Proc Natl Acad Sci U S A 99(24):15524–15529

Calin GA, Sevignani C, Dumitru CD, Hyslop T, Noch E, Yendamuri S et al (2004) Human microRNA genes are frequently located at fragile sites and genomic regions involved in cancers. Proc Natl Acad Sci U S A 101(9):2999–3004

Callegari E, Elamin BK, Giannone F, Milazzo M, Altavilla G, Fornari F et al (2012) Liver tumorigenicity promoted by microRNA-221 in a mouse transgenic model. Hepatology 56(3): 1025–1033

Catto JW, Miah S, Owen HC, Bryant H, Myers K, Dudziec E et al (2009) Distinct microRNA alterations characterize high- and low-grade bladder cancer. Cancer Res 69(21):8472–8481

Chan JA, Krichevsky AM, Kosik KS (2005) MicroRNA-21 is an antiapoptotic factor in human glioblastoma cells. Cancer Res 65(14):6029–6033

Chang TC, Wentzel EA, Kent OA, Ramachandran K, Mullendore M, Lee KH et al (2007) Transactivation of miR-34a by p53 broadly influences gene expression and promotes apoptosis. Mol Cell 26(6):745–752

Chang KH, Miller N, Kheirelseid EA, Ingoldsby H, Hennessy E, Curran CE et al (2011) MicroRNA-21 and PDCD4 expression in colorectal cancer. Eur J Surg Oncol 37(7):597–603

Chen Y, Liu W, Chao T, Zhang Y, Yan X, Gong Y et al (2008) MicroRNA-21 down-regulates the expression of tumor suppressor PDCD4 in human glioblastoma cell T98G. Cancer Lett 272(2):197–205

Chen X, Guo X, Zhang H, Xiang Y, Chen J, Yin Y et al (2009) Role of miR-143 targeting KRAS in colorectal tumorigenesis. Oncogene 28(10):1385–1392

Ciafrè SA, Galardi S, Mangiola A, Ferracin M, Liu CG, Sabatino G et al (2005) Extensive modulation of a set of microRNAs in primary glioblastoma. Biochem Biophys Res Commun 334(4): 1351–1358

Clapé C, Fritz V, Henriquet C, Apparailly F, Fernandez PL, Iborra F et al (2009) miR-143 interferes with ERK5 signaling, and abrogates prostate cancer progression in mice. PLoS One 4(10):e7542

Conti A, Aguennouz M, La Torre D, Tomasello C, Cardali S, Angileri FF et al (2009) miR-21 and 221 upregulation and miR-181b downregulation in human grade II-IV astrocytic tumors. J Neurooncol 93(3):325–332

Corsten MF, Miranda R, Kasmieh R, Krichevsky AM, Weissleder R, Shah K (2007) MicroRNA-21 knockdown disrupts glioma growth in vivo and displays synergistic cytotoxicity with neural precursor cell delivered S-TRAIL in human gliomas. Cancer Res 67(4):8994–9000

Costinean S, Zanesi N, Pekarsky Y, Tili E, Volinia S, Heerema N et al (2006) Pre-B cell proliferation and lymphoblastic leukemia/high-grade lymphoma in E(mu)-miR155 transgenic mice. Proc Natl Acad Sci U S A 103(18):7024–7029

Cui JG, Zhao Y, Sethi P, Li YY, Mahta A, Culicchia F et al (2010) Micro-RNA-128 (miRNA-128) down-regulation in glioblastoma targets ARP5 (ANGPTL6), Bmi-1 and E2F-3a, key regulators of brain cell proliferation. J Neurooncol 98(9):297–304

Di Leva G, Croce CM (2010) Roles of small RNAs in tumor formation. Trends Mol Med 16(6):257–267

Di Leva G, Gasparini P, Piovan C, Ngankeu A, Garofalo M, Taccioli C et al (2010) MicroRNA cluster 221–222 and estrogen receptor alpha interactions in breast cancer. J Natl Cancer Inst 102(10):706–721

Dillhoff M, Liu J, Frankel W, Croce C, Bloomston M (2008) MicroRNA-21 is overexpressed in pancreatic cancer and a potential predictor of survival. J Gastrointest Surg 12(12): 2171–2176

Dumitrescu RG (2009) Epigenetic targets in cancer epidemiology. Methods Mol Biol 471(4): 457–467

Dyson N, Howley PM, Munger K, Harlow E (1989) The human papilloma virus-16 E7 oncoprotein is able to bind to the retinoblastoma gene product. Science 243(4893):934–937

Feber A, Xi L, Luketich JD, Pennathur A, Landreneau RJ, Wu M et al (2008) MicroRNA expression profiles of esophageal cancer. J Thorac Cardiovasc Surg 135(2):255–260

Feber A, Xi L, Pennathur A, Gooding WE, Bandla S, Wu M et al (2011) MicroRNA prognostic signature for nodal metastases and survival in esophageal adenocarcinoma. Ann Thorac Surg 91(5):1523–1530

Ferretti E, De Smaele E, Po A, Di Marcotullio L, Tosi E, Espinola MS et al (2009) MicroRNA profiling in human medulloblastoma. Int J Cancer 124(3):568–577

Fornari F, Gramantieri L, Ferracin M, Veronese A, Sabbioni S, Calin GA et al (2008) MiR-221 controls CDKN1C/p57 and CDKN1B/p27 expression in human hepatocellular carcinoma. Oncogene 27(43):5651–5661

Frankel LB, Christoffersen NR, Jacobsen A, Lindow M, Krogh A, Lund AH (2008) Programmed cell death 4 (PDCD4) is an important functional target of the microRNA miR-21 in breast cancer cells. J Biol Chem 283(2):1026–1033

Gabriely G, Wurdinger T, Kesari S, Esau CC, Burchard J, Linsley PS et al (2008) MicroRNA 21 promotes glioma invasion by targeting matrix metalloproteinase regulators. Mol Cell Biol 28(17):5369–5380

Galardi S, Mercatelli N, Giorda E, Massalini S, Frajese GV, Ciafre SA et al (2007) miR-221 and miR-222 expression affects the proliferation potential of human prostate carcinoma cell lines by targeting p27(Kip1). J Biol Chem 282(32):23716–23724

Garofalo M, Croce CM (2011) microRNAs: master regulators as potential therapeutics in cancer. Annu Rev Pharmacol Toxicol 51:25–43

Garofalo M, Di Leva G, Romano G, Nuovo G, Suh SS, Ngankeu A et al (2009) miR-221&222 regulate TRAIL resistance and enhance tumorigenicity through PTEN and TIMP3 downregulation. Cancer Cell 16(6):498–509

Garofalo M, Quintavalle C, Romano G, Croce CM, Condorelli G (2012) miR221/222 in cancer: their role in tumor progression and response to therapy. Curr Mol Med 12(1):27–33

Godlewski J, Nowicki MO, Bronisz A, Williams S, Otsuki A, Nuovo G et al (2008) Targeting of the Bmi-1 oncogene/stem cell renewal factor by microRNA-128 inhibits glioma proliferation and self-renewal. Cancer Res 68(22):9125–9130

Gong J, Zhang JP, Li B, Zeng C, You K, Chen MX et al (2013) MicroRNA-125b promotes apoptosis by regulating the expression of Mcl-1, Bcl-w and IL-6R. Oncogene 32(25):3071–3079

Gregersen LH, Jacobsen AB, Frankel LB, Wen J, Krogh A, Lund AH (2010) MicroRNA-145 targets YES and STAT1 in colon cancer cells. PLoS One 5(1):e8836

Greither T, Grochola LF, Udelnow A, Lautenschläger C, Würl P, Taubert H (2010) Elevated expression of microRNAs 155, 203, 210 and 222 in pancreatic tumors is associated with poorer survival. Int J Cancer 126(1):73–80

He L, He X, Lim LP, de Stanchina E, Xuan Z, Liang Y et al (2007) A microRNA component of the p53 tumour suppressor network. Nature 447(7148):1130–1134

Hu Y, Correa AM, Hoque A, Guan B, Ye F, Huang J et al (2011) Prognostic significance of differentially expressed miRNAs in esophageal cancer. Int J Cancer 128(1):132–143

Huang Q, Gumireddy K, Schrier M, le Sage C, Nagel R, Nair S et al (2008) The microRNAs miR-373 and miR-520c promote tumour invasion and metastasis. Nat Cell Biol 10(2):202–210

Huang X, Liang M, Dittmar R, Wang L (2013) Extracellular MicroRNAs in urologic malignancies: chances and challenges. Int J Mol Sci 14(7):14785–14799

Iliopoulos D, Lindahl-Allen M, Polytarchou C, Hirsch HA, Tsichlis PN, Struhl K (2010) Loss of miR-200 inhibition of Suz12 leads to polycomb-mediated repression required for the formation and maintenance of cancer stem cells. Mol Cell 39(5):761–772

Iorio MV, Ferracin M, Liu CG, Veronese A, Spizzo R, Sabbioni S et al (2005) MicroRNA gene expression deregulation in human breast cancer. Cancer Res 65(16):7065–7070

Ito M, Kato T, Wakabayashi T (2009) The modulation of microRNAs by type I IFN through the activation of signal transducers and activators of transcription 3 in human glioma. Mol Cancer Res 7(12):2022–2030

Iwaya T, Yokobori T, Nishida N, Kogo R, Sudo T, Tanaka F et al (2012) Downregulation of miR-144 is associated with colorectal cancer progression via activation of mTOR signaling pathway. Carcinogenesis 33(14):2391–2397

Jabbar SF, Abrams L, Glick A, Lambert PF (2009) Persistence of high-grade cervical dysplasia and cervical cancer requires the continuous expression of the human papillomavirus type 16 E7 oncogene. Cancer Res 69(10):4407–4414

Johnson SM, Grosshans H, Shingara J, Byrom M, Jarvis R, Cheng A et al (2005) RAS is regulated by the let-7 microRNA family. Cell 120(5):635–647

Jones PA, Baylin SB (2007) The epigenomics of cancer. Cell 128(4):683–692

Kasinski AL, Slack FJ (2011) Epigenetics and genetics. MicroRNAs en route to the clinic: progress in validating and targeting microRNAs for cancer therapy. Nat Rev Cancer 11(12):849–864

Kent OA, Chivukula RR, Mullendore M, Wentzel EA, Feldmann G, Lee KH et al (2010) Repression of the miR-143/145 cluster by oncogenic Ras initiates a tumor-promoting feed-forward pathway. Genes Dev 24(24):2754–2759

Klein U, Lia M, Crespo M, Siegel R, Shen Q, Mo T et al (2010) The DLEU2/miR-15a/16-1 cluster controls B cell proliferation and its deletion leads to chronic lymphocytic leukemia. Cancer Cell 17(1):28–40

Kong D, Li Y, Wang Z, Banerjee S, Ahmad A, Kim HR et al (2009) miR-200 regulates PDGF-D-mediated epithelial-mesenchymal transition, adhesion, and invasion of prostate cancer cells. Stem Cells 27(8):1712–1721

Kong D, Banerjee S, Ahmad A, Li Y, Wang Z, Sethi S et al (2010) Epithelial to mesenchymal transition is mechanistically linked with stem cell signatures in prostate cancer cells. PLoS One 5(8):e12445

Kong KL, Kwong DL, Chan TH, Law SY, Chen L, Li Y et al (2012) MicroRNA-375 inhibits tumour growth and metastasis in oesophageal squamous cell carcinoma through repressing insulin-like growth factor 1 receptor. Gut 61(1):33–42

Kota J, Chivukula RR, O'Donnell KA, Wentzel EA, Montgomery CL, Hwang HW et al (2009) Therapeutic microRNA delivery suppresses tumorigenesis in a murine liver cancer model. Cell 137(6):1005–1017

Kurashige J, Kamohara H, Watanabe M, Tanaka Y, Kinoshita K, Saito S et al (2012) Serum microRNA-21 is a novel biomarker in patients with esophageal squamous cell carcinoma. J Surg Oncol 106(2):188–192

le Sage C, Nagel R, Egan DA, Schrier M, Mesman E, Mangiola A et al (2007) Regulation of the p27(Kip1) tumor suppressor by miR-221 and miR-222 promotes cancer cell proliferation. EMBO J 26(15):3699–3708

Lee YS, Dutta A (2007) The tumor suppressor microRNA let-7 represses the HMGA2 oncogene. Genes Dev 21(9):1025–1030

Lee EJ, Gusev Y, Jiang J, Nuovo GJ, Lerner MR, Frankel WL et al (2007) Expression profiling identifies microRNA signature in pancreatic cancer. Int J Cancer 120(5):1046–1054

Lee JW, Choi CH, Choi JJ, Kim WY, Kim TJ, Lee JH et al (2008) Altered MicroRNA expression in cervical carcinomas. Clin Cancer Res 14(9):2535–2542

Li T, Li D, Sha J, Sun P, Huang Y (2009) MicroRNA-21 directly targets MARCKS and promotes apoptosis resistance and invasion in prostate cancer cells. Biochem Biophys Res Commun 383(3):280–285

Li J, Zhang Y, Zhao J, Kong F, Chen Y (2011a) Overexpression of miR-22 reverses paclitaxel-induced chemoresistance through activation of PTEN signaling in p53-mutated colon cancer cells. Mol Cell Biochem 357(1–2):31–38

Li X, Lin R, Li J (2011b) Epigenetic silencing of microRNA-375 regulates PDK1 expression in esophageal cancer. Dig Dis Sci 56(10):2849–2856

Lin SL, Chiang A, Chang D, Ying SY (2008) Loss of mir-146a function in hormone-refractory prostate cancer. RNA 14(3):417–424

Liu M, Tang Q, Qiu M, Lang N, Li M, Zheng Y et al (2011) miR-21 targets the tumor suppressor RhoB and regulates proliferation, invasion and apoptosis in colorectal cancer cells. FEBS Lett 585(19):2998–3005

Liu R, Liao J, Yang M, Sheng J, Yang H, Wang Y et al (2012) The cluster of miR-143 and miR-145 affects the risk for esophageal squamous cell carcinoma through co-regulating fascin homolog 1. PLoS One 7(3):e33987

Liu H, Du L, Wen Z, Yang Y, Li J, Wang L et al (2013) Up-regulation of miR-182 expression in colorectal cancer tissues and its prognostic value. Int J Colorectal Dis 28(5):697–703

Louis DN, Ohgaki H, Wiestler OD, Cavenee WK, Burger PC, Jouvet A et al (2007) The 2007 WHO classification of tumours of the central nervous system. Acta Neuropathol 114(2): 97–109

Lu J, Getz G, Miska EA, Alvarez-Saavedra E, Lamb J, Peck D et al (2005) MicroRNA expression profiles classify human cancers. Nature 435(7043):834–838

Lu Q, Lu C, Zhou GP, Zhang W, Xiao H, Wang XR (2010) MicroRNA-221 silencing predisposed human bladder cancer cells to undergo apoptosis induced by TRAIL. Urol Oncol 28(6): 635–641

Lui WO, Pourmand N, Patterson BK, Fire A (2007) Patterns of known and novel small RNAs in human cervical cancer. Cancer Res 67(13):6031–6043
Luo X, Burwinkel B, Tao S, Brenner H (2011) MicroRNA signatures: novel biomarker for colorectal cancer? Cancer Epidemiol Biomarkers Prev 20(7):1272–1286
Ma L, Teruya-Feldstein J, Weinberg RA (2007) Tumour invasion and metastasis initiated by microRNA-10b in breast cancer. Nature 449(7163):682–688
Ma K, He Y, Zhang H, Fei Q, Niu D, Wang D et al (2012) DNA methylation-regulated miR-193a-3p dictates resistance of hepatocellular carcinoma to 5-fluorouracil via repression of SRSF2 expression. J Biol Chem 287(8):5639–5649
Majid S, Dar AA, Saini S, Shahryari V, Arora S, Zaman MS et al (2012) MicroRNA-1280 inhibits invasion and metastasis by targeting ROCK1 in bladder cancer. PLoS One 7(10):e46743
Martinez I, Gardiner AS, Board KF, Monzon FA, Edwards RP, Khan SA (2008) Human papillomavirus type 16 reduces the expression of microRNA-218 in cervical carcinoma cells. Oncogene 27(18):2575–2582
Medina PP, Nolde M, Slack FJ (2010) OncomiR addiction in an in vivo model of microRNA-21-induced pre-B-cell lymphoma. Nature 467(7311):86–90
Melar-New M, Laimins LA (2010) Human papillomaviruses modulate expression of microRNA 203 upon epithelial differentiation to control levels of p63 proteins. J Virol 84(10):5212–5221
Mendell JT, Olson EN (2012) MicroRNAs in stress signaling and human disease. Cell 148(6):1172–1187
Meng F, Henson R, Wehbe-Janek H, Ghoshal K, Jacob ST, Patel T (2007) MicroRNA-21 regulates expression of the PTEN tumor suppressor gene in human hepatocellular cancer. Gastroenterology 133(2):647–658
Mercatelli N, Coppola V, Bonci D, Miele F, Costantini A, Guadagnoli M et al (2008) The inhibition of the highly expressed miR-221 and miR-222 impairs the growth of prostate carcinoma xenografts in mice. PLoS One 3(12):e4029
Michael MZ, O'Connor SM, van Holst Pellekaan NG, Young GP, James RJ (2003) Reduced accumulation of specific microRNAs in colorectal neoplasia. Mol Cancer Res 1(12):882–891
Miller TE, Ghoshal K, Ramaswamy B, Roy S, Datta J, Shapiro CL et al (2008) MicroRNA-221/222 confers tamoxifen resistance in breast cancer by targeting p27Kip1. J Biol Chem 283(44): 29897–29903
Mukherji S, Ebert MS, Zheng GX, Tsang JS, Sharp PA, van Oudenaarden A (2011) MicroRNAs can generate thresholds in target gene expression. Nat Genet 43(9):854–859
Murakami Y, Yasuda T, Saigo K, Urashima T, Toyoda H, Okanoue T et al (2006) Comprehensive analysis of microRNA expression patterns in hepatocellular carcinoma and non-tumorous tissues. Oncogene 25(17):2537–2545
Musumeci M, Coppola V, Addario A, Patrizii M, Maugeri-Saccà M, Memeo L et al (2011) Control of tumor and microenvironment cross-talk by miR-15a and miR-16 in prostate cancer. Oncogene 30(41):4231–4242
Nagel R, le Sage C, Diosdado B, van der Waal M, Oude Vrielink JA, Bolijn A, Meijer GA, Agami R (2008) Regulation of the adenomatous polyposis coli gene by the miR-135 family in colorectal cancer. Cancer Res 68(14):5795–5802
Ni Y, Meng L, Wang L, Dong W, Shen H, Wang G et al (2013) MicroRNA-143 functions as a tumor suppressor in human esophageal squamous cell carcinoma. Gene 517(2):197–204
Nishida N, Yamashita S, Mimori K, Sudo T, Tanaka F, Shibata K et al (2012) MicroRNA-10b is a prognostic indicator in colorectal cancer and confers resistance to the chemotherapeutic agent 5-fluorouracil in colorectal cancer cells. Ann Surg Oncol 19(9):3065–3071
Nishimura J, Handa R, Yamamoto H, Tanaka F, Shibata K, Mimori K et al (2012) microRNA-181a is associated with poor prognosis of colorectal cancer. Oncol Rep 28(6):2221–2226
Noguchi S, Yasui Y, Iwasaki J, Kumazaki M, Yamada N, Naito S et al (2013) Replacement treatment with microRNA-143 and -145 induces synergistic inhibition of the growth of human bladder cancer cells by regulating PI3K/Akt and MAPK signaling pathways. Cancer Lett 328(2):353–361
Nuovo GJ, Wu X, Volinia S, Yan F, Di Leva G, Chin N et al (2010) Strong inverse correlation between microRNA-125b and human papillomavirus DNA in productive infection. Diagn Mol Pathol 19(3):135–143

Olaru AV, Yamanaka S, Vazquez C, Mori Y, Cheng Y, Abraham JM et al (2013) MicroRNA-224 negatively regulates p21 expression during late neoplastic progression in inflammatory bowel disease. Inflamm Bowel Dis 19(3):471–480

Pang Y, Young CY, Yuan H (2010) MicroRNAs and prostate cancer. Acta Biochim Biophys Sin (Shanghai) 42(6):363–369

Park SM, Gaur AB, Lengyel E, Peter ME (2008) The miR-200 family determines the epithelial phenotype of cancer cells by targeting the E-cadherin repressors ZEB1 and ZEB2. Genes Dev 22(7):894–907

Park JK, Lee EJ, Esau C, Schmittgen TD (2009) Antisense inhibition of microRNA-21 or -221 arrests cell cycle, induces apoptosis, and sensitizes the effects of gemcitabine in pancreatic adenocarcinoma. Pancreas 38(7):e190–e199

Park JK, Kogure T, Nuovo GJ, Jiang J, He L, Kim JH et al (2011) miR-221 silencing blocks hepatocellular carcinoma and promotes survival. Cancer Res 71(24):7608–7616

Parkin DM, Bray F (2006) Chapter 2: the burden of HPV-related cancers. Vaccine 24:S3/11–25

Pierson J, Hostager B, Fan R, Vibhakar R (2008) Regulation of cyclin dependent kinase 6 by microRNA 124 in medulloblastoma. J Neurooncol 90(1):1–7

Piperi C, Vlastos F, Farmaki E, Martinet N, Papavassiliou AG (2008) Epigenetic effects of lung cancer predisposing factors impact on clinical diagnosis and prognosis. J Cell Mol Med 12(5A):1495–1501

Prat A, Perou CM (2011) Deconstructing the molecular portraits of breast cancer. Mol Oncol 5(1):5–23

Qi L, Bart J, Tan LP, Platteel I, Sluis T, Huitema S et al (2009) Expression of miR-21 and its targets (PTEN, PDCD4, TM1) in flat epithelial atypia of the breast in relation to ductal carcinoma in situ and invasive carcinoma. BMC Cancer 9:163

Rao X, Di Leva G, Li M, Fang F, Devlin C, Hartman-Frey C et al (2011) MicroRNA-221/222 confers breast cancer fulvestrant resistance by regulating multiple signaling pathways. Oncogene 30(9):1082–1097

Raver-Shapira N, Marciano E, Meiri E, Spector Y, Rosenfeld N, Moskovits N et al (2007) Transcriptional activation of miR-34a contributes to p53-mediated apoptosis. Mol Cell 26(5):731–743

Ribas J, Ni X, Haffner M, Wentzel EA, Salmasi AH, Chowdhury WH et al (2009) miR-21: an androgen receptor-regulated microRNA that promotes hormone-dependent and hormone-independent prostate cancer growth. Cancer Res 69(18):7165–7169

Roldo C, Missiaglia E, Hagan JP, Falconi M, Capelli P, Bersani S et al (2006) MicroRNA expression abnormalities in pancreatic endocrine and acinar tumors are associated with distinctive pathologic features and clinical behavior. J Clin Oncol 24(29):4677–4684

Sachdeva M, Zhu S, Wu F, Wu H, Walia V, Kumar S et al (2009) p53 represses c-Myc through induction of the tumor suppressor miR-145. Proc Natl Acad Sci U S A 106(9):3207–3212

Sachdeva M, Wu H, Ru P, Hwang L, Trieu V, Mo YY (2011) MicroRNA-101-mediated Akt activation and estrogen-independent growth. Oncogene 30(7):822–831

Sato F, Hatano E, Kitamura K, Myomoto A, Fujiwara T, Takizawa S et al (2011) MicroRNA profile predicts recurrence after resection in patients with hepatocellular carcinoma within the Milan Criteria. PLoS One 6(1):e16435

Sayed D, Rane S, Lypowy J, He M, Chen IY, Vashistha H et al (2008) MicroRNA-21 targets Sprouty2 and promotes cellular outgrowths. Mol Biol Cell 19(8):3272–3282

Scheffner M, Werness BA, Huibregtse JM, Levine AJ, Howley PM (1990) The E6 oncoprotein encoded by human papillomavirus types 16 and 18 promotes the degradation of p53. Cell 63(6):1129–1136

Selcuklu SD, Donoghue MT, Spillane C (2009) miR-21 as a key regulator of oncogenic processes. Biochem Soc Trans 37(Pt 4):918–925

Shi XB, Xue L, Yang J, Ma AH, Zhao J, Xu M et al (2007) An androgen regulated miRNA suppresses Bak1 expression and induces androgen-independent growth of prostate cancer cells. Proc Natl Acad Sci U S A 104(50):1998–19988

Shi L, Chen J, Yang J, Pan T, Zhang S, Wang Z (2010) MiR-21 protected human glioblastoma U87MG cells from chemotherapeutic drug temozolomide induced apoptosis by decreasing Bax/Bcl-2 ratio and caspase-3 activity. Brain Res 1352(13):255–264

Shibata K, Tanaka S, Shiraishi T, Kitano S, Mori M (1999) G-protein g7 is down-regulated in cancers and associated with p 27kip1-induced growth arrest. Cancer Res 59(5):1096–1101

Shimono Y, Zabala M, Cho RW, Lobo N, Dalerba P, Qian D et al (2009) Downregulation of miRNA-200c links breast cancer stem cells with normal stem cells. Cell 138(3):592–603

Si ML, Zhu S, Wu H, Lu Z, Wu F, Mo YY (2007) miR-21-mediated tumor growth. Oncogene 26(19):2799–2803

Siegel R, Ward E, Brawley O, Jemal A (2011) Cancer statistics, 2011: the impact of eliminating socioeconomic and racial disparities on premature cancer deaths. CA Cancer J Clin 61(4):212–236

Sikand K, Slaibi JE, Singh R, Slane SD, Shukla GC (2010) miR 488* inhibits androgen receptor expression in prostate carcinoma cells. Int J Cancer 129(4):810–819

Song B, Wang Y, Xi Y, Kudo K, Bruheim S, Botchkina GI et al (2009) Mechanism of chemoresistance mediated by miR-140 in human osteosarcoma and colon cancer cells. Oncogene 28(46):4065–4074

Sun T, Wang Q, Balk S, Brown M, Lee GS, Kantoff P (2009) The role of microRNA-221 and microRNA-222 in androgen-independent prostate cancer cell lines. Cancer Res 69(8):3356–3363

Tao J, Lu Q, Wu D, Li P, Xu B, Qing W et al (2011) microRNA-21 modulates cell proliferation and sensitivity to doxorubicin in bladder cancer cells. Oncol Rep 25(6):1721–1729

Tavazoie SF, Alarcon C, Oskarsson T, Padua D, Wang Q, Bos PD et al (2008) Endogenous human microRNAs that suppress breast cancer metastasis. Nature 451(7175):147–152

Torrisani J, Bournet B, du Rieu MC, Bouisson M, Souque A, Escourrou J et al (2009) let-7 MicroRNA transfer in pancreatic cancer-derived cells inhibits in vitro cell proliferation but fails to alter tumor progression. Hum Gene Ther 20(8):831–844

Uziel T, Karginov FV, Xie S, Parker JS, Wang YD, Gajjar A et al (2009) The miR-17 92 cluster collaborates with the Sonic Hedgehog pathway in medulloblastoma. Proc Natl Acad Sci U S A 106(8):2812–2817

Valeri N, Gasparini P, Braconi C, Paone A, Lovat F, Fabbri M et al (2010) MicroRNA-21 induces resistance to 5-fluorouracil by down-regulating human DNA MutS homolog 2 (hMSH2). Proc Natl Acad Sci U S A 107(49):21098–21103

Varambally S, Cao Q, Mani RS, Shankar S, Wang X, Ateeq B et al (2008) Genomic loss of microRNA-101 leads to overexpression of histone methyltransferase EZH2 in cancer. Science 322(5908):1695–1699

Vinall RL, Ripoll AZ, Wang S, Pan CX, deVere White RW (2012) MiR-34a chemosensitizes bladder cancer cells to cisplatin treatment regardless of p53-Rb pathway status. Int J Cancer 130(11):2526–2538

Volanis D, Kadiyska T, Galanis A, Delakas D, Logotheti S, Zoumpourlis V (2010) Environmental factors and genetic susceptibility promote urinary bladder cancer. Toxicol Lett 193(2):131–137

Volinia S, Croce CM (2013) Prognostic microRNA/mRNA signature from the integrated analysis of patients with invasive breast cancer. Proc Natl Acad Sci U S A 110(18):7413–7417

Volinia S, Calin GA, Liu CG, Ambs S, Cimmino A, Petrocca F et al (2006) A microRNA expression signature of human solid tumors defines cancer gene targets. Proc Natl Acad Sci U S A 103(7):2257–2261

Volinia S, Galasso M, Sana ME, Wise TF, Palatini J, Huebner K et al (2012) Breast cancer signatures for invasiveness and prognosis defined by deep sequencing of microRNA. Proc Natl Acad Sci U S A 109(8):3024–3029

Walboomers JM, Jacobs MV, Manos MM, Bosch FX, Kummer JA, Shah KV et al (1999) Human papillomavirus is a necessary cause of invasive cervical cancer worldwide. J Pathol 189(1): 12–19

Wang X, Tang S, Le SY, Lu R, Rader JS, Meyers C et al (2008) Aberrant expression of oncogenic and tumor-suppressive microRNAs in cervical cancer is required for cancer cell growth. PLoS One 3(7):e2557

Wang J, Chen J, Chang P, LeBlanc A, Li D, Abbruzzesse JL et al (2009a) MicroRNAs in plasma of pancreatic ductal adenocarcinoma patients as novel blood-based biomarkers of disease. Cancer Prev Res (Phila) 2(9):807–813

Wang P, Zou F, Zhang X, Li H, Dulak A, Tomko RJ Jr et al (2009b) microRNA-21 negatively regulates Cdc25A and cell cycle progression in colon cancer cells. Cancer Res 69(20):8157–8165

Weiss FU, Marques IJ, Woltering JM, Vlecken DH, Aghdassi A, Partecke LI et al (2009) Retinoic acid receptor antagonists inhibit miR-10a expression and block metastatic behavior of pancreatic cancer. Gastroenterology 137(6):2136–2145

Wong CM, Wong CC, Lee JM, Fan DN, Au SL, Ng IO (2012) Sequential alterations of microRNA expression in hepatocellular carcinoma development and venous metastasis. Hepatology 55(5):1453–1461

Xiong B, Cheng Y, Ma L, Zhang C (2013) MiR-21 regulates biological behavior through the PTEN/PI-3 K/Akt signaling pathway in human colorectal cancer cells. Int J Oncol 42(1):219–228

Xu N, Papagiannakopoulos T, Pan G, Thomson JA, Kosik KS (2009) MicroRNA-145 regulates OCT4, SOX2, and KLF4 and represses pluripotency in human embryonic stem cells. Cell 137(4):647–658

Xu B, Niu X, Zhang X, Tao J, Wu D, Wang Z et al (2011) miR-143 decreases prostate cancer cells proliferation and migration and enhances their sensitivity to docetaxel through suppression of KRAS. Mol Cell Biochem 350(1–2):207–213

Xu K, Liang X, Shen K, Sun L, Cui D, Zhao Y et al (2012a) MiR-222 modulates multidrug resistance in human colorectal carcinoma by down-regulating ADAM-17. Exp Cell Res 318(17):2168–2177

Xu N, Shen C, Luo Y, Xia L, Xue F, Xia Q et al (2012b) Upregulated miR-130a increases drug resistance by regulating RUNX3 and Wnt signaling in cisplatin-treated HCC cell. Biochem Biophys Res Commun 425(2):468–472

Xu C, Zeng Q, Xu W, Jiao L, Chen Y, Zhang Z et al (2013) miRNA-100 inhibits human bladder urothelial carcinogenesis by directly targeting mTOR. Mol Cancer Ther 12(2):207–219

Yao Q, Xu H, Zhang QQ, Zhou H, Qu LH (2009) MicroRNA-21 promotes cell proliferation and downregulates the expression of programmed cell death 4 (PDCD4) in HeLa cervical carcinoma cells. Biochem Biophys Res Commun 388(3):539–542

Yin Y, Yan ZP, Lu NN, Xu Q, He J, Qian X et al (2013) Downregulation of miR-145 associated with cancer progression and VEGF transcriptional activation by targeting N-RAS and IRS1. Biochim Biophys Acta 1829(2):239–247

Yu F, Yao H, Zhu P, Zhang X, Pan Q, Gong C et al (2007) Let-7 regulates self renewal and tumorigenicity of breast cancer cells. Cell 131(6):1109–1123

Yugawa T, Kiyono T (2009) Molecular mechanisms of cervical carcinogenesis by high-risk human papillomaviruses: novel functions of E6 and E7 oncoproteins. Rev Med Virol 19(2):97–113

Zhang C, Kang C, You Y, Pu P, Yang W, Zhao P et al (2009a) Co-suppression of miR-221/222 cluster suppresses human glioma cell growth by targeting p27(kip1) in vitro and in vivo. Int J Oncol 34(6):1653–1660

Zhang Y, Chao T, Li R, Liu W, Chen Y, Yan X et al (2009b) MicroRNA-128 inhibits glioma cells proliferation by targeting transcription factor E2F3a. J Mol Med 87(1):43–51

Zhang CZ, Zhang JX, Zhang AL, Shi ZD, Han L, Jia ZF et al (2010a) MiR-221 and miR-222 target PUMA to induce cell survival in glioblastoma. Mol Cancer 9:229

Zhang J, Han L, Ge Y, Zhou X, Zhang A, Zhang C et al (2010b) miR-221/222 promote malignant progression of glioma through activation of the Akt pathway. Int J Oncol 36(4):913–920

Zhang GJ, Xiao HX, Tian HP, Liu ZL, Xia SS, Zhou T (2013) Upregulation of microRNA-155 promotes the migration and invasion of colorectal cancer cells through the regulation of claudin-1 expression. Int J Mol Med 31(6):1375–1380

Zhao JJ, Lin J, Yang H, Kong W, He L, Ma X et al (2008) MicroRNA-221/222 negatively regulates estrogen receptor alpha and is associated with tamoxifen resistance in breast cancer. J Biol Chem 283(45):31079–31086

Zheng C, Yinghao S, Li J (2012) MiR-221 expression affects invasion potential of human prostate carcinoma cell lines by targeting DVL2. Med Oncol 29(2):815–822

Zhi F, Chen X, Wang S, Xia X, Shi Y, Guan W et al (2010) The use of hsa-miR-21, hsa-miR-181b and hsa-miR-106a as prognostic indicators of astrocytoma. Eur J Cancer 46(9):1640–1649

Chapter 6
MicroRNAs and Blood Cancers

Charles H. Lawrie

Contents

Abstract Despite having been only formally recognized for just over 10 years, microRNAs (miRNAs) have become one of the trendiest topics in biology. It is now clear that dysfunctional expression of miRNAs is a characteristic of many, if not all, hematological malignancies. Many of the miRNAs aberrantly expressed in hematological malignancies also play a crucial regulatory role in normal hematopoietic

C.H. Lawrie (✉)
Oncology Research Area, Biodonostia Research Institute, San Sebastián, Spain

Ikerbasque, Basque Foundation for Science, Bilbao, Spain

Nuffield Department of Clinical Laboratory Sciences, University of Oxford, Oxford, UK
e-mail: charles.lawrie@biodonostia.org; charles.lawrie@ndcls.ox.ac.uk

S. Babashah (ed.), *MicroRNAs: Key Regulators of Oncogenesis*,
DOI 10.1007/978-3-319-03725-7_6, © Springer International Publishing Switzerland 2014

function. In this chapter we review the evidence for this assertion in particular focusing on the use of miRNAs as novel tools for biomarkers and therapeutic agents against hematological malignancies.

Keywords Hematological malignancies • Leukemias • Lymphomas • MicroRNA • Biomarker

1 Introduction

MicroRNAs (miRNAs) are a recently discovered class of naturally occurring short non-coding (18–24 nt) RNA molecules that regulate eukaryotic gene expression post-transcriptionally. There are now nearly 2,000 human miRNAs that have been identified through cloning and/or sequence analysis (http://www.mirbase.org) (Griffiths-Jones et al. 2006), and it is believed some 60 % of all human genes are a target for miRNA regulation (Friedman et al. 2009). MiRNAs have been shown to play key regulatory roles in virtually every aspect of biology including developmental timing, cell differentiation, apoptosis, cell proliferation, metabolism organ development, and hematopoiesis (Kim 2005). The potential importance of miRNAs in cancer is implied by the finding that the majority of human miRNAs are located at cancer-associated genomic regions (Calin et al. 2004b). The first indication that dysregulation of miRNAs was associated with hematological malignancies came from the seminal publication by Calin et al in 2002 (Calin et al. 2002), that made the connection between 13q14, a frequently deleted locus in chronic lymphocytic leukemia (CLL), and down-regulation of the miR-15a/16 cluster that is encoded within this region. There is now overwhelming evidence that dysfunctional expression of miRNAs is a common, if not ubiquitous, hallmark of cancer in general including haematological malignancies (Lawrie 2008; Iorio and Croce 2009).

The reasons for aberrant expression of miRNAs in hematological malignancies (and other cancers) are numerous, and can include chromosomal aberrations, epigenetic deregulation, aberrant expression of transcription factors that regulate promoter regions of miRNAs, and factors that change miRNA biosynthesis or function (Croce 2009; Babashah et al. 2012). Many of the miRNAs that are aberrantly expressed in hematological malignancies are also key regulators of hematopoiesis (Table 6.1, Fig. 6.1), and in order to fully understand the role of these miRNAs in the pathogenesis of these cancers it is crucial that we first understand their role under physiological conditions.

2 Role of MicroRNAs in Hematopoiesis

The process by which hematopoietic stem cells (HSC) maintain their pluripotency whilst at the same time responding to lineage determining signals to differentiate into the various hematopoietic lineages is a finely balanced process tightly

Table 6.1 Major microRNAs involved in hematological malignancies showing known roles in hematopoiesis, validated target genes and representative animal model

MicroRNA	Hematopoiesis	Malignancy	Target	Animal model
miR-155	B/T-cell development	DLBCL, BL, AML	HGAL, RTKN2 (Dagan et al. 2012); SMAD5 (Rai et al. 2010); PIK3R1 (Huang et al. 2012); SHIP (Pedersen et al. 2009); PU.1, CD10 (Thompson et al. 2011); AID (de Yebenes et al. 2008; Teng et al. 2008)	Eμ-enhancer expression → lymphoma (Costinean et al. 2006); Inducible miR-155 knock-in model (Babar et al. 2012) Engraftment of miR-155 HSCs → myeloid (O'Connell et al. 2008)
miR-17-92	B-cell development	B-cell lymphoma, CML, ALL, MCL	BIM, PTEN (Xiao et al. 2008); E2F1 (Woods et al. 2007); PP2A, PRKAA1 (Xiao et al. 2008; Ventura et al. 2008; Mu et al. 2009; Mavrakis et al. 2010); PHLPP2 (Rao et al. 2012); CCND1 (Chen et al. 2008; Deshpande et al. 2009)	miR-17-92 expression alone → lymphoproliferative/autoimmunity (Xiao et al. 2008) miR-17-92 along with MYC (Eμ-MYC) → lymphoma (He et al. 2005)
miR-34a	B-cell development	B-cell lymphoma	SIRT1 (Yamakuchi and Lowenstein 2009); MYC (Sotillo et al. 2011); FOXP1 (Craig et al. 2011a)	Xenotransplant model of ABC-type DLBCL inoculated with synthetic miR-34a (Craig et al. 2012)
miR-125	Myelopoiesis	DLBCL, ALL, MDS, AML	IRF4, PRDM1 (Malumbres et al. 2009); TNFAIP3 (Kim et al. 2012); LIN28A(Chaudhuri et al. 2012)	Eμ-enhancer expression → lymphoblastic leukemia/lymphoma (Costinean et al. 2006)
miR-223	Myelopoiesis, erythropoiesis, granulopoiesis, B-cell development	MALT, SzS	MEF2C (Johnnidis et al. 2008); LMO2 (Malumbres et al. 2009); NFIA, CEBPA (Fazi et al. 2005); E2A(Liu et al. 2010)	
miR-150	Myeolopoiesis, Megakaryopoiesis, B/T-cell development, NK cell development	MDS, MALT, NK/T-cell lymphoma	MYB (Zhou et al. 2007; Monticelli et al. 2005) NOTCH3 (Ghisi et al. 2011); DKC1, AKT2 (Watanabe et al. 2011)	

(continued)

Table 6.1 (continued)

MicroRNA	Hematopoiesis	Malignancy	Target	Animal model
miR-181	B/T-cell development	AML, CLL	AID, BCL2, CD69, TCRα, DUSP5, SHP2, PTPN22 (de Yebenes et al. 2008; Teng et al. 2008; Neilson et al. 2007)	
miR-221/222	Erythropoiesis		KIT (Felli et al. 2005)	
miR-15a/16-1		CLL, B-ALL, ALCL	BCL2, MCL1, CDK6 (Cimmino et al. 2005); TP53 (Fabbri et al. 2011); HIF-1A (Dejean et al. 2011); CCND1(Chen et al. 2008; Deshpande et al. 2009)	NZB strain (Raveche et al. 2007); MDR and miR-15a/16-1 deletions (Klein et al. 2010); CDR deletion (Lia et al. 2012)
miR-21		B-cell lymphoma, NK/T lymphoma, SzS	PDCD4, PTEN (Yamanaka et al. 2009); STAT3 (van der Fits et al. 2011); ANP32A, SMARCA4 (Schramedei et al. 2011)	Inducible miR-21 knock-in model (Medina et al. 2010)
miR-29a	Early HSC development	AML, CLL, MCL	HBP1 (Han et al. 2010); TCL1 (Pekarsky et al. 2006)	
miR-124a	Myelopoiesis	AML, ALL	EVI1, CEBPA (Hackanson et al. 2008); CDK6 (Agirre et al. 2012)	

Abbreviations: *AID* activation-induced cytidine deaminase, *ALCL* anaplastic large cell lymphoma, *ALL* acute lymphoblastic leukemia, *AML* acute myeloid leukemia, *ANP32A* acidic nuclear phosphoprotein 32 family member A, *BCL-2* B-cell leukemia/lymphoma-2, *BL* Burkitt lymphoma, *CCND1* Cyclin D1, *CDK6* Cyclin-dependent protein kinase 6, *CEBPA* CCAAT/enhancer binding protein alpha, *CLL* chronic lymphocytic leukemia, *CML* chronic myeloid leukemia, *DKC1* dyskeratosis congenita gene 1, *DLBCL* diffuse large B-cell lymphoma, *DUSP5* dual-specificity phosphatase 5, *E2F1* E2 transcription factor family-1, *EVI1* ecotropic virus integration 1, *FOXP1* Forkhead Box P1, *HBP1* HMG box-containing protein 1, *HGAL* human germinal-center associated lymphoma, *HIF-1A* hypoxia-induced factor 1α, *HSC* hematopoietic stem cells, *IRF-4* interferon regulatory factor 4, *LIN28A* Lin-28 Homologue A, *LMO2* LIM-only protein 2, *MALT* mucosa-associated lymphoid tissue, *MCL1* myeloid cell leukemia 1, *MDS* myelodysplastic syndrome, *MEF2C* myocyte enhancer factor 2C, *MYB* myeloblastosis, *NFIA* nuclear factor I-A, *NK* natural killer, *NZB strain* New Zealand Black strain, *PDCD4* programmed cell death 4, *PHLPP* PH domain leucine-rich repeat protein phosphatase 2, *PIK3R1* phosphatidylinositol 3-kinase regulatory subunit alpha, *PP2A* protein phosphatase 2, *PRDM1* PR domain zinc finger protein 1, *PRKAA1* AMP-activated protein kinase catalytic subunit alpha-1, *PTPN22* protein tyrosine phosphatase, non-receptor type 22, *PTEN* phosphatase and tensin homolog, *RTKN2* rabbit polyclonal anti–Rhotekin 2, *SHIP* the SH2-domain-containing inositol 5-phosphatase, *SIRT1* Sirtuin 1, *SMARCA4* SWI/SNF-related, matrix-associated, actin-dependent regulator chromatin, subfamily A, member 4, *STAT3* signal transducer and activator of transcription 3, *SzS* sézary syndrome, *TCL1* T-cell leukemia 1, *TCRα* T cell receptor alpha, *TNFAIP3* tumor necrosis factor-alpha-induced protein 3

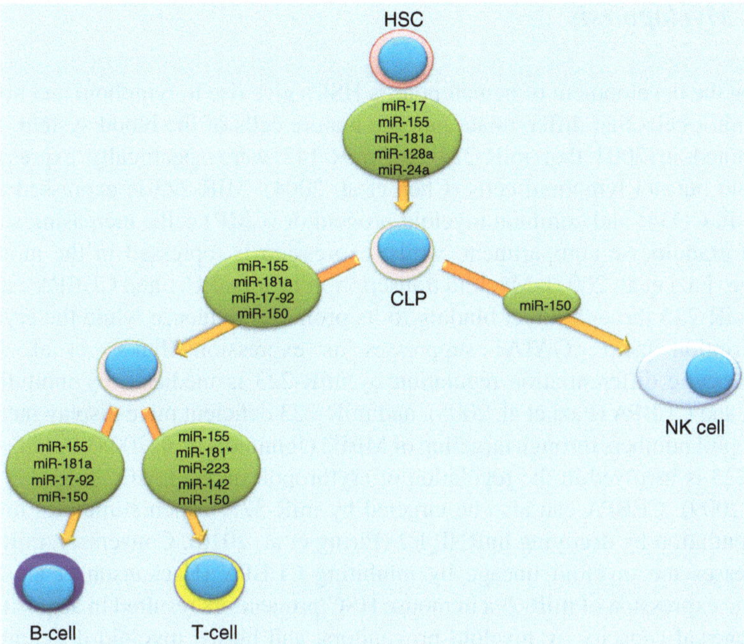

Fig. 6.1 Role of miRNAs in lymphopoiesis. For more details, see the text. *HSC* hematopoietic stem cell, *CLP* common lymphoid progenitor

controlled by a complex network of extrinsic and intrinsic stimuli, signaling pathways, growth factors, cytokines, transcription factors and other molecular components. MiRNAs have can target a great many of these factors and more generally can determine HSC fate, differentiation state and function, self-renewal ability, apoptosis levels as well as the balance of myeloid and lymphoid progenitor cells (Georgantas et al. 2007).

The general necessity of miRNAs in hematopoiesis has been demonstrated by multiple animal models. For example, deletion of Dicer in the thymus caused severe block in peripheral CD8⁺ development and reduced numbers of CD4⁺ cells which when stimulated underwent increased apoptosis and proliferated poorly (Cobb et al. 2005; Muljo et al. 2005). In contrast, when Dicer was deleted in HSCs this led to an increase in apoptosis and a reduction in hematopoietic ability (Guo et al. 2010). At the early progenitor stage, CD34⁺ HSCs derived either from bone marrow or peripheral blood compartments were found to commonly express 33 different miRNAs (Georgantas et al. 2007). Based on the predicted target genes for these miRNAs, as well as functional validation of 18 of these genes, it was proposed that miR-17, miR-24, miR-146, miR-155, miR-128 and miR-181 prevent the differentiation of early stage progenitor cells, whilst miR-16, miR-103 and miR-107 act later on, and miR-221, miR-222 and miR-223 control the terminal stages of hematopoietic development (Fig. 6.1).

2.1 Myelopoiesis

During the development of hematopoiesis HSCs give rise to lymphoid and myeloid progenitor cells that differentiate into the mature cells of the blood system. It was first noted in 2004 that miR-223 and miR-142 were specifically expressed in myeloid but not lymphoid cells (Chen et al. 2004). MiR-223 is expressed at low levels in $CD34^+$ and common myeloid progenitor (CMP) cells, increasing steadily in the granulocyte compartment, whilst expression is repressed in the monocyte lineage (Fazi et al. 2007). Myeloid transcription factors, PU.1 and CEBPA can activate miR-223 through direct binding to its promoter sequence while the erythroid transcription factor, GATA1 suppresses its expression (Fukao et al. 2007). Granulocytic differentiation regulation by miR-223 is mediated by inhibition of NFIA and CEBPA (Fazi et al. 2005), and miR-223 deficient mice display increased neutrophil numbers through targeting of MEF2 (Johnnidis et al. 2008). Additionally, miR-223 is involved in the regulation of erythropoiesis via LMO2 targeting (Felli et al. 2009). CEBPA can also be targeted by miR-328, which stimulates myeloid differentiation by decoying hnRNP E2 (Eiring et al. 2010). Conversely, miR-124a suppresses the myeloid lineage by inhibiting CEBPA (Hackanson et al. 2008). Ectopic expression of miR-29a in mouse HSC/progenitors resulted in acquisition of self-renewal capacity by myeloid progenitors, and biased myeloid differentiation (Han et al. 2010).

MiR-125 is highly expressed in human myeloblasts and promyelocytes, myeloid progenitor cell line 32D (Gerrits et al. 2012), and bone marrow-derived monocytes in response to antigen stimulation (Monk et al. 2010), but down-regulated in neutrophils (Sun et al. 2011). When miR-125b was inhibited in murine HSCs, lower levels of myeloid cells resulted and this effect could be phenocopied by inhibiting LIN28A (Chaudhuri et al. 2012).

The role of miRNA expression in erythropoiesis first came from a study of Choong et al who identified changes associated with progressing erythroid maturation in $CD34^+$ and K562 models (Choong et al. 2007). Afterwards many miRNA profiling studies have carried out [reviewed in Lawrie (2010)], however little consensus has been reached, primarily due to the many differing models of erythropoiesis used between studies. Felli and colleagues identified miR-221 and miR-222 as being highly expressed in $CD34^+$ cells but down-regulated in response to unilineage erythroid differentiation and targeting KIT (Felli et al. 2005). When miR-150 was expressed in $CD34^+$ cells cultured in the presence of thrombopoietin and Epo, an eight-fold increase of megakaryocytes was observed and transplanted bone marrow cells expressing miR-150 increased levels of megakaryocyte- colony forming units (CFUs) coupled with a decrease in erythroid-CFUs (Lu et al. 2008).

The miR-451/miR-144 cluster has been found to play a crucial role in erythropoiesis, with silencing of mir-451 in zebrafish causing a significant decrease in levels of erythroid markers coupled with severe anemia (Dore et al. 2008). MiR-451 is specifically expressed in erythrocytes and reticulocytes (Merkerova et al. 2008), and is the most highly expressed miRNA in mature erythrocytes, increasing >270-fold during erythroid-CFU culturing (Masaki et al. 2007). Expression of this cluster is directly regulated by GATA1 binding (Dore et al. 2008). In turn, miR-451 up-regulates levels of β-globin (HBB) (Zhan et al. 2007), and down-regulates GATA2 (Pase et al. 2009).

2.2 Lymphopoiesis

The first indication that miRNAs were important to lymphopoiesis came from Chen et al. in 2004 who showed that reconstitution of lethally irradiated mice with miR-181 expressing HSCs led to a significant increase in B-cells and cytotoxic $CD8^+$ T-cells (Chen et al. 2004). Subsequently, miR-181 has also been shown to regulate levels of CD69, BCL2 and TCRα in T cell development (Neilson et al. 2007), as well as being responsible for T-cell receptor sensitivity (Li et al. 2007).

MiR-155 deletion in mice caused them to become immunodeficient, with B cells that produced reduced levels of immunoglobulins in response to antigen treatment, and T cells that produced decreased levels of IL2 and IFNG. Both of these effects were caused by PU.1 targeting (Rodriguez et al. 2007; Thai et al. 2007). *In vitro* activation of B cells or $CD4^+$ T cells, strongly up-regulates miR-155 expression, whilst miR-155 deficient activated B cells express a third of normal levels of TNF and lymphotoxin, and T cell differentiation becomes biased towards the T_h2 phenotype (Vigorito et al. 2007).

Another important regulator of lymphocyte differentiation is the miR-17-92 cluster whose targeted deletion leads to a blockage in pro- to pre-B cell development via BIM targeting (Ventura et al. 2008). As well as BIM, members of this cluster also target PTEN, PP2A and AMP-activated kinase (PRKAA1), all of which play important roles in immune cell development (Xiao et al. 2008; Ventura et al. 2008; Mu et al. 2009; Mavrakis et al. 2010). Similar to the miR-17-92 cluster, ectopic expression of miR-34a in HSCs inhibit the transition of pro- to pre-B cell by FOXP1 inhibition (Rao et al. 2010), as does miR-150 expression via MYB down-regulation (Zhou et al. 2007). miR-150 has also been shown to be down-regulated in response to T-cell stimulation by both T_h1 or T_h2 subsets (Monticelli et al. 2005), and linked to megakaryopoiesis, driving megakaryocyte-erythrocyte progenitor cells to differentiate into megakaryocytes instead of erythrocytes (Lu et al. 2008) (Fig. 6.1).

3 MicroRNA Expression in Myeloid Malignancies

The general importance of miRNA dysregulation to the pathogenesis of myeloid disorders is suggested by the fact that more than 70 % of all human miRNAs are encoded within regions of recurrent copy-number alterations in myelodysplastic syndrome (MDS) and acute myeloid leukemia (AML) cell lines (Starczynowski et al. 2011). Several mouse models have been developed that suggest a direct link between miRNA deregulation and myeloid disorders. Mice with deleted Dicer1 in osteoprogenitor cells developed abnormal hematopoiesis, MDS and eventual AML (Raaijmakers et al. 2010). In another study, two miRNAs (miR-145 and miR-146a) encoded on 5q, a frequently deleted locus in myeloid disorders, were deleted in murine HSCs. These mice developed mild neutropenia, megakaryocytic dysplasia and a subset progressed to a fatal myeloid malignancy (Starczynowski et al. 2010).

Many studies have demonstrated that miRNAs are abnormally expressed in myeloid malignancies compared to counterpart controls; most commonly in AML [reviewed by Marcucci et al. (2011)], but also in MDS [reviewed by Rhyasen and Starczynowski (2012)] and chronic myeloid leukemia (CML) (Agirre et al. 2008, Babashah et al. 2013).

In AML, in particular, differences in miRNA expression between common cytogenetic subtypes have been widely documented including those harboring favorable-risk abnormalities such as t(8;21) (Jongen-Lavrencic et al. 2008; Li et al. 2008; Dixon-McIver et al. 2008; Cammarata et al. 2010), inv(16), t(15;17) (Jongen-Lavrencic et al. 2008, Li et al. 2008, Dixon-McIver et al. 2008), and inv(16), and those with less favorable-risk subtypes such as t(11q23)/MLL (Jongen-Lavrencic et al. 2008; Li et al. 2008; Dixon-McIver et al. 2008; Garzon et al. 2008b) and trisomy 8 cases (Garzon et al. 2008b). Additionally, karyotype normal AML patients also have distinctive miRNA expression patterns associated with recurrent molecular abnormalities including FLT3-ITD (Jongen-Lavrencic et al. 2008; Cammarata et al. 2010; Garzon et al. 2008a, b) and MLL duplications, mutations in WT1, IDH1, IDH2, NPM1 (Jongen-Lavrencic et al. 2008; Garzon et al. 2008a) and CEPBA (Jongen-Lavrencic et al. 2008), as well as high expression of BAALC (Langer et al. 2008), ERG and MN1 (Langer et al. 2009) genes. Microarray analyses of samples from 122 AML patients were used to define signatures associated with cytogenetically favorable-risk groups (Garzon et al. 2008b). Importantly, these findings were tested in an independent cohort of 60 AML patients using qRT-PCR and levels of miR-191 and miR-199a were found to be independent predictors of prognosis by multivariate analysis. Recently, up-regulation of miR-181a has been linked with a better prognostic outcome for karyotype normal AML patients (n = 187) (Schwind et al. 2010).

In MDS, 12 miRNAs were identified as being differentially expressed between high and low risk MDS patients (Erdogan et al. 2011). Another study found that high levels of miR-150 was associated with good cytogenetic-risk groups in MDS patients (n = 52) (Hussein et al. 2010). Up-regulation of the miR-17-92 cluster was found to be associated with chronic phase but not blast crisis CML patients (Venturini et al. 2007). A miRNA signature was described that could distinguish between CML patients that were responsive and those that were refractory to imatinib treatment (San Jose-Eneriz et al. 2009). miR-203, an epigenetically silenced miRNA in CML, can regulate BCL-ABL expression (Bueno et al. 2008), and miR-138, down-regulated in CML tumor cells but restored in response to imatinib treatment, was recently shown to act as a tumor suppressor in CML cells thorough targeting of BCR-ABL expression (Xu et al. 2012).

It has been reported that the miRNA expression profile of acute promyelocytic leukemia (APL) patients (and cell lines) treated with retinoic acid cause the up-regulation of several miRNAs including miR-223, miR-15a, miR-16, miR-342, miR-107 and let-7a (Garzon et al. 2007). Gauwerky et al. (1989) identified multiple chromosomal aberrations in a patient with APL including a t(8;17) translocation that fused the promoter and 5′ region of miR-142 to a truncated MYC gene.

4 MicroRNA Expression in Lymphoid Malignancies

4.1 Lymphoid Leukemias

The first report of miRNA dysregulation in CLL, or indeed any cancer, came in 2002 from the laboratory of Carlo Croce who made the observation that CLL patients that harbored a 13q14 deletion, also had lower levels of miR-15a and miR-16, miRNAs that are encoded at this locus (Calin et al. 2002). Subsequent research has demonstrated that down-regulation of these miRNAs results in a myriad of cellular responses mediated by targeting of BCL2, MCL1 and CDK6 (Cimmino et al. 2005). Additionally, a circuit between miR-15a/16-1, miR-34b/c, TP53 and ZAP-70 has been proposed to operate in CLL recently, as TP53 can induce expression these miRNAs, and in turn miR-15a/16-1 targets TP53, whilst ZAP-70 is inhibited by miR-34b/c (Fabbri et al. 2011).

The first *in vivo* functional evidence that CLL was linked to miRNA dysregulation came from the New Zealand Black (NZB) mouse strain that has a propensity to develop a disease akin to human CLL, and consequently was the subject of a genome-wide linkage program in order to identify disease-associated loci (Scaglione et al. 2007). A locus on chromosome 14 that was linked to lymphoproliferative disease in the mice was found to be synteneous with the human 13q14 locus, and to encode for the mir-15a/16-1 cluster. Moreover, levels of miR-16 were decreased in the lymphoid tissue of these mice. Sequencing of this region identified a recurrent point mutation, present in the NZB strain but not other closely-related strains, in a flanking region of the mir-15a/16-1 cluster that was very similar to a mutation previously identified in two CLL patients with a familial history of CLL and breast cancer (Calin et al. 2005).

In order to better define the role of the miR-15a/16-1 cluster *in vivo* mouse models have been created with deletions covering either the miR-15a/16-1 cluster alone, or also encompassing the adjacent DLEU2 gene (Klein et al. 2010). Both sets of homozygous mice developed a disease similar to human CLL, but those mice with the larger deletion had a more aggressive disease phenotype suggesting that components other than miR-15a/16-1 cluster contribute to CLL pathogenesis. Consistent with this idea, when the deletion was extended to include adjacent DLEU7 and RNASEH2B genes, mice developed an even more aggressive disease phenotype (Lia et al. 2012). Recently it has been shown that the miR-15a/16-1 cluster is also epigenetically silenced in 30–35 % of CLL samples, associated with an over-expression of HDAC1-3 perhaps suggesting that these patients could benefit from HDAC-inhibitor based therapies (Sampath et al. 2012).

Many miRNA signatures have been linked to CLL diagnosis and prognosis. For example, a 13 miRNA signature was described that could distinguish between aggressive and indolent forms of CLL (Calin et al. 2004a), and in another study a 32 gene signature was able to discriminate between the most common CLL cytogenetic subgroups (Visone et al. 2009). More recently high miR-21 levels has been associated with poor prognosis in CLL patients harboring the 17p deletion (Rossi et al. 2010), and levels of miR-181b were inversely correlated with disease severity over time leading to its proposal as a potential treatment biomarker for CLL progression (Visone et al. 2011). It has also been suggested that a miRNA signature can be used to predict those CLL patients that are refractory to Fludarabine treatment (Ferracin et al. 2010), and that these patients are characterized by low miR-34a levels even in the absence of p53 alterations (Zenz et al. 2009).

Similar to other leukemias, B-cell acute lymphoblastic leukemia (ALL) tumor cells have a distinct miRNA expression profile from their healthy counterpart ($CD33^+$) cells (Schotte et al. 2011). High levels of miR-128 are associated with ALL and can distinguish between AML and ALL with >98 % accuracy (Mi et al. 2007). Low levels of miR-16 were linked with better prognosis in pediatric ALL cases (Kaddar et al. 2009). Various cytogenetic subgroups of B-ALL including patients with 11q23/MLL, TEL-AML1, BCR-ABL and E2A-PBX1 translocations, and hyperdiploid patients also have distinct miRNA profiles. Let-7b for example, is down-regulated while the miR-17-92 cluster is over-expressed in B-ALL patients with MLL translocations (Schotte et al. 2011; Mi et al. 2010). Differentially expressed miRNAs have also been identified between T-ALL leukemic cells and healthy thymocytes (Schotte et al. 2011), and mice over-expressing miR-19 (part of the miR-17-92 cluster) in concert with Notch1, developed T-ALL faster than those mice expressing Notch1 alone (Mavrakis et al. 2010).

4.2 Lymphomas

4.2.1 Diffuse Large B-Cell Lymphoma

Diffuse Large B-Cell Lymphoma (DLBCL) was one of the first lymphomas to be linked with aberrant miRNA expression, with several reports observing over-expression of miR-155 in this malignancy (Kluiver et al. 2005; Lawrie et al. 2007; Eis et al. 2005). Forced over-expression of miR-155 in mice caused the development of a high grade B-cell lymphoma similar to DLBCL (Costinean et al. 2006); this oncogenicity was mediated by SHIP1 and C/EBPβ targeting (O'Connell et al. 2009; Pedersen et al. 2009; Yamanaka et al. 2009). When miR-155 was expressed using an inducible rather than a constitutive promoter again mice developed lymphoma, however when miR-155 expression was inhibited the tumor quickly receded and after 1 week remarkably mice had no detectable disease manifestations at all (Babar et al. 2012). Ectopic expression of miR-155 suppresses the *in vitro* growth-inhibitory effects of TGF-β1 and BMP2/4 in DLBCL cells via SMAD5 inhibition (Rai et al. 2010), and can regulate the phosphatidylinositol 3-kinase (PI3K)-AKT pathway via targeting of PIK3R1 in DLBCL (Huang et al. 2012). Furthermore, over-expression of this miRNA promotes TNFα-dependent growth of DLBCL cells *in vivo* in xenotransplant models (Pedersen et al. 2009). Interestingly it has recently been shown that SHIP1 is differentially expressed between the two molecular subtypes of DLBCL [activated B cell-like (ABC) and germinal center B cell-like (GC)] (Alizadeh et al. 2000), consistent with previous studies that identified differences in miR-155 expression levels between ABC- and GC-type DLBCL (Eis et al. 2005; Lawrie et al. 2007). Additionally, CD10 a marker of GC-type DLBCL (Hans et al. 2004), and constitutive expression of NF-κB, a hallmark of ABC-type DLBCL (Compagno et al. 2009), have been linked via the miR-155/PU.1 pathway (Thompson et al. 2011). When mice were inoculated with U2932, an ABC-type DLBCL cell line, treatment with exogenous miR-34a reduced tumor growth via targeting of Foxp1 (Craig et al. 2012), a molecule associated with ABC-type DLBCL (Choi et al. 2009), and also linked to high grade transformation of lymphoma via myc-mediated miR-34a repression (Craig et al. 2011a).

MiR-34a is a well described tumor suppressor miRNA that is closely connected with the p53 network in solid tumors (He et al. 2007), and a positive feedback loop exists whereby p53 induces miR-34a expression and in turn miR-34a activates p53 through SIRT1 inhibition (Yamakuchi and Lowenstein 2009).

Over-expression of the miR-17-92 cluster in conjunction with MYC accelerates lymphoma development and increased tumor aggressiveness (He et al. 2005; Tagawa et al. 2007). This effect is believed to be the result of the MYC/miR-17-92/E2F circuit (O'Donnell et al. 2005). MYC up-regulates the miR-17-92 cluster which targets E2F1, whilst conversely pro-proliferative E2F3 regulates the miR-17-92 cluster (Woods et al. 2007). Recently, miR-19 was identified as the key oncogenic component of the miR-17-92 cluster and in the Eμ-myc model was shown activate the Akt-mTOR pathway via antagonizing PTEN leading to promotion of cell survival (Olive et al. 2009).

4.2.2 Follicular Lymphoma

Despite being the most common form of indolent lymphoma, there are relatively few studies dedicated to miRNA expression in follicular lymphoma (FL). The expression levels of 153 miRNAs were measured in 46 FL samples compared to normal lymph nodes or DLBCL cases (Roehle et al. 2008). Our group looked at expression levels of 464 miRNAs in eighteen FL cases in comparison with 80 DLBCL cases and derived a 26-miRNA signature that could differentiate between FL cases and de novo cases of DLBCL (Lawrie et al. 2009). We also identified six miRNAs (miR-223, miR-217, miR-222, miR-221, let-7i and let-7b) that could distinguish FL cases that underwent high grade histological transformation from those that did not.

The miR-17-92 cluster has recently been suggested to be a useful diagnostic differentiator between the potentially confounding diagnostic classifications of GC-DLBCL and grade 3 FL cases (Fassina et al. 2012). Other studies have compared FL with nodal marginal zone lymphoma (NMZL) (Arribas et al. 2012), and follicular hyperplasia patients (Wang et al. 2012). The latter study also identified miRNAs associated with FL patients responsive to PACE chemotherapy, and demonstrated that p21 and SOCS2 were regulated by miR-20a/b and miR-194 in FL cell lines contributing to cell proliferation.

Although most FL cases harbor the t(14;18) translocation, others (~10 %) do not. Recently a study identified 17 miRNAs that distinguished between t(14;18) positive and negative cases including miR-16, miR-26a, miR-101, miR-29c and miR-138 that were associated with increased expression of CHEK1 and decreased expression of TCL1 in t(14;18) negative cases suggesting a late GC-type phenotype for this form of FL (Leich et al. 2011).

4.2.3 Mantle Cell Lymphoma

A number of miRNA signatures have now been described for mantle cell lymphoma (MCL) (Zhao et al. 2010; Di Lisio et al. 2010; Navarro et al. 2009a; Iqbal et al. 2012). The loss of potential miRNA target sites for miR-15/16 and members of the miR-17-92 cluster in the 3′UTR of CCND1 have been suggested to contribute to the pathogenic over-expression of Cyclin D1 in MCL (Chen et al. 2008; Deshpande et al. 2009). Over-expression of members of the miR-17-92 cluster has been associated with high MYC levels in aggressive MCL (Navarro et al. 2009a), and high proliferation gene signature (Iqbal et al. 2012), as well as activation of the PI3K/AKT pathway, and inhibition of chemotherapy-induced apoptosis in MCL cell lines (Chaudhuri et al. 2012). PHLPP2, an important regulator of the PI3K/Akt pathway, was also shown to be a target of the miR-17-92 cluster in addition to PTEN and BIM in MCL (Rao et al. 2012). Inhibition of miR-17-92 expression in an MCL xenotransplant model suppressed the PI3K/Akt pathway and resulted in decreased tumor growth. Also down-regulation of miR-29 was shown to activate CDK4/CDK6, and serve as a potential prognostic marker for this malignancy (Zhao et al. 2010).

4.2.4 Burkitt Lymphoma

The importance of miRNAs to the pathogenesis of Burkitt lymphoma (BL) is suggested by the fact that mice carrying a mutation in the 3′-UTR binding sequence for miR-155 in the activation-induced cytidine deaminase (AID) gene have increased levels of MYC-IgH (t(8;14)) translocations; the characteristic genetic hallmark of BL (Dorsett et al. 2008). MYC regulates and is itself regulated by a large set of miRNAs, leading to a complex regulatory loop that can contribute to lymphomagenesis (Gao et al. 2009; Chang et al. 2008; Bueno et al. 2011). It has even been suggested that MYC over-expression in BL cases that lack the classical t(8;14) translocation could be the result of miRNA deregulation (Leucci et al. 2008). In addition to a functional role, miRNAs may also be useful for a more accurate classification of the group of B-cell lymphomas with intermediate features between DLBCL and BL (Leucci et al. 2008). It has also recently been shown that the different epidemiologic subtypes of BL share a homogenous miRNA profile distinct from that of DLBCL (Lenze et al. 2011). It has been reported that miR-155 is not over-expressed in adult BL cases (Kluiver et al. 2005, 2006; van den Berg et al. 2003), but is in paediatric cases of BL (Metzler et al. 2004). Over-expression of miR-155, in Burkitt lymphoma and post-transplantation lymphoproliferative disorder (PTLD) at least, was proposed to be associated with EBV latency type-III infections (Jiang et al. 2006; Kluiver et al. 2006).

4.2.5 Hodgkin Lymphoma

A number of functional experiments have demonstrated the potential involvement of miRNAs to the pathogenesis of Hodgkin lymphoma (HL). A ribonucleoprotein chromatin immunoprecipitation (RIP-ChIP) approach was used to identify the target genes of aberrantly expressed miRNAs in HL cell lines, and identified an over-representation of genes associated with cell proliferation, apoptosis and the p53 pathway (Tan et al. 2009). In a more focused example, it was demonstrated that JAK2 is directly targeted by miR-135a, and that over-expression of this miRNA in HL cell lines increases apoptotic levels and decreases cell growth via Bcl-xL inhibition (Navarro et al. 2009b). Furthermore, patients with low miR-135a levels were found to have significantly poorer prognostic outcome. Inhibition of let-7 and miR-9 in HL cell lines resulted in reduced levels of PRDM1/BLIMP1 preventing plasma cell differentiation (Nie et al. 2008). miR-9 was also shown to target Dicer and HuR in HL, and inhibition of this miRNA led to a decrease in cytokine production and an impaired ability to attract inflammatory cells (Leucci et al. 2012). Ectopic administration of a miR-9 antagomir led to decreased tumor growth in a xenotransplant model.

4.2.6 Other B-Cell Lymphomas

A number of studies have recently investigated the role of miRNAs in mucosa-associated lymphoid tissue (MALT) lymphoma. For example, a 27-miRNA signature was able to distinguish gastric DLBCL from MALT lymphoma (Craig et al. 2011a), and the same researchers also proposed that transformation from gastritis to MALT lymphoma is epigenetically regulated by methylation of miR-203, consequently suggesting ABL1 as a potential target for treatment of this malignancy (Craig et al. 2011b). In another study a set of five miRNAs (miR-150, miR-550, miR-124a, miR-518b and miR-539) were identified as being differentially expressed in gastritis as opposed to MALT lymphoma (Thorns et al. 2012), and finally high miR-223 levels were found to correlate with increased E2A expression in gastric MALT lymphoma (Liu et al. 2010). Recently a miRNA signature of splenic marginal zone lymphoma (SMZL) cases was reported (Bouteloup et al. 2012).

4.2.7 T-Cell Lymphomas

Compared with B-cell lymphomas, relatively little is known of the role of miRNAs in T-cell lymphoma. Our group provided the first evidence for a functional role of miRNAs in T-cell lymphomas in a study that identified >100 aberrantly expressed miRNAs in the cutaneous T-cell lymphoma (CTCL), Sézary syndrome (SzS) (Ballabio et al. 2010). We identified miR-223 as a potential diagnostic marker (>85 % accuracy) by qRT-PCR for SzS, in both the training cohort and a separate validation cohort that contained patients with a confounding diagnosis [i.e. non-erythrodermic mycosis fungoides (MF)]. We also demonstrated a potential role for miR-342 in the pathogenesis of SzS through its targeting of RANKL which was associated with the protection of SzS cells from apoptosis. The identity of several of these aberrantly expressed miRNAs has now been validated in other independent studies (Narducci et al. 2011; Qin et al. 2012). Our group have also more recently carried out profiling studies on tumor stage MF (van Kester et al. 2011), and cutaneous anaplastic large cell lymphoma (cALCL) (Benner et al. 2012). Subsequently a qRT-PCR based classifier (miR-155, miR-203 and miR-205) has been proposed that can distinguish between the various forms of cutaneous T-cell lymphomas and related benign disorders (Ralfkiaer et al. 2011). Importantly both training (n = 90) and blinded test (n = 58) sets were used in this study. Besides its role in B-cell development, miR-150 also regulates natural killer (NK) cells via Myb targeting, (Bezman et al. 2011) as well as other T-cell subsets via NOTCH3 inhibition (Ghisi et al. 2011). Transduction of miR-150 into NK/T cell lymphoma cells increased apoptotic levels and decreased cell proliferation, effects mediated via targeting of DKC1 and AKT2, leading to a decrease in levels of BIM, p53 and phosphorylated AKT levels. Additionally, over-expression of miR-21 and miR-155, has been demonstrated to activate the PI3K-Akt pathway in NK/T-cell lymphomas (Yamanaka et al. 2009), and over-expression of miR-122 in CTCL induced AKT phosphorylation coupled with a decreased sensitivity to chemotherapy-induced apoptosis as well as inhibition of p53 (Manfe et al. 2012). Additionally miRNAs have been found to be encoded in the avian T cell lymphoma-causing virus, Marek's disease virus (MDV) (Yao et al. 2007; Burnside et al. 2006).

5 Summary and Future Directions

The miRNA field continues to grow at a phenomenal rate and new biological roles for miRNAs are constantly being uncovered. Whilst great effort has been put into identifying and cataloguing aberrantly expressed miRNAs in disease, very little is known about the functional consequences of this dysregulation, and understanding the biological function of identified miRNAs is perhaps the biggest challenge facing the miRNA field at the moment. The primary reason for this is a lack of knowledge about which genes are actually targeted by individual miRNAs and which of these genes are functionally important in specific cellular settings. With very few functionally annotated exceptions, current approaches to this problem primarily rely upon the use of the many predictive computational algorithms available. However, these algorithms typically predict hundreds or even thousands of target genes for each miRNA and in reality perform very poorly. When the most widely used algorithms were tested against experimentally validated miRNA-target gene interactions, sensitivity ranged from just 1.3 to 48.8 % (Sethupathy et al. 2006). Additionally, the degree of overlap between predictions (three algorithms) was found to range from 3.6 to 28.6 % and surprisingly no commonly predicted genes were identified at all when the five most-commonly used algorithms were compared. Importantly, this study showed that even when all five algorithms were used in union only 72 % of experimentally validated miRNA-target gene interactions were predicted. For example, KRAS and HRAS targeting by let-7 (Johnson et al. 2005), or E2F2 and MYC targeting by miR-24 (Lal et al. 2009) are not predicted targets of these algorithms. To compound matters further the function of a particular miRNA is dependent upon cellular context. Indeed, the same miRNA can act as both tumor suppressor and oncogene depending upon the cell type. For example, miR-222 is over-expressed in hepatocarcinoma where it targets tumor suppressor PTEN (Garofalo et al. 2009), but is down-regulated in erythroblastic leukemias where it targets the KIT oncogene (Felli et al. 2005). Consequently much effort has been expended to resolve this issue including techniques to directly measure the miRNA:target gene interface, the so-called 'targetome', in cells under physiologically relevant conditions. Particularly promising is the 'PAR-CLIP' technique (Hafner et al. 2010), that has now been used by several groups to elucidate the targetomes of multiple cellular systems (Skalsky et al. 2012; Lebedeva et al. 2011; Gottwein et al. 2011), however it has not yet been extensively applied to hematology.

MiRNAs show perhaps their greatest and certainly most immediate clinical potential, as novel biomarkers. MiRNA expression profiling can distinguish cancers according to diagnosis and developmental stage of the tumor to a greater degree of accuracy than traditional gene expression analysis, even discriminating between cancers that are poorly separated histologically (Lu et al. 2005). An especially useful characteristic of miRNAs is their remarkable stability which means that they can be robustly measured from routinely prepared formalin-fixed paraffin embedded (FFPE) biopsy material (Lawrie et al. 2007). A further manifestation of their stability is their presence in extracellular biological fluids including blood (Lawrie et al. 2008). We demonstrated for the first time that miRNAs were present in the blood (serum/plasma) of lymphoma patients at differential levels from healthy controls (Lawrie et al. 2008). The speed of miRNA biomarker discovery has been quite astonishing with over 5,000 publications in the last 5 years. However, it should be noted that the reliability of much of this data remains contentious and should be treated with some caution as the degree of discordancy between seemingly identical studies is worrisome, and in reality very few of the biomarkers studies published will ever make it into clinical practice. These discrepancies are probably due to the use of different control populations (unsorted cell types, populations sorted in a differing manner etc.), as well as technical variability (e.g. differing array platforms, statistical analyses and varying cytogenetic/molecular profiling techniques) between studies. Consequently there is a clear need for a standardized approach to be taken in future miRNA biomarker studies in order to rationalize these confounding factors. In particular, a systematic approach should be taken in a similar fashion to that achieved for other 'omic disciplines' (i.e. transcriptomics and genomics).

Perhaps the most promising clinical aspect of miRNAs is their potential as novel therapeutic molecules, either as a tool to modulate target genes associated with disease or by correcting dysfunctional expression of the miRNAs themselves (Babashah and Soleimani 2011). The former approach is particularly attractive in that a single agent (i.e. a miRNA) can be used against multiple targets in a disease pathway or even against the whole pathway (Bui and Mendell 2010). There are two major strategies to therapeutically modulate dysregulated miRNAs in disease; using miRNA mimics to restore physiological levels of miRNAs that are down-regulated (e.g. tumor suppressor miRNAs such as let-7 or miR-34), or the use of miRNA inhibitors targeted against over-expressed miRNAs (e.g. oncomiRs such as miR-21 or miR-155). In addition, indirect methods such as the use of epigenetic drugs like DNA-demethylating agents and histone deacetylase inhibitors may be of potential therapeutic use in re-expressing epigenetically silenced miRNAs (Agirre et al. 2009).

There is now a wealth of *in vivo* animal experiments that have established the proof-of-principle for the therapeutic efficacy of miRNAs in disease, however at present all but a couple of these studies are still at the pre-clinical stage. The major hurdles still to be resolved include the effective targeting of therapy (e.g. tissue-specific delivery, dosage and pharmacodynamics) and safety concerns (e.g. off-target effects, RNA-mediated immunostimulation and the use of viral vectors). That said, this is an area very much still in its infancy that is almost certain to flourish in the near future, and promises to add to the current arsenal of therapies available to the hematologist in their continual fight against disease. Whatever happens, the future for miRNAs in hematology is very promising, and we should remember that we are only at the very beginning of our understanding of non-coding (nc) RNA and that in reality miRNAs represent the tip of the ncRNA 'iceberg'. Indeed, although ~75 % of the human genome is transcribed (Djebali et al. 2012), the protein-encoding portion of the genome only accounts for 1.5 % (Alexander et al. 2010), whilst miRNAs represent another 1.8 % (Djebali et al. 2012). There is now emerging evidence that ncRNA species other than miRNAs are essential for both physiological function and development, as well as playing a fundamental role in disease (Mercer et al. 2009; Esteller 2011). Although relative to miRNAs, the study of other ncRNA molecules is very limited, many classes of ncRNAs are now recognized including short ncRNAs such as miRNAs, piRNAs and tiRNAs; mid-size ncRNAs such as snoRNAs, PASRs, TSSa-RNAs and PROMPTs; and long ncRNAs (lncRNAs) (Harries 2012; Esteller 2011). In particular, lncRNAs have been found to be more cell-type and tissue-type specific than protein coding genes and miRNAs (Cabili et al. 2011). It therefore seems highly likely that these molecules will become the next frontier of ncRNA discovery.

Acknowledgements I apologize to the authors of the many studies that were not included in this review due to space limitations. The author is supported by funding from the Starmer-Smith memorial lymphoma fund, and Ikerbasque, the Basque Foundation for Science.

References

Agirre X, Jimenez-Velasco A, San Jose-Eneriz E, Garate L, Bandres E, Cordeu L et al (2008) Down-regulation of hsa-miR-10a in chronic myeloid leukemia CD34+ cells increases USF2-mediated cell growth. Mol Cancer Res 6(12):1830–1840

Agirre X, Vilas-Zornoza A, Jimenez-Velasco A, Martin-Subero JI, Cordeu L, Garate L et al (2009) Epigenetic silencing of the tumor suppressor microRNA Hsa-miR-124a regulates CDK6 expression and confers a poor prognosis in acute lymphoblastic leukemia. Cancer Res 69(10):4443–4453

Agirre X, Martinez-Climent JA, Odero MD, Prosper F (2012) Epigenetic regulation of miRNA genes in acute leukemia. Leukemia 26(3):395–403
Alexander RP, Fang G, Rozowsky J, Snyder M, Gerstein MB (2010) Annotating non-coding regions of the genome. Nat Rev Genet 11(8):559–571
Alizadeh AA, Eisen MB, Davis RE, Ma C, Lossos IS, Rosenwald A et al (2000) Distinct types of diffuse large B-cell lymphoma identified by gene expression profiling. Nature 403(6769):503–511
Arribas AJ, Campos-Martin Y, Gomez-Abad C, Algara P, Sanchez-Beato M, Rodriguez-Pinilla MS et al (2012) Nodal marginal zone lymphoma: gene expression and miRNA profiling identify diagnostic markers and potential therapeutic targets. Blood 119(3):e9–e21
Babar IA, Cheng CJ, Booth CJ, Liang X, Weidhaas JB, Saltzman WM et al (2012) Nanoparticle-based therapy in an in vivo microRNA-155 (miR-155)-dependent mouse model of lymphoma. Proc Natl Acad Sci U S A 109(26):E1695–E1704
Babashah S, Soleimani M (2011) The oncogenic and tumour suppressive roles of microRNAs in cancer and apoptosis. Eur J Cancer 47(8):1127–1137
Babashah S, Sadeghizadeh M, Tavirani MR, Farivar S, Soleimani M (2012) Aberrant microRNA expression and its implications in the pathogenesis of leukemias. Cell Oncol (Dordr) 35(5):317–334
Babashah S, Sadeghizadeh M, Hajifathali A, Tavirani MR, Zomorod MS, Ghadiani M et al (2013) Targeting of the signal transducer Smo links microRNA-326 to the oncogenic Hedgehog pathway in CD34+ CML stem/progenitor cells. Int J Cancer 133(3):579–589
Ballabio E, Mitchell T, van Kester MS, Taylor S, Dunlop HM, Chi J et al (2010) MicroRNA expression in Sezary syndrome: identification, function, and diagnostic potential. Blood 116(7):1105–1113
Benner MF, Ballabio E, van Kester MS, Saunders NJ, Vermeer MH, Willemze R et al (2012) Primary cutaneous anaplastic large cell lymphoma shows a distinct miRNA expression profile and reveals differences from tumor-stage mycosis fungoides. Exp Dermatol 21(8):632–634
Bezman NA, Chakraborty T, Bender T, Lanier LL (2011) miR-150 regulates the development of NK and iNKT cells. J Exp Med 208(13):2717–2731
Bouteloup M, Verney A, Rachinel N, Callet-Bauchu E, Ffrench M, Coiffier B et al (2012) MicroRNA expression profile in splenic marginal zone lymphoma. Br J Haematol 156(2):279–281
Bueno MJ, Perez de Castro I, Gomez de Cedron M, Santos J, Calin GA, Cigudosa JC et al (2008) Genetic and epigenetic silencing of microRNA-203 enhances ABL1 and BCR-ABL1 oncogene expression. Cancer Cell 13(6):496–506
Bueno MJ, Gomez de Cedron M, Gomez-Lopez G, Perez de Castro I, Di Lisio L, Montes-Moreno S et al (2011) Combinatorial effects of microRNAs to suppress the Myc oncogenic pathway. Blood 117(23):6255–6266
Bui TV, Mendell JT (2010) Myc: maestro of microRNAs. Genes Cancer 1(6):568–575
Burnside J, Bernberg E, Anderson A, Lu C, Meyers BC, Green PJ et al (2006) Marek's disease virus encodes MicroRNAs that map to meq and the latency-associated transcript. J Virol 80(17):8778–8786
Cabili MN, Trapnell C, Goff L, Koziol M, Tazon-Vega B, Regev A et al (2011) Integrative annotation of human large intergenic noncoding RNAs reveals global properties and specific subclasses. Genes Dev 25(18):1915–1927
Calin GA, Dumitru CD, Shimizu M, Bichi R, Zupo S, Noch E et al (2002) Frequent deletions and down-regulation of micro- RNA genes miR15 and miR16 at 13q14 in chronic lymphocytic leukemia. Proc Natl Acad Sci U S A 99(24):15524–15529
Calin GA, Liu CG, Sevignani C, Ferracin M, Felli N, Dumitru CD et al (2004a) MicroRNA profiling reveals distinct signatures in B cell chronic lymphocytic leukemias. Proc Natl Acad Sci U S A 101(32):11755–11760
Calin GA, Sevignani C, Dumitru CD, Hyslop T, Noch E, Yendamuri S et al (2004b) Human microRNA genes are frequently located at fragile sites and genomic regions involved in cancers. Proc Natl Acad Sci U S A 101(9):2999–3004
Calin GA, Ferracin M, Cimmino A, Di Leva G, Shimizu M, Wojcik SE et al (2005) A microRNA signature associated with prognosis and progression in chronic lymphocytic leukemia. N Engl J Med 353(17):1793–1801

Cammarata G, Augugliaro L, Salemi D, Agueli C, La Rosa M, Dagnino L et al (2010) Differential expression of specific microRNA and their targets in acute myeloid leukemia. Am J Hematol 85(5):331–339
Chang TC, Yu D, Lee YS, Wentzel EA, Arking DE, West KM et al (2008) Widespread microRNA repression by Myc contributes to tumorigenesis. Nat Genet 40(1):43–50
Chaudhuri AA, So AY, Mehta A, Minisandram A, Sinha N, Jonsson VD et al (2012) Oncomir miR-125b regulates hematopoiesis by targeting the gene Lin28A. Proc Natl Acad Sci U S A 109(11):4233–4238
Chen CZ, Li L, Lodish HF, Bartel DP (2004) MicroRNAs modulate hematopoietic lineage differentiation. Science 303(5654):83–86
Chen RW, Bemis LT, Amato CM, Myint H, Tran H, Birks DK et al (2008) Truncation in CCND1 mRNA alters miR-16-1 regulation in mantle cell lymphoma. Blood 112(3):822–829
Choi WW, Weisenburger DD, Greiner TC, Piris MA, Banham AH, Delabie J et al (2009) A new immunostain algorithm classifies diffuse large B-cell lymphoma into molecular subtypes with high accuracy. Clin Cancer Res 15(17):5494–5502
Choong ML, Yang HH, McNiece I (2007) MicroRNA expression profiling during human cord blood-derived CD34 cell erythropoiesis. Exp Hematol 35(4):551–564
Cimmino A, Calin GA, Fabbri M, Iorio MV, Ferracin M, Shimizu M et al (2005) miR-15 and miR-16 induce apoptosis by targeting BCL2. Proc Natl Acad Sci U S A 102(39):13944–13949
Cobb BS, Nesterova TB, Thompson E, Hertweck A, O'Connor E, Godwin J et al (2005) T cell lineage choice and differentiation in the absence of the RNase III enzyme Dicer. J Exp Med 201(9):1367–1373
Compagno M, Lim WK, Grunn A, Nandula SV, Brahmachary M, Shen Q et al (2009) Mutations of multiple genes cause deregulation of NF-kappaB in diffuse large B-cell lymphoma. Nature 459(7247):717–721
Costinean S, Zanesi N, Pekarsky Y, Tili E, Volinia S, Heerema N et al (2006) Pre-B cell proliferation and lymphoblastic leukemia/high-grade lymphoma in E(mu)-miR155 transgenic mice. Proc Natl Acad Sci U S A 103(18):7024–7029
Craig VJ, Cogliatti SB, Imig J, Renner C, Neuenschwander S, Rehrauer H et al (2011a) Myc-mediated repression of microRNA-34a promotes high-grade transformation of B-cell lymphoma by dysregulation of FoxP1. Blood 117(23):6227–6236
Craig VJ, Cogliatti SB, Rehrauer H, Wundisch T, Muller A (2011b) Epigenetic silencing of microRNA-203 dysregulates ABL1 expression and drives Helicobacter-associated gastric lymphomagenesis. Cancer Res 71(10):3616–3624
Craig VJ, Tzankov A, Flori M, Schmid CA, Bader AG, Muller A (2012) Systemic microRNA-34a delivery induces apoptosis and abrogates growth of diffuse large B-cell lymphoma in vivo. Leukemia 26(11):2421–2424
Croce CM (2009) Causes and consequences of microRNA dysregulation in cancer. Nat Rev Genet 10(10):704–714
Dagan LN, Jiang X, Bhatt S, Cubedo E, Rajewsky K, Lossos IS (2012) miR-155 regulates HGAL expression and increases lymphoma cell motility. Blood 119(2):513–520
de Yebenes VG, Belver L, Pisano DG, Gonzalez S, Villasante A, Croce C et al (2008) miR-181b negatively regulates activation-induced cytidine deaminase in B cells. J Exp Med 205(10):2199–2206
Dejean E, Renalier MH, Foisseau M, Agirre X, Joseph N, de Paiva GR et al (2011) Hypoxia-microRNA-16 downregulation induces VEGF expression in anaplastic lymphoma kinase (ALK)-positive anaplastic large-cell lymphomas. Leukemia 25(12):1882–1890
Deshpande A, Pastore A, Deshpande AJ, Zimmermann Y, Hutter G, Weinkauf M et al (2009) 3′UTR mediated regulation of the cyclin D1 proto-oncogene. Cell Cycle 8(21):3584–3592
Di Lisio L, Gomez-Lopez G, Sanchez-Beato M, Gomez-Abad C, Rodriguez ME, Villuendas R et al (2010) Mantle cell lymphoma: transcriptional regulation by microRNAs. Leukemia 24(7):1335–1342
Dixon-McIver A, East P, Mein CA, Cazier JB, Molloy G, Chaplin T et al (2008) Distinctive patterns of microRNA expression associated with karyotype in acute myeloid leukaemia. PLoS One 3(5):e2141

Djebali S, Davis CA, Merkel A, Dobin A, Lassmann T, Mortazavi A et al (2012) Landscape of transcription in human cells. Nature 489(7414):101–108
Dore LC, Amigo JD, Dos Santos CO, Zhang Z, Gai X, Tobias JW et al (2008) A GATA-1-regulated microRNA locus essential for erythropoiesis. Proc Natl Acad Sci U S A 105(9):3333–3338
Dorsett Y, McBride KM, Jankovic M, Gazumyan A, Thai TH, Robbiani DF et al (2008) MicroRNA-155 suppresses activation-induced cytidine deaminase-mediated Myc-Igh translocation. Immunity 28(5):630–638
Eiring AM, Harb JG, Neviani P, Garton C, Oaks JJ, Spizzo R et al (2010) miR-328 functions as an RNA decoy to modulate hnRNP E2 regulation of mRNA translation in leukemic blasts. Cell 140(5):652–665
Eis PS, Tam W, Sun L, Chadburn A, Li Z, Gomez MF et al (2005) Accumulation of miR-155 and BIC RNA in human B cell lymphomas. Proc Natl Acad Sci U S A 102(10):3627–3632
Erdogan B, Facey C, Qualtieri J, Tedesco J, Rinker E, Isett RB et al (2011) Diagnostic microRNAs in myelodysplastic syndrome. Exp Hematol 39(9):915–926, e2
Esteller M (2011) Non-coding RNAs in human disease. Nat Rev Genet 12(12):861–874
Fabbri M, Bottoni A, Shimizu M, Spizzo R, Nicoloso MS, Rossi S et al (2011) Association of a microRNA/TP53 feedback circuitry with pathogenesis and outcome of B-cell chronic lymphocytic leukemia. JAMA 305(1):59–67
Fassina A, Marino F, Siri M, Zambello R, Ventura L, Fassan M et al (2012) The miR-17-92 microRNA cluster: a novel diagnostic tool in large B-cell malignancies. Lab Invest 92(11):1574–1582
Fazi F, Rosa A, Fatica A, Gelmetti V, De Marchis ML, Nervi C et al (2005) A minicircuitry comprised of microRNA-223 and transcription factors NFI-A and C/EBPalpha regulates human granulopoiesis. Cell 123(5):819–831
Fazi F, Racanicchi S, Zardo G, Starnes LM, Mancini M, Travaglini L et al (2007) Epigenetic silencing of the myelopoiesis regulator microRNA-223 by the AML1/ETO oncoprotein. Cancer Cell 12(5):457–466
Felli N, Fontana L, Pelosi E, Botta R, Bonci D, Facchiano F et al (2005) MicroRNAs 221 and 222 inhibit normal erythropoiesis and erythroleukemic cell growth via kit receptor down-modulation. Proc Natl Acad Sci U S A 102(50):18081–18086
Felli N, Pedini F, Romania P, Biffoni M, Morsilli O, Castelli G et al (2009) MicroRNA 223-dependent expression of LMO2 regulates normal erythropoiesis. Haematologica 94(4):479–486
Ferracin M, Zagatti B, Rizzotto L, Cavazzini F, Veronese A, Ciccone M et al (2010) MicroRNAs involvement in fludarabine refractory chronic lymphocytic leukemia. Mol Cancer 9:123
Friedman RC, Farh KK, Burge CB, Bartel DP (2009) Most mammalian mRNAs are conserved targets of microRNAs. Genome Res 19(1):92–105
Fukao T, Fukuda Y, Kiga K, Sharif J, Hino K, Enomoto Y et al (2007) An evolutionarily conserved mechanism for microRNA-223 expression revealed by microRNA gene profiling. Cell 129(3):617–631
Gao P, Tchernyshyov I, Chang TC, Lee YS, Kita K, Ochi T et al (2009) c-Myc suppression of miR-23a/b enhances mitochondrial glutaminase expression and glutamine metabolism. Nature 458(7239):762–765
Garofalo M, Di Leva G, Romano G, Nuovo G, Suh SS, Ngankeu A et al (2009) miR-221&222 regulate TRAIL resistance and enhance tumorigenicity through PTEN and TIMP3 downregulation. Cancer Cell 16(6):498–509
Garzon R, Pichiorri F, Palumbo T, Visentini M, Aqeilan R, Cimmino A et al (2007) MicroRNA gene expression during retinoic acid-induced differentiation of human acute promyelocytic leukemia. Oncogene 26(28):4148–4157
Garzon R, Garofalo M, Martelli MP, Briesewitz R, Wang L, Fernandez-Cymering C et al (2008a) Distinctive microRNA signature of acute myeloid leukemia bearing cytoplasmic mutated nucleophosmin. Proc Natl Acad Sci U S A 105(10):3945–3950
Garzon R, Volinia S, Liu CG, Fernandez-Cymering C, Palumbo T, Pichiorri F et al (2008b) MicroRNA signatures associated with cytogenetics and prognosis in acute myeloid leukemia. Blood 111(6):3183–3189

Gauwerky CE, Huebner K, Isobe M, Nowell PC, Croce CM (1989) Activation of MYC in a masked t(8;17) translocation results in an aggressive B-cell leukemia. Proc Natl Acad Sci U S A 86(22):8867–8871

Georgantas RW 3rd, Hildreth R, Morisot S, Alder J, Liu CG, Heimfeld S et al (2007) CD34+ hematopoietic stem-progenitor cell microRNA expression and function: a circuit diagram of differentiation control. Proc Natl Acad Sci U S A 104(8):2750–2755

Gerrits A, Walasek MA, Olthof S, Weersing E, Ritsema M, Zwart E et al (2012) Genetic screen identifies microRNA cluster 99b/let-7e/125a as a regulator of primitive hematopoietic cells. Blood 119(2):377–387

Ghisi M, Corradin A, Basso K, Frasson C, Serafin V, Mukherjee S et al (2011) Modulation of microRNA expression in human T-cell development: targeting of NOTCH3 by miR-150. Blood 117(26):7053–7062

Gottwein E, Corcoran DL, Mukherjee N, Skalsky RL, Hafner M, Nusbaum JD et al (2011) Viral microRNA targetome of KSHV-infected primary effusion lymphoma cell lines. Cell Host Microbe 10(5):515–526

Griffiths-Jones S, Grocock RJ, van Dongen S, Bateman A, Enright AJ (2006) miRBase: microRNA sequences, targets and gene nomenclature. Nucleic Acids Res 34(Database issue):D140–D144

Guo S, Lu J, Schlanger R, Zhang H, Wang JY, Fox MC et al (2010) MicroRNA miR-125a controls hematopoietic stem cell number. Proc Natl Acad Sci U S A 107(32):14229–14234

Hackanson B, Bennett KL, Brena RM, Jiang J, Claus R, Chen SS et al (2008) Epigenetic modification of CCAAT/enhancer binding protein alpha expression in acute myeloid leukemia. Cancer Res 68(9):3142–3151

Hafner M, Landthaler M, Burger L, Khorshid M, Hausser J, Berninger P et al (2010) Transcriptome-wide identification of RNA-binding protein and microRNA target sites by PAR-CLIP. Cell 141(1):129–141

Han YC, Park CY, Bhagat G, Zhang J, Wang Y, Fan JB et al (2010) microRNA-29a induces aberrant self-renewal capacity in hematopoietic progenitors, biased myeloid development, and acute myeloid leukemia. J Exp Med 207(3):475–489

Hans CP, Weisenburger DD, Greiner TC, Gascoyne RD, Delabie J, Ott G et al (2004) Confirmation of the molecular classification of diffuse large B-cell lymphoma by immunohistochemistry using a tissue microarray. Blood 103(1):275–282

Harries LW (2012) Long non-coding RNAs and human disease. Biochem Soc Trans 40(4):902–906

He L, Thomson JM, Hemann MT, Hernando-Monge E, Mu D, Goodson S et al (2005) A microRNA polycistron as a potential human oncogene. Nature 435(7043):828–833

He L, He X, Lim LP, de Stanchina E, Xuan Z, Liang Y et al (2007) A microRNA component of the p53 tumour suppressor network. Nature 447(7148):1130–1134

Huang X, Shen Y, Liu M, Bi C, Jiang C, Iqbal J et al (2012) Quantitative proteomics reveals that miR-155 regulates the PI3K-AKT pathway in diffuse large B-cell lymphoma. Am J Pathol 181(1):26–33

Hussein K, Theophile K, Busche G, Schlegelberger B, Gohring G, Kreipe H et al (2010) Significant inverse correlation of microRNA-150/MYB and microRNA-222/p27 in myelodysplastic syndrome. Leuk Res 34(3):328–334

Iorio MV, Croce CM (2009) MicroRNAs in cancer: small molecules with a huge impact. J Clin Oncol 27(34):5848–5856

Iqbal J, Shen Y, Liu Y, Fu K, Jaffe ES, Liu C et al (2012) Genome-wide miRNA profiling of mantle cell lymphoma reveals a distinct subgroup with poor prognosis. Blood 119(21):4939–4948

Jiang J, Lee EJ, Schmittgen TD (2006) Increased expression of microRNA-155 in Epstein-Barr virus transformed lymphoblastoid cell lines. Genes Chromosomes Cancer 45(1):103–106

Johnnidis JB, Harris MH, Wheeler RT, Stehling-Sun S, Lam MH, Kirak O et al (2008) Regulation of progenitor cell proliferation and granulocyte function by microRNA-223. Nature 451(7182):1125–1129

Johnson SM, Grosshans H, Shingara J, Byrom M, Jarvis R, Cheng A et al (2005) RAS is regulated by the let-7 microRNA family. Cell 120(5):635–647

Jongen-Lavrencic M, Sun SM, Dijkstra MK, Valk PJ, Lowenberg B (2008) MicroRNA expression profiling in relation to the genetic heterogeneity of acute myeloid leukemia. Blood 111(10):5078–5085

Kaddar T, Chien WW, Bertrand Y, Pages MP, Rouault JP, Salles G et al (2009) Prognostic value of miR-16 expression in childhood acute lymphoblastic leukemia relationships to normal and malignant lymphocyte proliferation. Leuk Res 33(9):1217–1223

Kim VN (2005) MicroRNA biogenesis: coordinated cropping and dicing. Nat Rev Mol Cell Biol 6(5):376–385

Kim SW, Ramasamy K, Bouamar H, Lin AP, Jiang D, Aguiar RC (2012) MicroRNAs miR-125a and miR-125b constitutively activate the NF-kappaB pathway by targeting the tumor necrosis factor alpha-induced protein 3 (TNFAIP3, A20). Proc Natl Acad Sci U S A 109(20):7865–7870

Klein U, Lia M, Crespo M, Siegel R, Shen Q, Mo T et al (2010) The DLEU2/miR-15a/16-1 cluster controls B cell proliferation and its deletion leads to chronic lymphocytic leukemia. Cancer Cell 17(1):28–40

Kluiver J, Poppema S, de Jong D, Blokzijl T, Harms G, Jacobs S et al (2005) BIC and miR-155 are highly expressed in Hodgkin, primary mediastinal and diffuse large B cell lymphomas. J Pathol 207(2):243–249

Kluiver J, Haralambieva E, de Jong D, Blokzijl T, Jacobs S, Kroesen BJ et al (2006) Lack of BIC and microRNA miR-155 expression in primary cases of Burkitt lymphoma. Genes Chromosomes Cancer 45(2):147–153

Lal A, Navarro F, Maher CA, Maliszewski LE, Yan N, O'Day E et al (2009) miR-24 Inhibits cell proliferation by targeting E2F2, MYC, and other cell-cycle genes via binding to "seedless" 3′UTR microRNA recognition elements. Mol Cell 35(5):610–625

Langer C, Radmacher MD, Ruppert AS, Whitman SP, Paschka P, Mrozek K et al (2008) High BAALC expression associates with other molecular prognostic markers, poor outcome, and a distinct gene-expression signature in cytogenetically normal patients younger than 60 years with acute myeloid leukemia: a Cancer and Leukemia Group B (CALGB) study. Blood 111(11):5371–5379

Langer C, Marcucci G, Holland KB, Radmacher MD, Maharry K, Paschka P et al (2009) Prognostic importance of MN1 transcript levels, and biologic insights from MN1-associated gene and microRNA expression signatures in cytogenetically normal acute myeloid leukemia: a cancer and leukemia group B study. J Clin Oncol 27(19):3198–3204

Lawrie CH (2008) microRNA expression in lymphoid malignancies: new hope for diagnosis and therapy? J Cell Mol Med 12(5A):1432–1444

Lawrie CH (2010) microRNA expression in erythropoiesis and erythroid disorders. Br J Haematol 150(2):144–151

Lawrie CH, Soneji S, Marafioti T, Cooper CD, Palazzo S, Paterson JC et al (2007) MicroRNA expression distinguishes between germinal center B cell-like and activated B cell-like subtypes of diffuse large B cell lymphoma. Int J Cancer 121(5):1156–1161

Lawrie CH, Gal S, Dunlop HM, Pushkaran B, Liggins AP, Pulford K et al (2008) Detection of elevated levels of tumour-associated microRNAs in serum of patients with diffuse large B-cell lymphoma. Br J Haematol 141(5):672–675

Lawrie CH, Chi J, Taylor S, Tramonti D, Ballabio E, Palazzo S et al (2009) Expression of microRNAs in diffuse large B cell lymphoma is associated with immunophenotype, survival and transformation from follicular lymphoma. J Cell Mol Med 13(7):1248–1260

Lebedeva S, Jens M, Theil K, Schwanhausser B, Selbach M, Landthaler M et al (2011) Transcriptome-wide analysis of regulatory interactions of the RNA-binding protein HuR. Mol Cell 43(3):340–352

Leich E, Zamo A, Horn H, Haralambieva E, Puppe B, Gascoyne RD et al (2011) MicroRNA profiles of t(14;18)-negative follicular lymphoma support a late germinal center B-cell phenotype. Blood 118(20):5550–5558

Lenze D, Leoncini L, Hummel M, Volinia S, Liu CG, Amato T et al (2011) The different epidemiologic subtypes of Burkitt lymphoma share a homogenous micro RNA profile distinct from diffuse large B-cell lymphoma. Leukemia 25(12):1869–1876

Leucci E, Cocco M, Onnis A, De Falco G, van Cleef P, Bellan C et al (2008) MYC translocation-negative classical Burkitt lymphoma cases: an alternative pathogenetic mechanism involving miRNA deregulation. J Pathol 216(4):440–450

Leucci E, Zriwil A, Gregersen LH, Jensen KT, Obad S, Bellan C et al (2012) Inhibition of miR-9 de-represses HuR and DICER1 and impairs Hodgkin lymphoma tumour outgrowth in vivo. Oncogene 31(49):5081–5089

Li QJ, Chau J, Ebert PJ, Sylvester G, Min H, Liu G et al (2007) miR-181a is an intrinsic modulator of T cell sensitivity and selection. Cell 129(1):147–161

Li Z, Lu J, Sun M, Mi S, Zhang H, Luo RT et al (2008) Distinct microRNA expression profiles in acute myeloid leukemia with common translocations. Proc Natl Acad Sci U S A 105(40):15535–15540

Lia M, Carette A, Tang H, Shen Q, Mo T, Bhagat G et al (2012) Functional dissection of the chromosome 13q14 tumor-suppressor locus using transgenic mouse lines. Blood 119(13):2981–2990

Liu TY, Chen SU, Kuo SH, Cheng AL, Lin CW (2010) E2A-positive gastric MALT lymphoma has weaker plasmacytoid infiltrates and stronger expression of the memory B-cell-associated miR-223: possible correlation with stage and treatment response. Mod Pathol 23(11):1507–1517

Lu J, Getz G, Miska EA, Alvarez-Saavedra E, Lamb J, Peck D et al (2005) MicroRNA expression profiles classify human cancers. Nature 435(7043):834–838

Lu J, Guo S, Ebert BL, Zhang H, Peng X, Bosco J et al (2008) MicroRNA-mediated control of cell fate in megakaryocyte-erythrocyte progenitors. Dev Cell 14(6):843–853

Malumbres R, Sarosiek KA, Cubedo E, Ruiz JW, Jiang X, Gascoyne RD et al (2009) Differentiation stage-specific expression of microRNAs in B lymphocytes and diffuse large B-cell lymphomas. Blood 113(16):3754–3764

Manfe V, Biskup E, Rosbjerg A, Kamstrup M, Skov AG, Lerche CM et al (2012) miR-122 regulates p53/Akt signalling and the chemotherapy-induced apoptosis in cutaneous T-cell lymphoma. PLoS One 7(1):e29541

Marcucci G, Mrozek K, Radmacher MD, Garzon R, Bloomfield CD (2011) The prognostic and functional role of microRNAs in acute myeloid leukemia. Blood 117(4):1121–1129

Masaki S, Ohtsuka R, Abe Y, Muta K, Umemura T (2007) Expression patterns of microRNAs 155 and 451 during normal human erythropoiesis. Biochem Biophys Res Commun 364(3):509–514

Mavrakis KJ, Wolfe AL, Oricchio E, Palomero T, de Keersmaecker K, McJunkin K et al (2010) Genome-wide RNA-mediated interference screen identifies miR-19 targets in Notch-induced T-cell acute lymphoblastic leukaemia. Nat Cell Biol 12(4):372–379

Medina PP, Nolde M, Slack FJ (2010) OncomiR addiction in an in vivo model of microRNA-21-induced pre-B-cell lymphoma. Nature 467(7311):86–90

Mercer TR, Dinger ME, Mattick JS (2009) Long non-coding RNAs: insights into functions. Nat Rev Genet 10(3):155–159

Merkerova M, Belickova M, Bruchova H (2008) Differential expression of microRNAs in hematopoietic cell lineages. Eur J Haematol 81(4):304–310

Metzler M, Wilda M, Busch K, Viehmann S, Borkhardt A (2004) High expression of precursor microRNA-155/BIC RNA in children with Burkitt lymphoma. Genes Chromosomes Cancer 39(2):167–169

Mi S, Lu J, Sun M, Li Z, Zhang H, Neilly MB et al (2007) MicroRNA expression signatures accurately discriminate acute lymphoblastic leukemia from acute myeloid leukemia. Proc Natl Acad Sci U S A 104(50):19971–19976

Mi S, Li Z, Chen P, He C, Cao D, Elkahloun A et al (2010) Aberrant overexpression and function of the miR-17-92 cluster in MLL-rearranged acute leukemia. Proc Natl Acad Sci U S A 107(8):3710–3715

Monk CE, Hutvagner G, Arthur JS (2010) Regulation of miRNA transcription in macrophages in response to Candida albicans. PLoS One 5(10):e13669

Monticelli S, Ansel KM, Xiao C, Socci ND, Krichevsky AM, Thai TH et al (2005) MicroRNA profiling of the murine hematopoietic system. Genome Biol 6(8):R71

Mu P, Han YC, Betel D, Yao E, Squatrito M, Ogrodowski P et al (2009) Genetic dissection of the miR-17 92 cluster of microRNAs in Myc-induced B-cell lymphomas. Genes Dev 23(24):2806–2811

Muljo SA, Ansel KM, Kanellopoulou C, Livingston DM, Rao A, Rajewsky K (2005) Aberrant T cell differentiation in the absence of Dicer. J Exp Med 202(2):261–269
Narducci MG, Arcelli D, Picchio MC, Lazzeri C, Pagani E, Sampogna F et al (2011) MicroRNA profiling reveals that miR-21, miR486 and miR-214 are upregulated and involved in cell survival in Sezary syndrome. Cell Death Dis 2:e151
Navarro A, Bea S, Fernandez V, Prieto M, Salaverria I, Jares P et al (2009a) MicroRNA expression, chromosomal alterations, and immunoglobulin variable heavy chain hypermutations in Mantle cell lymphomas. Cancer Res 69(17):7071–7078
Navarro A, Diaz T, Martinez A, Gaya A, Pons A, Gel B et al (2009b) Regulation of JAK2 by miR-135a: prognostic impact in classic Hodgkin lymphoma. Blood 114(14):2945–2951
Neilson JR, Zheng GX, Burge CB, Sharp PA (2007) Dynamic regulation of miRNA expression in ordered stages of cellular development. Genes Dev 21(5):578–589
Nie K, Gomez M, Landgraf P, Garcia JF, Liu Y, Tan LH et al (2008) MicroRNA-mediated down-regulation of PRDM1/Blimp-1 in Hodgkin/Reed-Sternberg cells: a potential pathogenetic lesion in Hodgkin lymphomas. Am J Pathol 173(1):242–252
O'Connell RM, Rao DS, Chaudhuri AA, Boldin MP, Taganov KD, Nicoll J et al (2008) Sustained expression of microRNA-155 in hematopoietic stem cells causes a myeloproliferative disorder. J Exp Med 205(3):585–594
O'Connell RM, Chaudhuri AA, Rao DS, Baltimore D (2009) Inositol phosphatase SHIP1 is a primary target of miR-155. Proc Natl Acad Sci U S A 106(17):7113–7118
O'Donnell KA, Wentzel EA, Zeller KI, Dang CV, Mendell JT (2005) c-Myc-regulated microRNAs modulate E2F1 expression. Nature 435(7043):839–843
Olive V, Bennett MJ, Walker JC, Ma C, Jiang I, Cordon-Cardo C et al (2009) miR-19 is a key oncogenic component of mir-17-92. Genes Dev 23(24):2839–2849
Pase L, Layton JE, Kloosterman WP, Carradice D, Waterhouse PM, Lieschke GJ (2009) miR-451 regulates zebrafish erythroid maturation in vivo via its target gata2. Blood 113(8):1794–1804
Pedersen IM, Otero D, Kao E, Miletic AV, Hother C, Ralfkiaer E et al (2009) Onco-miR-155 targets SHIP1 to promote TNFalpha-dependent growth of B cell lymphomas. EMBO Mol Med 1(5):288–295
Pekarsky Y, Santanam U, Cimmino A, Palamarchuk A, Efanov A, Maximov V et al (2006) Tcl1 expression in chronic lymphocytic leukemia is regulated by miR-29 and miR-181. Cancer Res 66(24):11590–11593
Qin Y, Buermans HP, van Kester MS, van der Fits L, Out-Luiting JJ, Osanto S et al (2012) Deep-sequencing analysis reveals that the miR-199a2/214 cluster within DNM3os represents the vast majority of aberrantly expressed microRNAs in Sezary syndrome. J Invest Dermatol 132(5):1520–1522
Raaijmakers MH, Mukherjee S, Guo S, Zhang S, Kobayashi T, Schoonmaker JA et al (2010) Bone progenitor dysfunction induces myelodysplasia and secondary leukaemia. Nature 464(7290):852–857
Rai D, Kim SW, McKeller MR, Dahia PL, Aguiar RC (2010) Targeting of SMAD5 links microRNA-155 to the TGF-beta pathway and lymphomagenesis. Proc Natl Acad Sci U S A 107(7):3111–3116
Ralfkiaer U, Hagedorn PH, Bangsgaard N, Lovendorf MB, Ahler CB, Svensson L et al (2011) Diagnostic microRNA profiling in cutaneous T-cell lymphoma (CTCL). Blood 118(22):5891–5900
Rao DS, O'Connell RM, Chaudhuri AA, Garcia-Flores Y, Geiger TL, Baltimore D (2010) MicroRNA-34a perturbs B lymphocyte development by repressing the forkhead box transcription factor Foxp1. Immunity 33(1):48–59
Rao E, Jiang C, Ji M, Huang X, Iqbal J, Lenz G et al (2012) The miRNA-17 approximately 92 cluster mediates chemoresistance and enhances tumor growth in mantle cell lymphoma via PI3K/AKT pathway activation. Leukemia 26(5):1064–1072
Raveche ES, Salerno E, Scaglione BJ, Manohar V, Abbasi F, Lin YC et al (2007) Abnormal microRNA-16 locus with synteny to human 13q14 linked to CLL in NZB mice. Blood 109(12):5079–5086
Rhyasen GW, Starczynowski DT (2012) Deregulation of microRNAs in myelodysplastic syndrome. Leukemia 26(1):13–22

Rodriguez A, Vigorito E, Clare S, Warren MV, Couttet P, Soond DR et al (2007) Requirement of bic/microRNA-155 for normal immune function. Science 316(5824):608–611
Roehle A, Hoefig KP, Repsilber D, Thorns C, Ziepert M, Wesche KO et al (2008) MicroRNA signatures characterize diffuse large B-cell lymphomas and follicular lymphomas. Br J Haematol 142(5):732–744
Rossi S, Shimizu M, Barbarotto E, Nicoloso MS, Dimitri F, Sampath D et al (2010) microRNA fingerprinting of CLL patients with chromosome 17p deletion identify a miR-21 score that stratifies early survival. Blood 116(6):945–952
Sampath D, Liu C, Vasan K, Sulda M, Puduvalli VK, Wierda WG et al (2012) Histone deacetylases mediate the silencing of miR-15a, miR-16, and miR-29b in chronic lymphocytic leukemia. Blood 119(5):1162–1172
San Jose-Eneriz E, Roman-Gomez J, Jimenez-Velasco A, Garate L, Martin V, Cordeu L et al (2009) MicroRNA expression profiling in Imatinib-resistant Chronic Myeloid Leukemia patients without clinically significant ABL1-mutations. Mol Cancer 8:69
Scaglione BJ, Salerno E, Balan M, Coffman F, Landgraf P, Abbasi F et al (2007) Murine models of chronic lymphocytic leukaemia: role of microRNA-16 in the New Zealand Black mouse model. Br J Haematol 139(5):645–657
Schotte D, De Menezes RX, Akbari Moqadam F, Khankahdani LM, Lange-Turenhout E, Chen C et al (2011) MicroRNA characterize genetic diversity and drug resistance in pediatric acute lymphoblastic leukemia. Haematologica 96(5):703–711
Schramedei K, Morbt N, Pfeifer G, Lauter J, Rosolowski M, Tomm JM et al (2011) MicroRNA-21 targets tumor suppressor genes ANP32A and SMARCA4. Oncogene 30(26):2975–2985
Schwind S, Maharry K, Radmacher MD, Mrozek K, Holland KB, Margeson D et al (2010) Prognostic significance of expression of a single microRNA, miR-181a, in cytogenetically normal acute myeloid leukemia: a Cancer and Leukemia Group B study. J Clin Oncol 28(36):5257–5264
Sethupathy P, Megraw M, Hatzigeorgiou AG (2006) A guide through present computational approaches for the identification of mammalian microRNA targets. Nat Methods 3(11):881–886
Skalsky RL, Corcoran DL, Gottwein E, Frank CL, Kang D, Hafner M et al (2012) The viral and cellular microRNA targetome in lymphoblastoid cell lines. PLoS Pathog 8(1):e1002484
Sotillo E, Laver T, Mellert H, Schelter JM, Cleary MA, McMahon S et al (2011) Myc overexpression brings out unexpected antiapoptotic effects of miR-34a. Oncogene 30(22):2587–2594
Starczynowski DT, Kuchenbauer F, Argiropoulos B, Sung S, Morin R, Muranyi A et al (2010) Identification of miR-145 and miR-146a as mediators of the 5q- syndrome phenotype. Nat Med 16(1):49–58
Starczynowski DT, Morin R, McPherson A, Lam J, Chari R, Wegrzyn J et al (2011) Genome-wide identification of human microRNAs located in leukemia-associated genomic alterations. Blood 117(2):595–607
Sun SM, Dijkstra MK, Bijkerk AC, Brooimans RA, Valk PJ, Erkeland SJ et al (2011) Transition of highly specific microRNA expression patterns in association with discrete maturation stages of human granulopoiesis. Br J Haematol 155(3):395–398
Tagawa H, Karube K, Tsuzuki S, Ohshima K, Seto M (2007) Synergistic action of the microRNA-17 polycistron and Myc in aggressive cancer development. Cancer Sci 98(9):1482–1490
Tan LP, Seinen E, Duns G, de Jong D, Sibon OC, Poppema S et al (2009) A high throughput experimental approach to identify miRNA targets in human cells. Nucleic Acids Res 37(20):e137
Teng G, Hakimpour P, Landgraf P, Rice A, Tuschl T, Casellas R et al (2008) MicroRNA-155 is a negative regulator of activation-induced cytidine deaminase. Immunity 28(5):621–629
Thai TH, Calado DP, Casola S, Ansel KM, Xiao C, Xue Y et al (2007) Regulation of the germinal center response by microRNA-155. Science 316(5824):604–608
Thompson RC, Herscovitch M, Zhao I, Ford TJ, Gilmore TD (2011) NF-kappaB down-regulates expression of the B-lymphoma marker CD10 through a miR-155/PU.1 pathway. J Biol Chem 286(3):1675–1682
Thorns C, Kuba J, Bernard V, Senft A, Szymczak S, Feller AC et al (2012) Deregulation of a distinct set of microRNAs is associated with transformation of gastritis into MALT lymphoma. Virchows Arch 460(4):371–377

van den Berg A, Kroesen BJ, Kooistra K, de Jong D, Briggs J, Blokzijl T et al (2003) High expression of B-cell receptor inducible gene BIC in all subtypes of Hodgkin lymphoma. Genes Chromosomes Cancer 37(1):20–28

van der Fits L, van Kester MS, Qin Y, Out-Luiting JJ, Smit F, Zoutman WH et al (2011) MicroRNA-21 expression in CD4+ T cells is regulated by STAT3 and is pathologically involved in Sezary syndrome. J Invest Dermatol 131(3):762–768

van Kester MS, Ballabio E, Benner MF, Chen XH, Saunders NJ, van der Fits L et al (2011) miRNA expression profiling of mycosis fungoides. Mol Oncol 5(3):273–280

Ventura A, Young AG, Winslow MM, Lintault L, Meissner A, Erkeland SJ et al (2008) Targeted deletion reveals essential and overlapping functions of the miR-17 through 92 family of miRNA clusters. Cell 132(5):875–886

Venturini L, Battmer K, Castoldi M, Schultheis B, Hochhaus A, Muckenthaler MU et al (2007) Expression of the miR-17-92 polycistron in chronic myeloid leukemia (CML) CD34+ cells. Blood 109(10):4399–4405

Vigorito E, Perks KL, Abreu-Goodger C, Bunting S, Xiang Z, Kohlhaas S et al (2007) microRNA-155 regulates the generation of immunoglobulin class-switched plasma cells. Immunity 27(6):847–859

Visone R, Rassenti LZ, Veronese A, Taccioli C, Costinean S, Aguda BD et al (2009) Karyotype-specific microRNA signature in chronic lymphocytic leukemia. Blood 114(18):3872–3879

Visone R, Veronese A, Rassenti LZ, Balatti V, Pearl DK, Acunzo M et al (2011) miR-181b is a biomarker of disease progression in chronic lymphocytic leukemia. Blood 118(11):3072–3079

Wang W, Corrigan-Cummins M, Hudson J, Maric I, Simakova O, Neelapu SS et al (2012) MicroRNA profiling of follicular lymphoma identifies microRNAs related to cell proliferation and tumor response. Haematologica 97(4):586–594

Watanabe A, Tagawa H, Yamashita J, Teshima K, Nara M, Iwamoto K et al (2011) The role of microRNA-150 as a tumor suppressor in malignant lymphoma. Leukemia 25(8):1324–1334

Woods K, Thomson JM, Hammond SM (2007) Direct regulation of an oncogenic micro-RNA cluster by E2F transcription factors. J Biol Chem 282(4):2130–2134

Xiao C, Srinivasan L, Calado DP, Patterson HC, Zhang B, Wang J et al (2008) Lymphoproliferative disease and autoimmunity in mice with increased miR-17-92 expression in lymphocytes. Nat Immunol 9(4):405–414

Xu C, Fu H, Gao L, Wang L, Wang W, Li J et al (2012) BCR-ABL/GATA1/miR-138 mini circuitry contributes to the leukemogenesis of chronic myeloid leukemia. Oncogene. doi:10.1038/onc.2012.557. [Epub ahead of print]

Yamakuchi M, Lowenstein CJ (2009) MiR-34, SIRT1 and p53: the feedback loop. Cell Cycle 8(5):712–715

Yamanaka Y, Tagawa H, Takahashi N, Watanabe A, Guo YM, Iwamoto K et al (2009) Aberrant overexpression of microRNAs activate AKT signaling via down-regulation of tumor suppressors in natural killer-cell lymphoma/leukemia. Blood 114(15):3265–3275

Yao Y, Zhao Y, Xu H, Smith LP, Lawrie CH, Sewer A et al (2007) Marek's disease virus type 2 (MDV-2)-encoded microRNAs show no sequence conservation to those encoded by MDV-1. J Virol 81(13):7164–7170

Zenz T, Habe S, Denzel T, Mohr J, Winkler D, Buhler A et al (2009) Detailed analysis of p53 pathway defects in fludarabine-refractory chronic lymphocytic leukemia (CLL): dissecting the contribution of 17p deletion, TP53 mutation, p53-p21 dysfunction, and miR34a in a prospective clinical trial. Blood 114(13):2589–2597

Zhan M, Miller CP, Papayannopoulou T, Stamatoyannopoulos G, Song CZ (2007) MicroRNA expression dynamics during murine and human erythroid differentiation. Exp Hematol 35(7):1015–1025

Zhao JJ, Lin J, Lwin T, Yang H, Guo J, Kong W et al (2010) microRNA expression profile and identification of miR-29 as a prognostic marker and pathogenetic factor by targeting CDK6 in mantle cell lymphoma. Blood 115(13):2630–2639

Zhou B, Wang S, Mayr C, Bartel DP, Lodish HF (2007) miR-150, a microRNA expressed in mature B and T cells, blocks early B cell development when expressed prematurely. Proc Natl Acad Sci U S A 104(17):7080–7085 van den Berg A, Kroesen BJ, Kooistra K, de Jong D, Briggs J, Blokzijl T et al (2003) High expression of B-cell receptor inducible gene BIC in all subtypes of Hodgkin lymphoma. Genes Chromosomes Cancer 37:20–28

[illegible]

Chapter 7
MicroRNAs and Oncogenic Human Viruses

Muriel Thirion, Teru Kanda, Yoshiki Murakami, Takahiro Ochiya, and Hisashi Iizasa

Contents

M. Thirion • T. Ochiya
Division of Molecular and Cellular Medicine, National Cancer Center Research Institute, Tokyo, Japan

T. Kanda
Division of Virology, Aichi Cancer Center Research Institute, Nagoya, Japan

Y. Murakami
Department of Hepatology, Graduate School of Medicine, Osaka City University, Osaka, Japan

H. Iizasa (✉)
Division of Stem Cell Biology, Institute for Genetic Medicine, Hokkaido University, Sapporo, Japan
e-mail: iizasah@igm.hokudai.ac.jp

S. Babashah (ed.), *MicroRNAs: Key Regulators of Oncogenesis*,
DOI 10.1007/978-3-319-03725-7_7, © Springer International Publishing Switzerland 2014

Abstract MicroRNAs (miRNAs) are small, non-coding RNAs that regulate mRNA expression by post-transcriptional mechanism in eukaryotic cells. Some viruses also encode primary transcripts containing miRNA-like structures, and such transcripts are subjected to host miRNA processing pathway to generate viral miRNAs. Viral miRNAs derived from oncogenic viruses are often associated with tumor progression. Moreover, infections with oncogenic viruses alter the expression of host miRNAs, increasing the risk of tumor progression and viral escape from the host immune mechanism. In this chapter, we discuss the roles of virally-regulated cellular miRNAs in the respective viral life-cycles and in virus-related tumors.

Keywords microRNA • Oncogenic viruses • Tumorigenesis • Immune system

1 Introduction

1.1 Discovery of Viral miRNAs

MicroRNAs (miRNAs) are 18–25 nucleotides (nt) non-coding small RNAs derived from double-stranded RNAs, and play an important role in eukaryotic cells by post-transcriptional repression of mRNAs. It has been shown that some viruses encode primary transcripts containing miRNA-like structures. In 2004, Pfeffer et al. (2004) reported that Epstein-Barr virus (EBV) strain B95-8 encodes 5 viral pre-miRNA-like structures, and that viral miRNAs were detected in infected cells. Moreover, Cai et al. (2006) reported that wild-type EBV encodes 13 more pre-miRNAs than EBV B95-8 strain as the 12 kb region that is deleted in EBV B95-8 strain is rich in pre-miRNA genes. The expression of viral miRNAs are very common in cells that are infected with other Herpesviruses (Pfeffer et al. 2005) (http://www.mirbase.org) including Kaposi's sarcoma-associated herpesvirus (KSHV), human cytomegalovirus, herpes simplex viruses (HSVs), and also observed in simian virus 40-infected cells. It is speculated that viral miRNAs may suppress viral transcripts or host-specific genes. However, the pathophysiological role of viral miRNAs is not clearly understood.

1.2 Viral Infection and miRNAs

Oncogenic viral infections induce the expression of several miRNAs that are associated with cancer progression. Virally induced miRNAs play the role of oncogenes when they target tumor suppressor genes. Moreover, when viral infections involve regulation of oncogenes, they repress some host miRNAs with tumor suppressive functions. Some herpesviruses such as EBV and KSHV encode pri-miRNA-like structures that are tolerated as self-entities by the host machinery. Virally derived-factors repress host miRNA cascade and are called "RNA-silencing suppressors" (RSSs) (de Vries and Berkhout 2008). RSSs were originally identified in plant viruses and oncogenic viruses origin interact with miRNA pathway (de Vries and Berkhout 2008).

2 MicroRNAs in Epstein-Barr Virus; Expression, Regulation and Function Epstein-Barr Virus

2.1 EBV Encoded miRNAs

EBV is a ubiquitous human herpesvirus that establishes life-long latent infection in human B lymphocytes and pharyngeal epithelial cells (kieff 2007). EBV has quite a large genome (~170 kb) and encodes >70 open reading frames. While many of the virally encoded proteins are immunogenic in the human body, miRNAs can affect gene expression in the host without stimulating an immune response. Therefore, encoding miRNAs work to the advantage of the virus. EBV miRNAs were the first virally encoded miRNAs to be identified (Pfeffer et al. 2004). A Burkitt's lymphoma cell line harboring EBV B95-8 strain, a laboratory strain with 12 kb deletion in its genome, (Baer et al. 1984) was used as a source of RNA. Five miRNAs were identified in the study. Later studies revealed that there are far more miRNAs in the wild-type EBV, and the region deleted in the EBV B95-8 strain is rich in pre-miRNA genes (Lo et al. 2012) (Fig. 7.1). Currently, 44 mature miRNAs that are encoded at two different loci in the EBV genome have been identified: 4 mature miRNAs encoded at the BHRF1 locus and 40 mature miRNAs encoded at the BART locus (Pfeffer et al. 2004; Cai et al. 2006; Grundhoff et al. 2006; Zhu

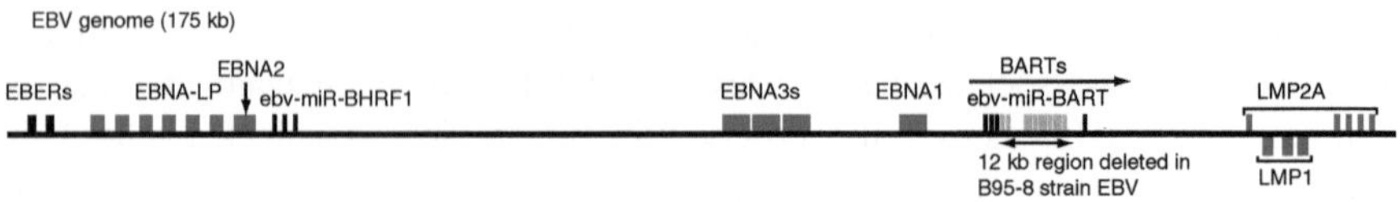

Fig. 7.1 Schematic illustration of EBV genome. The positions of EBV miRNA genes are indicated together with those of EBV latent genes, including BARTs. The 12-kb region missing in the EBV B95-8 strain is also indicated

et al. 2009) (Fig. 7.1). The presence of such a high number of miRNAs in EBV indicates the evolutionary selection of these miRNAs. A complete listing of EBV miRNAs (ebv-miR-BHRF1 and ebv-miR-BART) with both mature and precursor sequences can be found at www.mirbase.org. EBV miRNAs have no notable sequence similarity with known host (human) cell miRNAs, and no orthologous miRNAs are identified in other human Herpesviruses (Pfeffer et al. 2005). In comparison with Rhesus lymphocryptovirus, it is apparent that many of EBV miRNAs are evolutionarily conserved (Cai et al. 2006).

The expression levels of EBV miRNAs in various EBV-infected cells have been examined using various strategies, including the stem-loop PCR method (Amoroso et al. 2011; Chen et al. 2005; Cosmopoulos et al. 2009; Pratt et al. 2009) and direct sequencing of small RNA libraries, either through traditional or high-throughput sequencing method (Lung et al. 2009; Zhu et al. 2009; Chen et al. 2010). The results revealed that EBV miRNAs are expressed at markedly different levels among cell lines (Pratt et al. 2009). Four miRNAs encoded within the BHRF1 locus (hereafter referred to as miR-BHRFs) are highly expressed in cells with latency type III (Xia et al. 2008; Cai et al. 2006) [expressing all EBNAs (EBNA1, 2, 3A-C), LMP1, LMP2A, EBERs, and BARTs (BamHI A Rightward Transcripts)]. The miR-BHRFs are also highly expressed in primary EBV-associated AIDS-related diffuse large B-cell lymphomas (DLBCL) (Xia et al. 2008), but they are undetectable in B cells or epithelial cells with latency type I (expressing EBNA1, EBER, and BARTs) or latency type II (expressing EBNA1, LMP1 and LMP2A, EBER, and BARTs). On the other hand, miR-BART miRNAs (miR-BARTs) are expressed not only in B cells with type III latency, but also in epithelial cells with latency type I or type II (Cai et al. 2006). The miR-BARTs are of particular interest as they are highly expressed in nasopharyngeal carcinomas (Zhu et al. 2009; Cosmopoulos et al. 2009), gastric carcinoma cells (Kim do et al. 2007), and NK/T lymphomas-derived cell lines (Ramakrishnan et al. 2011). Therefore, it is likely that miR-BARTs somehow contribute to the tumorigenesis (Lo et al. 2012; Marquitz and Raab-Traub 2012; Raab-Traub 2012). Transcripts now referred to as BARTs originally identified from nasopharyngeal carcinoma cells (Hitt et al. 1989), have remained enigmatic for many years. However, it is now clear that BARTs most likely serve as primary transcripts that are processed to generate miR-BARTs. Interestingly, the currently identified all miR-BARTs are encoded in the introns of the transcripts of BART, and are subject to highly complicated splicing (Edwards et al. 2008).

Table 7.1 Targeting genes of EBV encoded miRNAs

Target genes	EBV miRNAs	Reference
Viral target genes		
BALF5 (DNA polymerase)	miR-BART2-5p	Barth et al. (2008)
LMP1	miR-BART1-5p, -16, -17-5p	Lo et al. (2007)
	miR-BART9	Ramakrishnan et al. (2011)
	miR-BART19-5p, -5-5p	Riley et al. (2012)
LMP2A	miR-BART22	Lung et al. (2009)
BHRF1	miR-BART10-3p	Riley et al. (2012)
Cellular target genes		
Bim	miR-BART cluster 1 and cluster 2	Marquitz et al. (2011)
PUMA	miR-BART5	Choy et al. (2008)
DICER1	miR-BART6-5p	Iizasa et al. (2010)
CXCL-11	miR-BHRF1-3	Xia et al. (2008)
IPO7	miR-BART3	Vereide et al. (2013)
CASP3	miR-BART16	Vereide et al. (2013)
GUF1, SCRN1	miR-BHRF1-1	Skalsky et al. (2012)
CAPRIN2	miR-BART13-3p	Riley et al. (2012)

2.2 *Pathophysiological Roles of EBV Encoded miRNAs*

Viral miRNAs can either target other EBV transcripts or cellular transcripts. The viral and cellular targets of EBV miRNAs so far identified are listed in Table 7.1. MiR-BART2-5p, which is located directly antisense to the 3′-UTR of BALF5 (a viral polymerase) can down-regulate the expression of BLAF5, inhibiting the transition from latent to lytic viral replication (Barth et al. 2008). Several miR-BARTs suppress the expression of viral oncoproteins LMP1 (Riley et al. 2012; Lo et al. 2007; Ramakrishnan et al. 2011) and LMP2A (Lung et al. 2009). Cellular targets of EBV miRNAs so far identified include proapoptotic proteins Bim (Marquitz et al. 2011) and BBC3/PUMA (Choy et al. 2008), a Dicer (Iizasa et al. 2010), an interferon-inducible T-cell-attracting chemokine CXCL-11/I-TAC (Xia et al. 2008), IPO7, and CASP3 (Vereide et al. 2013). Genome-wide searches for the targets of EBV miRNAs (miRNA targetome) have been conducted using either human Burkitt's lymphoma cell lines (Dolken et al. 2010), primary effusion lymphoma cell lines (co-infected with EBV and KSHV) (Gottwein et al. 2011), or EBV-transformed lymphoblastoid cell lines (Skalsky et al. 2012; Riley et al. 2012).

It is now technically feasible to utilize recombinant viruses, having miRNA genes either deleted or restored in the EBV genome, to clarify the biological significance of viral miRNAs. It was shown, by two independent studies, that disruption of genes encoding miR-BHRF1 results in slightly attenuated outgrowth of infected primary B cells (Feederle et al. 2011; Seto et al. 2010). The EBV B95-8 strain lacks 17 pre-miRNAs of miR-BARTs. Research groups attempted to reconstitute the expression of all EBV-encoded miR-BARTs by ectopically inserting the missing pre-miRNA genes that were driven by heterologous promoters (Vereide et al. 2013; Seto et al. 2010). However, the displaced miR-BARTs were not expressed as efficiently as the endogenous miRNAs (Seto et al. 2010). The efficient expression of miR-BARTs may require primary transcripts under the control of native BART promoter, followed by proper processing of the primary transcripts.

It was also shown that EBV miRNAs were secreted from infected B cells and that they were functional upon transfer via exosomes in primary monocyte-derived dendritic cells (Pegtel et al. 2010). Another study recently demonstrated that certain plasma EBV miRNAs did not copurify with exosomes, implicating non-exosomal transport of miRNAs into plasma (Gourzones et al. 2013). Further studies are required to clarify the functional significance of viral miRNAs secreted into plasma via exosomal or non-exosomal mechanisms.

2.3 Alteration of Human miRNA Pathway by EBV Infection

Regulating host gene expression is crucial for viruses to survive in host cells, and it is now becoming apparent that viral miRNAs significantly contribute to such regulations, especially in latently infected cells where a few viral proteins are expressed. Viral miRNAs can affect the expression of cellular miRNAs. Specific cellular miRNAs, namely, miR-21, miR-155, and miR-146a, were found to be up-regulated in B lymphocytes transformed by EBV B95-8 strain (Godshalk et al. 2008; Mrazek et al. 2007), while other cellular miRNAs were dramatically down-regulated following EBV infection of primary B cells (Godshalk et al. 2008). It is tempting to speculate that the up-regulation of miR-21 plays critical roles in EBV-mediated transformation, as miR-21 is a well-characterized oncomir (Gabriely et al. 2008). Therefore, it appears that viral and cellular miRNA regulatory networks affect each other, and virus-host interactions are apparently far more complicated than previously thought.

3 MicroRNAs in Kaposi's Sarcoma-Associated Herpesvirus; Expression, Regulation and Function

3.1 KSHV Encoded miRNAs

KSHV belongs to the human herpesvirus family and is implicated in human diseases such as Kaposi's sarcoma (KS), AIDS-related primary effusion lymphoma (PEL), and multicentric castle-man's disease (Boshoff and Weiss 2002). KSHV exists as a latent or lytic infection in host cells. Pfeffer et al. and other groups discovered

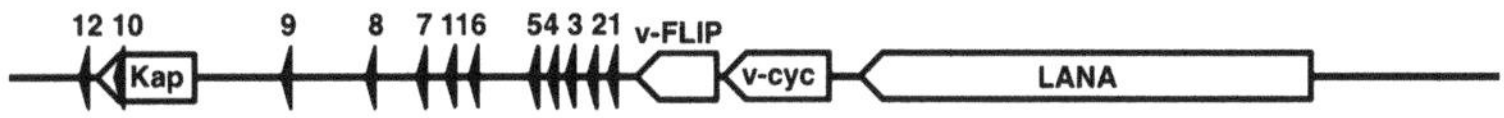

Fig. 7.2 Location of KSHV-encoded miRNAs in KSHV genome *Black triangle*: miRNA, *Kap*: kaposin

KHSV-derived miRNAs in latently infected cells (Pfeffer et al. 2004, 2005; Cai et al. 2005; Samols et al. 2005). KSHV encodes 12 miR-K12 pre-miRNAs (24 miRNAs) and A-to-I RNA edited mir-K12-10a is registered as a mir-K12-10b on miRBase (http://www.mirbase.org) (Pfeffer et al. 2005; Umbach and Cullen 2010; Lin et al. 2010). Most miR-K12s are localized in the intron of K12 (Kaposin) and two pre-miRNAs are localized in the protein- coding region and 3′-UTR of K12, respectively (Fig. 7.2).

3.2 Pathophysiological Roles of KSHV Encoded miRNAs

MiR-K12s are expressed in latently infected cells; however their role in the viral life cycle is largely unknown. MiR-K12-9 suppresses the expression of RTA, which is an essential transcription factor for KSHV lytic infection (Bellare and Ganem 2009; Lin et al. 2011). Transfection of miR-K12-7 or miR-K12-5 also represses RTA-expression (Lin et al. 2011; Lu et al. 2010). Moreover, mutated KSHV that lacks miR-K12s, except miR-K12-10 and miR-K12-12, increased lytic protein expression by enhancing NF-κB activation (Lei et al. 2010). These reports indicate that miR-K12s suppress lytic reactivation and maintain latent infection in host cells.

Seed sequences of miR-K12s are similar to human miRNAs (KSHV-K12-11 and human miR-155, miR-K12-6-5p and human miR-15a and miR-16) (Skalsky et al. 2007; Gottwein et al. 2007). These reports suggest that miR-K12s may target human genes to maintain latent infections. MiR-K12s repressed thrombospondin1 (THBS1), a tumor suppressor, via inhibition of angiogenesis and down-regulation of THBS1 expression, was also previously observed in KS lesion (Samols et al. 2007; Taraboletti et al. 1999). MiR-K12-5, -9, -10a, and -10b repress Bcl-2-associated transcription factor 1 (BCLAF1), which is a repressor of Bcl2 family and induces apoptosis (Ziegelbauer et al. 2009). MiR-K12-11 targets the xCT-negative regulator BACH-1 (Qin et al. 2010a). xCT is an amino acid transporter that protects cells from environmental oxidative stress. KS legions show high expression of xCT (Qin et al. 2010a), and interestingly, xCT is reported to be a regulator of cancer stem cells (Ishimoto et al. 2011). MiR-K12-1 represses cyclin-dependent kinase inhibitor p21. Inhibition of miR-K12-1 results in cell cycle arrest by p53 activation (Gottwein and Cullen 2010). These miR-K12-targeting genes are related to the pathogenesis of KSHV-associated diseases.

3.3 Immune Defense and KSHV Encoded miRNAs

Latent infection of KSHV were observed in 40–50 % of the population in south part of Africa, 10 % of the north American population and 4 % of the Japanese population; however, most of these people were kept healthy (Fujii et al. 1999). MiR-K12s are expressed in latently infected cells. To escape from host immune mechanism, miR-K12s may regulate this mechanism. MiR-K12-7 inhibits the expression of MHC class I polypeptide-related sequence B, which is recognized by NK cells (Nachmani et al. 2009). miR-K12-10a represses the tumor necrosis factor receptor superfamily member 12A, which regulates apoptosis and inflammatory response (Abend et al. 2010). MiR-K12s also alter human cytokine expression via targeting of the cytokine repressor C/EBPβ p20 (Qin et al. 2010b). C/EBPβ p20 is a repressor of IL-6 and IL-10. MiR-K12s induce the expression of these cytokines in murine macrophages (Qin et al. 2010b). A bioinformatics sequence analysis revealed that this could be attributed to viral miRNA-mediated expression of a known repressor of these cytokines.

To identify target genes of miR-K12s, a bioinformatics approach was used. Identifying target genes of viral miRNAs is much more difficult than identifying those of mammalian miRNAs, because species conservation of 3′-UTR miRNA targeting site is not useful for viral miRNAs. Recently, high-throughput sequencing of RNA isolated by cross-linking immunoprecipitation (HITS-CLIP) was developed to identify miRNA target gene by immunoprecipitation of Ago2-miRNAs and associated mRNAs. This method recovered approximately 1,000 cellular targeting genes of miR-K12s, including THBS1, BACH1, and C/EBPβin PEL cell lines (Haecker et al. 2012). Interestingly, HITS-CLIP revealed that the miR-K12s predominate Ago2-associated miRNAs and miR-K12smay contribute to global alteration of the human miRNAs pathway in KSHV-infected cells (Haecker et al. 2012). KSHV-positive KS is derived from endothelial cells and the mRNA expression profile of endothelial cells is not to the same as that of PEL. In the near future, the pathophysiological role of miR-K12s in endothelial cells will be identified using the HITS-CLIP method.

4 MicroRNAs in High-Risk Human Papillomavirus; Expression, Regulation and Function

4.1 Basic Knowledge of HPV Infection

Human papillomaviruses (HPVs) have small double-stranded circular genomic DNA that encode early genes (E1, E2, E4, E5, E6, and E7) and late genes (L1 and L2) (Zheng and Baker 2006). HPVs infect squamous epithelium, and then integrate into the epithelial stem cells on the basal membrane. HPV early genes are expressed in the epithelial stem cells; however, expression of viral late genes and viral DNA replication are observed in differentiated epithelial layers. The E6 and E7 genes of high-risk HPVs (HPV16 and HPV18) have oncogenic activity and inactivate p53 and pRb, respectively (Scheffner et al. 1990; Dyson et al. 1989; Gonzalez et al. 2001).

4.2 MiRNAs Expression Profile of HPV Infected Cells and Pathophysiological Role of miRNAs in HPV Infection

DNA viruses encode viral miRNAs and therefore are able to regulate viral life cycle or human immune defense. However, HPVs do not have viral miRNAs because they have small genome (size, 8 kb). Infection with high-risk HPVs leads to tumorigenesis in the epithelial stem cells by the inactivation of tumor suppressive factors. The component p53 interacts with the Drosha-DGCR8 complex component p68 and regulates part of miRNA processing (Suzuki et al. 2009). The effect of high-risk HPVs infection in miRNA expression is largely unknown.

Dreher et al. (2011) reported that the expression level of miR-145 was increased in high-risk- HPV infected cells compared to low-risk HPV-infected cells. High-risk HPV component E6 is one of the key factors for tumorigenesis, it suppresses the expression of miR-145 (Shi et al. 2012; Gunasekharan and Laimins 2013), miR-218 (Martinez et al. 2008), miR-34a (Wang et al. 2009b; Xie et al. 2013) and miR-23b (Au Yeung et al. 2011). Previously, miR-145 was identified as a tumor suppressive miRNA (called "anti-oncomir") (Cho et al. 2009) as it suppresses the expression of c-Myc (Sachdeva et al. 2009), MUC1 (Sachdeva and Mo 2010) and stem cell-related transcription factors (Xu et al. 2009). Tumor-related target genes of miR-218 are Robo1 (Tie et al. 2010), survivin (Alajez et al. 2011), Runx2 (Zhang et al. 2011b), and the mTOR component Rictor (Uesugi et al. 2011). Moreover, miR-34a and miR-23b repress c-Myc (Christoffersen et al. 2010; Gao et al. 2009) and these miRNAs are induced by p53. These reports suggest that the high-risk HPV E6 gene represses tumor suppressive miRNAs via p53 inactivation.

High-risk HPV E7 can inactivate pRb and induce the activation of the transcription factor E2F (Scheffner et al. 1990; Dyson et al. 1989; Gonzalez et al. 2001). Interestingly, E7 suppresses miR-203 expression and induces p63 expression; p63 is an enhancer of cancer stem cells (Melar-New and Laimins 2010; Keyes et al. 2011). MiR-203 is a repressor of dermal stem cells, but the molecular mechanism underlying the transcriptional regulation of miR-203 is unknown (Yi et al. 2008). Moreover, high-risk HPV E5 regulates miR-146a, miR-203, and miR-324-5p (Greco et al. 2011). High-risk HPV infection itself may regulate cell differentiation by repressing the expression of human miRNAs.

5 MicroRNAs in Hepatitis C Virus; Expression, Regulation and Function

5.1 Basic Knowledge of HCV

Hepatitis C virus (HCV) has a 9.6 kb genome that encodes a single positive-strand polyprotein, which is organized in structural and the non-structural (NS-) replication proteins. The open reading frame is flanked by the 5′- and 3′ -UTRs that contain the cis-signals for the translation and replication of the viral RNA. The structural proteins, which form the viral particle, include the core protein and the envelope glycoproteins E1 and E2. The non-structural proteins include p7 ion channel, NS2-3 protease, NS3 serine protease and RNA helicase, NS4A polypeptide, NS4B and NS5A proteins, and NS5B RNA-dependent RNA polymerase (RdRp) (Appel et al. 2006; Moradpour et al. 2007) (Fig. 7.3). HCV infection is a cause of chronic hepatitis, liver cirrhosis, and hepatocellular carcinoma (HCC) (Wasley and Alter 2000).

5.2 The Expression and Role of miRNAs in Liver

The miRNA expression pattern differs dramatically among internal organs. miR-122 constitutes ~70 % of the hepatic miRNAs (Landgraf et al. 2007), and its function in the liver is varied. MiR-122 maintains the hepatic function by down-regulating genes involved in cholesterol synthesis like HMG-CoA reductase, amongst others (Esau et al. 2006). The serum lipid profiles of both liver-specific and germline knockouts of miR-122 induced a 30 % reduction in the total cholesterol, LDL, HDL, and serum triglyceride level. Against expectations, the knockout miR-122 mice had progressive steatohepatitis (Hsu et al. 2012; Tsai et al. 2012). MiR-122 can also regulate lipid synthesis in the liver by controlling expression of *Agpat1* and *Cidec* (Hsu et al. 2012; Tsai et al. 2012). These genes are a part of the triglyceride biosynthesis pathway (Kim et al. 2008). Gatfield et al. (2009) showed that miR-122 is associated with circadian rhythm as the circadian metabolic regulators of the PPAR family are regulated by the miR-122-mediated metabolic control.

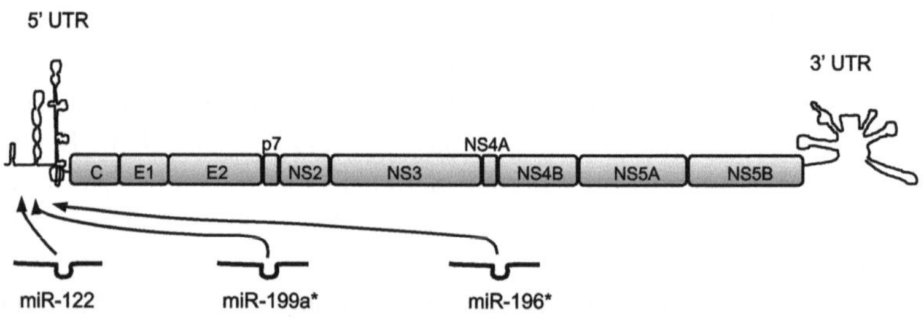

Fig. 7.3 Structure of the HCV genome. The recognition sites of miR-122, miR-199a*, and miR-196* on the HCV genome and in miRNAs are shown. This confirms the *in vitro* replication of HCV

In miR-122-knock out animals, steatohepatitis and liver fibrosis were observed (Hsu et al. 2012; Tsai et al. 2012; Gatfield et al. 2009). MiR-122 is also related to liver inflammation (Lanford et al. 2010). Several researchers showed that the expression level of miR-122 is reduced in experimental models and clinical samples of HCC, and loss of miR-122 is associated with tumor invasiveness and cancer progression (Hsu et al. 2012; Tsai et al. 2012; Wu et al. 2009; Coulouarn et al. 2009; Bai et al. 2009; Cheung et al. 2008; Wang et al. 2012).

5.3 *MiRNA and HCV Associated Liver Disease*

It was demonstrated that HCV replication is controlled by miR-122 (Jopling et al. 2005), and since then, the function of miR-122 in the hepatic tissue is mostly analyzed in relation to HCV replication. The reasons why HCV replication is controlled by suppressing the function of miR-122 are (1) The binding site of miR-122 is downstream to the internal ribosomal entry site, which controls duplication in the early stages of HCV infection (Henke et al. 2008), (2) It is possible that the isoprenoid biosynthetic pathway, controlled by miR-122, regulates HCV replication.miR-122 can directly regulate HCV replication when used as a target gene (Henke et al. 2008), (3) The recognition site of miR-122 in HCV is located in both the 5′-UTR and 3′-UTR domains. miR-122 forms an oligomeric complex in which one miR-122 molecule binds to the 5′-terminus of the HCV RNA, masking the 5′-terminal sequences of the HCV genome while the 3′ nucleotides are overhanging (Machlin et al. 2011). We highlight the most recent findings regarding the role of miRNAs in viral hepatitis, liver fibrosis, and HCC by analyzing the possible mechanisms by which they contribute to the progression of chronic liver disease. MiR-122, which is liver-tropic, can control HCV by stimulating and accelerating translation during replication of HCV (Jopling et al. 2005) and inhibition of miR-122 can block HCV replication.

Lohmann et al. (1999) have developed the HCV subgenomic replicon system, in which an HCV subgenomic replicon autonomously replicates in Huh-7 cells (HCV replicon cells). This technology has contributed greatly to the development of anti viral agents, and helped us to monitor the effect of miRNA on the replication of HCV. The algorithms that search miRNAs responsible for HCV-targets were demonstrated. MiRNAs, except miR-122, can also control the replication of HCV (Hsu et al. 2007) (Fig. 7.4). MiR-199a* can recognize the 5′-UTR region so over-expressing or inhibiting miR-199a* can respectively suppress or enhance HCV replication (Murakami et al. 2009). MiR-196, a HCV protein repressor, can recognize HCV genome as target gene (Hou et al. 2010) (Fig. 7.3).

MiR-130a expression was significantly higher in HCV-infected hepatocytes and liver biopsy specimens than in controls. MiR-130a can regulate interferon-induced trans-membrane 1 (IFITM1). Up-regulation of miR-130a in HCV infections reduces the expression level of IFITM1. This can inhibit HCV replication (Bhanja Chowdhury et al. 2012). The hepatic miRNA expression pattern that exists in

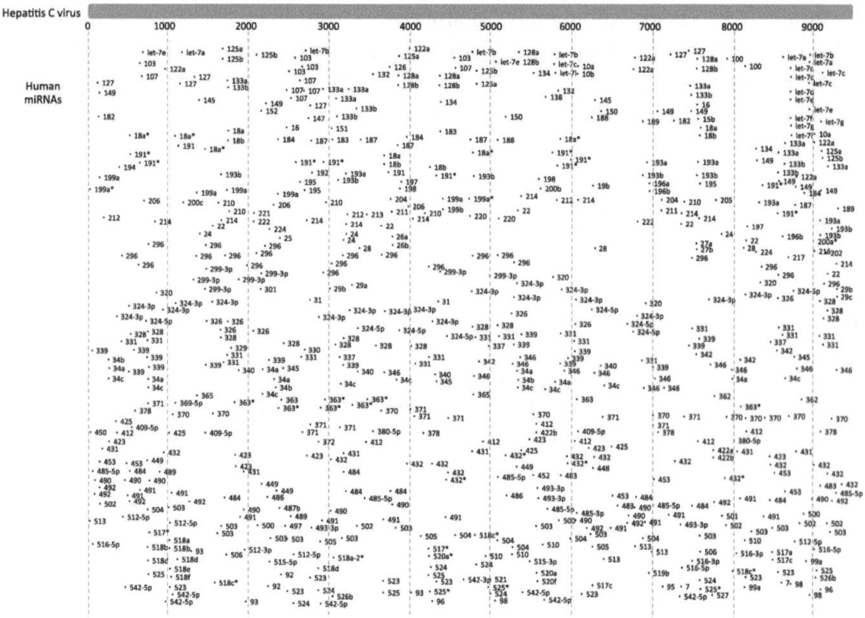

Fig. 7.4 Hypothetical miRNAs target sites on HCV genome

chronic hepatitis c (CHC) patients before pegylated interferon and ribavirin combination therapy is associated with their therapeutic outcome. The expression level of nine miRNAs was significantly different in the sustained virological response (SVR) and non-responder (NR) groups. The accuracy of this diagnosis is 70.5 % (Murakami et al. 2010). Viral species may have different expression patterns for miRNA; for example, expression patterns of miRNAs are unique in HBV and HCV infections and are closely related to liver disease progression. When seventeen miRNAs are down-regulated in HCC, cancer –associated pathways such as cell cycle, adhesion, proteolysis, transcription, and translation are enhanced. However, when miRNAs are up-regulated in HCC, the anti-tumor immune response is suppressed (Murakami et al. 2010).

The miRNAs can recognize HCV genome by using the *in silico* target search algorithm (ViTa: http://vita.mbc.nctu.edu.tw). The number above the bar indicates the nucleotide number.

The paragraphs written above summarize the close relationship between miRNA and HCV infection and chronic liver disease. The accumulated information between expression pattern of miRNAs and HCV infection can pave the way for clinical application. This knowledge has opened the path to clinical applications of miRNA analysis. Many researchers have attempted to diagnose cancer using the miRNA expression in serum or plasma (Kosaka et al. 2010). Expression pattern in circulating miRNAs were used to diagnose chronic liver disease (Bihrer et al. 2011; Cermelli et al. 2011; van der Meer et al. 2013; Shrivastava et al. 2013; Murakami et al. 2012). The second phase of the clinical trial for chronic hepatitis C involves a complementary based sequence of miR-122 and the uses of locked nucleic acid-miR-122 to control the function of liver miR-122 (Janssen et al. 2013). Biomarkers and nucleic acid drugs will be applied clinically in the near future.

6 MicroRNAs in Hepatitis B Virus; Expression, Regulation and Function

6.1 *Basic Knowledge of HBV Infection*

The hepatitis B virus (HBV) is a small enveloped partially double-stranded DNA virus that belongs to the *Hepadnaviridae* family. This virus primarily infects hepatocytes and causes acute and chronic liver disease. Among the 2,000 million people worldwide infected with HBV, more than 350 million remain chronically infected and become carriers of the virus (Ganem and Prince 2004). Epidemiological studies have revealed that chronic HBV infection is the major etiological factor in the development of HCC. Indeed, more than a half of the HCC patients are HBV carriers (Parkin 2006). Despite the availability of an efficacious vaccine, persistent HBV infection remains a challenging global health issue that requires a better understanding of the virus biology and pathogenesis for improved control and treatment.

The life cycle of HBV is complex (Fig. 7.5a). The initial stages of the acute HBV infection, including virion attachment, uncoating and nucleocapsid transport to the cell nucleus, are still poorly understood (Seeger and Mason 2000; Yan et al. 2012). Once delivered into the nucleus, the 3.2 kb relaxed circular DNA genome is converted into a covalently closed circular DNA (cccDNA) from which all the viral RNAs are transcribed. These transcripts include the pregenomic RNA (pgRNA) that will serve as template for reverse transcription and the subgenomic mRNAs that derive from the four overlapping gene sequences composing the viral genome. These sequences comprise the pre-S and surface genes, the precore and core genes, the polymerase gene, and the X gene. The newly formed nucleocapsids can either assemble with envelope proteins in the endoplasmic reticulum and form mature virion that will be secreted, or return to the nucleus to maintain the cccDNA amplification. When the immune system fails to clear the virus, the HBV infection becomes chronic (Fig. 7.5b). Eventually, the viral genetic material or sequences can integrate into the host cellular DNA. The integration has been frequently observed and is associated with HCC (Brechot et al. 1980; Paterlini-Brechot et al. 2003).

MiRNAs play key roles in the regulation of almost every cellular process in all multicellular eukaryotes (Bartel 2009). As intracellular pathogens, viruses are affected by these post-transcriptional modulators and have found a way to subvert their effects. Several viruses, especially the herpesviruses, encode for miRNAs that increase their replication potential and/or allow the evasion from the innate immune system (Skalsky and Cullen 2010). This chapter will outline the implication of miRNAs in the HBV biology and the associated pathogenesis, including HCC development. We will also outline the present and future miRNA-based strategies for the diagnosis, prognosis and treatment of the HBV-related HCC.

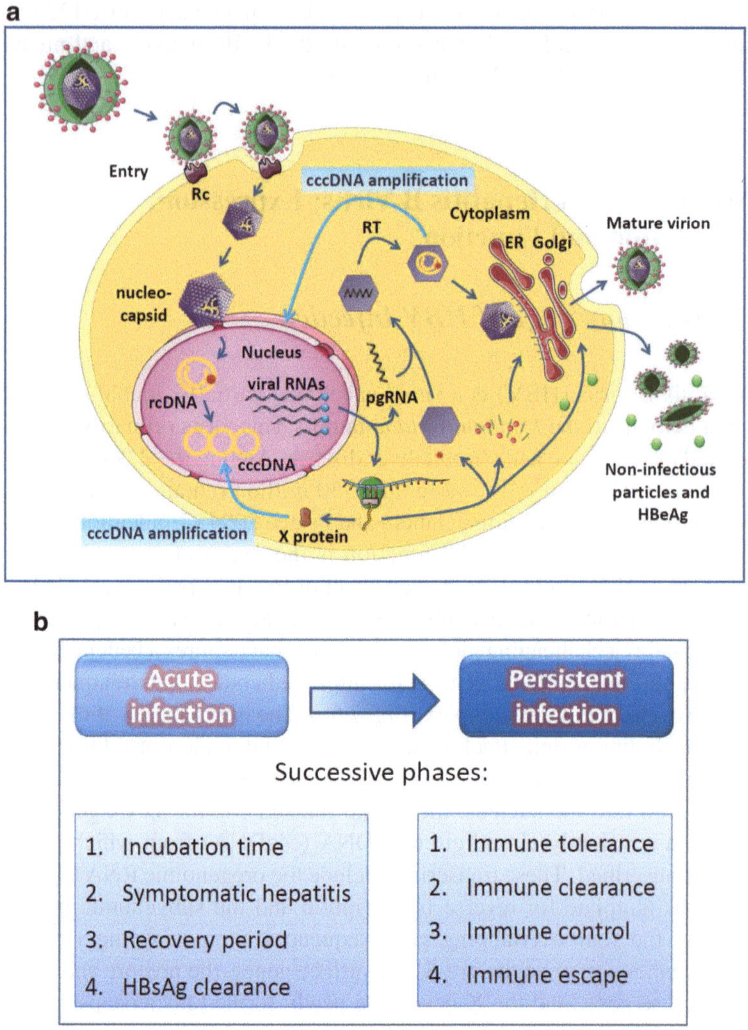

Fig. 7.5 The HBV infection (**a**) Schematic representation of the HBV life cycle. (**b**) HBV natural history of infection. Abbreviations: *cccDNA* covalently closed circular DNA, *ER* endoplasmic reticulum, *HBeAg* hepatitis B extracellular "e" antigen, *HBsAg* HBV surface antigen, *pgRNA* pregenomic RNA, *Rc* receptor, *rcDNA* relaxed circular DNA, *RT* reverse transcription

6.2 MiRNAs Involved in the Regulation of HBV Gene Expression, Replication and Effects on the Carcinogenesis

Viruses, nuclear DNA viruses in particular, need some time to complete their life cycle. During this period, the host cell can develop defense mechanisms such as cell cycle arrest and viral clearance. By taking advantage of the cellular miRNA machinery,

Table 7.2 Cellular miRNAs and their effects on HBV biology, pathogenesis or related-HCC HBV (↑): Promotes HBV replication, HBV (↓): Inhibits HBV replication, HCC (↑): Development and/or growth of HCC

Target genes	miRNAs	miRNA expressions	HBV or HCC status	Reference
Viral target genes				
HBsAg	miR-199-3p	Up	HBV(↓)	Zhang et al. (2010)
HBVpre-S1	miR-210	Up	HBV(↓)	Zhang et al. (2010)
Cellular target genes				
HDAC4	miR-1	Up	HBV (↑)	Zhang et al. (2011a)
c-myb	miR-15a	Down	HCC (↑)	Liu et al. (2009)
E2F1 (c-myc repressor)	miR-17-92 cluster	Up	HCC (↑)	Connolly et al. (2008)
PTEN (?)	miR-21	Up	HCC (↑)	Connolly et al. (2008)
cyclin G1 (p53 modulator)	miR-122	Down	HBV (↑), HCC (↑)	Wang et al. (2012)
DNMT1	miR-152	Down	HBV (↓)	Huang et al. (2010)
SOCS1 (STAT inhibitor)	miR-155	Up	HBV(↓)	Su et al. (2011)
HLA-A (miR-181)	miR-181a, -181b, 200b	Up	HBV (↑)	Liu et al. (2009)
NFIB	miR-372,-373	Up	HBV (↑)	Guo et al. (2011)
STAT3	let-7 family	Down	HBV (↑?), HCC (↑)	Wang et al. (2010)

these viruses can more easily and efficiently help to promote a favorable cellular environment for viral replication and achievement of the life cycle (Skalsky and Cullen 2010). The modulation of the machinery could be made by direct action on the cellular miRNAs (Backes et al. 2012; Jopling et al. 2005) (inhibition or up-regulation) or by expression of their own miRNAs that will mimic their cellular counterparts (Gottwein et al. 2007; Lu and Cullen 2004). Despite the fact that HBV is a nuclear DNA virus, none viral-encoded miRNA has been identified so far. Only one putative HBV-miRNA, with hypothetical regulation role on its own genome, was deduced by computational approach (Jin et al. 2007). However, several cellular miRNAs are involved in the HBV viral replication. They are presented here above and summarized in Table 7.2.

6.2.1 Cellular miRNAs That Promote HBV Replication

MiR-1 can enhance the HBV core promoter transcription and thus increase the viral replication by modulating the expression of several host genes such as transcription factors (Zhang et al. 2011a). The report has confirmed that the histone deacetylase 4 (HDAC4) expression is down-regulated by miR-1. Knowing that the cccDNA amplification is controlled by epigenetic regulation (Pollicino et al. 2006), miR-1 could act in complementarity with the nuclear HBV X protein (HBx) in order to induce these modifications (Belloni et al. 2009). However, miR-1 can also inhibit the cell proliferation and even induce a reverse cancer cell phenotype (Zhang et al. 2011a). The roles of miR-1 in the cell proliferation and hepatocellular carcinogenesis (Datta et al. 2008) seem to be contradictory with the viral replication and with the characteristics of oncogenic virus but must represent benefit for HBV survival.

Another miRNA, miR-501, has also been suggested to work together with HBx for the benefit of viral replication (Jin et al. 2013). HBx itself has also the ability to deregulate the cellular miRNAs expression. This small protein is a key regulator of HBV infection. It is usually over-expressed in HCC and accumulated evidence indicates that HBx can promote hepatocarcinogenesis by disrupting the normal physiologic mechanisms of the host cell (Chirillo et al. 1997; Lee et al. 2005; Tian et al. 2013). The let-7 family of miRNAs has been demonstrated to be negatively regulated by HBx (Wang et al. 2010). This miRNA family is often observed down-regulated in many cancers including HCC (Guo et al. 2006; Johnson et al. 2005; Yu et al. 2007). The consequence of this down-regulation is the increase activity of that signal transducer and activator of transcription 3 (STAT3) that supports the cell proliferation, and potentially the hepatocarcinogenesis.

Finally, the miRNAs can promote the viral replication by the indirect stimulation the HBV enhancer element I or II. It is the case for the CCAAT/enhancer binding protein that binds and activates the HBV enhancer II in a dose-dependent manner (Lopez-Cabrera et al. 1991). miR-372, together with miR-373, targets the nuclear factor I/B, an important regulator of several viruses (Nagata et al. 1983), and so supports the HBV expression (Guo et al. 2011).

6.2.2 Cellular miRNAs That Prevent HBV Replication

One of the best studied miRNAs in liver-related diseases is miR-122. This liver-specific miRNA is expressed at high levels in normal hepatocytes (about 70 % of the total miRNA population in the adult liver) (Lagos-Quintana et al. 2002) and is pivotal in numerous aspects of the liver function such as lipid metabolism, liver development, differentiation, growth and neoplastic transformation (Girard et al. 2008). The essential role of mir-122 in the HCV replication reflects furthermore the importance of this miRNA in the infection process (Jopling et al. 2005). While the loss of miR-122 expression is impeding HCV replication, it is enhancing the replication in the circumstance of HBV infection (Wang et al. 2012). In fact, miR-122 can negatively regulate the viral gene expression and replication by direct binding to a highly conserved sequence of HBV (Chen et al. 2011). This repression effect can apparently be impeded by a negative feedback loop involving the Heme oxygenase-1 (Qiu et al. 2010). A recent study has reported the indirect implication of HBx in miR-122 deregulation (Song et al. 2013) that could, at least partially, explain the difference observed between the two viruses. Knowing that miR-122 expression is low in HBV and HCC tissues (Wang et al. 2012; Kutay et al. 2006) and that HBV replication is usually low or absent in HCC cells (Wong et al. 2006), miR-122 is a highly potential linker between HBV infection and liver carcinogenesis (Wang et al. 2012; Fan et al. 2011) and therefore a predilected target for future clinical applications.

The miR-17-92 cluster is also important in the HBV-associated HCC. This polycistron includes six miRNAs (miR-17-5p, miR-18a, miR-19a, miR-19b, miR-20a and miR-92a-1) and its up-regulated expression is associated with malignancies (Hayashita et al. 2005). By using human HBV-positive human HCC tissues, hepatoma cell lines and woodchuck hepatitis virus -induced HCC animal model (Popper et al. 1987), Connolly and colleagues were able to demonstrate the elevated expression of miR-17-92 cluster and its implication in the malignant phenotype (Connolly et al. 2008). The expression could be amplified by c-myc activation (He et al. 2005), under HBx control (Terradillos et al. 1997), to contribute to HBV latency state (Jung et al. 2013). The consequence is the induction of liver oncogenesis. Since the RNA intermediates of HBV (pgRNA and transcripts) are good targets of miRNA action, it is not surprising to observe several cellular miRNAs with different binding sites. So, in addition to miR-122 that targets the polymerase region (Chen et al. 2011), the mir-199a-3p and mir-210 can repress the S and pre-S1 regions, respectively (Zhang et al. 2010).

All the examples illustrating cellular miRNAs as inhibitors of the viral replication are a bit difficult to comprehend initially because of their obvious negative effect on HBV infection. However, it can be understood by keeping in mind the survival of the virus into the host organism. The natural history of HBV infection shows often a transition from acute to chronic infection, especially in young children. This step corresponds to a failure of the immune system to eradicate the virus (Fig. 7.5b). One of the escape pathways is the successful adaptation to the immune-induced down-regulation of replication. The virus could evade the immune system by reaching a dormant state into the infected hepatocytes, under the cccDNA form, and survive until its eventual life cycle reactivation (Ganem and Prince 2004; Belloni et al. 2009, 2012; Huang et al. 2010). The study of Huang and colleagues reports the CpG islands methylation of the cccDNA by the DNA methyltransferase 1 (DNMT1) to prevent the viral gene expression and therefore the viral antigen presentation. DNMT1 over-expression is induced by a decrease of miR-152, under the effect of HBx (Huang et al. 2010).

6.3 MiRNAs in the Modulation of the Immune System and Effects on the Carcinogenesis

HBV must adapt to a very complex network in order to survive. It has to cope with the modification of homeostasis, the cell cycle arrest, the apoptosis and the destruction of the host cell by the immune cells. MiRNAs are also important in the development and function of immune system (Baltimore et al. 2008). Some miRNAs in particular are crucial for modulating innate and adaptive immune responses. MiR-155 has multi-roles during an innate immune response such as the regulation of the acute inflammatory response after recognition of pathogens by the toll-like

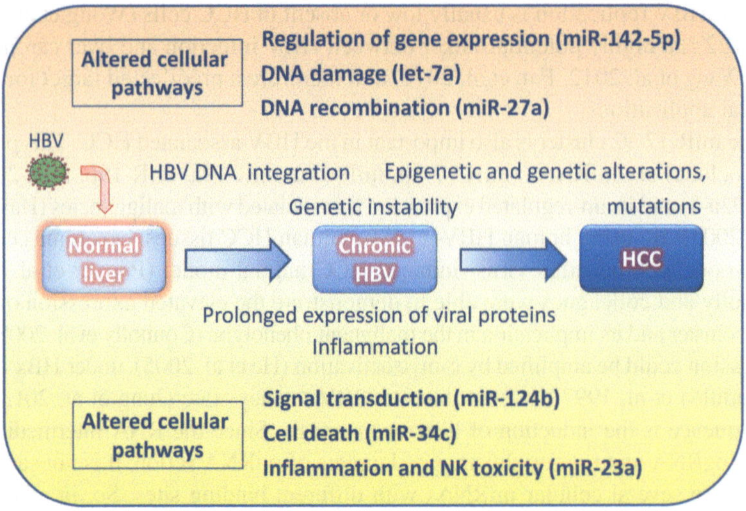

Fig. 7.6 Chronology of events from the HBV infection until HCC development. The indicated altered miRNAs and related pathways are based on the results from Ura et al. (2009)

receptors (O'Connell et al. 2007; Tili et al. 2007). The up-regulation of miR-155 can lead to prolonged exposure to inflammation, a well-known causal agent to cancers like HCC (Berasain et al. 2009). Two recent studies suggest a role of miR-155 in hepatocarcinogenesis and HBV infection (Table 7.2). Using HCC-induced mouse model, Wang and collaborators have demonstrated an oncogenic role of miR-155 at the early stages of the tumorigenesis (Wang et al. 2009a). On the other hand, the ectopic expression of miR-155 in human hepatoma cells enhances the innate immunity through promotion of the JAK/STAT pathway and down-regulates HBx expression (Su et al. 2011).

A study analyzing the modified expression profiles of miRNAs in a stable HBV-expressing cell line revealed the up-regulation of miR-181a (Liu et al. 2009) (Table 7.2). The deregulation of this miRNA in liver cell might participate to the establishment of HBV persistence through inhibition of the human leukocyte antigen A (HLA-A) -dependent HBV antigen presentation. To date, it is unclear if miRNAs altered in the host cell, like miR-181a and miR-146a also present in Liu's study, miRNAs involved in ubiquitous and cell-specific regulatory functions, could affect directly the immune cells. The presence of circulating miRNAs, as well as the existence of intercellular nanovesicle-mediated miRNA transfer and its impact on the environmental modulation, could potentially support that hypothesis (Arataki et al. 2013; Waidmann et al. 2012; Li et al. 2010, 2012; Zhou et al. 2011; Kogure et al. 2011). The current knowledge shows an altered miRNA profile expression between normal and HCC liver at the different stages, and between the HBV and HCV-induced HCC (Murakami et al. 2006; Li et al. 2008; Budhu et al. 2008; Ura et al. 2009). For the latest one, this reflects the variation in the cellular pathways that are modulated as a consequence of the viral infection (Fig. 7.6).

6.4 *MiRNAs as Biomarkers and Treatment-Based Strategies for HBV Infection and HBV-Induced HCC*

It is important to know the precise mechanisms, the cellular pathways that the viral infection or cancer cells alter in the different steps of the infection and/or tumor evolution. The knowledge will allow developing powerful targeted therapeutical strategies. The significance of miRNAs in antiviral immunity and liver carcinogenesis emphasizes their values as therapeutic targets for HBV infection and HBV-induced HCC. MiR-122 and miR-18a are of particular interest. They are both released in the blood and could be used as potential non-invasive biomarkers for HBV-related HCC screening (Liu et al. 2009; Waidmann et al. 2012; Li et al. 2012). Some other reports suggest using a miRNA panel in order to improve the specificity of the test (Li et al. 2010; Zhou et al. 2011). In addition with the current routinely used markers such as HBV surface antigen, HBV extracellular antigen and alanine aminotransferase, the circulating miRNAs represent a significant clinical value for better evaluation of the HBV-infection status, liver injury and early diagnosis of HCC.

In the therapeutic perspective, the work of Ura's group is valuable. They analyzed the livers of HBV and HCV positive patients with HCC to identify the miRNAs that are differentially expressed. Nineteen miRNAs were clearly differentiated between HBV and HCV groups, six specific for HBV and thirteen specific for HCV. Based on the miRNAs profile, they made a pathway analysis of candidate targeted genes and were also able to distinguish the cellular mechanisms altered in HBV or HCV-infected livers (Ura et al. 2009). The HBV infection alters mostly the pathways related to signal transduction, inflammation and natural killer toxicity, DNA damage, recombination, and cell death (Fig. 7.6), while HCV infection modifies those involved in immune response involving antigen presentation, cell cycle and cell adhesion (Ura et al. 2009).

Moreover, technological advances in the delivery of miRNA and RNA interference enable safe and efficient *in vivo* miRNA gene therapy, as exemplify by the recent study from Kota and collaborators on the liver cancer (Kota et al. 2009). They used an adeno-associated virus to deliver miR-26a in a mouse model of HCC. This resulted in the successful inhibition of the cancer cell proliferation, induction of the tumor-specific apoptosis, and protection from disease progression without toxicity.

7 Concluding Remarks

MiRNAs have emerged as novel key players in the control of gene expression in cells. Investigations of their profiling have unveiled specific miRNA deregulations in tumors and in condition of viral infection. On the viral point of view, the deregulated pathways mirror the strategies of the virus to allow its replication and evade the host defense mechanisms to survive. On the cellular point of view, they mirror the immune response that is trying to get rid of the intruder and that become deregulated. In both cases, the viral infection leads to the alteration of miRNA expression by RSSs that can trigger tumorigenesis. Several oncogenic viruses, especially herpesviruses like EBV and KSHV, encode their own miRNAs to modify both cellular and viral gene expression (Pfeffer et al. 2004). This step is crucial for their latency phase. On the other hand, HPV, HBV and HCV do not express viral miRNAs but can affect the host miRNA pathway. The present and future knowledge about miRNA will broaden our understanding of the pathogenesis of oncogenic viruses and most certainly allow developing efficient oncogenic viral therapies.

Acknowledgements The authors would like to thank Servier Medical Art for their image bank used to create the illustrations.

References

Abend JR, Uldrick T, Ziegelbauer JM (2010) Regulation of tumor necrosis factor-like weak inducer of apoptosis receptor protein (TWEAKR) expression by Kaposi's sarcoma-associated herpesvirus microRNA prevents TWEAK-induced apoptosis and inflammatory cytokine expression. J Virol 84(23):12139–12151

Alajez NM, Lenarduzzi M, Ito E, Hui AB, Shi W, Bruce J et al (2011) MiR-218 suppresses nasopharyngeal cancer progression through downregulation of survivin and the SLIT2-ROBO1 pathway. Cancer Res 71(6):2381–2391

Amoroso R, Fitzsimmons L, Thomas WA, Kelly GL, Rowe M, Bell AI (2011) Quantitative studies of Epstein-Barr virus-encoded microRNAs provide novel insights into their regulation. J Virol 85(2):996–1010

Appel N, Schaller T, Penin F, Bartenschlager R (2006) From structure to function: new insights into hepatitis C virus RNA replication. J Biol Chem 281(15):9833–9836

Arataki K, Hayes CN, Akamatsu S, Akiyama R, Abe H, Tsuge M et al (2013) Circulating microRNA-22 correlates with microRNA-122 and represents viral replication and liver injury in patients with chronic hepatitis B. J Med Virol 85(5):789–798

Au Yeung CL, Tsang TY, Yau PL, Kwok TT (2011) Human papillomavirus type 16 E6 induces cervical cancer cell migration through the p53/microRNA-23b/urokinase-type plasminogen activator pathway. Oncogene 30(21):2401–2410

Backes S, Shapiro JS, Sabin LR, Pham AM, Reyes I, Moss B et al (2012) Degradation of host microRNAs by poxvirus poly(A) polymerase reveals terminal RNA methylation as a protective antiviral mechanism. Cell Host Microbe 12(2):200–210

Baer R, Bankier AT, Biggin MD, Deininger PL, Farrell PJ, Gibson TJ et al (1984) DNA sequence and expression of the B95-8 Epstein-Barr virus genome. Nature 310(5974):207–211

Bai S, Nasser MW, Wang B, Hsu SH, Datta J, Kutay H et al (2009) MicroRNA-122 inhibits tumorigenic properties of hepatocellular carcinoma cells and sensitizes these cells to sorafenib. J Biol Chem 284(46):32015–32027

Baltimore D, Boldin MP, O'Connell RM, Rao DS, Taganov KD (2008) MicroRNAs: new regulators of immune cell development and function. Nat Immunol 9(8):839–845

Bartel DP (2009) MicroRNAs: target recognition and regulatory functions. Cell 136(2):215–233

Barth S, Pfuhl T, Mamiani A, Ehses C, Roemer K, Kremmer E et al (2008) Epstein-Barr virus-encoded microRNA miR-BART2 down-regulates the viral DNA polymerase BALF5. Nucleic Acids Res 36(2):666–675

Bellare P, Ganem D (2009) Regulation of KSHV lytic switch protein expression by a virus-encoded microRNA: an evolutionary adaptation that fine-tunes lytic reactivation. Cell Host Microbe 6(6):570–575

Belloni L, Pollicino T, De Nicola F, Guerrieri F, Raffa G, Fanciulli M et al (2009) Nuclear HBx binds the HBV minichromosome and modifies the epigenetic regulation of cccDNA function. Proc Natl Acad Sci U S A 106(47):19975–19979

Belloni L, Allweiss L, Guerrieri F, Pediconi N, Volz T, Pollicino T et al (2012) IFN-alpha inhibits HBV transcription and replication in cell culture and in humanized mice by targeting the epigenetic regulation of the nuclear cccDNA minichromosome. J Clin Invest 122(2):529–537

Berasain C, Castillo J, Perugorria MJ, Latasa MU, Prieto J, Avila MA (2009) Inflammation and liver cancer: new molecular links. Ann N Y Acad Sci 1155:206–221

Bhanja Chowdhury J, Shrivastava S, Steele R, Di Bisceglie AM, Ray R, Ray RB (2012) Hepatitis C virus infection modulates expression of interferon stimulatory gene IFITM1 by upregulating miR-130A. J Virol 86(18):10221–10225

Bihrer V, Friedrich-Rust M, Kronenberger B, Forestier N, Haupenthal J, Shi Y et al (2011) Serum miR-122 as a biomarker of necroinflammation in patients with chronic hepatitis C virus infection. Am J Gastroenterol 106(9):1663–1669

Boshoff C, Weiss R (2002) AIDS-related malignancies. Nat Rev Cancer 2(5):373–382

Brechot C, Pourcel C, Louise A, Rain B, Tiollais P (1980) Presence of integrated hepatitis B virus DNA sequences in cellular DNA of human hepatocellular carcinoma. Nature 286(5772): 533–535

Budhu A, Jia HL, Forgues M, Liu CG, Goldstein D, Lam A et al (2008) Identification of metastasis-related microRNAs in hepatocellular carcinoma. Hepatology 47(3):897–907

Cai X, Lu S, Zhang Z, Gonzalez CM, Damania B, Cullen BR (2005) Kaposi's sarcoma-associated herpesvirus expresses an array of viral microRNAs in latently infected cells. Proc Natl Acad Sci U S A 102(15):5570–5575

Cai X, Schafer A, Lu S, Bilello JP, Desrosiers RC, Edwards R (2006) Epstein-Barr virus microRNAs are evolutionarily conserved and differentially expressed. PLoS Pathog 2(3):e23

Cermelli S, Ruggieri A, Marrero JA, Ioannou GN, Beretta L (2011) Circulating microRNAs in patients with chronic hepatitis C and non-alcoholic fatty liver disease. PLoS One 6(8):e23937

Chen C, Ridzon DA, Broomer AJ, Zhou Z, Lee DH, Nguyen JT et al (2005) Real-time quantification of microRNAs by stem-loop RT-PCR. Nucleic Acids Res 33(20):e179

Chen SJ, Chen GH, Chen YH, Liu CY, Chang KP, Chang YS et al (2010) Characterization of Epstein-Barr virus miRNAome in nasopharyngeal carcinoma by deep sequencing. PLoS One 5(9) pii: e12745

Chen Y, Shen A, Rider PJ, Yu Y, Wu K, Mu Y et al (2011) A liver-specific microRNA binds to a highly conserved RNA sequence of hepatitis B virus and negatively regulates viral gene expression and replication. FASEB J 25(12):4511–4521

Cheung O, Puri P, Eicken C, Contos MJ, Mirshahi F, Maher JW et al (2008) Nonalcoholic steatohepatitis is associated with altered hepatic microRNA expression. Hepatology 48(6):1810–1820

Chirillo P, Pagano S, Natoli G, Puri PL, Burgio VL, Balsano C et al (1997) The hepatitis B virus X gene induces p53-mediated programmed cell death. Proc Natl Acad Sci U S A 94(15): 8162–8167

Cho WC, Chow AS, Au JS (2009) Restoration of tumour suppressor hsa-miR-145 inhibits cancer cell growth in lung adenocarcinoma patients with epidermal growth factor receptor mutation. Eur J Cancer 45(12):2197–2206

Choy EY, Siu KL, Kok KH, Lung RW, Tsang CM, To KF et al (2008) An Epstein-Barr virus-encoded microRNA targets PUMA to promote host cell survival. J Exp Med 205(11):2551–2560

Christoffersen NR, Shalgi R, Frankel LB, Leucci E, Lees M, Klausen M et al (2010) p53-independent upregulation of miR-34a during oncogene-induced senescence represses MYC. Cell Death Differ 17(2):236–245

Connolly E, Melegari M, Landgraf P, Tchaikovskaya T, Tennant BC, Slagle BL et al (2008) Elevated expression of the miR-17-92 polycistron and miR-21 in hepadnavirus-associated hepatocellular carcinoma contributes to the malignant phenotype. Am J Pathol 173(3):856–864

Cosmopoulos K, Pegtel M, Hawkins J, Moffett H, Novina C, Middeldorp J et al (2009) Comprehensive profiling of Epstein-Barr virus microRNAs in nasopharyngeal carcinoma. J Virol 83(5):2357–2367

Coulouarn C, Factor VM, Andersen JB, Durkin ME, Thorgeirsson SS (2009) Loss of miR-122 expression in liver cancer correlates with suppression of the hepatic phenotype and gain of metastatic properties. Oncogene 28(40):3526–3536

Datta J, Kutay H, Nasser MW, Nuovo GJ, Wang B, Majumder S et al (2008) Methylation mediated silencing of MicroRNA-1 gene and its role in hepatocellular carcinogenesis. Cancer Res 68(13):5049–5058

de Vries W, Berkhout B (2008) RNAi suppressors encoded by pathogenic human viruses. Int J Biochem Cell Biol 40(10):2007–2012

Dolken L, Malterer G, Erhard F, Kothe S, Friedel CC, Suffert G et al (2010) Systematic analysis of viral and cellular microRNA targets in cells latently infected with human gamma-herpesviruses by RISC immunoprecipitation assay. Cell Host Microbe 7(4):324–334

Dreher A, Rossing M, Kaczkowski B, Andersen DK, Larsen TJ, Christophersen MK et al (2011) Differential expression of cellular microRNAs in HPV 11, -16, and -45 transfected cells. Biochem Biophys Res Commun 412(1):20–25

Dyson N, Howley PM, Munger K, Harlow E (1989) The human papilloma virus-16 E7 oncoprotein is able to bind to the retinoblastoma gene product. Science 243(4893):934–937

Edwards RH, Marquitz AR, Raab-Traub N (2008) Epstein-Barr virus BART microRNAs are produced from a large intron prior to splicing. J Virol 82(18):9094–9106

Esau C, Davis S, Murray SF, Yu XX, Pandey SK, Pear M et al (2006) miR-122 regulation of lipid metabolism revealed by in vivo antisense targeting. Cell Metab 3(2):87–98

Fan CG, Wang CM, Tian C, Wang Y, Li L, Sun WS et al (2011) miR-122 inhibits viral replication and cell proliferation in hepatitis B virus-related hepatocellular carcinoma and targets NDRG3. Oncol Rep 26(5):1281–1286

Feederle R, Haar J, Bernhardt K, Linnstaedt SD, Bannert H, Lips H et al (2011) The members of an Epstein-Barr virus microRNA cluster cooperate to transform B lymphocytes. J Virol 85(19):9801–9810

Fujii T, Taguchi H, Katano H, Mori S, Nakamura T, Nojiri N et al (1999) Seroprevalence of human herpesvirus 8 in human immunodeficiency virus 1-positive and human immunodeficiency virus 1-negative populations in Japan. J Med Virol 57(2):159–162

Gabriely G, Wurdinger T, Kesari S, Esau CC, Burchard J, Linsley PS et al (2008) MicroRNA 21 promotes glioma invasion by targeting matrix metalloproteinase regulators. Mol Cell Biol 28(17):5369–5380

Ganem D, Prince AM (2004) Hepatitis B virus infection–natural history and clinical consequences. N Engl J Med 350(11):1118–1129

Gao P, Tchernyshyov I, Chang TC, Lee YS, Kita K, Ochi T et al (2009) c-Myc suppression of miR-23a/b enhances mitochondrial glutaminase expression and glutamine metabolism. Nature 458(7239):762–765

Gatfield D, Le Martelot G, Vejnar CE, Gerlach D, Schaad O, Fleury-Olela F et al (2009) Integration of microRNA miR-122 in hepatic circadian gene expression. Genes Dev 23(11):1313–1326. 9

Girard M, Jacquemin E, Munnich A, Lyonnet S, Henrion-Caude A (2008) miR-122, a paradigm for the role of microRNAs in the liver. J Hepatol 48(4):648–656

Godshalk SE, Bhaduri-McIntosh S, Slack FJ (2008) Epstein-Barr virus-mediated dysregulation of human microRNA expression. Cell Cycle 7(22):3595–3600

Gonzalez SL, Stremlau M, He X, Basile JR, Munger K (2001) Degradation of the retinoblastoma tumor suppressor by the human papillomavirus type 16 E7 oncoprotein is important for functional inactivation and is separable from proteasomal degradation of E7. J Virol 75(16):7583–7591

Gottwein E, Cullen BR (2010) A human herpesvirus microRNA inhibits p21 expression and attenuates p21-mediated cell cycle arrest. J Virol 84(10):5229–5237

Gottwein E, Mukherjee N, Sachse C, Frenzel C, Majoros WH, Chi JT et al (2007) A viral microRNA functions as an orthologue of cellular miR-155. Nature 450(7172):1096–1099

Gottwein E, Corcoran DL, Mukherjee N, Skalsky RL, Hafner M, Nusbaum JD et al (2011) Viral microRNA targetome of KSHV-infected primary effusion lymphoma cell lines. Cell Host Microbe 10(5):515–526

Gourzones C, Ferrand FR, Amiel C, Verillaud B, Barat A, Guerin M et al (2013) Consistent high concentration of the viral microRNA BART17 in plasma samples from nasopharyngeal carcinoma patients – evidence of non-exosomal transport. Virol J 10:119
Greco D, Kivi N, Qian K, Leivonen SK, Auvinen P, Auvinen E (2011) Human papillomavirus 16 E5 modulates the expression of host microRNAs. PLoS One 6(7):e21646
Grundhoff A, Sullivan CS, Ganem D (2006) A combined computational and microarray-based approach identifies novel microRNAs encoded by human gamma-herpesviruses. RNA 12(5):733–750
Gunasekharan V, Laimins LA (2013) Human papillomaviruses modulate microRNA 145 expression to directly control genome amplification. J Virol 87(10):6037–6043
Guo Y, Chen Y, Ito H, Watanabe A, Ge X, Kodama T et al (2006) Identification and characterization of lin-28 homolog B (LIN28B) in human hepatocellular carcinoma. Gene 384:51–61
Guo H, Liu H, Mitchelson K, Rao H, Luo M, Xie L et al (2011) MicroRNAs-372/373 promote the expression of hepatitis B virus through the targeting of nuclear factor I/B. Hepatology 54(3):808–819
Haecker I, Gay LA, Yang Y, Hu J, Morse AM, McIntyre LM et al (2012) Ago HITS-CLIP expands understanding of Kaposi's sarcoma-associated herpesvirus miRNA function in primary effusion lymphomas. PLoS Pathog 8(8):e1002884
Hayashita Y, Osada H, Tatematsu Y, Yamada H, Yanagisawa K, Tomida S et al (2005) A polycistronic microRNA cluster, miR-17-92, is overexpressed in human lung cancers and enhances cell proliferation. Cancer Res 65(21):9628–9632
He L, Thomson JM, Hemann MT, Hernando-Monge E, Mu D, Goodson S et al (2005) A microRNA polycistron as a potential human oncogene. Nature 435(7043):828–833
Henke JI, Goergen D, Zheng J, Song Y, Schuttler CG, Fehr C et al (2008) microRNA-122 stimulates translation of hepatitis C virus RNA. EMBO J 27(24):3300–3310
Hitt MM, Allday MJ, Hara T, Karran L, Jones MD, Busson P et al (1989) EBV gene expression in an NPC-related tumour. EMBO J 8(9):2639–2651
Hou W, Tian Q, Zheng J, Bonkovsky HL (2010) MicroRNA-196 represses Bach1 protein and hepatitis C virus gene expression in human hepatoma cells expressing hepatitis C viral proteins. Hepatology 51(5):1494–1504
Hsu PW, Lin LZ, Hsu SD, Hsu JB, Huang HD (2007) ViTa: prediction of host microRNAs targets on viruses. Nucleic Acids Res 35(Database issue):D381–D385
Hsu SH, Wang B, Kota J, Yu J, Costinean S, Kutay H et al (2012) Essential metabolic, anti-inflammatory, and anti-tumorigenic functions of miR-122 in liver. J Clin Invest 122(8):2871–2883
Huang J, Wang Y, Guo Y, Sun S (2010) Down-regulated microRNA-152 induces aberrant DNA methylation in hepatitis B virus-related hepatocellular carcinoma by targeting DNA methyltransferase 1. Hepatology 52(1):60–70
Iizasa H, Wulff BE, Alla NR, Maragkakis M, Megraw M, Hatzigeorgiou A et al (2010) Editing of Epstein-Barr virus-encoded BART6 microRNAs controls their dicer targeting and consequently affects viral latency. J Biol Chem 285(43):33358–33370
Ishimoto T, Nagano O, Yae T, Tamada M, Motohara T, Oshima H et al (2011) CD44 variant regulates redox status in cancer cells by stabilizing the xCT subunit of system xc(−) and thereby promotes tumor growth. Cancer Cell 19(3):387–400
Janssen HL, Reesink HW, Lawitz EJ, Zeuzem S, Rodriguez-Torres M, Patel K et al (2013) Treatment of HCV infection by targeting microRNA. N Engl J Med 368(18):1685–1694
Jin WB, Wu FL, Kong D, Guo AG (2007) HBV-encoded microRNA candidate and its target. Comput Biol Chem 31(2):124–126
Jin J, Tang S, Xia L, Du R, Xie H, Song J et al (2013) MicroRNA-501 promotes HBV replication by targeting HBXIP. Biochem Biophys Res Commun 430(4):1228–1233
Johnson SM, Grosshans H, Shingara J, Byrom M, Jarvis R, Cheng A et al (2005) RAS is regulated by the let-7 microRNA family. Cell 120(5):635–647
Jopling CL, Yi M, Lancaster AM, Lemon SM, Sarnow P (2005) Modulation of hepatitis C virus RNA abundance by a liver-specific microRNA. Science 309(5740):1577–1581

Jung YJ, Kim JW, Park SJ, Min BY, Jang ES, Kim NY et al (2013) c-Myc-mediated overexpression of miR-17-92 suppresses replication of hepatitis B virus in human hepatoma cells. J Med Virol 85(6):969–978

Keyes WM, Pecoraro M, Aranda V, Vernersson-Lindahl E, Li W, Vogel H et al (2011) DeltaNp63alpha is an oncogene that targets chromatin remodeler Lsh to drive skin stem cell proliferation and tumorigenesis. Cell Stem Cell 8(2):164–176

Kieff EaR AB (2007) Epstein-Barr virus and its replication. In: Fields virology, 5th edn. Lippincott Williams & Wilkins, Philadelphia

Kim do N, Chae HS, Oh ST, Kang JH, Park CH, Park WS et al (2007) Expression of viral microRNAs in Epstein-Barr virus-associated gastric carcinoma. J Virol 81(2):1033–1036

Kim YJ, Cho SY, Yun CH, Moon YS, Lee TR, Kim SH (2008) Transcriptional activation of Cidec by PPARgamma2 in adipocyte. Biochem Biophys Res Commun 377(1):297–302

Kogure T, Lin WL, Yan IK, Braconi C, Patel T (2011) Intercellular nanovesicle-mediated microRNA transfer: a mechanism of environmental modulation of hepatocellular cancer cell growth. Hepatology 54(4):1237–1248

Kosaka N, Iguchi H, Ochiya T (2010) Circulating microRNA in body fluid: a new potential biomarker for cancer diagnosis and prognosis. Cancer Sci 101(10):2087–2092

Kota J, Chivukula RR, O'Donnell KA, Wentzel EA, Montgomery CL, Hwang HW et al (2009) Therapeutic microRNA delivery suppresses tumorigenesis in a murine liver cancer model. Cell 137(6):1005–1017

Kutay H, Bai S, Datta J, Motiwala T, Pogribny I, Frankel W et al (2006) Downregulation of miR-122 in the rodent and human hepatocellular carcinomas. J Cell Biochem 99(3):671–678

Lagos-Quintana M, Rauhut R, Yalcin A, Meyer J, Lendeckel W, Tuschl T (2002) Identification of tissue-specific microRNAs from mouse. Curr Biol 12(9):735–739

Landgraf P, Rusu M, Sheridan R, Sewer A, Iovino N, Aravin A et al (2007) A mammalian microRNA expression atlas based on small RNA library sequencing. Cell 129(7):1401–1414

Lanford RE, Hildebrandt-Eriksen ES, Petri A, Persson R, Lindow M, Munk ME et al (2010) Therapeutic silencing of microRNA-122 in primates with chronic hepatitis C virus infection. Science 327(5962):198–201

Lee AT, Ren J, Wong ET, Ban KH, Lee LA, Lee CG (2005) The hepatitis B virus X protein sensitizes HepG2 cells to UV light-induced DNA damage. J Biol Chem 280(39):33525–33535

Lei X, Bai Z, Ye F, Xie J, Kim CG, Huang Y et al (2010) Regulation of NF-kappaB inhibitor IkappaBalpha and viral replication by a KSHV microRNA. Nat Cell Biol 12(2):193–199

Li W, Xie L, He X, Li J, Tu K, Wei L et al (2008) Diagnostic and prognostic implications of microRNAs in human hepatocellular carcinoma. Int J Cancer 123(7):1616–1622

Li LM, Hu ZB, Zhou ZX, Chen X, Liu FY, Zhang JF et al (2010) Serum microRNA profiles serve as novel biomarkers for HBV infection and diagnosis of HBV-positive hepatocarcinoma. Cancer Res 70(23):9798–9807

Li L, Guo Z, Wang J, Mao Y, Gao Q (2012) Serum miR-18a: a potential marker for hepatitis B virus-related hepatocellular carcinoma screening. Dig Dis Sci 57(11):2910–2916

Lin YT, Kincaid RP, Arasappan D, Dowd SE, Hunicke-Smith SP, Sullivan CS (2010) Small RNA profiling reveals antisense transcription throughout the KSHV genome and novel small RNAs. RNA 16(8):1540–1558

Lin X, Liang D, He Z, Deng Q, Robertson ES, Lan K (2011) miR-K12-7-5p encoded by Kaposi's sarcoma-associated herpesvirus stabilizes the latent state by targeting viral ORF50/RTA. PLoS One 6(1):e16224

Liu Y, Zhao JJ, Wang CM, Li MY, Han P, Wang L et al (2009) Altered expression profiles of microRNAs in a stable hepatitis B virus-expressing cell line. Chin Med J 122(1):10–14

Lo AK, To KF, Lo KW, Lung RW, Hui JW, Liao G et al (2007) Modulation of LMP1 protein expression by EBV-encoded microRNAs. Proc Natl Acad Sci U S A 104(41):16164–16169

Lo AK, Dawson CW, Jin DY, Lo KW (2012) The pathological roles of BART miRNAs in nasopharyngeal carcinoma. J Pathol 227(4):392–403

Lohmann V, Korner F, Koch J, Herian U, Theilmann L, Bartenschlager R (1999) Replication of subgenomic hepatitis C virus RNAs in a hepatoma cell line. Science 285(5424):110–113

Lopez-Cabrera M, Letovsky J, Hu KQ, Siddiqui A (1991) Transcriptional factor C/EBP binds to and transactivates the enhancer element II of the hepatitis B virus. Virology 183(2):825–829
Lu S, Cullen BR (2004) Adenovirus VA1 noncoding RNA can inhibit small interfering RNA and microRNA biogenesis. J Virol 78(23):12868–12876
Lu F, Stedman W, Yousef M, Renne R, Lieberman PM (2010) Epigenetic regulation of Kaposi's sarcoma-associated herpesvirus latency by virus-encoded microRNAs that target Rta and the cellular Rbl2-DNMT pathway. J Virol 84(6):2697–2706
Lung RW, Tong JH, Sung YM, Leung PS, Ng DC, Chau SL et al (2009) Modulation of LMP2A expression by a newly identified Epstein-Barr virus-encoded microRNA miR-BART22. Neoplasia 11(11):1174–1184
Machlin ES, Sarnow P, Sagan SM (2011) Masking the 5′ terminal nucleotides of the hepatitis C virus genome by an unconventional microRNA-target RNA complex. Proc Natl Acad Sci U S A 108(8):3193–3198
Marquitz AR, Raab-Traub N (2012) The role of miRNAs and EBV BARTs in NPC. Semin Cancer Biol 22(2):166–172
Marquitz AR, Mathur A, Nam CS, Raab-Traub N (2011) The Epstein-Barr virus BART microRNAs target the pro-apoptotic protein Bim. Virology 412(2):392–400
Martinez I, Gardiner AS, Board KF, Monzon FA, Edwards RP, Khan SA (2008) Human papillomavirus type 16 reduces the expression of microRNA-218 in cervical carcinoma cells. Oncogene 27(18):2575–2582
Melar-New M, Laimins LA (2010) Human papillomaviruses modulate expression of microRNA 203 upon epithelial differentiation to control levels of p63 proteins. J Virol 84(10):5212–5221
Moradpour D, Penin F, Rice CM (2007) Replication of hepatitis C virus. Nat Rev Microbiol 5(6):453–463
Mrazek J, Kreutmayer SB, Grasser FA, Polacek N, Huttenhofer A (2007) Subtractive hybridization identifies novel differentially expressed ncRNA species in EBV-infected human B cells. Nucleic Acids Res 35(10):e73
Murakami Y, Yasuda T, Saigo K, Urashima T, Toyoda H, Okanoue T et al (2006) Comprehensive analysis of microRNA expression patterns in hepatocellular carcinoma and non-tumorous tissues. Oncogene 25(17):2537–2545
Murakami Y, Aly HH, Tajima A, Inoue I, Shimotohno K (2009) Regulation of the hepatitis C virus genome replication by miR-199a. J Hepatol 50(3):453–460
Murakami Y, Tanaka M, Toyoda H, Hayashi K, Kuroda M, Tajima A et al (2010) Hepatic microRNA expression is associated with the response to interferon treatment of chronic hepatitis C. BMC Med Genomics 3:48
Murakami Y, Toyoda H, Tanahashi T, Tanaka J, Kumada T, Yoshioka Y et al (2012) Comprehensive miRNA expression analysis in peripheral blood can diagnose liver disease. PLoS One 7(10):e48366
Nachmani D, Stern-Ginossar N, Sarid R, Mandelboim O (2009) Diverse herpesvirus microRNAs target the stress-induced immune ligand MICB to escape recognition by natural killer cells. Cell Host Microbe 5(4):376–385
Nagata K, Guggenheimer RA, Hurwitz J (1983) Specific binding of a cellular DNA replication protein to the origin of replication of adenovirus DNA. Proc Natl Acad Sci U S A 80(20):6177–6181
O'Connell RM, Taganov KD, Boldin MP, Cheng G, Baltimore D (2007) MicroRNA-155 is induced during the macrophage inflammatory response. Proc Natl Acad Sci U S A 104(5):1604–1609
Parkin DM (2006) The global health burden of infection-associated cancers in the year 2002. Int J Cancer 118(12):3030–3044
Paterlini-Brechot P, Saigo K, Murakami Y, Chami M, Gozuacik D, Mugnier C et al (2003) Hepatitis B virus-related insertional mutagenesis occurs frequently in human liver cancers and recurrently targets human telomerase gene. Oncogene 22(25):3911–3916
Pegtel DM, Cosmopoulos K, Thorley-Lawson DA, van Eijndhoven MA, Hopmans ES, Lindenberg JL et al (2010) Functional delivery of viral miRNAs via exosomes. Proc Natl Acad Sci U S A 107(14):6328–6333

Pfeffer S, Zavolan M, Grasser FA, Chien M, Russo JJ, Ju J et al (2004) Identification of virus-encoded microRNAs. Science 304(5671):734–736

Pfeffer S, Sewer A, Lagos-Quintana M, Sheridan R, Sander C, Grasser FA et al (2005) Identification of microRNAs of the herpesvirus family. Nat Methods 2(4):269–276

Pollicino T, Belloni L, Raffa G, Pediconi N, Squadrito G, Raimondo G et al (2006) Hepatitis B virus replication is regulated by the acetylation status of hepatitis B virus cccDNA-bound H3 and H4 histones. Gastroenterology 130(3):823–837

Popper H, Roth L, Purcell RH, Tennant BC, Gerin JL (1987) Hepatocarcinogenicity of the woodchuck hepatitis virus. Proc Natl Acad Sci U S A 84(3):866–870

Pratt ZL, Kuzembayeva M, Sengupta S, Sugden B (2009) The microRNAs of Epstein-Barr virus are expressed at dramatically differing levels among cell lines. Virology 386(2):387–397

Qin Z, Freitas E, Sullivan R, Mohan S, Bacelieri R, Branch D et al (2010a) Upregulation of xCT by KSHV-encoded microRNAs facilitates KSHV dissemination and persistence in an environment of oxidative stress. PLoS Pathog 6(1):e1000742

Qin Z, Kearney P, Plaisance K, Parsons CH (2010b) Pivotal advance: Kaposi's sarcoma-associated herpesvirus (KSHV)-encoded microRNA specifically induce IL-6 and IL-10 secretion by macrophages and monocytes. J Leukoc Biol 87(1):25–34

Qiu L, Fan H, Jin W, Zhao B, Wang Y, Ju Y et al (2010) miR-122-induced down-regulation of HO-1 negatively affects miR-122-mediated suppression of HBV. Biochem Biophys Res Commun 398(4):771–777

Raab-Traub N (2012) Novel mechanisms of EBV-induced oncogenesis. Curr Opin Virol 2(4): 453–458

Ramakrishnan R, Donahue H, Garcia D, Tan J, Shimizu N, Rice AP et al (2011) Epstein-Barr virus BART9 miRNA modulates LMP1 levels and affects growth rate of nasal NK T cell lymphomas. PLoS One 6(11):e27271

Riley KJ, Rabinowitz GS, Yario TA, Luna JM, Darnell RB, Steitz JA (2012) EBV and human microRNAs co-target oncogenic and apoptotic viral and human genes during latency. EMBO J 31(9):2207–2221

Sachdeva M, Mo YY (2010) MicroRNA-145 suppresses cell invasion and metastasis by directly targeting mucin 1. Cancer Res 70(1):378–387

Sachdeva M, Zhu S, Wu F, Wu H, Walia V, Kumar S et al (2009) p53 represses c-Myc through induction of the tumor suppressor miR-145. Proc Natl Acad Sci U S A 106(9):3207–3212

Samols MA, Hu J, Skalsky RL, Renne R (2005) Cloning and identification of a microRNA cluster within the latency-associated region of Kaposi's sarcoma-associated herpesvirus. J Virol 79(14):9301–9305

Samols MA, Skalsky RL, Maldonado AM, Riva A, Lopez MC, Baker HV et al (2007) Identification of cellular genes targeted by KSHV-encoded microRNAs. PLoS Pathog 3(5):e65

Scheffner M, Werness BA, Huibregtse JM, Levine AJ, Howley PM (1990) The E6 oncoprotein encoded by human papillomavirus types 16 and 18 promotes the degradation of p53. Cell 63(6):1129–1136

Seeger C, Mason WS (2000) Hepatitis B virus biology. Microbiol Mol Biol Rev MMBR 64(1):51–68

Seto E, Moosmann A, Gromminger S, Walz N, Grundhoff A, Hammerschmidt W (2010) Micro RNAs of Epstein-Barr virus promote cell cycle progression and prevent apoptosis of primary human B cells. PLoS Pathog 6(8):e1001063

Shi M, Du L, Liu D, Qian L, Hu M, Yu M et al (2012) Glucocorticoid regulation of a novel HPV-E6-p53-miR-145 pathway modulates invasion and therapy resistance of cervical cancer cells. J Pathol 228(2):148–157

Shrivastava S, Petrone J, Steele R, Lauer GM, Bisceglie AM, Ray RB (2013) Upregulation of circulating miR-20a is correlated with hepatitis C virus mediated liver disease progression. Hepatology 58(3):863–871

Skalsky RL, Cullen BR (2010) Viruses, microRNAs, and host interactions. Annu Rev Microbiol 64:123–141

Skalsky RL, Samols MA, Plaisance KB, Boss IW, Riva A, Lopez MC et al (2007) Kaposi's sarcoma-associated herpesvirus encodes an ortholog of miR-155. J Virol 81(23):12836–12845

Skalsky RL, Corcoran DL, Gottwein E, Frank CL, Kang D, Hafner M et al (2012) The viral and cellular microRNA targetome in lymphoblastoid cell lines. PLoS Pathog 8(1):e1002484

Song K, Han C, Zhang J, Lu D, Dash S, Feitelson M et al (2013) Epigenetic regulation of miR-122 by PPARgamma and hepatitis B virus X protein in hepatocellular carcinoma cells. Hepatology. doi:10.1002/hep.26514. [Epub ahead of print]

Su C, Hou Z, Zhang C, Tian Z, Zhang J (2011) Ectopic expression of microRNA-155 enhances innate antiviral immunity against HBV infection in human hepatoma cells. Virol J 8:354

Suzuki HI, Yamagata K, Sugimoto K, Iwamoto T, Kato S, Miyazono K (2009) Modulation of microRNA processing by p53. Nature 460(7254):529–533

Taraboletti G, Benelli R, Borsotti P, Rusnati M, Presta M, Giavazzi R et al (1999) Thrombospondin-1 inhibits Kaposi's sarcoma (KS) cell and HIV-1 Tat-induced angiogenesis and is poorly expressed in KS lesions. J Pathol 188(1):76–81

Terradillos O, Billet O, Renard CA, Levy R, Molina T, Briand P et al (1997) The hepatitis B virus X gene potentiates c-myc-induced liver oncogenesis in transgenic mice. Oncogene 14(4):395–404

Tian Y, Yang W, Song J, Wu Y, Ni B (2013) Hepatitis B virus x protein-induced aberrant epigenetic modifications contributing to human hepatocellular carcinoma pathogenesis. Mol Cell Biol 33(15):2810–2816

Tie J, Pan Y, Zhao L, Wu K, Liu J, Sun S et al (2010) MiR-218 inhibits invasion and metastasis of gastric cancer by targeting the Robo1 receptor. PLoS Genet 6(3):e1000879

Tili E, Michaille JJ, Cimino A, Costinean S, Dumitru CD, Adair B et al (2007) Modulation of miR-155 and miR-125b levels following lipopolysaccharide/TNF-alpha stimulation and their possible roles in regulating the response to endotoxin shock. J Immunol 179(8):5082–5089

Tsai WC, Hsu SD, Hsu CS, Lai TC, Chen SJ, Shen R et al (2012) MicroRNA-122 plays a critical role in liver homeostasis and hepatocarcinogenesis. J Clin Invest 122(8):2884–2897

Uesugi A, Kozaki K, Tsuruta T, Furuta M, Morita K, Imoto I et al (2011) The tumor suppressive microRNA miR-218 targets the mTOR component Rictor and inhibits AKT phosphorylation in oral cancer. Cancer Res 71(17):5765–5778

Umbach JL, Cullen BR (2010) In-depth analysis of Kaposi's sarcoma-associated herpesvirus microRNA expression provides insights into the mammalian microRNA-processing machinery. J Virol 84(2):695–703

Ura S, Honda M, Yamashita T, Ueda T, Takatori H, Nishino R et al (2009) Differential microRNA expression between hepatitis B and hepatitis C leading disease progression to hepatocellular carcinoma. Hepatology 49(4):1098–1112

van der Meer AJ, Farid WR, Sonneveld MJ, de Ruiter PE, Boonstra A, van Vuuren AJ et al (2013) Sensitive detection of hepatocellular injury in chronic hepatitis C patients with circulating hepatocyte-derived microRNA-122. J Viral Hepat 20(3):158–166

Vereide DT, Seto E, Chiu YF, Hayes M, Tagawa T, Grundhoff A et al (2013) Epstein-Barr virus maintains lymphomas via its miRNAs. Oncogene. doi:10.1038/onc.2013.71. [Epub ahead of print]

Waidmann O, Bihrer V, Pleli T, Farnik H, Berger A, Zeuzem S et al (2012) Serum microRNA-122 levels in different groups of patients with chronic hepatitis B virus infection. J Viral Hepat 19(2):e58–e65

Wang B, Majumder S, Nuovo G, Kutay H, Volinia S, Patel T et al (2009a) Role of microRNA-155 at early stages of hepatocarcinogenesis induced by choline-deficient and amino acid-defined diet in C57BL/6 mice. Hepatology 50(4):1152–1161

Wang X, Wang HK, McCoy JP, Banerjee NS, Rader JS, Broker TR et al (2009b) Oncogenic HPV infection interrupts the expression of tumor-suppressive miR-34a through viral oncoprotein E6. RNA 15(4):637–647. d

Wang Y, Lu Y, Toh ST, Sung WK, Tan P, Chow P et al (2010) Lethal-7 is down-regulated by the hepatitis B virus x protein and targets signal transducer and activator of transcription 3. J Hepatol 53(1):57–66

Wang S, Qiu L, Yan X, Jin W, Wang Y, Chen L et al (2012) Loss of microRNA 122 expression in patients with hepatitis B enhances hepatitis B virus replication through cyclin G(1)-modulated P53 activity. Hepatology 55(3):730–741

Wasley A, Alter MJ (2000) Epidemiology of hepatitis C: geographic differences and temporal trends. Semin Liver Dis 20(1):1–16

Wong DK, Yuen MF, Poon RT, Yuen JC, Fung J, Lai CL (2006) Quantification of hepatitis B virus covalently closed circular DNA in patients with hepatocellular carcinoma. J Hepatol 45(4):553–559

Wu X, Wu S, Tong L, Luan T, Lin L, Lu S et al (2009) miR-122 affects the viability and apoptosis of hepatocellular carcinoma cells. Scand. J Gastroenterol 44(11):1332–1339

Xia T, O'Hara A, Araujo I, Barreto J, Carvalho E, Sapucaia JB et al (2008) EBV microRNAs in primary lymphomas and targeting of CXCL-11 by ebv-mir-BHRF1-3. Cancer Res 68(5):1436–1442

Xie X, Piao L, Bullock BN, Smith A, Su T, Zhang M et al (2013) Targeting HPV16 E6-p300 interaction reactivates p53 and inhibits the tumorigenicity of HPV-positive head and neck squamous cell carcinoma. Oncogene. doi:10.1038/onc.2013.25. [Epub ahead of print]

Xu N, Papagiannakopoulos T, Pan G, Thomson JA, Kosik KS (2009) MicroRNA-145 regulates OCT4, SOX2, and KLF4 and represses pluripotency in human embryonic stem cells. Cell 137(4):647–658

Yan H, Zhong G, Xu G, He W, Jing Z, Gao Z et al (2012) Sodium taurocholate cotransporting polypeptide is a functional receptor for human hepatitis B and D virus. eLife 1:e00049

Yi R, Poy MN, Stoffel M, Fuchs E (2008) A skin microRNA promotes differentiation by repressing 'stemness'. Nature 452(7184):225–229

Yu F, Yao H, Zhu P, Zhang X, Pan Q, Gong C et al (2007) let-7 regulates self renewal and tumorigenicity of breast cancer cells. Cell 131(6):1109–1123

Zhang GL, Li YX, Zheng SQ, Liu M, Li X, Tang H (2010) Suppression of hepatitis B virus replication by microRNA-199a-3p and microRNA-210. Antivir Res 88(2):169–175

Zhang X, Zhang E, Ma Z, Pei R, Jiang M, Schlaak JF et al (2011a) Modulation of hepatitis B virus replication and hepatocyte differentiation by MicroRNA-1. Hepatology 53(5):1476–1485

Zhang Y, Xie RL, Croce CM, Stein JL, Lian JB, van Wijnen AJ et al (2011b) A program of microRNAs controls osteogenic lineage progression by targeting transcription factor Runx2. Proc Natl Acad Sci U S A 108(24):9863–9868

Zheng ZM, Baker CC (2006) Papillomavirus genome structure, expression, and post-transcriptional regulation. Front Biosci 11:2286–2302

Zhou J, Yu L, Gao X, Hu J, Wang J, Dai Z et al (2011) Plasma microRNA panel to diagnose hepatitis B virus-related hepatocellular carcinoma. J Clin Oncol 29(36):4781–4788

Zhu JY, Pfuhl T, Motsch N, Barth S, Nicholls J, Grasser F et al (2009) Identification of novel Epstein-Barr virus microRNA genes from nasopharyngeal carcinomas. J Virol 83(7):3333–3341

Ziegelbauer JM, Sullivan CS, Ganem D (2009) Tandem array-based expression screens identify host mRNA targets of virus-encoded microRNAs. Nat Genet 41(1):130–134

Chapter 8
MicroRNAs Linking Cancer and Inflammation: Focus on Liver Cancer

Angélique Gougelet and Sabine Colnot

Contents

Abstract Due to their great variety of targets, microRNAs (miRNAs) play a key role in number of physiological processes and in oncogenesis. The identification of specific miRNA signatures in various types of tumors, including hepatocellular carcinoma (HCC), highlighted the dual role of miRNAs, both oncogenic and tumor suppressive. HCC is a cancer of poor prognosis that mainly develops on an injured

A. Gougelet (✉) • S. Colnot
Oncogenesis of Digestive Epithelia, Inserm, U1016, Institut Cochin, Paris 75014, France

Cnrs, UMR8104, Paris 75014, France

Université Paris Descartes, Sorbonne Paris Cité, Paris 75006, France
e-mail: angelique.gougelet@inserm.fr

S. Babashah (ed.), *MicroRNAs: Key Regulators of Oncogenesis*, DOI 10.1007/978-3-319-03725-7_8,

liver. Here, we will review the current knowledge concerning the deregulation of miRNA expression at all stages of hepatocarcinogenesis, including the underlying liver disease, in particular steatohepatitis and fibrosis. We will develop the data concerning the identification of specific miRNA signatures in hepatocellular carcinoma, either in tumor or in sera. Since HCC appears in a context of chronic inflammation after recurrent liver injury, it also constitutes one of the best example in which miRNAs are key mediators for the communication between tumor cells and tumor microenvironment. The role of miRNAs in liver inflammation will be detailed in this chapter. To conclude, all studies focusing on miRNAs in HCC argue for their possible use as diagnostic and prognostic biomarkers. Promisingly, miRNAs appear as potent therapeutic targets to improve HCC treatment, for which surgery remains the most frequent therapeutic option.

Keywords Liver disease • Hepatocellular carcinoma • Inflammation • MicroRNA signature • MicroRNA-based therapies

1 Liver Architecture

By its central position, liver plays a crucial metabolic function in the body. The liver is divided into lobules centered onto the central vein and surrounded by six portal triads, composed of a branch of the hepatic artery, the portal vein and a biliary duct (Fig. 8.1).

Blood flow within the liver determines the organization around the sinusoids of the liver cell plate, the anatomical unit of the hepatic parenchyma. The hepatocyte population represents 60 % to the total number of liver cells and assumes the main liver functions. According to their localization near the portal vein or around the central vein, the hepatocytes show different metabolic properties due to the expression of different enzymes. This phenomenon was defined by Jungermann as the liver metabolic zonation (Katz and Jungermann 1976). Nonparenchymal liver cells also populate liver in the sinusoidal compartment of the lobule. They regulate hepatocyte function and protect them from liver injuries through the secretion of various molecules [(Bouwens et al. 1992) for a review]. Three main cell types composed the wall of hepatic sinusoid: the sinusoidal endothelial cells (SECs), the Kupffer cells, and the hepatic stellate cells (HSCs, also known as Ito cells). Sinusoidal endothelial cells constitute the wall of the hepatic sinusoid, in which they exert filtrating and endocytic activity. SECs also present original markers (i.e. the immunoglobulin receptors or CD4), which facilitate their interaction with Kupffer cells to clear immune complexes and to favor a T-cell tolerance. SECs are also able to secrete cytokines, eicosanoids, and some components of the extracellular matrix (ECM). Kupffer cells are the liver macrophages. Their phagocytic activity is required to control damages induced by dietary and bacterial metabolites arriving from the gut. They also secrete immune mediators in response to bacterial products like

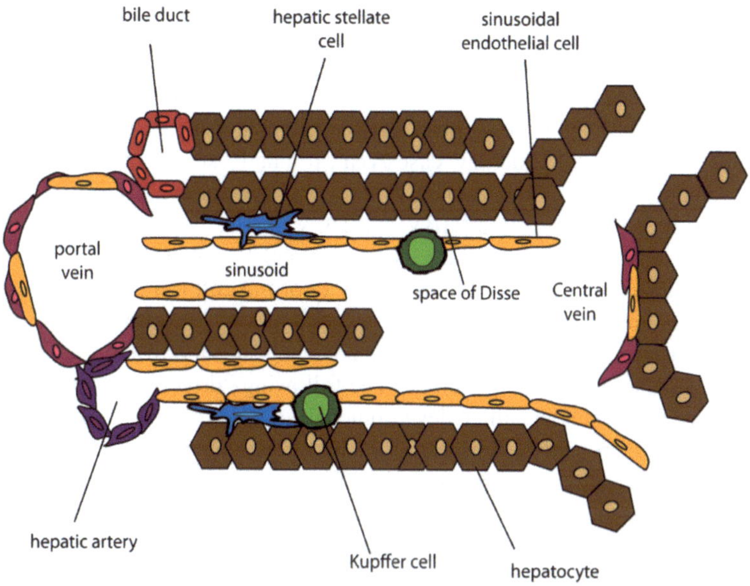

Fig. 8.1 Liver lobule organization. Liver parenchyma is organized in cell plates distributed around the sinusoids. The sinusoid wall is composed of endothelial cells, hepatic stellate cells and Kupffer cells. The hepatic parenchyma is divided into two distinct populations of hepatocytes according to their localization around the central vein or the portal triad composed of the portal vein, the hepatic artery and a bile duct

lipopolysaccharide (LPS) to promote an oral tolerance to bacterial antigens (Crispe 2009). During liver injury, Kupffer cells are activated, secrete inflammatory mediators [i.e. tumor necrosis factor α (TNFα), interleukin-10, and reactive oxygen species] to control the early phase of liver inflammation and to favor innate immune defense (Seki and Brenner 2008). Hepatic stellate cells are perisinusoidal cells involved in collagen synthesis and vitamin A storage. During liver injury, HSCs transdifferentiate into myofibroblast-like cells; this constitutes the crucial event for inflammatory fibrotic process. In response to hepatocyte damages, HSCs secrete cytokines, chemokines and growth factors, and participate to ECM remodeling (Brenner et al. 2000). Finally, liver immunosurveillance is also assumed by intrahepatic lymphocytes called Pit cells, which reside in the sinusoid lumen after they have crossed the sinusoid endothelium [(Lalor et al. 2002) for a review]. In response to different damages, a panel of inflammatory cytokines and chemokines are secreted in liver (i.e. IFNγ, IP-10, and MIP-1α), which results in the recruitment of various immune cells. Different types of lymphocytes can be recruited in liver: CD4 and CD8 T cells, dendritic cells, liver-specific natural killer (NK) or NKT cells. All these cells are required for a correct protection against viral infections and tumor cells [(Parker and Picut 2012) for a review]. Following repeated insults to liver, an exacerbation of the inflammatory response can occur and promote tumor development.

2 Liver Cancer

Liver cancer constitutes the third cause of cancer-related death worldwide. The most frequent form of primary liver cancer is hepatocellular carcinoma (HCC) with a prevalence of 500,000 new cases/year. It mainly affects men with cirrhosis. HCC frequently appears in a context of chronic liver disease caused by infection with hepatitis B and C viruses (HBV and HCV, respectively), alcohol consumption, obesity or genotoxic exposure [(El-Serag and Rudolph 2007) for a review]. The classical view of carcinogenesis is that a normal cell is submitted to the first "gatekeeping" mutation (i.e. single-base substitutions, deletion, insertion, and translocation) leading to a selective growth advantage to survive and proliferate, and to finally become a microscopic clone. The following "driver" mutations facilitate cell growth, and result in tumor development (Vogelstein et al. 2013). HCC is a complex and heterogeneous disease due to different genetic alterations, whose sequential appearance during the evolution from benign to malignant lesions is still poorly known. At the end of the 2000s, the global transcriptomic analysis performed on HCC samples established a molecular classification of these tumors (Boyault et al. 2007). The most frequently signaling pathways altered in HCC are the β-catenin pathway (15–40 %), the p53 pathway (26 %), and the Ras pathway (5–10 %) (de La Coste et al. 1998; Guichard et al. 2012). In brief, the subgroups of HCC mutated for the β-catenin, appear in an etiological context of HCV virus infection or alcohol abuse. They present a lower proliferation rate, are more genetically stable, well differentiated, cholestatic and of relative better prognosis (Audard et al. 2007); inversely, the other subgroups of HCC appeared after hepatitis B virus infection. They are characterized by a genetic instability, are poorly differentiated, steatotic and of bad prognosis (Fig. 8.2).

Liver cancer is also a perfect example of the relationship between tumor microenvironment and tumor development and/or progression. Over the past decades, the studies focusing on cancer do not consider so far the tumor cells as an unique entity but as an organ, whose development is orchestrated by surrounding cells, which could be immune cells, inflammatory cells, endothelial cells or fibroblasts. The tumor microenvironment was in consequence defined as a hallmark of cancer by Hanahan and Weinberg, and the tumor-promoting inflammation was defined these recent years as one of its emerging characteristic (Hanahan and Weinberg 2011). To prevent an exacerbation of liver chronic inflammation, immune cells dedicated to protect the hepatic parenchyma drain the adult liver and their recruitment and/or activation influence hepatocarcinogenesis (Anson et al. 2012; He and Karin 2011).

Although numerous efforts have been conducted to improve the comprehension and the treatment of liver cancer, the median life expectancy after HCC diagnosis does not go beyond 6 months. In consequence, it appears crucial to identify new molecular actors involved in liver tumor progression, which could be useful for its diagnosis and its treatment. In consideration with the key role of microRNAs

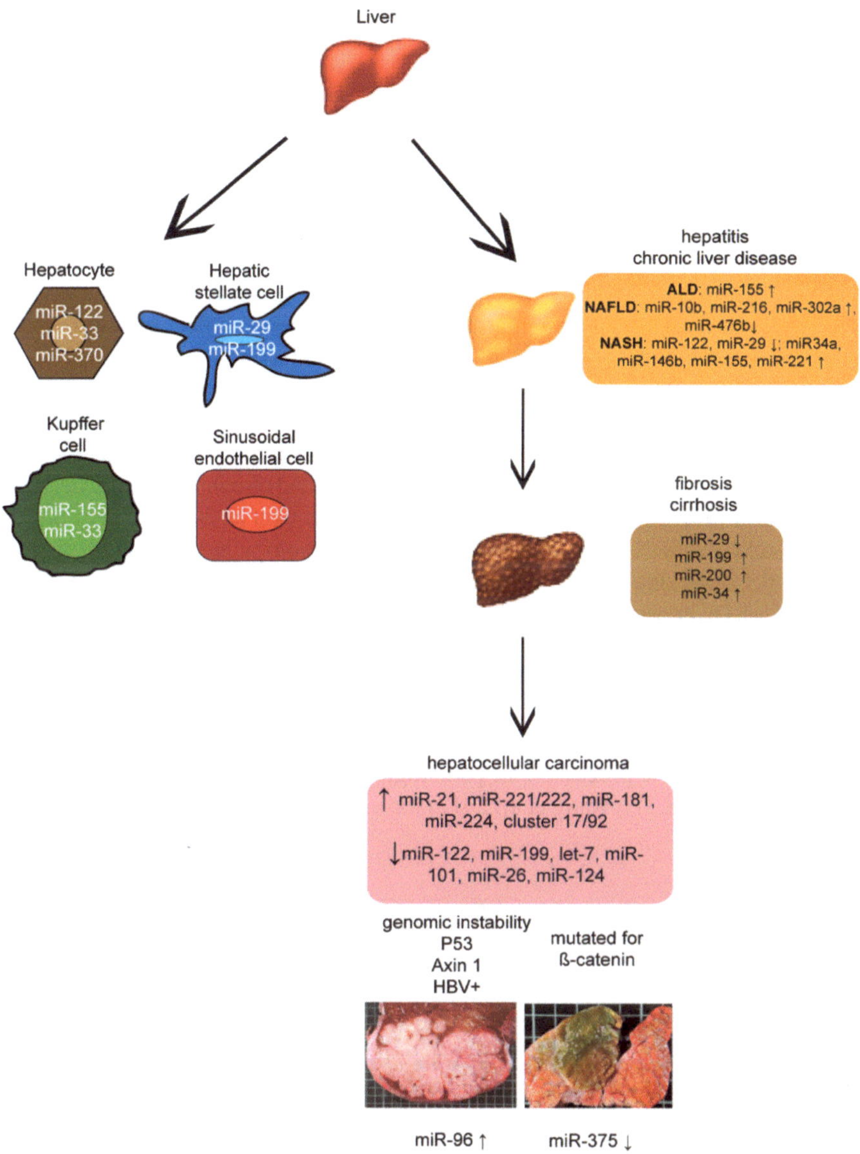

Fig. 8.2 MicroRNA expression in physiological and pathological conditions. For more details, see the text. *ALD* alcohol-liver disease, *NAFLD* non-fatty acid liver disease, *NASH* non alcoholic steatohepatitis

(miRNAs) in metabolism, immunity, inflammation and cancer, these small RNAs appear as promising targets. In this chapter, we review the main data concerning the impact of miRNAs in HCC progression and the therapies based on miRNAs, which proceed from this molecular deciphering.

3 MicroRNAs in Liver

Since the discovery of lin-4 in Caenorhabditis elegans in 1993 (Lee et al. 1993), miRNAs emerge as promising prognostic and diagnostic biomarkers, in particular in cancer. More than half of the genes encoding miRNAs are located in cancer-associated regions or fragile sites (Calin et al. 2004). Their abnormal expression contributes to the progression of tumors, in which they play an ambivalent role of tumor suppressor or oncogene (Babashah and Soleimani 2011). MiRNA expression is submitted to a great variety of genetic regulations, similarly to more "classical" genes, i.e. gene amplification or deletion, promoter hypermethylation, translocation, single nucleotide polymorphism. A number of works based on a global analysis of miRNA expression in different types of tumors highlighted specific miRNA signatures according to the cellular and tissular origin of the tumors (Lu et al. 2005; Volinia et al. 2006; Babashah et al. 2012). These recent years, a number of studies have shown the key role of miRNAs in liver organogenesis and also in liver tumor development and/or progression (Ladeiro et al. 2008; Murakami et al. 2006) (Fig. 8.3).

The role of miRNAs in liver physiology and pathology has been supported by two works of Sekine et al., which showed by a transgenic mouse model that the deletion of Dicer, the crucial protein involved in miRNA maturation, results in a loss of liver zonation and the appearance of tumors (Sekine et al. 2009a, b). Thereafter, a number of studies has highlighted the preponderant role of a panel of miRNAs in embryonic liver development, whose the great majority are deregulated in hepatic diseases like alcoholic liver disease (ALD), non-alcoholic steatohepatitis (NASH) and fibrotic processes, all pathologies predisposing to HCC, and in HCC themselves (Fig. 8.3) [(Wang et al. 2012) for a review].

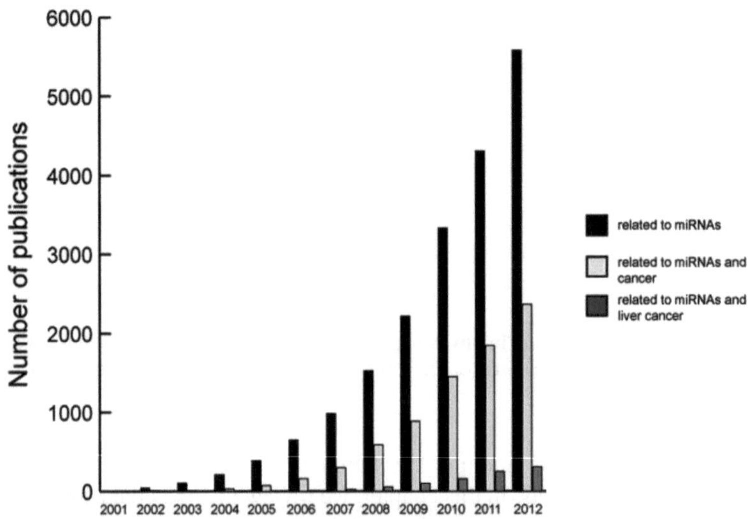

Fig. 8.3 Number of publications concerning liver cancer and miRNAs referenced in PUBMED in February 2013

4 Main MicroRNAs Involved in Liver Metabolism

MicroRNA expression in the liver changes during its development: in the adult liver in addition to miR-122, other miRNAs such as miR-21, miR-29a, miR-194 and the let-7 family are also abundant, while miR-92a and miR-483 are expressed in the fetal liver (Girard et al. 2008). Due to the central metabolic function of liver, it is not surprising that hepatic miRNAs mainly target metabolic enzymes and in consequence finely regulate these processes.

4.1 MiR-122

MiR-122 is the most abundant miRNA in liver (70 % of liver miRNAs). It is a target of the transcription factor HNF-4α, a master regulator of hepatocyte differentiation. MiR-122 is involved in iron and lipid metabolisms and cholesterol synthesis [(Vaulont 2011; Wen and Friedman 2012) for reviews]. This miRNA targets cyclin G1, a protein involved in G2/M cell cycle arrest in response to DNA damages (Gramantieri et al. 2007), and the anti-apoptotic molecule Bcl-w (Lin et al. 2008). In addition, miR-122 is used by hepatitis C virus for its replication. That is the reason why the first therapeutic strategy based on miRNA inhibition in liver was the anti-miR-122, Miravirsen, which succeeds in the treatment of chronic hepatitis C virus in primates (Lanford et al. 2010) and in humans (Janssen et al. 2013).

4.2 MiR-370

MiR-370 has been also described as a key regulator of cholesterol and lipid metabolism. Iliopoulos et al. firstly showed that the over-expression of a dominant negative form of c-Jun leads to an up-regulation of miR-370 correlated to a down-regulation of carnitine palmitoyl transferase-1 (CPT-1), which ends to impair fatty acid oxidation (Iliopoulos et al. 2010). The authors also demonstrate in hepatic cell culture that miR-370 up-regulates miR-122 and SREBP-1c expression.

4.3 MiR-33

MiR-33a/b are also key regulators of lipid metabolism and cholesterol trafficking in cooperation with SREBP. By different approaches, three studies showed that miR-33 regulates cholesterol efflux and high-density lipoprotein formation through the targeting of Abca1 and Abcg1 (Marquart et al. 2010; Najafi-Shoushtari et al. 2010; Rayner et al. 2010). MiR33a/b also regulated the β-oxidation of fatty acids by the binding to CPT-1 (Gerin et al. 2010). In addition, miR-33 can target AMPKα and thus increases the level of cholesterol, fatty acids and lipids in hepatocytes (Davalos et al. 2011).

5 Main MicroRNAs Deregulated in Liver Pathologies

Due to their essential role in metabolic processes, it appears intuitive that the expression of the miRNAs described in the last paragraph is altered in chronic liver diseases. In particular, a prolonged decrease in miR-122 expression is associated with non-alcoholic fatty liver disease (NAFLD), NASH and fibrosis (Alisi et al. 2011; Cheung et al. 2008; Tsai et al. 2012). In addition, a miRNA signature has been identified for the different forms of liver diseases (see Fig. 8.2).

5.1 MiR-155 in Alcoholic Liver Disease

MiR-155 is mainly expressed by Kupffer cells. Its over-expression in these cells, and also in hepatocytes, is observed in alcohol-fed mice. This induction is responsible for the highest production of TNFα by Kupffer cells observed in patients with ALD (Bala et al. 2011). Recently, a study performed by the same team highlighted the presence of miR-155 into exosomes, a multiprotein structure secreted by mammalian cells, in a mouse model of ALD (Bala et al. 2012). This miRNA targets SOCS1, an inhibitor of signal transducer and activator of transcription 3 (STAT3)/ matrix metalloproteinase 9 (MMP9) cascade (Yan et al. 2013).

5.2 MiRNA Deregulation in NAFLD and NASH

NAFLD is the most prevalent chronic liver disease in industrialized countries in expansion these recent years with the increase of obesity, leading to hepatic accumulation of triglycerides in the absence of alcohol consumption. This pathology could be characterized by a simple steatosis but could also result to NASH. The impact of miRNAs in NAFLD has been firstly assessed on steatotic human L02 hepatocytes cultured with high concentration of free fatty acids by Zheng et al. They correlate miR-10b over-expression with the triglyceride levels in L02 cells due to the targeting of the peroxisome proliferator-activated receptor α (PPARα) (Zheng et al. 2010). In mice fed with high-fat diet, Ahn et al. shows that a loss of miR-476b is associated with an increase in the lipoprotein lipase level (Ahn et al. 2011). Finally, a recent study performed on mice depleted for the low-density lipoprotein receptor fed with a high fat/western-type diet correlates the decrease in miR-216 and miR-302a expression with NAFLD development (Hoekstra et al. 2012).

Concerning the alteration of miRNAs in NASH, a first study of Cheung et al. on human samples reveals that miR-146b and miR-34a are induced in the liver of NASH subjects (Cheung et al. 2008). MiR-34a is a well-documented miRNA, which is a target of p53 and a key regulator of the NAD-dependent deacetylase Sirtuin-1. Sirtuin-1 inhibition leads to AMP Kinase dephosphorylation, HMG-CoA reductase phosphorylation and hepatic cholesterol accumulation (Min et al. 2012). MiR-34a and miR-122 deregulation is also confirmed in mice fed with methyl-deficient diet. In this model, miR-155, miR-200b and miR-221 are found over-expressed, while miR-29c, miR-192 and miR-203 are down-regulated (Pogribny et al. 2010).

5.3 *MiRNAs in Fibrosis*

NAFLD pathology can also progress to complications like liver fibrosis. Liver fibrosis consists to a remodeling of liver architecture mainly orchestrated by hepatic stellate cells. HSCs switch from a quiescent stage to myofibroblast-like cells. Activated HSC secrete ECM proteins, collagen and inflammatory mediators in response to liver injury. HSC express two crucial miRNAs, miR-29 and miR-199. MiR-29 is a target of two inflammatory pathways in liver, the NFkB and TGFβ pathway. It targets the anti-apoptotic bcl-2 and Mcl-1 molecules (Law and Wong 2011). In a mouse model of liver fibrosis induced by carbon tetrachloride, the family miR-29a/b/c is lost (Roderburg et al. 2011). MiR-199 is found in HSCs, and also in sinusoidal endothelial cells. Its expression is up-regulated during fibrosis. In particular, its expression and those of miR-200a/b are highly increased in human samples and in a mouse liver fibrosis model (Murakami et al. 2011). The miR-34 family is also a crucial actor of liver fibrosis. Its expression is down-regulated in fibrotic liver induced by dimethylnitrosamine in rats (Li et al. 2011b).

To conclude, these miRNAs deregulated in all hepatic disease predisposing to cancer, are also affected in liver tumors and in particular in hepatocellular carcinoma. These miRNAs could constitute potent early biomarkers for the initiation and progression of liver tumors. They could also be promising targets for the prevention and the treatment of HCC.

6 MicroRNA Signature in Liver Tumors

6.1 *MiRNA Alteration in Hepatocellular Carcinoma*

Besides the divergence in the results obtained by the authors due to the differences in samples, miRNA measurement and statistical analyses, the global analysis of the miRNAs expressed in liver tumors highlighted a miRNA signature common to all types of HCC. As mentioned previously, the great majority of miRNAs found deregulated in chronic liver diseases are also found deregulated in HCC. A loss of miR-122 (Ladeiro et al. 2008), miR-29 (Xiong et al. 2010), and miR-199 (Hou et al. 2011) has been identified in HCC (Fig. 8.2). In addition, other miRNAs with a tumor suppressive role have also been largely described. We could cite: Let-7 (Lan et al. 2011), miR-101 (Su et al. 2009) and miR-124 (Furuta et al. 2010). Inversely, some oncogenic miRNAs have been also identified: miR-155 (Yan et al. 2013), miR-21 (Meng et al. 2007), miR-221/miR-222 (Pineau et al. 2010), miR-181 (Wang et al. 2010), miR-224 (Wang et al. 2008), and the miRNA cluster miR-17/92 (Yang et al. 2010) (Fig. 8.2) [(Huang and He 2011; Law and Wong 2011) for reviews].

All these miRNAs participate to apoptosis (Table 8.1). The mitochondrial apoptotic pathway is particularly induced through the increase expression of anti-apoptotic actors [i.e. Mcl-1 targeted by miR-101 (Su et al. 2009), and miR-29 (Xiong et al. 2010), bcl-w targeted by miR-122 (Lin et al. 2008), bcl-2 by miR-29 (Xiong et al. 2010)], and inversely through the inhibition of pro-apoptotic mediators [i.e. Bmf inhibited by miR-221 (Gramantieri et al. 2009), and Bim by miR-25 (Li et al. 2009c)]. MiR-221/222 can also confer resistance to the extrinsic apoptotic pathway dependent on TNF-related apoptosis induced ligand (TRAIL) (Garofalo et al. 2009). These miRNAs also regulate the tumor growth through the modulation of many cell cycle actors, in particular miR-122 targets cyclin G1 (Gramantieri et al. 2007), miR-221 regulates p27 (le Sage et al. 2007) and p57 (Fornari et al. 2008), miR-124 represses the cyclin-dependent kinase 6 (Furuta et al. 2010). Finally, all these miRNAs can also alter the survival of tumor cells by the targeting of receptor tyrosine kinase pathways like the ERK pathway targeted by miR-101 (Li et al. 2009b), and let-7c (Shah et al. 2007), the c-MET-activated pathway by miR-34a (Li et al. 2009a), or the PI3K/AKT pathway repressed by miR-21 (Meng et al. 2007), and miR-221/222 (Garofalo et al. 2009; Wong et al. 2010).

MiR-221 is one of the miRNAs, for which the role in the development and the progression of liver cancer is the best-described. As mentioned before, this miRNA is induced during fibrosis and chronic liver disease. Number of studies showed its induction in HCC samples (Fornari et al. 2008; Garofalo et al. 2009; Gramantieri et al. 2009; le Sage et al. 2007; Pineau et al. 2010; Wong et al. 2010). MiR-221 is also reduced in cholangiocarcinoma, epithelial tumors emerging from biliary cells (Karakatsanis et al. 2013). The main targets of miR-221/222 in HCC are the tumor suppressor p27, p57, PTEN and the inhibitor of metalloproteinase TIMP-3. *In vivo* experiments are currently performed to test its invalidation in mice; this aspect will be developed at the end of this chapter. Another well-studied miRNA in HCC is miR-21. This miRNA is induced in tumor cells (Karakatsanis et al. 2013; Zhu et al. 2012), as well as in liver cancer stem cells (Li et al. 2010b), and in cholangiocarcinoma samples (Karakatsanis et al. 2013; Selaru et al. 2009). Interestingly, miR-21 can be detected in patient sera, a part that we will detail in the next paragraph. It thus constitutes a potent biomarker for HCC diagnosis (Xu et al. 2011a; Zhou et al. 2011); in particular, miR-21 is a marker of necroinflammation for HCC developed during hepatitis C virus infection (Bihrer et al. 2011).

Table 8.1 Role of microRNAs deregulated in hepatocellular carcinoma

Proliferation		Apoptosis		Migration/invasion		Serum
Pro-	Anti-	Pro-	Anti-	Pro-	Anti-	
miR-18a (ERα) (Liu et al. 2009)	Let-7 (c-myc) (Lan et al. 2011)	Let-7 (bcl-xL) (Shimizu et al. 2010)	**miR-221 (Bmf)** (Gramantieri et al. 2009)	miR-143 (FNDC3B) (Zhang et al. 2009)	Let-7 (COLA12) (Ji et al. 2010)	Let-7c (Li et al. 2010a)
miR-21 (PTEN) (Meng et al. 2007)	miR-1 (FOXP1, HDAC4) (Datta et al. 2008)	miR-101 (Mcl-1) (Su et al. 2009)	miR-224 (API-5) (Wang et al. 2008)	miR-151 (RhoG DIA) (Ding et al. 2010)	miR-1 (c-MET) (Datta et al. 2008)	miR-17-5p (Zheng et al. 2012)
miR-93 (E2F1) (Li et al. 2009c)	miR-122 (cyclin G1,IGFR1) (Gramantieri et al. 2007; Lanford et al. 2010)	miR-122 (bcl-w) (Lin et al. 2008)	miR-25 (Bim) (Li et al. 2009c)	miR-181b (TIMP3)(Wang et al. 2010)	miR-9-2 (e-cadherin) (Tan et al. 2010)	miR-18a (Li et al. 2012)
miR-106b (p21) (Li et al. 2009c)	miR-124 (SMYD3,cdk6) (Furuta et al. 2010; Hatziapostolou et al. 2011)	miR-203 (ABCE1) (Furuta et al. 2010)		miR-21 (PTEN) (Meng et al. 2007)	miR-101 (FOS) (Li et al. 2009b)	miR-122 (Xu et al. 2011a,b)
miR-221/222 (p27, p57) (Fornari et al. 2008; Park et al. 2011)	miR-26 (cyclin D2, E2) (Kota et al. 2009)	miR-29 (Mcl-1, bcl-2) (Xionget al. 2010)		**miR-221/222 (PPP2R2A)** (Garofalo et al. 2009; Wonget al. 2010)	miR-122 (ADAM10,17) (Tsai et al. 2012)	miR-15b (Liu et al. 2012)
	miR-195 (cyclin D, cdk4/6, E2F1) (Xu et al. 2009)			miR-30d (Gαi2) (Yao et al. 2010)	miR-124 (VIM) (Furuta et al. 2010)	miR-130b (Liu et al. 2012)
	miR-199 (mTOR) (Fornari et al. 2010)			miR-17-5p (p38) (Yang et al. 2010)	miR-199 (c-MET) (Fornari et al. 2010)	miR-183 (Liu et al. 2012)
	miR-223 (Stathmin1) (Wong et al. 2008)				miR-23b (uPA, c-MET) (Salvi et al. 2009)	miR-21 (Bihrer et al. 2011; Xu et al. 2011a)
	miR-375 (YAP) (Liu et al. 2010)				miR-34a (c-MET) (Li et al. 2009a)	miR-25 (Li et al. 2010a)
					miR-375 (YAP) (Liu et al. 2010)	**miR-221** (Li et al. 2011a)
						miR-223 (Xu et al. 2011a)
						miR-224 (Li et al. 2011a)
						miR-375(Li et al. 2010a)
						miR-34a (Cermelli et al. 2011)

Above table compiled all the data concerning the miRNAs described to interfere with proliferation, apoptosis, migration and invasion of HCC tumors. The miRNAs listed here are at least cited by two independent works and their targets are clearly identified. The miRNAs, which could be detected in the patient serum, are listed in the last column. The miRNAs written in bold are those for which a therapeutic strategy anti-miRNA has been tested in HCC; the miRNAs written in red are those for which a mimic strategy has been administered to animals

Finally, Cairo et al. (2010) have identified two clusters of miRNAs (miR-100/let-7a-2/miR-125b-1 and the cluster miR-371-3), which are targets for the oncogene MYC, as a bad prognosis factor in liver tumors. By the measurement of the four miRNAsmiR-100, miR-371, miR-373 and let-7a, the authors are able to discriminate on a large cohort of patients, the tumor samples from HCC patients with tumor properties very closed to hepatoblastoma, the most common primitive liver tumor in child; this subtype of HCC are less differentiated and more invasive. This work supports the existence of a miRNA signature common to hepatoblastoma and HCC, and highlights the great interest of this signature of four miRNAs for the prognosis and the treatment of both tumors.

Despite the identification of a common miRNA signature for HCC samples, other miRNAs are specific of a class of tumor. A loss of miR-375 is characteristic of tumors mutated for the β-catenin, while miR-126* is lost in HCC developed after alcohol abuse, and miR-96 is associated to hepatitis B virus infection (Zucman-Rossi 2010). The miRNA miR-18a is also a very interesting marker in HCC. Indeed, HCC is more prevalent in men (by 2–6-fold), and the women with HCC have a longer life expectancy. This suggests the importance of sex-dependent molecular mechanisms in HCC progression, which could be notably linked to estrogens. MiR-18a is, in fact, a regulator of the estrogen receptor α (ERα), which is over-expressed in HCC samples from woman patients. It thus appears that the over-expression of miR-18a is associated with a loss of the protective effects of estrogen for these women (Liu et al. 2009). In addition, this miRNA has been recently described as a serum biomarker for HCC due to HBV infection (Li et al. 2012).

Finally, some works also showed that this miRNA panel could constitute a predictive tool to follow the tumor response to surgery or chemotherapeutic treatments. In this way, miR-122 sensitizes HCC cells to adriamycin and vincristine (Xu et al. 2011b), while miR-199a-3p affects cell sensitivity to doxorubicin and 5-fluorouracil (Fornari et al. 2010).

6.2 MiRNA Expression in HCC Patient Sera

As we suggested before, miRNAs not only constitute potent biomarkers in tumor cells, but they could also be detectable in various biological fluids like serum, plasma, urine or saliva. Circulating miRNAs could be associated to proteins or encapsulated into exosomes. Conde-Vancells et al. (2008) deciphered the proteome of exosomes secreted by hepatocytes. Many of the proteins were previously described to be present in exosomes secreted by various cells (i.e. CD81 or CD63), while others are only expressed in hepatocytes-derived exosomes, and involved in the liver-specific functions of lipid, carbohydrate and xenobiotic metabolism. To date, there are few data concerning the miRNA content of exosomes derived from any type of liver cells, or the characterization of their functional impact on the recipient hepatic cells. In 2009, a work showed that exosomes purified from bile are able to stimulate the growth of SEC cells, due to the presence of hedgehog in these exosomes (Witek et al. 2009; Yang et al. 2008). For various types of tumors, it has been shown that miRNA expression could be altered in patient sera and in particular for HCC. Although there are some divergence concerning the discriminating miRNAs identified in patient sera, some miRNAs deregulated in HCC are over-expressed in HCC patient sera: miR-21, miR-221/222 and miR-122 (Li et al. 2011a; Xu et al. 2011a; Zhou et al. 2011). The reason why miR-122 is decreased in liver tumors and increased in patient sera is not elucidated yet. Recently, Liu et al. argue for the measurement of miR-15b and miR-130b in serum for the diagnosis of HCC (Liu et al. 2012). By this method, the authors could detect early stage HCC that could not be discriminated by α-fœtoprotein measurement. The detection of miRNAs in body fluids could be a great advantage for the diagnosis and the treatment of HCC, and also for other pathologies, since it could constitute a non-invasive tool.

6.3 MiRNA Expression in Metastatic HCC

In addition to the identification a specific miRNA signature in tumors, a number of works aimed to identify a miRNA signature in metastatic HCC. Divergent results were obtained probably due to divergence in the sampling and the data analysis. Although a recent work of Wong et al. (2012) is in favor of a more pronounced alteration of the same panel of miRNAs in metastases as compared to the tumors, a study of Budhu et al. (2008) showed the existence of a miRNA signature of 20 metastatic miRNAs. Numbers of studies concerning only one miRNA have now enriched the deep-sequencing data, and it appears that metastatic HCC express some miRNAs found deregulated in HCC like miR-122 (Tsai et al. 2012), miR-124 (Budhu et al. 2008), miR-17 (Yang et al. 2010), let-7g (Ji et al. 2010), miR-21 (Zhu et al. 2012), miR-221 (Wong et al. 2010) or miR-143 (Zhang et al. 2009), but the authors also identified an alteration of other miRNAs, i.e. miR-9-2 (Tan et al. 2010), miR-148a, miR-125b, miR-30a (Budhu et al. 2008), or miR-34a (Li et al. 2009a) (Table 8.1). Some of these miRNAs have been described as promoters of metastasis. As we mentioned previously, miR-221 and miR-222 target the tumor suppressor TIMP3, and thus activate the AKT pathway and metalloproteinases to induce HCC invasion and metastasis (Garofalo et al. 2009; Wong et al. 2010). TIMP3 is also a target of miR-181b, which is induced by TGFβ in HCC (Wang et al. 2010). MiR-181b induction results in MMP2 and MMP9 activation, also induced by miR-21, another miRNA induced in metastatic HCC. More recently, it has been shown that miR-143 exerts a pro-metastatic activity *in vitro* and *in vivo*. MiR-143b targets the fibronectin type III domain containing 3B (FNDC3B) (Zhang et al. 2009). This miRNA is induced by NFκB in metastatic HBV-related HCC. Inversely, there are also anti-metastatic miRNAs, i.e. miR-122, which suppresses intrahepatic metastasis through the inhibition of disintegrin and metalloprotease ADAM10 and ADAM17 (Bai et al. 2009). Let-7g was also found impaired in metastatic HCC. Transfection of let-7g inhibited cell migration and this antimetastatic effect is reduced by the addition of type I collagen alpha2 (Ji et al. 2010). Finally, the main signaling pathway targeted by miRNAs currently deregulated in HCC is the c-MET activated pathway. c-MET activation stimulates cell motility and invasion, and protects cells from apoptosis. c-MET is regulated by miR-1, miR-23b, miR-34a and miR-199a-3p, four miRNAs altered in HCC. Silencing of miR-1 inhibits HCC growth and invasion by the targeting of c-MET (Datta et al. 2008). MiR-34a can decrease c-MET mRNA and protein level and impair the invasive capacity of HepG2 cells (Li et al. 2009a). Finally, miR-23b is also able to decrease c-MET and the urokinase-type plasminogen activator (uPA) to reduce the proliferation and migration properties of HCC cells (Salvi et al. 2009).

7 MicroRNAs and HCC Microenvironment: Role in Inflammation

The tumor microenvironment is an expending domain, which not only considers the behavior of cancer cells alone, but also integrates the surrounding cells. This dynamic system is largely orchestrated by inflammatory cells like immune cells, fibroblasts and myofibroblasts, the cytokines and chemokines they secrete and also the extracellular matrix. Liver cancer constitutes a good example in which the microenvironment plays a crucial role in the progression of the disease due to its chronic inflammation. The HCC microenvironment is a complex and intricate system composed of various cells, i.e. cancer-associated fibroblasts, hepatic stellate cells, tumor-infiltrating leukocytes, endothelial cells, hepatoma cells and tumor associated macrophages [(Leonardi et al. 2012) for a review]. All these cells secrete a cocktail of growth factors (HGF, EGF, FGF, TGFβ), cytokines (IL-6, IL-8, IL-4, IL-10, TNFα) and chemokines (CCL17, CCL22, CCL24), which all orchestrate immune response to liver injury (Fig. 8.4).

The transforming growth factor β plays a key and paradoxical role in carcinogenesis. In liver, it has been shown that TGFβ plays crucial roles at different steps of liver pathogenesis including cirrhosis (Matsuzaki 2009), HBV and HCV infection (Marotta et al. 2004) and tumorigenesis (Massague 2008). TGFβ exerts a tumor suppressive activity in the premalignant state but could also promote tumor progression, through the repression of immune surveillance and through increase in tumor cell proliferation, migration and invasion. As mentioned before, TGFβ induces miR-181b, a process blocked by SMAD4 silencing, and this promotes cell survival, mobility and invasion (Wang et al. 2010). TGFβ also induces the cluster of miRNAs, miR-23a, miR-27a and miR-24 to promote cell proliferation and invasion (Huang et al. 2008). A recent work of Yang et al. (2012) has shown that TGFβ represses the expression of miR-34a resulting in CCL22 production to recruit of lymphocyte T regulators leading to immune escape in HBV-related HCC. The axis TGFβ/miR-34a/CCL22 thus promotes venous metastasis in this type of HCC. TGFβ is also required for HSC activation during fibrogenesis (Liu et al. 2006). Recently, miR-146a has been described as a modulator of TGFβ-1 dependent activation of HSC. The interleukin-6 is also a crucial mediator in liver cancer progression. Its level is increased during cirrhosis (Tilg et al. 1992), and a high serum level of IL-6 is a risk factor and a factor of bad prognosis for HCC (Nakagawa et al. 2009;

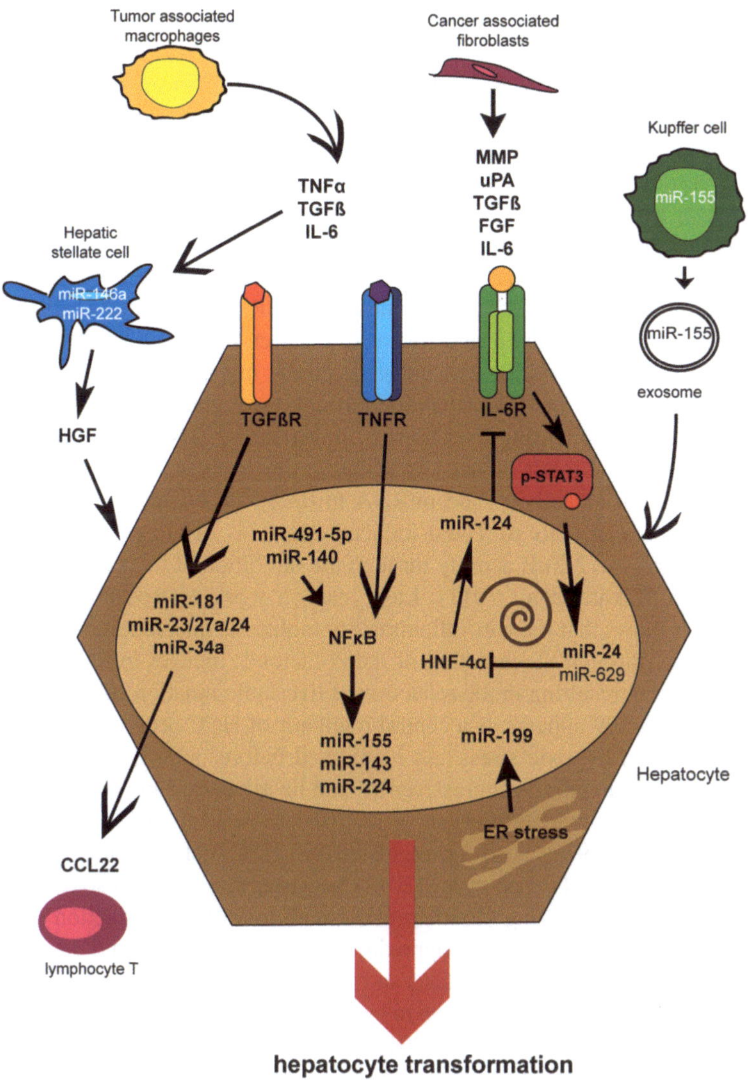

Fig. 8.4 MicroRNA network involved in liver inflammation. For more details, see the text

Wong et al. 2009).A recent work of Iliopoulos' lab highlights a circuit orchestrated by miRNAs that results in liver tumorigenesis and inflammation. This feedback loop, governed by miR-24 and miR-629, promotes a HNF-4α transient inhibition resulting in miR-124 induction and STAT3 activation. The authors described the interconnection between the trio miR-24/miR-629/miR-124 and HNF-4α as a key regulator of HCC progression by limiting the tumor inflammatory environment through the inhibition of the IL-6/STAT3 pathway, predisposing a patient to liver cancer development and immune escape (Hatziapostolou et al. 2011). The level of interleukin-6 has also been inversely correlated to miR-26 expression in HCC (Ji et al. 2009). Patients with a low level of miR-26 and inversely a high level of IL-6 have a lower survival but are also better responders to IFNα treatment, suggesting that this miRNA could be a predictive marker for patient response to IFNα treatment. In this study, NFκB expression has also been inversely correlated to miR-26 level. NFκB is a transcription factor induced in response to TNFα. It plays an essential role in liver homeostasis by preventing hepatocyte death. As mentioned before, NFκB regulates a number of miRNA deregulated in HCC i.e. miR-155 and miR-143 (Zhang et al. 2009). NFκB has also been demonstrated as a regulator of miR-224, a miRNA up-regulated in HCC. NFκB binding on miR-224 promoter is significantly induced in response to TNFα and LPS (Scisciani et al. 2012). The IKK inhibitor BMS-345541 blocks pre-miR-224-induced cellular migration and invasion. NFκB is also a key regulator of HSC activation through the regulation of miR-222. MiR-222 induction in HSC during liver fibrosis in response to TNFα and TGFβ is abrogated by a NFκB inhibitor (Ogawa et al. 2012). Inversely, some miRNAs regulate NFκB expression or activity: the miRNA miR-491-5p controls NFκB level and sensitizes hepatoma cells to TNFα-induced apoptosis (Yoon et al. 2010), while miR-140 suppresses NFκB activity through the inhibition of its coactivators NRP1 and NCoA1 (Takata et al. 2011). Last years, a work of Boesch-Saadatmandi et al. (2012) shows that the anti-inflammatory molecule, querceptin, which reduces NFκB activity through the targeting of redox factor 1, induces miR-122 and miR-125b expression, resulting in the reduction of liver inflammation in mice.

Finally, miR-199 appears as a central regulator of HCC progression but also of liver fibrosis and hypoxic stress. As mentioned before, miR-199 is decreased in HCC in correlation with its aggressiveness and invasiveness, and thus constitutes a potential predictive marker (Fornari et al. 2010; Hou et al. 2011). As previously mentioned, miR-199 is also up-regulated in HSC cells during fibrogenesis (Murakami et al. 2011). Lee et al. (2012) suggests that the farnesoid X receptor FXR decrease in liver during fibrotic process leads to miR-199 induction and LKB1 reduction.miR-199 also controls hypoxia, which affects tumor and stromal cells, and enhances proliferation, metastasis and resistance to treatment of HCC. The loss of miR-199b has been directly linked to hypoxia-induced factor 1α (HIF-1α) activation in HCC samples (Wang et al. 2011). MiR-199b transfection into HCC cells inhibits cell proliferation during hypoxia and restores their radiosensitivity. Recently, the cluster miR-199a/miR-214 has also been identified as a key orchestrator of HCC progression in response to endoplasmic reticulum stress (Duan et al. 2012). The authors showed that the expression of these miRNAs is suppressed in response to unfolded protein response-induced hypoxia, and this is repressed by NFκB inhibition.

To conclude, as supported by all these data, miRNAs constitute potent biomarkers, which can be very useful to improve the classification of HCC, its diagnosis and the prediction of its response to treatment. A non-invasive diagnostic tool based on miRNAs is also possible, by blood or urinary sampling, even if an improvement of their detection in body fluids is required. Finally, tumoral miRNAs also emerge as promising therapeutic targets, as we will develop in the next part of this chapter.

8 MicroRNAs in Liver: A New Therapeutic Approach

Although numerous efforts have been conducted to improve HCC treatment, surgery is still nowadays the most frequent therapeutic option. Sorafenib, an inhibitor of receptors tyrosine kinase, is the unique drug accepted by FDA for the treatment of liver cancer. It has only enhanced the patient life expectancy of 3 months. MiRNAs appear as promising therapeutic targets since they could target different signaling pathways by their multiplicity of action. Above all, miRNA-based therapy could target in the same time the metabolic, inflammatory and immune counterparts of liver tumors responsible for the development and progression of HCC.

Following the identification of a panel of microRNAs deregulated in HCC, a number of works have been conducted to restore the expression of these miRNAs, initially *in vitro* and then *in vivo* (Table 8.1). The *in vitro* experiments on cell lines gave encouraging results, which support the potent effect of an administration of miRNA-based therapies to mice. A number of mouse models mimicking liver disease, and particularly HCC, have been developed these recent years (Heindryckx et al. 2009). Chemically induced tumor models, i.e. following diethylnitrosamine (Hacker et al. 1991) or carbon tetrachloride injection (Weisburger 1977) are frequently used. Some teams also work on xenograft models following ectopic or orthotopic implantation. Alternatively, various mouse models have been genetically created to model the different subtypes of HCC, following the introduction of a viral gene (Araki et al. 1989), or the activation of the oncogenes myc (Thorgeirsson and Santoni-Rugiu 1996) or β-catenin (Colnot et al. 2004). More recently, these miRNA-based therapies have been successfully tested in primates and in patients.

As we mentioned at the beginning of this chapter, an anti-miR-122, called Miravirsen, based on the locked nucleic acid technology (LNA), was the first therapy to enter into clinical trials in 2008 to treat HCV. A LNA is a nucleic acid chemically modified by the addition of a methylene bridge in the ribose. This modification increases the LNA affinity for its target and also its stability. The LNA becomes resistant to nuclease degradation. The first works performed in mouse and primates infected with hepatitis C virus suggest an absence of toxicity for Miravirsen. This anti-miR-122 successfully impairs the level of viral RNAs in plasma, since HCV uses liver miR-122 for its replication (Lanford et al. 2010). Recently, encouraging data were obtained in patients infected with chronic HCV. Miravirsen successfully decreases the level of virus RNAs and of plasmatic cholesterol (Janssen et al. 2013). Another type of inhibitors has also been tested *in vivo*, the antagomiRs. An inhibitor against miR-221 has been tested in a mouse model, and successfully delays HCC development (Park et al. 2011). An antagomiR presents a sequence complementary to the whole sequence of the miRNA of interest, in contrast to a LNA, which only hybridizes with a part of the miRNA sequence. It induces the sequestration or the degradation of its miRNA target. To improve the antagomiR stability *in vivo*, some chemical modifications have been added, i.e. an addition of 2′-O-methyl or 2′-O-methoxyethyl (MOE) group, or creation of a phosphorothioate backbone. In 2010, Rosetta Genomics Ltd. developed a systemic 2′MOE-antagomiR against miR-191, frequently over-expressed in HCC. The orthotopic xenograft mouse model study demonstrates that a marked suppression of a microRNA, miR-191, leads to a significant two-fold decrease of tumor mass due to reduction of cancer-cell proliferation and enhanced apoptosis (Elyakim et al. 2010). These recent years, a new class of inhibitors appears the tiny-LNAs. Due to their shorter sequence, these LNAs have a better tissue distribution and can inhibit an entire family of miRNAs. A tiny-LNA directed against miR-122 showed a better efficiency in liver cells and *in vivo* as compared to LNAs (Obad et al. 2011).

The second approach based on miRNA consists to restore miRNA expression thanks to a double strand RNA, mimicking the activity of the endogenous miRNA, and, so, called "mimic". The administration of a mimic of miR-26 with an adenovirus vehicle in a mouse model of HCC mutated for the oncogene MYC, successfully abrogates cancer-cell proliferation and tumor development in correlation with a loss of cyclin E_2 and D_2 in tumors without toxic effect in the remaining liver (Kota et al. 2009). In the same way, the work of Hatziapostolou et al. (2011), supporting the crucial role of the trio miR-24/miR-629/miR-124 in IL-6-dependent inflammation in HCC induced by diethylnitrosamine, shows the efficiency of an administration of a mimic of miR-124 encapsulated into liposomes in this mouse model of HCC. Finally, the miRNA therapeutics company, dedicated to develop mimic-based therapies, has created liposomes containing a mimic of miR-34a, called miRX34. This mimic formulation is currently in clinical trials for solid tumors, and in particular for liver tumors. They showed that in a model of human HCC grafted to mouse, miRX34 significantly improved mouse survival. This company also begins the preclinical trials for a mimic let-7g in liver tumors (www.mirnatherapeutics.com).

Despite promising results obtained for miRNA-based therapies, a number of challenges remained to date to improve the efficiency of this type of treatments. Their inconvenient are similar to those which have delayed the use of therapies based on small interfering RNAs: improvement of their stability, free administration or delivery encapsulated into nanoparticules or exosomes, problems of specificity, off-targets or targeting of non-identified mRNAs, tissue distribution, response persistence, and secondary effects.

9 Conclusions

Since their discovery in 1993, the study of miRNAs is an expanding domain that allows the identification of a miRNA signature for a great variety of tumors and in particular for liver tumors. The better knowledge of the panel of miRNAs deregulated in liver disease and notably in hepatocellular carcinoma is in favor of the use of miRNAs as potent tools for diagnosis, prognosis and treatment of HCC patients. Interestingly, miRNAs integrate all the counterparts of liver tumorigenesis, whether inflammation, hypoxia, immune escape as well as tumor cell proliferation, migration and invasion. Although a great number of studies help to decipher the different targets of one miRNA of interest, the main challenge for the future remains to better understand how the different targets of each miRNA interact in one pathology and how a mRNA, targeted by different miRNAs, behave in this context.

Acknowledgements This work has been supported by the Ligue Nationale Contre le Cancer (Labellisation Programme 2011–2013) and the Agence Nationale de la Recherche (WNT-METABOLIV, 2011–2013).

References

Ahn J, Lee H, Chung CH, Ha T (2011) High fat diet induced downregulation of microRNA-467b increased lipoprotein lipase in hepatic steatosis. Biochem Biophys Res Commun 414(4):664–669

Alisi A, Da Sacco L, Bruscalupi G, Piemonte F, Panera N, De Vito R et al (2011) Mirnome analysis reveals novel molecular determinants in the pathogenesis of diet-induced nonalcoholic fatty liver disease. Lab Invest 91(2):283–293

Anson M, Crain-Denoyelle AM, Baud V, Chereau F, Gougelet A, Terris B et al (2012) Oncogenic beta-catenin triggers an inflammatory response that determines the aggressiveness of hepatocellular carcinoma in mice. J Clin Invest 122(2):586–599

Araki K, Miyazaki J, Hino O, Tomita N, Chisaka O, Matsubara K et al (1989) Expression and replication of hepatitis B virus genome in transgenic mice. Proc Natl Acad Sci U S A 86(1):207–211

Audard V, Grimber G, Elie C, Radenen B, Audebourg A, Letourneur F et al (2007) Cholestasis is a marker for hepatocellular carcinomas displaying beta-catenin mutations. J Pathol 212(3):345–352

Babashah S, Soleimani M (2011) The oncogenic and tumour suppressive roles of microRNAs in cancer and apoptosis. Eur J Cancer 47(8):1127–1137

Babashah S, Sadeghizadeh M, Tavirani MR, Farivar S, Soleimani M (2012) Aberrant microRNA expression and its implications in the pathogenesis of leukemias. Cell Oncol (Dordr) 35(5):317–334

Bai S, Nasser MW, Wang B, Hsu SH, Datta J, Kutay H et al (2009) MicroRNA-122 inhibits tumorigenic properties of hepatocellular carcinoma cells and sensitizes these cells to sorafenib. J Biol Chem 284(46):32015–32027

Bala S, Marcos M, Kodys K, Csak T, Catalano D, Mandrekar P et al (2011) Up-regulation of microRNA-155 in macrophages contributes to increased tumor necrosis factor alpha (TNF{alpha}) production via increased mRNA half-life in alcoholic liver disease. J Biol Chem 286(2):1436–1444

Bala S, Petrasek J, Mundkur S, Catalano D, Levin I, Ward J et al (2012) Circulating microRNAs in exosomes indicate hepatocyte injury and inflammation in alcoholic, drug-induced, and inflammatory liver diseases. Hepatology 56(5):1946–1957

Bihrer V, Waidmann O, Friedrich-Rust M, Forestier N, Susser S, Haupenthal J et al (2011) Serum microRNA-21 as marker for necroinflammation in hepatitis C patients with and without hepatocellular carcinoma. PLoS One 6(10):e26971

Boesch-Saadatmandi C, Wagner AE, Wolffram S, Rimbach G (2012) Effect of quercetin on inflammatory gene expression in mice liver in vivo – role of redox factor 1, miRNA-122 and miRNA-125b. Pharmacol Res 65(5):523–530

Bouwens L, De Bleser P, Vanderkerken K, Geerts B, Wisse E (1992) Liver cell heterogeneity: functions of non-parenchymal cells. Enzyme 46(1–3):155–168

Boyault S, Rickman DS, de Reynies A, Balabaud C, Rebouissou S, Jeannot E et al (2007) Transcriptome classification of HCC is related to gene alterations and to new therapeutic targets. Hepatology 45(1):42–52

Brenner DA, Waterboer T, Choi SK, Lindquist JN, Stefanovic B, Burchardt E et al (2000) New aspects of hepatic fibrosis. J Hepatol 32(1 Suppl):32–38

Budhu A, Jia HL, Forgues M, Liu CG, Goldstein D, Lam A et al (2008) Identification of metastasis-related microRNAs in hepatocellular carcinoma. Hepatology 47(3):897–907

Cairo S, Wang Y, de Reynies A, Duroure K, Dahan J, Redon MJ et al (2010) Stem cell-like micro-RNA signature driven by Myc in aggressive liver cancer. Proc Natl Acad Sci U S A 107(47):20471–20476

Calin GA, Sevignani C, Dumitru CD, Hyslop T, Noch E, Yendamuri S et al (2004) Human microRNA genes are frequently located at fragile sites and genomic regions involved in cancers. Proc Natl Acad Sci U S A 101(9):2999–3004

Cermelli S, Ruggieri A, Marrero JA, Ioannou GN, Beretta L (2011) Circulating microRNAs in patients with chronic hepatitis C and non-alcoholic fatty liver disease. PLoS One 6(8):e23937

Cheung O, Puri P, Eicken C, Contos MJ, Mirshahi F, Maher JW et al (2008) Nonalcoholic steatohepatitis is associated with altered hepatic microRNA expression. Hepatology 48(6):1810–1820

Colnot S, Decaens T, Niwa-Kawakita M, Godard C, Hamard G, Kahn A et al (2004) Liver-targeted disruption of Apc in mice activates beta-catenin signaling and leads to hepatocellular carcinomas. Proc Natl Acad Sci U S A 101(49):17216–17221

Conde-Vancells J, Rodriguez-Suarez E, Embade N, Gil D, Matthiesen R, Valle M et al (2008) Characterization and comprehensive proteome profiling of exosomes secreted by hepatocytes. J Proteome Res 7(12):5157–5166

Crispe IN (2009) The liver as a lymphoid organ. Annu Rev Immunol 27:147–163

de La Coste A, Romagnolo B, Billuart P, Renard CA, Buendia MA, Soubrane O et al (1998) Somatic mutations of the beta-catenin gene are frequent in mouse and human hepatocellular carcinomas. Proc Natl Acad Sci U S A 95(15):8847–8851

Datta J, Kutay H, Nasser MW, Nuovo GJ, Wang B, Majumder S et al (2008) Methylation mediated silencing of microRNA-1 gene and its role in hepatocellular carcinogenesis. Cancer Res 68(13):5049–5058

Davalos A, Goedeke L, Smibert P, Ramirez CM, Warrier NP, Andreo U et al (2011) miR-33a/b contribute to the regulation of fatty acid metabolism and insulin signaling. Proc Natl Acad Sci U S A 108(22):9232–9237

Ding J, Huang S, Wu S, Zhao Y, Liang L, Yan M et al (2010) Gain of miR-151 on chromosome 8q24.3 facilitates tumour cell migration and spreading through downregulating RhoGDIA. Nat Cell Biol 12(4):390–399

Duan Q, Wang X, Gong W, Ni L, Chen C, He X et al (2012) ER stress negatively modulates the expression of the miR-199a/214 cluster to regulates tumor survival and progression in human hepatocellular cancer. PLoS One 7(2):e31518

El-Serag HB, Rudolph KL (2007) Hepatocellular carcinoma: epidemiology and molecular carcinogenesis. Gastroenterology 132(7):2557–2576

Elyakim E, Sitbon E, Faerman A, Tabak S, Montia E, Belanis L et al (2010) hsa-miR-191 is a candidate oncogene target for hepatocellular carcinoma therapy. Cancer Res 70(20):8077–8087

Fornari F, Gramantieri L, Ferracin M, Veronese A, Sabbioni S, Calin GA et al (2008) MiR-221 controls CDKN1C/p57 and CDKN1B/p27 expression in human hepatocellular carcinoma. Oncogene 27(43):5651–5661

Fornari F, Milazzo M, Chieco P, Negrini M, Calin GA, Grazi GL et al (2010) MiR-199a-3p regulates mTOR and c-Met to influence the doxorubicin sensitivity of human hepatocarcinoma cells. Cancer Res 70(12):5184–5193

Furuta M, Kozaki KI, Tanaka S, Arii S, Imoto I, Inazawa J (2010) miR-124 and miR-203 are epigenetically silenced tumor-suppressive microRNAs in hepatocellular carcinoma. Carcinogenesis 31(5):766–776

Garofalo M, Di Leva G, Romano G, Nuovo G, Suh SS, Ngankeu A et al (2009) miR-221&222 regulate TRAIL resistance and enhance tumorigenicity through PTEN and TIMP3 downregulation. Cancer Cell 16(6):498–509

Gerin I, Clerbaux LA, Haumont O, Lanthier N, Das AK, Burant CF et al (2010) Expression of miR-33 from an SREBP2 intron inhibits cholesterol export and fatty acid oxidation. J Biol Chem 285(44):33652–33661

Girard M, Jacquemin E, Munnich A, Lyonnet S, Henrion-Caude A (2008) miR-122, a paradigm for the role of microRNAs in the liver. J Hepatol 48(4):648–656

Gramantieri L, Ferracin M, Fornari F, Veronese A, Sabbioni S, Liu CG et al (2007) Cyclin G1 is a target of miR-122a, a microRNA frequently down-regulated in human hepatocellular carcinoma. Cancer Res 67(13):6092–6099

Gramantieri L, Fornari F, Ferracin M, Veronese A, Sabbioni S, Calin GA et al (2009) MicroRNA-221 targets Bmf in hepatocellular carcinoma and correlates with tumor multifocality. Clin Cancer Res 15(16):5073–5081

Guichard C, Amaddeo G, Imbeaud S, Ladeiro Y, Pelletier L, Maad IB et al (2012) Integrated analysis of somatic mutations and focal copy-number changes identifies key genes and pathways in hepatocellular carcinoma. Nat Genet 44(6):694–698

Hacker HJ, Mtiro H, Bannasch P, Vesselinovitch SD (1991) Histochemical profile of mouse hepatocellular adenomas and carcinomas induced by a single dose of diethylnitrosamine. Cancer Res 51(7):1952–1958

Hanahan D, Weinberg RA (2011) Hallmarks of cancer: the next generation. Cell 144(5):646–674

Hatziapostolou M, Polytarchou C, Aggelidou E, Drakaki A, Poultsides GA, Jaeger SA et al (2011) An HNF4alpha-miRNA inflammatory feedback circuit regulates hepatocellular oncogenesis. Cell 147(6):1233–1247

He G, Karin M (2011) NF-kappaB and STAT3 – key players in liver inflammation and cancer. Cell Res 21(1):159–168

Heindryckx F, Colle I, Van Vlierberghe H (2009) Experimental mouse models for hepatocellular carcinoma research. Int J Exp Pathol 90(4):367–386

Hoekstra M, van der Sluis RJ, Kuiper J, Van Berkel TJ (2012) Nonalcoholic fatty liver disease is associated with an altered hepatocyte microRNA profile in LDL receptor knockout mice. J Nutr Biochem 23(6):622–628

Hou J, Lin L, Zhou W, Wang Z, Ding G, Dong Q et al (2011) Identification of miRNomes in human liver and hepatocellular carcinoma reveals miR-199a/b-3p as therapeutic target for hepatocellular carcinoma. Cancer Cell 19(2):232–243

Huang S, He X (2011) The role of microRNAs in liver cancer progression. Br J Cancer 104(2):235–240

Huang S, He X, Ding J, Liang L, Zhao Y, Zhang Z et al (2008) Upregulation of miR-23a approximately 27a approximately 24 decreases transforming growth factor-beta-induced tumor-suppressive activities in human hepatocellular carcinoma cells. Int J Cancer 123(4):972–978

Iliopoulos D, Drosatos K, Hiyama Y, Goldberg IJ, Zannis VI (2010) MicroRNA-370 controls the expression of microRNA-122 and Cpt1alpha and affects lipid metabolism. J Lipid Res 51(6):1513–1523

Janssen HL, Reesink HW, Lawitz EJ, Zeuzem S, Rodriguez-Torres M, Patel K et al (2013) Treatment of HCV infection by targeting microRNA. N Engl J Med 368(18):1685–1694

Ji J, Shi J, Budhu A, Yu Z, Forgues M, Roessler S et al (2009) MicroRNA expression, survival, and response to interferon in liver cancer. N Engl J Med 361(15):1437–1447

Ji J, Zhao L, Budhu A, Forgues M, Jia HL, Qin LX et al (2010) Let-7g targets collagen type I alpha2 and inhibits cell migration in hepatocellular carcinoma. J Hepatol 52(5):690–697

Karakatsanis A, Papaconstantinou I, Gazouli M, Lyberopoulou A, Polymeneas G, Voros D (2013) Expression of microRNAs, miR-21, miR-31, miR-122, miR-145, miR-146a, miR-200c, miR-221, miR-222, and miR-223 in patients with hepatocellular carcinoma or intrahepatic cholangiocarcinoma and its prognostic significance. Mol Carcinog 52(4):297–303

Katz N, Jungermann K (1976) Autoregulatory shift from fructolysis to lactate gluconeogenisis in rat hepatocyte suspensions. The problem of metabolic zonation of liver parenchyma. Hoppe Seylers Z Physiol Chem 357(3):359–375

Kota J, Chivukula RR, O'Donnell KA, Wentzel EA, Montgomery CL, Hwang HW et al (2009) Therapeutic microRNA delivery suppresses tumorigenesis in a murine liver cancer model. Cell 137(6):1005–1017

Ladeiro Y, Couchy G, Balabaud C, Bioulac-Sage P, Pelletier L, Rebouissou S et al (2008) MicroRNA profiling in hepatocellular tumors is associated with clinical features and oncogene/tumor suppressor gene mutations. Hepatology 47(6):1955–1963

Lalor PF, Shields P, Grant A, Adams DH (2002) Recruitment of lymphocytes to the human liver. Immunol Cell Biol 80(1):52–64

Lan FF, Wang H, Chen YC, Chan CY, Ng SS, Li K et al (2011) Hsa-let-7g inhibits proliferation of hepatocellular carcinoma cells by downregulation of c-Myc and upregulation of p16(INK4A). Int J Cancer 128(2):319–331

Lanford RE, Hildebrandt-Eriksen ES, Petri A, Persson R, Lindow M, Munk ME et al (2010) Therapeutic silencing of microRNA-122 in primates with chronic hepatitis C virus infection. Science 327(5962):198–201

Law PT, Wong N (2011) Emerging roles of microRNA in the intracellular signaling networks of hepatocellular carcinoma. J Gastroenterol Hepatol 26(3):437–449

le Sage C, Nagel R, Egan DA, Schrier M, Mesman E, Mangiola A et al (2007) Regulation of the p27(Kip1) tumor suppressor by miR-221 and miR-222 promotes cancer cell proliferation. EMBO J 26(15):3699–3708

Lee RC, Feinbaum RL, Ambros V (1993) The C. elegans heterochronic gene lin-4 encodes small RNAs with antisense complementarity to lin-14. Cell 75(5):843–854

Lee CG, Kim YW, Kim EH, Meng Z, Huang W, Hwang SJ et al (2012) Farnesoid X receptor protects hepatocytes from injury by repressing miR-199a-3p, which increases levels of LKB1. Gastroenterology 142(5):1206–1217, e1207

Leonardi GC, Candido S, Cervello M, Nicolosi D, Raiti F, Travali S et al (2012) The tumor microenvironment in hepatocellular carcinoma (review). Int J Oncol 40(6):1733–1747

Li N, Fu H, Tie Y, Hu Z, Kong W, Wu Y et al (2009a) miR-34a inhibits migration and invasion by down-regulation of c-Met expression in human hepatocellular carcinoma cells. Cancer Lett 275(1):44–53

Li S, Fu H, Wang Y, Tie Y, Xing R, Zhu J et al (2009b) MicroRNA-101 regulates expression of the v-fos FBJ murine osteosarcoma viral oncogene homolog (FOS) oncogene in human hepatocellular carcinoma. Hepatology 49(4):1194–1202

Li Y, Tan W, Neo TW, Aung MO, Wasser S, Lim SG et al (2009c) Role of the miR-106b-25 microRNA cluster in hepatocellular carcinoma. Cancer Sci 100(7):1234–1242

Li LM, Hu ZB, Zhou ZX, Chen X, Liu FY, Zhang JF et al (2010a) Serum microRNA profiles serve as novel biomarkers for HBV infection and diagnosis of HBV-positive hepatocarcinoma. Cancer Res 70(23):9798–9807

Li R, Qian N, Tao K, You N, Wang X, Dou K (2010b) MicroRNAs involved in neoplastic transformation of liver cancer stem cells. J Exp Clin Cancer Res 29:169

Li J, Wang Y, Yu W, Chen J, Luo J (2011a) Expression of serum miR-221 in human hepatocellular carcinoma and its prognostic significance. Biochem Biophys Res Commun 406(1):70–73

Li WQ, Chen C, Xu MD, Guo J, Li YM, Xia QM et al (2011b) The rno-miR-34 family is upregulated and targets ACSL1 in dimethylnitrosamine-induced hepatic fibrosis in rats. FEBS J 278(9):1522–1532

Li L, Guo Z, Wang J, Mao Y, Gao Q (2012) Serum miR-18a: a potential marker for hepatitis B virus-related hepatocellular carcinoma screening. Dig Dis Sci 57(11):2910–2916

Lin CJ, Gong HY, Tseng HC, Wang WL, Wu JL (2008) miR-122 targets an anti-apoptotic gene, Bcl-w, in human hepatocellular carcinoma cell lines. Biochem Biophys Res Commun 375(3):315–320

Liu X, Hu H, Yin JQ (2006) Therapeutic strategies against TGF-beta signaling pathway in hepatic fibrosis. Liver Int 26(1):8–22

Liu WH, Yeh SH, Lu CC, Yu SL, Chen HY, Lin CY et al (2009) MicroRNA-18a prevents estrogen receptor-alpha expression, promoting proliferation of hepatocellular carcinoma cells. Gastroenterology 136(2):683–693

Liu AM, Poon RT, Luk JM (2010) MicroRNA-375 targets Hippo-signaling effector YAP in liver cancer and inhibits tumor properties. Biochem Biophys Res Commun 394(3):623–627

Liu AM, Yao TJ, Wang W, Wong KF, Lee NP, Fan ST et al (2012) Circulating miR-15b and miR-130b in serum as potential markers for detecting hepatocellular carcinoma: a retrospective cohort study. BMJ Open 2(2):e000825

Lu J, Getz G, Miska EA, Alvarez-Saavedra E, Lamb J, Peck D et al (2005) MicroRNA expression profiles classify human cancers. Nature 435(7043):834–838
Marotta F, Vangieri B, Cecere A, Gattoni A (2004) The pathogenesis of hepatocellular carcinoma is multifactorial event. Novel immunological treatment in prospect. Clin Ter 155(5):187–199
Marquart TJ, Allen RM, Ory DS, Baldan A (2010) miR-33 links SREBP-2 induction to repression of sterol transporters. Proc Natl Acad Sci U S A 107(27):12228–12232
Massague J (2008) TGFbeta in cancer. Cell 134(2):215–230
Matsuzaki K (2009) Modulation of TGF-beta signaling during progression of chronic liver diseases. Front Biosci 14:2923–2934
Meng F, Henson R, Wehbe-Janek H, Ghoshal K, Jacob ST, Patel T (2007) MicroRNA-21 regulates expression of the PTEN tumor suppressor gene in human hepatocellular cancer. Gastroenterology 133(2):647–658
Min HK, Kapoor A, Fuchs M, Mirshahi F, Zhou H, Maher J et al (2012) Increased hepatic synthesis and dysregulation of cholesterol metabolism is associated with the severity of nonalcoholic fatty liver disease. Cell Metab 15(5):665–674
Murakami Y, Yasuda T, Saigo K, Urashima T, Toyoda H, Okanoue T et al (2006) Comprehensive analysis of microRNA expression patterns in hepatocellular carcinoma and non-tumorous tissues. Oncogene 25(17):2537–2545
Murakami Y, Toyoda H, Tanaka M, Kuroda M, Harada Y, Matsuda F et al (2011) The progression of liver fibrosis is related with overexpression of the miR-199 and 200 families. PLoS One 6(1):e16081
Najafi-Shoushtari SH, Kristo F, Li Y, Shioda T, Cohen DE, Gerszten RE et al (2010) MicroRNA-33 and the SREBP host genes cooperate to control cholesterol homeostasis. Science 328(5985):1566–1569
Nakagawa H, Maeda S, Yoshida H, Tateishi R, Masuzaki R, Ohki T et al (2009) Serum IL-6 levels and the risk for hepatocarcinogenesis in chronic hepatitis C patients: an analysis based on gender differences. Int J Cancer 125(10):2264–2269
Obad S, dos Santos CO, Petri A, Heidenblad M, Broom O, Ruse C et al (2011) Silencing of microRNA families by seed-targeting tiny LNAs. Nat Genet 43(4):371–378
Ogawa T, Enomoto M, Fujii H, Sekiya Y, Yoshizato K, Ikeda K et al (2012) MicroRNA-221/222 upregulation indicates the activation of stellate cells and the progression of liver fibrosis. Gut 61(11):1600–1609
Park JK, Kogure T, Nuovo GJ, Jiang J, He L, Kim JH et al (2011) miR-221 silencing blocks hepatocellular carcinoma and promotes survival. Cancer Res 71(24):7608–7616
Parker GA, Picut CA (2012) Immune functioning in non lymphoid organs: the liver. Toxicol Pathol 40(2):237–247
Pineau P, Volinia S, McJunkin K, Marchio A, Battiston C, Terris B et al (2010) miR-221 overexpression contributes to liver tumorigenesis. Proc Natl Acad Sci U S A 107(1):264–269
Pogribny IP, Starlard-Davenport A, Tryndyak VP, Han T, Ross SA, Rusyn I et al (2010) Difference in expression of hepatic microRNAs miR-29c, miR-34a, miR-155, and miR-200b is associated with strain-specific susceptibility to dietary nonalcoholic steatohepatitis in mice. Lab Invest 90(10):1437–1446
Rayner KJ, Suarez Y, Davalos A, Parathath S, Fitzgerald ML, Tamehiro N et al (2010) MiR-33 contributes to the regulation of cholesterol homeostasis. Science 328(5985):1570–1573
Roderburg C, Urban GW, Bettermann K, Vucur M, Zimmermann H, Schmidt S et al (2011) MicroRNA profiling reveals a role for miR-29 in human and murine liver fibrosis. Hepatology 53(1):209–218
Salvi A, Sabelli C, Moncini S, Venturin M, Arici B, Riva P et al (2009) MicroRNA-23b mediates urokinase and c-met downmodulation and a decreased migration of human hepatocellular carcinoma cells. FEBS J 276(11):2966–2982
Scisciani C, Vossio S, Guerrieri F, Schinzari V, De Iaco R, D'Onorio de Meo P et al (2012) Transcriptional regulation of miR-224 upregulated in human HCCs by NFkappaB inflammatory pathways. J Hepatol 56(4):855–861

Seki E, Brenner DA (2008) Toll-like receptors and adaptor molecules in liver disease: update. Hepatology 48(1):322–335

Sekine S, Ogawa R, Ito R, Hiraoka N, McManus MT, Kanai Y et al (2009a) Disruption of Dicer1 induces dysregulated fetal gene expression and promotes hepatocarcinogenesis. Gastroenterology 136(7):2304–2315, e2301–2304

Sekine S, Ogawa R, McManus MT, Kanai Y, Hebrok M (2009b) Dicer is required for proper liver zonation. J Pathol 219(3):365–372

Selaru FM, Olaru AV, Kan T, David S, Cheng Y, Mori Y et al (2009) MicroRNA-21 is overexpressed in human cholangiocarcinoma and regulates programmed cell death 4 and tissue inhibitor of metalloproteinase 3. Hepatology 49(5):1595–1601

Shah YM, Morimura K, Yang Q, Tanabe T, Takagi M, Gonzalez FJ (2007) Peroxisome proliferator-activated receptor alpha regulates a microRNA-mediated signaling cascade responsible for hepatocellular proliferation. Mol Cell Biol 27(12):4238–4247

Shimizu S, Takehara T, Hikita H, Kodama T, Miyagi T, Hosui A et al (2010) The let-7 family of microRNAs inhibits Bcl-xL expression and potentiates sorafenib-induced apoptosis in human hepatocellular carcinoma. J Hepatol 52(5):698–704

Su H, Yang JR, Xu T, Huang J, Xu L, Yuan Y et al (2009) MicroRNA-101, down-regulated in hepatocellular carcinoma, promotes apoptosis and suppresses tumorigenicity. Cancer Res 69(3):1135–1142

Takata A, Otsuka M, Kojima K, Yoshikawa T, Kishikawa T, Yoshida H et al (2011) MicroRNA-22 and microRNA-140 suppress NF-kappaB activity by regulating the expression of NF-kappaB coactivators. Biochem Biophys Res Commun 411(4):826–831

Tan HX, Wang Q, Chen LZ, Huang XH, Chen JS, Fu XH et al (2010) MicroRNA-9 reduces cell invasion and E-cadherin secretion in SK-Hep-1 cell. Med Oncol 27(3):654–660

Thorgeirsson SS, Santoni-Rugiu E (1996) Transgenic mouse models in carcinogenesis: interaction of c-myc with transforming growth factor alpha and hepatocyte growth factor in hepatocarcinogenesis. Br J Clin Pharmacol 42(1):43–52

Tilg H, Wilmer A, Vogel W, Herold M, Nolchen B, Judmaier G et al (1992) Serum levels of cytokines in chronic liver diseases. Gastroenterology 103(1):264–274

Tsai WC, Hsu SD, Hsu CS, Lai TC, Chen SJ, Shen R et al (2012) MicroRNA-122 plays a critical role in liver homeostasis and hepatocarcinogenesis. J Clin Invest 122(8):2884–2897

Vaulont S (2011) miR-122, a microRNA gatekeeper of iron homeostasis. Med Sci M/S 27(8–9):704–706

Vogelstein B, Papadopoulos N, Velculescu VE, Zhou S, Diaz LA Jr, Kinzler KW (2013) Cancer genome landscapes. Science 339(6127):1546–1558

Volinia S, Calin GA, Liu CG, Ambs S, Cimmino A, Petrocca F et al (2006) A microRNA expression signature of human solid tumors defines cancer gene targets. Proc Natl Acad Sci U S A 103(7):2257–2261

Wang Y, Lee AT, Ma JZ, Wang J, Ren J, Yang Y et al (2008) Profiling microRNA expression in hepatocellular carcinoma reveals microRNA-224 up-regulation and apoptosis inhibitor-5 as a microRNA-224-specific target. J Biol Chem 283(19):13205–13215

Wang B, Hsu SH, Majumder S, Kutay H, Huang W, Jacob ST et al (2010) TGFbeta-mediated upregulation of hepatic miR-181b promotes hepatocarcinogenesis by targeting TIMP3. Oncogene 29(12):1787–1797

Wang C, Song B, Song W, Liu J, Sun A, Wu D et al (2011) Underexpressed microRNA-199b-5p targets hypoxia-inducible factor-1alpha in hepatocellular carcinoma and predicts prognosis of hepatocellular carcinoma patients. J Gastroenterol Hepatol 26(11):1630–1637

Wang XW, Heegaard NH, Orum H (2012) MicroRNAs in liver disease. Gastroenterology 142(7):1431–1443

Weisburger EK (1977) Carcinogenicity studies on halogenated hydrocarbons. Environ Health Perspect 21:7–16

Wen J, Friedman JR (2012) miR-122 regulates hepatic lipid metabolism and tumor suppression. J Clin Invest 122(8):2773–2776

Witek RP, Yang L, Liu R, Jung Y, Omenetti A, Syn WK et al (2009) Liver cell-derived microparticles activate hedgehog signaling and alter gene expression in hepatic endothelial cells. Gastroenterology 136(1):320–330, e322
Wong QW, Lung RW, Law PT, Lai PB, Chan KY, To KF et al (2008) MicroRNA-223 is commonly repressed in hepatocellular carcinoma and potentiates expression of Stathmin1. Gastroenterology 135(1):257–269
Wong VW, Yu J, Cheng AS, Wong GL, Chan HY, Chu ES et al (2009) High serum interleukin-6 level predicts future hepatocellular carcinoma development in patients with chronic hepatitis B. Int J Cancer 124(12):2766–2770
Wong QW, Ching AK, Chan AW, Choy KW, To KF, Lai PB et al (2010) MiR-222 overexpression confers cell migratory advantages in hepatocellular carcinoma through enhancing AKT signaling. Clin Cancer Res 16(3):867–875
Wong CM, Wong CC, Lee JM, Fan DN, Au SL, Ng IO (2012) Sequential alterations of microRNA expression in hepatocellular carcinoma development and venous metastasis. Hepatology 55(5):1453–1461
Xiong Y, Fang JH, Yun JP, Yang J, Zhang Y, Jia WH et al (2010) Effects of microRNA-29 on apoptosis, tumorigenicity, and prognosis of hepatocellular carcinoma. Hepatology 51(3):836–845
Xu T, Zhu Y, Xiong Y, Ge YY, Yun JP, Zhuang SM (2009) MicroRNA-195 suppresses tumorigenicity and regulates G1/S transition of human hepatocellular carcinoma cells. Hepatology 50(1):113–121
Xu J, Wu C, Che X, Wang L, Yu D, Zhang T et al (2011a) Circulating microRNAs, miR-21, miR-122, and miR-223, in patients with hepatocellular carcinoma or chronic hepatitis. Mol Carcinog 50(2):136–142
Xu Y, Xia F, Ma L, Shan J, Shen J, Yang Z et al (2011b) MicroRNA-122 sensitizes HCC cancer cells to adriamycin and vincristine through modulating expression of MDR and inducing cell cycle arrest. Cancer Lett 310(2):160–169
Yan XL, Jia YL, Chen L, Zeng Q, Zhou JN, Fu CJ et al (2013) Hepatocellular carcinoma-associated mesenchymal stem cells promote hepatocarcinoma progression: role of the S100A4-miR155-SOCS1-MMP9 axis. Hepatology 57(6):2274–2286
Yang L, Wang Y, Mao H, Fleig S, Omenetti A, Brown KD et al (2008) Sonic hedgehog is an autocrine viability factor for myofibroblastic hepatic stellate cells. J Hepatol 48(1):98–106
Yang F, Yin Y, Wang F, Wang Y, Zhang L, Tang Y et al (2010) miR-17-5p Promotes migration of human hepatocellular carcinoma cells through the p38 mitogen-activated protein kinase-heat shock protein 27 pathway. Hepatology 51(5):1614–1623
Yang P, Li QJ, Feng Y, Zhang Y, Markowitz GJ, Ning S et al (2012) TGF-beta-miR-34a-CCL22 signaling-induced Treg cell recruitment promotes venous metastases of HBV-positive hepatocellular carcinoma. Cancer Cell 22(3):291–303
Yao J, Liang L, Huang S, Ding J, Tan N, Zhao Y et al (2010) MicroRNA-30d promotes tumor invasion and metastasis by targeting Galphai2 in hepatocellular carcinoma. Hepatology 51(3):846–856
Yoon S, Kim TH, Natarajan A, Wang SS, Choi J, Wu J et al (2010) Acute liver injury upregulates microRNA-491-5p in mice, and its overexpression sensitizes Hep G2 cells for tumour necrosis factor-alpha-induced apoptosis. Liver Int 30(3):376–387
Zhang X, Liu S, Hu T, Liu S, He Y, Sun S (2009) Up-regulated microRNA-143 transcribed by nuclear factor kappa B enhances hepatocarcinoma metastasis by repressing fibronectin expression. Hepatology 50(2):490–499
Zheng L, Lv GC, Sheng J, Yang YD (2010) Effect of miRNA-10b in regulating cellular steatosis level by targeting PPAR-alpha expression, a novel mechanism for the pathogenesis of NAFLD. J Gastroenterol Hepatol 25(1):156–163
Zheng J, Dong P, Gao S, Wang N, Yu F (2012) High expression of serum miR-17-5p associated with poor prognosis in patients with hepatocellular carcinoma. Hepatogastroenterology 60(123):549–552

Zhou J, Yu L, Gao X, Hu J, Wang J, Dai Z et al (2011) Plasma microRNA panel to diagnose hepatitis B virus-related hepatocellular carcinoma. J Clin Oncol 29(36):4781–4788

Zhu Q, Wang Z, Hu Y, Li J, Li X, Zhou L et al (2012) miR-21 promotes migration and invasion by the miR-21-PDCD4-AP-1 feedback loop in human hepatocellular carcinoma. Oncol Rep 27(5):1660–1668

Zucman-Rossi J (2010) Molecular classification of hepatocellular carcinoma. Dig Liver Dis 42(Suppl 3):S235–S241

Chapter 9
MicroRNAs and Regulatory Pathways in Tumorigenesis

Jun Wei, Yong Li, Xiaoming Liu, and William C. Cho

Contents

Abstract The signaling pathways have been demonstrated to invariably alter in controlling cell growth and differentiation in cancers. The understanding of complex signalings in tumorigenesis has led to the development of tumor-specific, molecularly targeted agents against a wide variety of human cancers. Recently, numerous microRNAs (miRNAs) has been identified as having an oncogenic or tumor suppressor property, shown to be involved in cell proliferation, differentiation, apoptosis, and

---

Jun Wei and Yong Li contributed equally in this work.

J. Wei
Institute of Stem Cell Research, General Hospital of Ningxia Medical University, Yinchuan, Ningxia 750004, China

Y. Li
College of Life Science, Ningxia University, Yinchuan, Ningxia 750021, China

X. Liu (✉)
Institute of Stem Cell Research, General Hospital of Ningxia Medical University, Yinchuan, Ningxia 750004, China

College of Life Science, Ningxia University, Yinchuan, Ningxia 750021, China
e-mail: lxm1966@nxu.edu.cn

W.C. Cho (✉)
Department of Clinical Oncology, Queen Elizabeth Hospital, Hong Kong, China
e-mail: williamcscho@gmail.com

S. Babashah (ed.), *MicroRNAs: Key Regulators of Oncogenesis*,
DOI 10.1007/978-3-319-03725-7_9, © Springer International Publishing Switzerland 2014

radio- or chemo-sensitivity in cancers. Importantly, a growing body of studies has demonstrated that some of the miRNAs function as important modulators in cellular pathways and they appear to play pivotal roles in tumorigenesis, as well as the radio-resistance and/or chemo-resistance to cancer therapy regimens, suggesting miRNAs may be potential novel anticancer targets for cancers. However, the mechanisms that miRNAs target these pathways remains fragmentary, it is therefore critical to understand the underlying mechanisms and to identify targets for therapeutic intervention. This chapter aims to discuss the current understanding of mechanisms that miRNAs integrate the activity of regulatory pathways in tumorigenesis, mainly focuses on the miRNAs those target several well-defined signaling pathways with potential targets for developing anticancer agents, including the Wnt/ß-catenin, mTOR/Akt/PI3K, EGFR, VEGF and KRAS pathways.

Keywords MicroRNA • Cancer target • Wnt signaling • mTOR pathway • EGFR • KRAS • VEGF

1 Introduction

Cell signaling is an important communicating system that governs basic cellular activities and coordinates cell actions. Alterations in the cellular signaling pathways have been suggested to be responsible for diseases including cancers. The cancer cells are able to gain functions to proliferate independently of exogenous signals and to evade mechanisms that limit cell proliferation, such as apoptosis and replicative senescence (Grant 2008; Martin 2003), which implies that targeting of cellular signals may be a novel strategy for developing anticancer agents. Indeed, numerous currently investigated potential targets for cancer therapy are signaling proteins of these pathways.

MicroRNAs (miRNAs, miRs) are a class of small non-coding RNA that impact gene expression by post-transcriptional arrest or direct degradation of mRNA. By using the deep sequencing technology, over thousands of miRNAs have been discovered during last decade (Schmitz and Wolkenhauer 2013). The miRNAs have been suggested to be involved in many unexpected biologic processes, ranging from development and homeostasis to diseases such as cancers. Since miRNAs have been found to frequently resided at fragile sites and genomic regions that are deleted or amplified in cancers, together with increasing number of studies that has manifested the alteration of miRNA expression profiles and dysregulation of miRNA expression in a broad range of cancers, implicating that they may act as key players in diverse processes of human tumorigenesis, from the oncogene expression to the metastasis of cancer, in which miRNAs acted as tumor suppressor or oncogene (Babashah and Soleimani 2011; Lu et al. 2005; Inui et al. 2010; Cho 2012; Hoshino and Matsubara 2013; Rossbach 2012).

miRNAs impact on mammalian cell apoptosis, proliferation, differentiation, development, and metabolism through a mechanism of regulating critical signaling proteins, the processes are dose-sensitive and the fine-tuning, suggesting miRNA are prime candidates for regulation of signaling pathways (Inui et al. 2010). The increasing evidence of cellular cofactors impacting miRNAs has made them as attractive diagnostic markers and therapeutic targets for cancers. Furthermore, the understanding of underlying mechanism of miRNA in cancer-related signaling pathways may also offer broader pathogenic insight into basic processes of tumorigenesis.

The dysregulation of several cellular signaling pathways, such as the Wnt/ß-catenin, mammalian target of rapamycin (mTOR)/protein kinase B (Akt)/phosphatidylinositol 3-kinase (PI3K), epidermal growth factor receptor (EGFR), vascular endothelial growth factor (VEGF) and KRAS pathways have been broadly recognized to play pivotal roles in human tumorigenesis, and increasing number of miRNAs have been defined as key players in the regulation of these signaling pathway in many types of cancer (Anton et al. 2011; Liu et al. 2011; Laplante and Sabatini 2012; Mabuchi et al. 2009; Hua et al. 2006; Sasahira et al. 2012; Mosakhani et al. 2012a, b; Weiss et al. 2008). In this chapter, we discuss the crosstalk between miRNAs and Wnt/ß-catenin, mTOR/Akt/PI3K, EGFR, VEGF, or KRAS pathways in tumorigenesis, uncovering the complex interplay between the miRNAs and their targeting signaling proteins will allow us to better understand the integrated functions of miRNAs in signal transduction cascades and develop novel targets for cancer therapy.

2 MicroRNAs and Wnt/ß-catenin Pathway

The Wnt signaling pathway is comprised of canonical and non-canonical signaling pathways. The canonical pathway is also referred as Wnt/ß-catenin signaling pathway, in which the Wnt ligands bind to a complex of FZD and LRP5/6 of cell membrane, and result in the formation of a multiprotein complex of adenomatous polyposis coli (APC), glycogen synthase kinase 3ß and axins, which in turn, prevent ß-catenin phosphorylation and degradation by proteasome in cytosol, thereby accumulating ß-catenin to translocate into the nucleus where it binds to the transcriptional factor, T-cell factor (TCF)/lymphocyte enhancer transcription factor-1 (Lef-1) to activate target genes involved in cell survival, proliferation, differentiation, apoptosis and/or invasion (Clevers 2006; Cselenyi et al. 2008). A large body of evidence demonstrates that Wnt/ß-catenin signaling plays a central role in tumorigenesis, epithelial-mesenchymal transition (EMT) and metastasis of many types of cancer (Zhang et al. 2013; Valenta et al. 2012). Accumulating evidence revealed that the alteration of miRNA expression was connected to dysregulation of several signal pathways, including the Wnt/ß-catenin pathway, which played key roles in the initiation and progression of cancers, the chemoresistance and EMT of cancer cells, as well as the fate determination of cancer stem cells (CSCs) (Chen et al. 2009a; Anton et al. 2011; Liu et al. 2011). Some of miRNAs targeting Wnt/ß-catenin signaling pathway and their potential impacts on this pathway are summarized in the Table 9.1.

In order to systematically identify miRNAs that regulate Wnt/ß-catenin pathway, Anton et al. identified 38 candidate miRNAs able to either activate or repress the Wnt/ß-catenin pathway from 470 miRNAs screened from human HEK293 cells. Interestingly, consistent with the results reported in literatures, the Wnt-repressing

Table 9.1 MicroRNAs that target Wnt/ß-catenin signaling pathway involved in carcinogenesis

MicroRNA	Potential functions	Reference(s)
miR-126	Promotes colorectal cancer	Schepeler et al. (2012)
miR125b	Activates Wnt/ß-catenin/TCF4 axis	Liu et al. (2013)
miR-27	Activates Wnt signaling to promote epithelial-mesenchymal transition of gastric cancer	Zhang et al. (2011d)
miR-130a	Directly represses RUNX3 and activates Wnt/ß-catenin signaling to increase drug resistance in HCC	Liu et al. (2011)
miR-145	Directly targets ß-catenin to regulate Wnt/ß-catenin signaling in human colon cancer cells	Yamada et al. (2013)
miR-181 family	Transcriptionally activated by Wnt signaling in HCC	Ji et al. (2011)
miR-374a	Constitutively activates Wnt/ß-catenin signaling in breast cancer	Cai et al. (2013)
miR-34	Directly targets Wnt signaling and links to p53 in various cancers	Cha et al. (2012) and Kim et al. (2011)
miR-21	Functionally activates ß-catenin/TCF4 signaling in cancer cell proliferation and invasion	Lan et al. (2012)
miR-200a	Directly targets ß-catenin in cancer cells	Saydam et al. (2009) and Su et al. (2012)
miR-218	Directs a Wnt signaling circuit to promote differentiation of osteoblasts and osteomimicry of cancer cells	Hassan et al. (2012)
miR-122	Directly targets Wnt/ß-catenin pathway to suppress HCC proliferation and induce apoptosis	Xu et al. (2012b)
miR-92b	Directly targets Nemo-like kinase to activate Wnt/ß-catenin signaling in glioma.	Wang et al. (2013a)
miR-101	Targets Wnt/ß-catenin signaling pathway in colon cancer cells	Strillacci et al. (2013)

HCC hepatocellular carcinoma

miRNAs tend to be anti-oncomiRs and down-regulated in cancers while Wnt-activating miRNAs tend to be oncomiRs and up-regulated during tumorigenesis (Anton et al. 2011). In this context, miRNAs could directly target and regulate Wnt/ß-catenin signaling. For instances, miR-200a was down-regulated in human meningioma, where it was found to directly targets ß-catenin mRNA, thereby inhibiting its translation and blocking Wnt/ß-catenin signaling (Saydam et al. 2009); similar finding was also demonstrated in human colon cancer cells, where miR-145 was down-regulated, and ectopic expression of miR-145 inhibited the growth of DLD-1 colon cancer cells by disturbing ß-catenin translocation into the nucleus, accordingly lead to the down-regulation of TCF/Lef-1 transcriptional target genes c-Myc and Cyclin D1. This finding suggested a pivotal role of miR-145 in modulating intracellular translocation of ß-catenin on Wnt/ß-catenin signaling pathway (Yamada et al. 2013). On the other hand, some of miRNAs, such as the miR-21 was up-regulated in many types of cancers, the miR-21 was found to directly transcriptionally activate TCF4 in colorectal cancer (CRC) cells (Lan et al. 2012). In contrast, the Wnt signaling was able to modulate the expression of miRNAs. For example, in hepatocellular carcinoma (HCC), miR-122 plays a negative role to suppresses cell proliferation and induces cell apoptosis by directly targeting Wnt/ß-catenin pathway (Xu et al. 2012b), while the expression of some miRNAs, such as miR-181 could be directly induced upon an activation of Wnt/ß-catenin signaling or inhibition of GSK3β (Ji et al. 2011). In addition to directly targeting Wnt/ß-catenin signaling, there is evidence that miRNAs are able to indirectly target Wnt signaling. Wang et al. found that miR-92b activated Wnt/ß-catenin signaling by indirectly targeting Nemo-like kinase (NLK) in glioma (Wang et al. 2013a).

Increasing evidence suggests the presence of highly resistant CSCs in various types of cancer, and the dysregulation of miRNAs and Wnt/ß-catenin pathway are involved in the chemoresistance for cancer therapy (Pan et al. 2013; Cui et al. 2012b; Vangipuram et al. 2012). Liu et al. recently uncovered a novel mechanism that up-regulation of miR-125b by Wnt signaling in Snail enriched CSCs. Enforced expression of Snail increases the expression of miR-125b through the activated Wnt/ß-catenin/TCF4 axis, which in turn, repress Bak1 to confer chemoresistance. The Snail-induced chemoresistance could be re-sensitized by either restoring the expression of Bak1 or depleting miR-125b. Additionally, ectopic expression of miR-125b leads to enrichment of the $CD24^-/CD44^+$ CSC population, while repression of miR-125b expression or restoration of the expression of Bak1 increases the non-stem cell population ($CD24^-/CD44^+$) in Snail-over-expressing cells (Liu et al. 2013). Cisplatin is one of the commonly used chemotherapeutic drugs for many types of cancer, previous study in the head and neck cancer cell line was associated with decreased expression of Wnt signaling inhibitor DKK1, and over-expression of DKK1 could partially reverse the resistance, suggesting the elevation of canonical Wnt signaling might correlated with cisplatin resistance (Gosepath et al. 2008). The members of miR-181 family could be transcriptionally activated by Wnt/ß-catenin signaling pathway in $EpCAM^+AFP^+$ hepatic CSCs (HepCSCs) in HCC (Ji et al. 2011). Xu et al. recently also found that that the expression of miR-130a was elevated in HCC patients treated with cisplatin based chemotherapy and cisplatin-resistant Huh7 cells. Mechanistically, they further demonstrated that up-regulation of miR-130a directly targeted tumor suppressor gene RUNX3, which in turn, activated Wnt/ß-catenin signaling and increased drug resistance, implying that miR-130a may offer a novel target for chemotherapy of HCC (Xu et al. 2012c).

Cancer metastasis involves a series of biological steps during which the cancer cells acquire the ability to invade surrounding tissues and survive outside the original tumor site by EMT, and Wnt/ß-catenin signaling is an important pathway to drive EMT and metastasis in cancers (Cong et al. 2013). In gastric cancers, miR-27 was up-regulated and paralleled the increased levels of the EMT-associated genes ZEB1, ZEB2, Slug, and Vimentin, as well as decreased E-cadherin levels. Such a miR-27 promoted EMT was through by a mechanism of activation of Wnt pathway, in which the APC gene of Wnt pathway was the direct and functional target of miR-27 (Zhang et al. 2011d). Additionally, members of miR-200 family are capable of directly targeting Wnt pathway, the down-regulation of miR-200s in gastric cancers were also correlated with EMT by activation of Wnt/ß-catenin signaling and increased expression of ZEB1 and ZEB2 (Su et al. 2012; Cong et al. 2013). Similarly, hyperactivated Wnt/ß-catenin signaling was detected in miR-374a expressing metastatic breast cancer (BC) cells. The over-expression of miR-374a enhanced the ability of EMT and metastasis in BC cells by directly targeting and inhibiting multiple negative regulators of the Wnt/ß-catenin signaling cascade, including Wnt inhibitory factor-1 (WIF1), Phosphatase and tensin homolog (PTEN), and Wnt5A (Cai et al. 2013).

3 MicroRNAs and mTOR/Akt/PI3K Pathway

mTOR is an atypical serine/threonine protein kinase, which exists in two functionally distinct complexes: mTORC1 (containing mTOR, Raptor, etc.) and mTORC2 (containing mTOR, Rictor, etc.). The mTOR pathway is activated by multiple extracellular signals, which, in turn, regulates protein translation, cell proliferation, cell differentiation, cell invasion, cell apoptosis and cell cycle (Li et al. 2011a; Laplante and Sabatini 2012). It has been demonstrated that mTOR/Akt/PI3K pathway is a key signaling system that frequently dysregulated in a wide variety of human cancers, in which the mTOR/Akt/PI3K pathway is constitutively activated and is associated with a poor prognosis, invasion, and metastasis. The mTOR inhibitors including rapamycin and rapamycin derivatives thus have recently been applied for anticancer treatments (Gulhati et al. 2011).

The potential role of miRNAs in the regulation of mTOR pathway has been investigated in a variety of cancer types (Alqurashi et al. 2013). Table 9.2 lists several miRNAs that target the mTOR/Akt/PI3K signal pathway. The mTOR signaling is highly activated with a down-regulation of its regulatory miRNAs in some cancers. The miR-99 family members, such as miR-99a/b and miR-100, are down-regulated in various malignant cancers, including clear cell ovarian cancer (Doghman et al. 2010) and esophageal squamous cell carcinoma (SCC) (Sun et al. 2013b). Enforced expression of miR-100 showed an inhibition of mTOR signaling and an enhanced sensitivity to a mTOR inhibitor, rapamycin analog RAD001 (Everolimus) (Nagaraja et al. 2010); over-expression of miR-99a/100 repressed the mTOR signaling by directly targeting its 3′ UTR of mTOR. Similarly, miR-199a-3p was also shown to be down-regulated and target mTOR in HCC cells (Fornari et al. 2010). Other identified miRNAs that able to directly target mTOR include miR-520c and miR-373 (Liu and Wilson 2012), miR-30a (Zhong et al. 2013) and miR-144 (Iwaya et al. 2012). These results suggested that mTOR targeted miRNAs may be potential therapeutic targets for inducing apoptosis to combat cancers. For instance, miR-126 was found to increase irradiation-induced non-small cell lung cancer (NSCLC) cell apoptosis and re-sensitize NSCLC cells to radiotherapy by targeting the PI3K/Akt pathway (Wang et al. 2011).

mTOR signaling is constitutively active with the down-regulation of miR-30a and miR-144 in CRC, in which miRNAs play a role of metastasis suppressor (Zhong

Table 9.2 MicroRNAs that target mTOR/Akt/PI3K signaling pathway involved in carcinogenesis

MicroRNA	Potential functions	Reference(s)
miR-101	Inhibits mTOR pathway	Nagaraja et al. (2010)
miR-99a/100	Represses mTOR signal pathway in various cancer cells	Nagaraja et al. (2010), Sun et al. (2013b), Oneyama et al. (2011) and Doghman et al. (2010)
miR-122	Directly targets IGF1R and regulates PI3K/Akt/mTOR/p70S6K pathway in breast cancer	Wang et al. (2012a)
miR-144	Directly targets mTOR in colorectal carcinoma cells.	Iwaya et al. (2012)
miR-520c/373	Directly targets mTOR to up-regulate matrix metalloproteinase 9 in fibrosarcoma cells	Liu and Wilson (2012)
miR-125b	Represses tumor growth activity by targeting the PI3K/Akt/mTOR signaling pathway	Cui et al. (2012a)
miR-30a	Inhibits colorectal carcinoma cell migration and invasion by targeting the PI3K/Akt/mTOR pathway	Zhong et al. (2013)
miR-199a-3p	Directly targets mTOR and leads to cell cycle arrest in hepatocellular carcinoma cells	Fornari et al. (2010)
miR-204	Inhibits Akt/mTOR signaling pathway by targeting brain-derived neurotrophic factor in cancer cells	Imam et al. (2012)
miR-218	Targets the mTOR component Rictor and inhibits Akt phosphorylation in oral cancer	Uesugi et al. (2011)

et al. 2013; Iwaya et al. 2012). The ectopic expression of miR-30a was found to down-regulate the expression of Akt and mTOR and their phosphorylated forms in CRC cells (Zhong et al. 2013). Clinicopathologic analysis further demonstrated that down-regulation of miR-144 was associated with enhanced malignant potential and poor prognosis *via* activation of the mTOR signaling pathway in CRC, although only high expression of a component of mTORC2, Rictor, was found to be associated with poor prognosis of CRC (Iwaya et al. 2012). The Rictor was able to directly regulate the phosphorylation of Akt at Ser-473, which was also identified as a target of miR-218 in oral squamous cell carcinoma cells (Uesugi et al. 2011). Moreover, *in silico* miRNA:target mRNA prediction analysis revealed a miR-144 binding site in its 3′ UTR region of both Rictor and mTOR (Iwaya et al. 2012).

Apart from direct target of mTOR pathway, miRNAs were found to indirectly regulate mTOR signaling by targeting other signaling pathways. The miR-204 was suggested as a potent tumor suppressor, which was frequently lost in multiple cancers. miR-204 exerts its function by targeting genes involved in tumorigenesis including brain-derived neurotrophic factor (BDNF), the loss of miR-204 elevates BDNF expression and subsequent activation of the small GTPase Rac1 and actin reorganization through the Akt/mTOR signaling pathway, that leads to cancer cell migration and invasion (Imam et al. 2012). Same as seen in the miR-204, tumor suppressor miR-122, was able to functionally inhibit the tumorigenesis of BC through targeting insulin-like growth factor 1 receptor (IGF1R), a receptor tyrosine kinase, and regulating PI3K/Akt/mTOR/p70S6K pathway, suggesting miR-122 may serve as a novel therapeutic or diagnostic/prognostic-target for BC therapy (Wang et al. 2012a).

In addition, mTOR has been demonstrated to most commonly induces autophagy, a major intracellular degradation pathway for long-lived proteins and organelles, by which the cell gains an ability to survive stressful conditions, and the abnormalities of autophagy contribute to many diseases including the cancers (Laplante and Sabatini 2012). miR-376b and miR-181a have been recently demonstrated to regulate rapamycin-induced autophagy in MCF-7 and Huh-7 cells, through a mechanism of directly targeting autophagy proteins ATG4C and ATG5, respectively (Korkmaz et al. 2012; Tekirdag et al. 2013).

4 MicroRNAs and EGFR Pathway

EGFR, a receptor tyrosine kinase associated with cell proliferation and survival, which is hyperactivated and is associated with disease progression, resistance to chemotherapy and radiotherapy, and poor prognosis in many epithelial originated cancers (Bianco et al. 2005). Given the importance of EGFR in cancer biology, the EGFR and its downstream signaling effectors are potential biomarkers of prognosis major targets for new therapeutics, such as monoclonal antibodies and tyrosine kinase inhibitors (TKIs) (Bianco et al. 2005; Samakoglu et al. 2012). As one of the extensively investigated cellular signaling pathways, EGFR signaling pathway is involved in cell migration, invasion, viability and proliferation in tumorigenesis of many types of cancer, and up-regulation of EGFR occurs in a range of cancers and correlates with high metastases and a poor prognosis (Perez et al. 2013; Teixeira et al. 2012). Accumulating evidence has suggested that miRNAs are key players in regulating EGFR signaling in different types of cancer, and they can be served as tumor suppressors and potential therapeutic targets, i.e. a miRNA that regulates EGFR may have therapeutic potential for cancer treatment (Barker et al. 2010). Several miRNAs which were known to regulate EGFR signaling are listed in the Table 9.3.

The discovery of EGFR signaling pathway has led to the development of two classes of drugs, oral EGFR-TKIs and monoclonal antibodies (mAbs) for the treatment of a range of EGFR-driven tumors, most notably lung, colon, pancreatic, and head and neck cancer. However, the benefits of these therapeutic agents to patients are ultimately limited by the emergence of mutations and other molecular mechanisms that eventually developing drug resistance (Politi et al. 2010). MET receptor has been reported to be involved in the acquired resistance of NSCLC to TKIs, and there are evidences show that miRNA may participate in the EGFR/MET network in lung cancer and provide a clue to overcoming EGFR-TKI resistance (Garofalo et al. 2012; Wang et al. 2012b). Garofalo et al. reported that miR-30b, miR-30c, miR-221 and miR-222, miR-103 and miR-203 were able to modulate EGFR and MET receptor, suggesting that they might have important roles in EGFR-TKI

Table 9.3 MicroRNAs that target EGFR signaling pathway involved in carcinogenesis

MicroRNA	Potential functions	Reference(s)
miR-200	Regulates epithelial-mesenchymal transition in anaplastic thyroid cancer cells, bladder cancer cells, reverses resistance of EGFR therapy	Zhang et al. (2012) and Adam et al. (2009)
miR-542-5p, 1203, 1237, 541, 1911	Down-regulates EGFR in human lung cancer cells	Yamaguchi et al. (2012) and Webster et al. (2009)
miR-7	Inhibits EGFR/PI3K/Akt signaling and reverses radio-resistance in various cancer cells	Kefas et al. (2008), Lee et al. (2011) and Webster et al. (2009)
miR-145	Negatively regulates EGFR expression in lung cancer cells.	Cho et al. (2011)
miR-21	Regulates the EGFR/Akt pathway in a PTEN independent manner	Zhou et al. (2010)
miR-214	Regulates acquired resistance to EGFR-TKIs in cancer cells through a PTEN/Akt pathway	Wang et al. (2012c)
miR-27a	Directly targets EGFR and contributes to mutant p53 gain-of-function.	Wang et al. (2013c) and Acunzo et al. (2013)
miR-146a	Inhibits EGFR in prostate and non-small cell lung cancer cells	Li et al. (2010), Xu et al. (2012a) and Chen et al. (2013)
miR-146b-5p	Suppresses EGFR expression in glioblastoma cell lines	Katakowski et al. (2010)
miR-133a	Directly targets EGFR and regulates its downstream signal molecule Akt in breast cancer	Cui et al. (2013)
miR-133b	Suppresses EGFR pathway signaling and enhances susceptibility to EGFR-TKI in lung cancer cells by directly targeting EGFR	Liu et al. (2012)
miR-128b	Directly regulates EGFR expression in non-small cell lung cancer	Weiss et al. (2008)

gefitinib-induced apoptosis of NSCLC cells (Garofalo et al. 2012). In agreement with this finding, Wang et al. found that miR-214 was dramatically elevated in a gefitinib resistant cell line HCC827/GR (with a 746E-750A in-frame deletion of EGFR gene), with an inversely expression of PTEN. Interestingly, knockdown miR-214 showed an increasing expression of PTEN and phosphorylated Akt (p-Akt), and re-sensitization of HCC827/GR to gefitinib, clearly indicating that targeting of miR-214 might reverse the acquired resistance to EGFR-TKIs therapy in NSCLC (Wang et al. 2012c).

Bioinformatic analysis has predicted 138 miRNAs that potentially target EGFR in NSCLC, and some of them have been confirmed experimentally (Chan et al. 2012). Using similar strategy, three miR-7 target sites can be predicted in human EGFR mRNA 3′ UTR. A number of studies have demonstrated the ability of miR-7 to down-regulate EGFR in different cancer cells (Kefas et al. 2008; Webster et al. 2009). In addition to directly targeting EGFR mRNA, miR-7 has been demonstrated to regulate the activity of downstream effectors of EGFR signaling pathway, including the Akt and ERK1/2, in multiple EGFR-expressing cancer cell lines (Webster et al. 2009). Yamaguchi et al. found that miR-542-5p was able to directly targets EGFR mRNA, by which it functionally down-regulated EGFR in human lung cancer cells. More importantly, the miR-542-5p expression was inversed with EGFR protein levels in human lung cancer tissue samples (Yamaguchi et al. 2012). In another study, Chen et al. demonstrated that miR-146a was capable of suppressing cell growth, inducing cellular apoptosis, inhibiting EGFR downstream signaling and the migratory capacity in various NSCLC cell lines (H358, H1650, H1975, HCC827 and H292), through an EGFR mutation status independent mechanism of directly targeting EGFR and NF-κB signaling. Furthermore, a combination of miR-146a mimic and EGFR-TKIs or anti-EGFR mAb, cetuximab, miR-146a exhibited a capacity to enhance the inhibition of cell proliferation in NSCLC cells (Chan et al. 2012). miR-27a (Acunzo et al. 2013), miR-145 (Cho et al. 2011) and miR-133b (Liu et al. 2012) were also identified to directly target EGFR in lung cancer. On the other hand, the aberrant expression and/or mutation(s) of EGFR may also alter the expression of miRNAs in cancers. For instance, a miRNA expression profiling study in lung cancer of never-smokers showed that the changes of expression of some miRNAs in the cases with EGFR mutations were more remarkable than those without these mutations. Moreover, the miR-21, one of the most studied miRNA species in human cancer, is elevated in most types of cancer. It was found to enhance further by the activated EGFR signaling pathway in this study (Seike et al. 2009), suggesting that EGFR signaling not only is regulated by tumor suppressive miRNAs, but also has potential to regulate some miRNAs acting as oncogene.

In order to unravel the coordination of miRNAs and EGFR signaling network on a global miRNA (miRNome) level in BC, Uhlmann et al. performed a large-scale miRNA screening approach with a combination of high-throughput proteomic read-out and network-based data analysis, and finally they identified and validated miR-124, miR-147 and miR-193a-3p as novel tumor suppressors that co-target EGFR-driven cell cycle network proteins and inhibit cell cycle progression and proliferation in BC (Uhlmann et al. 2012). Recent study also revealed that miR-133a was significantly down-regulated in BC, and ectopic expression of miR-133a in BC cells arrested the cell cycles. Mechanistic study further verified that miR-133a was able to directed target EGFR, and regulate EGFR signaling in BC through Akt signaling pathway (Cui et al. 2013), in which miR-133a-regulated EGFR expression was positively correlated with that of p-Akt, which was consisted with that in prostate cancer and glioblastoma (Gan et al. 2010; Kefas et al. 2008). These results demonstrate a tumor suppressor role of miR-133a in BC, in which miR-133a regulates the tumorigenesis of BC through targeting EGFR (Cui et al. 2013). Additionally, miR-146a and miR-146b-5p were also found to regulate EGFR signaling by directly targeting EGFR in cancer cells (Xu et al. 2012a; Katakowski et al. 2010).

5 MicroRNAs and VEGF Pathway

Angiogenesis is one of the key characteristics of malignant cancer, and VEGF is the most prominent driving force for angiogenesis and cancer progression, which is mostly up-regulated in many cancers, therefore it is recognized as an attractive target for both cancer diagnosis and therapy. There is an increasing evidence shows that VEGF is a potential target for some miRNAs in a wide range of cancers, several miRNAs have been demonstrated to be associated with vascular development. Hua et al. predicted and experimental confirmed that VEGF might be a potential target for a group of candidate miRNAs, including miR-16, let-7b, miR-17-5p, miR-27, miR-106, miR-107, miR-193, miR-210, miR-320 and miR-361 (Hua et al. 2006; Kulshreshtha et al. 2008). Roccaro et al. recently found that miR-15a/16 were down-regulated or undetectable in relapsed/refractory multiple myeloma (MM) patients, and miR-15a/16 inhibits MM cell-triggered endothelial cell growth and capillary formation *in vitro*. Enforced expression of miR-15a/16 in MM cells led to decreased pro-angiogenic activity of MM cells led to a dramatic inhibition of VEGF secretion, immunoblotting and luciferase reporter assays further revealed that VEGF-A was a direct target of miR-15a/16 (Roccaro et al. 2009; Sun et al. 2013a). Table 9.4 lists the potential miRNAs that target VEGF pathway.

Table 9.4 MicroRNAs that target VEGF signaling pathway involved in carcinogenesis

MicroRNA	Potential functions	Reference(s)
miR-126	Promotes or inhibits VEGF signaling in a tissue specific manner	Fish et al. (2008), Wang et al. (2008), Liu et al. (2009), Zhu et al. (2011), (2012) and Sasahira et al. (2012)
miR-20b	Down-regulates VEGF in breast cancer MCF-7 cells	Cascio et al. (2010)
miR-145	Directly targets VEGFA in breast cancer	Zou et al. (2012)
miR-125b	Represses VEGF expression in various cancers	Smits et al. (2012), Bi et al. (2012) and He et al. (2013)
miR-195	Directly inhibits VEGF expression in hepatocellular carcinoma	Wang et al. (2013b)
miR-199a	Inhibits VEGF expression in ovarian cancer	He et al. (2013)
miR-200	Directly targets Flt1/VEGFR1 in human lung adenocarcinoma	Roybal et al. (2011)
miR-205	Directly targets VEGFA in human glioblastoma cells	Yue et al. (2012)
miR-206	Inhibits VEGF expression in laryngal squamous cell carcinoma	Zhang et al. (2011a)
miR-361-5p	Directly targets VEGFA in human cutaneous squamous cell carcinoma	Kanitz et al. (2012)

miR-126 is an extensively studied human miRNA, it has been identified as a suppressor of tumor formation in various types of cancer by directly targeting several well-defined signaling including the VEGFA (Fish and Srivastava 2009; Liu et al. 2009; Sasahira et al. 2012). miR-126 defected mice showed a phenotype including loss of vascular integrity and defects in endothelial cell proliferation, migration and angiogenesis (Schepeler et al. 2012). Hence, the miR-126 is one of the most current investigated miRNAs in the angiogenesis of cancer. In this context, miR-126 is down-regulated in different malignancies (Li et al. 2011b). The down-regulation of miR-126 in tumors is correlated with an increased activity of the VEGF/PI3K/Akt signaling pathway and promotion of vascular formation and cancer metastasis, which was demonstrated in a study that introduction of miR-126 mimics into BC MCF-7 cancer cells (Zhu et al. 2011). Mechanistically, miR-126 has been found to be expressed specifically in epithelial cells where it regulated the expressions of sprouty-related protein 1 (SPRED1) and PI3K regulatory subunit 2 (PIK3R2) to activate VEGF signaling in response to angiogenic growth factors (Fish et al. 2008; Wang et al. 2008). Apart from its function of targeting VEGF signaling, miR-126 was found to be involved in regulating the response of NSCLC cells to cancer chemotherapy. Introduction of lung adenocarcinoma A459 cells with miR-126 mimics led an increased miR-126 level that was significantly associated with an enhanced drug sensitivity by decreasing the half maximal inhibitory concentration of adriamycin (ADM) and vincristine, and a down-regulation multidrug resistance-associated protein 1 (MRP1), as well as inactivation of the Akt signaling pathway (Zhu et al. 2012).

Glioblastomas characterized by the secretion of large amounts of the angiogenic VEGF are a supportive evidence for the therapeutic benefit of VEGF pathway targeting (Reardon et al. 2011). Smits et al. found that down-regulation of miR-125b in glioblastoma-associated endothelial cells was correlated with increased expression of transcription factor myc-associated zinc finger protein (MAZ) that regulates VEGF. Interestingly, endothelial cells exposed to VEGF showed a down-regulation of miR-125b and an increased MAZ expression. Additionally, elevated MAZ expression was observed in brain blood vessels of glioblastoma patients. This study suggested a functional feed-forward loop in glioblastoma related angiogenesis, in which elevated VEGF inhibits the expression of miR-125b, resulting in increased expression of MAZ, which in its turn causes transcriptional activation of VEGF (Smits et al. 2012).

Similarly, down-regulation of miR-125a and miR-195 were observed in HCC cells and tissues (Bi et al. 2012; Wang et al. 2013b). The alteration of miR-125a in HCC was inversely correlated with expression of matrix metalloproteinase 11 (MMP11) and VEGFA, both of them were direct targets of miR-125a (Bi et al. 2012). In addition to directly target VEGF, miR-195 also showed a function of directly inhibiting the expression of pro-angiogenic factors VAV2 and CDC42. Furthermore, the down-regulation of miR-195 led to enhanced VEGF levels in the HCC tissues, by which VEGFR2 signaling was subsequently activated in endothelial cells and thereby promoted angiogenesis (Wang et al. 2013b). Such anti-angiogenesis activity of miRNAs through targeting of VEGF signaling was also demonstrated in other miRNAs from various cancers, including the miR-145 in BC (Zou et al. 2012), and the miR-361-5p in SCC of the skin (Kanitz et al. 2012). These studies strongly evidenced that miRNAs targeting VEGF signaling hold a great promise for developing diagnostic markers and therapeutic agents for cancer treatment.

6 MicroRNAs and KRAS Pathway

The oncogene v-Ki-ras2 Kirsten rat sarcoma viral oncogene homolog also known as, KRAS or GTPase KRas, is a key molecule of EGFR/RAS/mitogen-activated protein kinase (MAPK) pathway. The EGFR/RAS/MAPK pathway regulates a variety of biological activities of malignant cells, including cell proliferation, migration, and chemosensitivity (Cai et al. 2009). As a downstream component of EGFR signaling pathway, KRAS plays an important role in transduction of extracellular signals from EGFR to downstream effectors involved in cell division, apoptosis, and differentiation, KRAS mutations therefore are predictive of resistance to anti-EGFR treatment that depends on the presence of wild-type KRAS, and patients with KRAS mutation show the least likely to benefit from anti-EGFR therapy in both CRC and lung cancer (Amado et al. 2008; De Roock et al. 2010). To identify the association of miRNA signatures KRAS status in CRC, Mosakhani et al. performed a miRNA microarray analysis on metastatic CRC tissues, and they found the KRAS mutation was associated with up-regulation of miR-127-3p, miR-92a, and miR-486-3p and down-regulation of miR-378 (Mosakhani et al. 2012b).

With the facts of complementary sites of let-7 miRNAs reside in the KRAS 3′ UTR region (Johnson et al. 2005), let-7 post-transcriptional regulates KRAS expression in CRC cells (Akao et al. 2006), the progresses of CRC are through an adenoma-carcinoma sequence and KRAS plays a major function in this process, and anti-EGFR treatments in CRC, it is therefore let-7 family of miRNAs in CRC and lung cancer have received more attention (Johnson et al. 2005; Zhang et al. 2011c; Ruzzo et al. 2012; He et al. 2010; Smits et al. 2011).

Several lines of evidence indicate a down-regulation of let-7 in a variety of cancers, including the CRC and lung cancer where the reduction of let-7 caused reactivation of some let-7 targeted oncogenes, such as KRAS and c-Myc. Recently, a functional single nucleotide polymorphism (SNP), rs61764370 has been defined in the KRAS 3′ UTR in the let-7 complementary site (LCS6), which consists of a T>G base change that alters the binding ability of the mature let-7 to the KRAS mRNA.

miRNA polymorphisms (miR-SNPs) analysis further revealed that KRAS expression was increased in lung cancer tumors with G variant allele (Chin et al. 2008), and the G allele was correlated with a better outcome in early stages of CRC patients, especially in those with the KRAS mutation (Smits et al. 2011). Of note, this polymorphic association may vary among cancer types and stages (Smits et al. 2011). For instance, the LCS6 G allele was found to be associated with reduced survival in patients with head and neck squamous cell carcinoma (HNSCC) (Christensen et al. 2009) and with a poor outcome in postmenopausal women with epithelial ovarian cancer (Ratner et al. 2012). Moreover, conflicting results were reported in the polymorphic studies on the involvement of KRAS let-7 LCS6 SNP in the response to different targeted therapies (Graziano et al. 2010; Zhang et al. 2011c). In a study conducted by Zhang et al. found that the LCS6 SNP was associated with object response rate (ORR) to cetuximab treatment in metastatic CRC (mCRC) patients with wild type KRAS, in which a variant G allele (TG or GG) had a 42 % ORR compared with a 9 % ORR in those with TT genotype. In addition, KRAS wild type patients with TG/GG genotypes had trend of longer median progression-free survival (PFS) (3.9 *vs* 1.3 months) and overall survival (OS) (10.7 *vs* 6.4 months) compared to those with TT genotypes (Zhang et al. 2011b). In contrast to this finding, however, Graziano et al. found that LCS6 G-allele carriers had worse OS (P=0.001) and PFS (P=0.004) than the T/T carriers in mCRC patients without BRAF V600E mutation who underwent salvage cetuximab-irinotecan therapy, and the G-allele genotype was more frequent in patients with a KRAS mutation in comparison with those who carried a wild type KRAS (P=0.004) (Graziano et al. 2010). This finding was supported by a recently study performed by Sebio et al. (2013), in which they found the LCS6 G-allele was significant associated with nonresponse to anti-EGFR-based treatment, mCRC patients (without BRAF V600E mutation) carrying G allele (TG or G/G genotypes) had nonresponse to anti-EGFR therapy regardless of chemotherapy backbone, in contrast, 31.9 % of patients with the T/T genotype presented a complete or a partial response (P=0.004) (Sebio et al. 2013). Such a differential response to anti-EGFR therapy may be attributed to the genetic variation of among ethnic populations (Patel et al. 2010). In an exploratory analysis, Ruzzo et al. quantified the let-7a in patients with mCRC who harboring KRAS mutation and underwent third-line therapy with cetuximab plus irinotecan, they found a positive effect of high let-7a level on both OS and PFS times in patients with KRAS wild type LCS6 (Ruzzo et al. 2012). The miRNAs regulated resistance against anti-EGFR agents was also recently found in HNSCC where miR-122 regulates resistance to mAb cetuximab (Hatakeyama et al. 2010), albeit the underlying mechanism requires further investigation. These results suggest that let-7 analysis may serve to identify subgroups of patients who may still benefit from anti-EGFR therapy, as well as to provide informative cues for designing alternative strategies treating patients with KRAS mutations.

In addition to let miRNA family, miR-143 has also been found to impact on CRC cell proliferation and survival by targeting KRAS mRNA (Chen et al. 2009b). In this context, miR-143 was down-regulated in CRC tissues, the low levels of miR-143 were an independent negative prognostic factor for cancer-specific survival (CSS) and was found to be associated with a shorter PFS in patients with KRAS wild-type CRC who were treated with EGFR-targeted agents, but it was not a useful predictive biomarker in KRAS wild-type CRC patients treated with EGFR-targeting agents, such as mAb cetuximab or panitumuma (Pichler et al. 2012). In addition, Xu et al. demonstrated an inverse correlation of miR-143 expression and KRAS protein in prostate cancer samples, ectopic expression of miR-143 arrested cell proliferation and migration in prostate cancer cells, and increased the sensitivity to docetaxel by

Table 9.5 MicroRNAs that target KRAS signaling pathway involved in carcinogenesis

MicroRNA	Potential functions	Reference(s)
miR-200	Accompany with KRAS mutation	Ota et al. (2012)
let-7 family	Negatively regulates let-60/RAS, and down-regulates KRAS with antitumor effects in the presence of activating KRAS mutations	Johnson et al. (2005), Zhang et al. (2011c), Ruzzo et al. (2012), He et al. (2010) and Smits et al. (2011)
miR-143	Directly targets to KRAS and severs as an independent prognostic biomarker for various cancers	Chen et al. (2009b), Pichler et al. (2012), Loboda et al. (2010), Gao et al. (2011) and Xu et al. (2011)
miR-126b	Suppresses the growth of nasopharyngeal carcinoma by targeting KRAS	Deng et al. (2011)
miR-96	Suppresses KRAS expression in pancreatic cancer	Yu et al. (2010)
miR-217	As a potential tumor suppressor by targeting KRAS in pancreatic ductal adenocarcinoma	Zhao et al. (2010)
miR-30c	Directly targets KRAS and inhibits the proliferation of breast cancer cells	Tanic et al. (2012)

targeting EGFR/RAS/MAPK pathway, suggesting that miR-143 may play a crucial role in tumorigenesis of prostate cancer and chemosensitivity by targeting KRAS and subsequent inactivation of MAPK pathway (Xu et al. 2011).

Apart from the let-7 and miR-143, miR-216b (Deng et al. 2011), miR-30c (Tanic et al. 2012), miR-96 (Yu et al. 2010) and miR-217 (Zhao et al. 2010) also shown potentials to target KRAS pathway in various cancers. Xu et al. found a down-regulation of miR-216b in nasopharyngeal carcinoma (NPC) cells and tissues. Moreover, the decreased miR-216b level was directly related to advanced clinical stage and lymph node metastasis, and an increased expression of KRAS protein. The miR-216b was able to directly target KRAS mRNA and suppress NPC cell proliferation, invasion and tumor growth through a mechanism of inhibition of the KRAS-related Akt and ERK pathways (Deng et al. 2011). Same as seen in the miR-216b, miR-30c (Tanic et al. 2012), miR-96 (Yu et al. 2010) and miR-217 (Zhao et al. 2010) were also found to be down-regulated in respective BC, pancreatic cancer and pancreatic ductal adenocarcinoma cells. It has been experimentally confirmed that they are capable of directly targeting KRAS mRNA, and suppressing cancer cell proliferation, invasion through inhibition of the KRAS signaling pathway. The miRNAs that target KRAS pathway is listed in Table 9.5.

7 Perspectives and Challenges

The signaling pathways play key roles in controlling cell growth and differentiation, as well as tissue homeostatic processes; however, they are commonly hijacked in oncogenesis. With the uncovering of the underlying mechanisms of signaling pathways, increasing numbers of tumor-specific, molecularly targeted agents are under developing. Cancers from diverse organs of origin with similar molecular traits should be managed together (Cortés et al. 2013). The emergence of miRNAs and the understanding of the underlying mechanisms of their targeted signaling open up new perspectives for discovery and development of targeted therapies against different cancers. The miRNAs have been linked to many critical biologic processes in the eukaryotic cell, in which they exert their functions by targeting various signaling molecules. Although the underlying mechanism of miRNA in the tumorigenesis is far from fully understood, miRNA-based therapeutics has shown a significant promise in developing novel molecular targeting agents for cancer treatment, and the miRNAs that target various tumorigenesis signaling are of particular interest. In this context, administration of miRNA mimics or miRNA expressing vectors could be employed to up-regulate miRNAs; and administration of anti-sense nucleotides could be used for down-regulation of miRNAs (oncomiRs). However, apart from the complex of signaling pathways that interact through crosstalk and feedback loops, and the unclear functional consequences of miRNA abnormalities in cancers, there are several other potential limitations in the development and testing of miRNA-based therapeutics deserved further investigation. These include, but are not limited to, the off-target effect, tissue specificity, internal nucleases avoidance, *in vivo* delivery, and toxicity for miRNA-based targeting therapy. Nevertheless, preclinical studies and early trials in humans have clearly demonstrate that there is potential for developing novel therapeutics based on miRNAs by targeting signaling pathways that involved in tumorigenesis.

References

Acunzo M, Romano G, Palmieri D, Lagana A, Garofalo M, Balatti V et al (2013) Cross-talk between MET and EGFR in non-small cell lung cancer involves miR-27a and Sprouty2. Proc Natl Acad Sci U S A 110(21):8573–8578

Adam L, Zhong M, Choi W, Qi W, Nicoloso M, Arora A et al (2009) miR-200 expression regulates epithelial-to-mesenchymal transition in bladder cancer cells and reverses resistance to epidermal growth factor receptor therapy. Clin Cancer Res 15(16):5060–5072

Akao Y, Nakagawa Y, Naoe T (2006) let-7 microRNA functions as a potential growth suppressor in human colon cancer cells. Biol Pharm Bull 29(5):903–906

Alqurashi N, Hashimi SM, Wei MQ (2013) Chemical inhibitors and microRNAs (miRNA) targeting the mammalian target of rapamycin (mTOR) pathway: potential for novel anticancer therapeutics. Int J Mol Sci 14(2):3874–3900

Amado RG, Wolf M, Peeters M, Van Cutsem E, Siena S, Freeman DJ et al (2008) Wild-type KRAS is required for panitumumab efficacy in patients with metastatic colorectal cancer. J Clin Oncol 26(10):1626–1634

Anton R, Chatterjee SS, Simundza J, Cowin P, Dasgupta R (2011) A systematic screen for micro-RNAs regulating the canonical Wnt pathway. PLoS One 6(10):e26257

Babashah S, Soleimani M (2011) The oncogenic and tumour suppressive roles of microRNAs in cancer and apoptosis. Eur J Cancer 47(8):1127–1137

Barker A, Giles KM, Epis MR, Zhang PM, Kalinowski F, Leedman PJ (2010) Regulation of ERBB receptor signalling in cancer cells by microRNA. Curr Opin Pharmacol 10(6):655–661

Bi Q, Tang S, Xia L, Du R, Fan R, Gao L et al (2012) Ectopic expression of miR-125a inhibits the proliferation and metastasis of hepatocellular carcinoma by targeting MMP11 and VEGF. PLoS One 7(6):e40169

Bianco R, Troiani T, Tortora G, Ciardiello F (2005) Intrinsic and acquired resistance to EGFR inhibitors in human cancer therapy. Endocr Relat Cancer 12(Suppl 1):S159–S171

Cai C, Portnoy DC, Wang H, Jiang X, Chen S, Balk SP (2009) Androgen receptor expression in prostate cancer cells is suppressed by activation of epidermal growth factor receptor and ERBB2. Cancer Res 69(12):5202–5209

Cai J, Guan H, Fang L, Yang Y, Zhu X, Yuan J et al (2013) MicroRNA-374a activates Wnt/beta-catenin signaling to promote breast cancer metastasis. J Clin Investig 123(2):566–579

Cascio S, D'Andrea A, Ferla R, Surmacz E, Gulotta E, Amodeo V et al (2010) miR-20b modulates VEGF expression by targeting HIF-1 alpha and STAT3 in MEF-7 breast cancer cells. J Cell Physiol 224(1):242–249

Cha YH, Kim NH, Park C, Lee I, Kim HS, Yook JI (2012) miRNA-34 intrinsically links p53 tumor suppressor and Wnt signaling. Cell Cycle 11(7):1273–1281

Chan LW, Wang FF, Cho WC (2012) Genomic sequence analysis of EGFR regulation by microRNAs in lung cancer. Curr Top Med Chem 12(8):920–926

Chen HC, Chen GH, Chen YH, Liao WL, Liu CY, Chang KP et al (2009a) MicroRNA deregulation and pathway alterations in nasopharyngeal carcinoma. Br J cancer 100(6): 1002–1011

Chen X, Guo X, Zhang H, Xiang Y, Chen J, Yin Y et al (2009b) Role of miR-143 targeting KRAS in colorectal tumorigenesis. Oncogene 28(10):1385–1392

Chen G, Umelo IA, Lv S, Teugels E, Fostier K, Kronenberger P et al (2013) miR-146a inhibits cell growth, cell migration and induces apoptosis in non-small cell lung cancer cells. PLoS One 8(3):e60317

Chin LJ, Ratner E, Leng S, Zhai R, Nallur S, Babar I et al (2008) A SNP in a let-7 microRNA complementary site in the KRAS 3′ untranslated region increases non-small cell lung cancer risk. Cancer Res 68(20):8535–8540

Cho WC (2012) MicroRNAs as therapeutic targets and their potential applications in cancer therapy. Expert Opin Ther Targets 16(8):747–759

Cho WC, Chow AS, Au JS (2011) miR-145 inhibits cell proliferation of human lung adenocarcinoma by targeting EGFR and NUDT1. RNA Biol 8(1):125–131

Christensen BC, Moyer BJ, Avissar M, Ouellet LG, Plaza SL, McClean MD et al (2009) A let-7 microRNA-binding site polymorphism in the KRAS 3′ UTR is associated with reduced survival in oral cancers. Carcinogenesis 30(6):1003–1007

Clevers H (2006) Wnt/beta-catenin signaling in development and disease. Cell 127(3):469–480

Cong N, Du P, Zhang A, Shen F, Su J, Pu P et al (2013) Downregulated microRNA-200a promotes emt and tumor growth through the Wnt/beta-catenin pathway by targeting the E-cadherin repressors ZEB1/ZEB2 in gastric adenocarcinoma. Oncol Rep 29(4):1579–1587

Cortés J, Calvo E, Vivancos A, Perez-Garcia J, Recio JA, Seoane J (2013) New approach to cancer therapy based on a molecularly defined cancer classification. CA Cancer J Clin. doi: 10.3322/caac.21211 [Epub ahead of print]

Cselenyi CS, Jernigan KK, Tahinci E, Thorne CA, Lee LA, Lee E (2008) LRP6 transduces a canonical Wnt signal independently of AXIN degradation by inhibiting GSK3's phosphorylation of beta-catenin. Proc Natl Acad Sci U S A 105(23):8032–8037

Cui F, Li X, Zhu X, Huang L, Huang Y, Mao C et al (2012a) miR-125b inhibits tumor growth and promotes apoptosis of cervical cancer cells by targeting phosphoinositide 3-kinase catalytic subunit delta. Cell Physiol Biochem 30(5):1310–1318

Cui J, Jiang W, Wang S, Wang L, Xie K (2012b) Role of Wnt/beta-catenin signaling in drug resistance of pancreatic cancer. Curr Pharm Des 18(17):2464–2471

Cui W, Zhang S, Shan C, Zhou L, Zhou Z (2013) MicroRNA-133a regulates cell cycle and proliferation of breast cancer cells by targeting epidermal growth factor receptor through EGFR/Akt signaling pathway. FEBS J 280(16):3962–3974

De Roock W, Claes B, Bernasconi D, De Schutter J, Biesmans B, Fountzilas G et al (2010) Effects of KRAS, BRAF, nRAS, and PIK3CA mutations on the efficacy of cetuximab plus chemotherapy in chemotherapy-refractory metastatic colorectal cancer: a retrospective consortium analysis. Lancet Oncol 11(8):753–762

Deng M, Tang H, Zhou Y, Zhou M, Xiong W, Zheng Y et al (2011) miR-216b suppresses tumor growth and invasion by targeting KRAS in nasopharyngeal carcinoma. J Cell Sci 124(Pt 17):2997–3005

Doghman M, El Wakil A, Cardinaud B, Thomas E, Wang J, Zhao W et al (2010) Regulation of insulin-like growth factor-mammalian target of rapamycin signaling by microRNA in childhood adrenocortical tumors. Cancer Res 70(11):4666–4675

Fish JE, Srivastava D (2009) MicroRNAs: opening a new vein in angiogenesis research. Sci Signal 2(52):pe1

Fish JE, Santoro MM, Morton SU, Yu S, Yeh RF, Wythe JD et al (2008) miR-126 regulates angiogenic signaling and vascular integrity. Dev Cell 15(2):272–284

Fornari F, Milazzo M, Chieco P, Negrini M, Calin GA, Grazi GL et al (2010) miR-199a-3p regulates mTOR and c-MET to influence the doxorubicin sensitivity of human hepatocarcinoma cells. Cancer Res 70(12):5184–5193

Gan Y, Shi C, Inge L, Hibner M, Balducci J, Huang Y (2010) Differential roles of ERK and Akt pathways in regulation of EGFR-mediated signaling and motility in prostate cancer cells. Oncogene 29(35):4947–4958

Gao JS, Zhang Y, Tang X, Tucker LD, Tarwater PM, Quesenberry PJ et al (2011) The evil, microRNA-143, k-RAS axis in colon cancer. FEBS Lett 585(4):693–699

Garofalo M, Romano G, Di Leva G, Nuovo G, Jeon YJ, Ngankeu A et al (2012) EGFR and MET receptor tyrosine kinase-altered microRNA expression induces tumorigenesis and gefitinib resistance in lung cancers. Nat Med 18(1):74–82

Gosepath EM, Eckstein N, Hamacher A, Servan K, von Jonquieres G, Lage H et al (2008) Acquired cisplatin resistance in the head-neck cancer cell line CA127 is associated with decreased DKK1 expression and can partially be reversed by overexpression of DKK1. Int J Cancer 123(9):2013–2019

Grant S (2008) Cotargeting survival signaling pathways in cancer. J Clin Investig 118(9): 3003–3006

Graziano F, Canestrari E, Loupakis F, Ruzzo A, Galluccio N, Santini D et al (2010) Genetic modulation of the let-7 microRNA binding to KRAS 3′-untranslated region and survival of metastatic colorectal cancer patients treated with salvage cetuximab-irinotecan. Pharmacogenomics J 10(5):458–464

Gulhati P, Bowen KA, Liu J, Stevens PD, Rychahou PG, Chen M et al (2011) MTORC1 and MTORC2 regulate EMT, motility, and metastasis of colorectal cancer via RHOA and RAC1 signaling pathways. Cancer Res 71(9):3246–3256

Hassan MQ, Maeda Y, Taipaleenmaki H, Zhang W, Jafferji M, Gordon JA et al (2012) miR-218 directs a Wnt signaling circuit to promote differentiation of osteoblasts and osteomimicry of metastatic cancer cells. J Biol Chem 287(50):42084–42092

Hatakeyama H, Cheng H, Wirth P, Counsell A, Marcrom SR, Wood CB et al (2010) Regulation of heparin-binding EGF-like growth factor by miR-212 and acquired cetuximab-resistance in head and neck squamous cell carcinoma. PLoS One 5(9):e12702

He XY, Chen JX, Zhang Z, Li CL, Peng QL, Peng HM (2010) The let-7a microRNA protects from growth of lung carcinoma by suppression of k-RAS and c-MYC in nude mice. J Cancer Res Clin Oncol 136(7):1023–1028

He J, Jing Y, Li W, Qian X, Xu Q, Li FS et al (2013) Roles and mechanism of miR-199a and miR-125b in tumor angiogenesis. PLoS One 8(2):e56647

Hoshino I, Matsubara H (2013) MicroRNAs in cancer diagnosis and therapy: from bench to bedside. Surg Today 43(5):467–478

Hua Z, Lv Q, Ye W, Wong CK, Cai G, Gu D et al (2006) miRNA-directed regulation of VEGF and other angiogenic factors under hypoxia. PLoS One 1:e116

Imam JS, Plyler JR, Bansal H, Prajapati S, Bansal S, Rebeles J et al (2012) Genomic loss of tumor suppressor miRNA-204 promotes cancer cell migration and invasion by activating Akt/mTOR/RAC1 signaling and actin reorganization. PLoS One 7(12):e52397

Inui M, Martello G, Piccolo S (2010) MicroRNA control of signal transduction. Nat Rev Mol Cell Biol 11(4):252–263

Iwaya T, Yokobori T, Nishida N, Kogo R, Sudo T, Tanaka F et al (2012) Downregulation of miR-144 is associated with colorectal cancer progression via activation of mTOR signaling pathway. Carcinogenesis 33(12):2391–2397

Ji J, Yamashita T, Wang XW (2011) Wnt/beta-catenin signaling activates microRNA-181 expression in hepatocellular carcinoma. Cell Biosci 1(1):4

Johnson SM, Grosshans H, Shingara J, Byrom M, Jarvis R, Cheng A et al (2005) Ras is regulated by the let-7 microRNA family. Cell 120(5):635–647

Kanitz A, Imig J, Dziunycz PJ, Primorac A, Galgano A, Hofbauer GF et al (2012) The expression levels of microRNA-361-5p and its target VEGFA are inversely correlated in human cutaneous squamous cell carcinoma. PLoS One 7(11):e49568

Katakowski M, Zheng X, Jiang F, Rogers T, Szalad A, Chopp M (2010) miR-146b-5p suppresses EGFR expression and reduces in vitro migration and invasion of glioma. Cancer Investig 28(10):1024–1030

Kefas B, Godlewski J, Comeau L, Li Y, Abounader R, Hawkinson M et al (2008) MicroRNA-7 inhibits the epidermal growth factor receptor and the Akt pathway and is down-regulated in glioblastoma. Cancer Res 68(10):3566–3572

Kim NH, Kim HS, Kim NG, Lee I, Choi HS, Li XY et al (2011) P53 and microRNA-34 are suppressors of canonical Wnt signaling. Sci Signal 4(197):ra71

Korkmaz G, le Sage C, Tekirdag KA, Agami R, Gozuacik D (2012) miR-376b controls starvation and mTOR inhibition-related autophagy by targeting ATG4C and BECN1. Autophagy 8(2):165–176

Kulshreshtha R, Davuluri RV, Calin GA, Ivan M (2008) A microRNA component of the hypoxic response. Cell Death Differ 15(4):667–671

Lan F, Yue X, Han L, Shi Z, Yang Y, Pu P et al (2012) Genome-wide identification of TCF7l2/TCF4 target mirnas reveals a role for miR-21 in Wnt-driven epithelial cancer. Int J Oncol 40(2):519–526

Laplante M, Sabatini DM (2012) mTOR signaling in growth control and disease. Cell 149(2): 274–293

Lee KM, Choi EJ, Kim IA (2011) MicroRNA-7 increases radiosensitivity of human cancer cells with activated EGFR-associated signaling. Radiother Oncol 101(1):171–176

Li Y, Vandenboom TG 2nd, Wang Z, Kong D, Ali S, Philip PA et al (2010) miR-146a suppresses invasion of pancreatic cancer cells. Cancer Res 70(4):1486–1495

Li D, Liu X, Lin L, Hou J, Li N, Wang C et al (2011a) MicroRNA-99a inhibits hepatocellular carcinoma growth and correlates with prognosis of patients with hepatocellular carcinoma. J Biol Chem 286(42):36677–36685

Li XM, Wang AM, Zhang J, Yi H (2011b) Down-regulation of miR-126 expression in colorectal cancer and its clinical significance. Med Oncol 28(4):1054–1057

Liu P, Wilson MJ (2012) miR-520c and miR-373 upregulate MMP9 expression by targeting mTOR and SIRT1, and activate the RAS/RAF/MEK/ERK signaling pathway and NF-KAPPAB factor in human fibrosarcoma cells. J Cell Physiol 227(2):867–876

Liu B, Peng XC, Zheng XL, Wang J, Qin YW (2009) miR-126 restoration down-regulate VEGF and inhibit the growth of lung cancer cell lines in vitro and in vivo. Lung Cancer 66(2):169–175

Liu Y, Huang T, Zhao X, Cheng L (2011) MicroRNAs modulate the Wnt signaling pathway through targeting its inhibitors. Biochem Biophys Res Commun 408(2):259–264

Liu L, Shao X, Gao W, Zhang Z, Liu P, Wang R et al (2012) MicroRNA-133b inhibits the growth of non-small-cell lung cancer by targeting the epidermal growth factor receptor. FEBS J 279(20):3800–3812

Liu Z, Liu H, Desai S, Schmitt DC, Zhou M, Khong HT et al (2013) miR-125b functions as a key mediator for snail-induced stem cell propagation and chemoresistance. J Biol Chem 288(6):4334–4345

Loboda A, Nebozhyn M, Klinghoffer R, Frazier J, Chastain M, Arthur W et al (2010) A gene expression signature of Ras pathway dependence predicts response to PI3K and Ras pathway inhibitors and expands the population of Ras pathway activated tumors. BMC Med Genomics 3:26

Lu J, Getz G, Miska EA, Alvarez-Saavedra E, Lamb J, Peck D et al (2005) MicroRNA expression profiles classify human cancers. Nature 435(7043):834–838

Mabuchi S, Kawase C, Altomare DA, Morishige K, Sawada K, Hayashi M et al (2009) mTOR is a promising therapeutic target both in cisplatin-sensitive and cisplatin-resistant clear cell carcinoma of the ovary. Clin Cancer Res 15(17):5404–5413

Martin GS (2003) Cell signaling and cancer. Cancer Cell 4(3):167–174

Mosakhani N, Lahti L, Borze I, Karjalainen-Lindsberg ML, Sundstrom J, Ristamaki R et al (2012a) MicroRNA profiling predicts survival in anti-EGFR treated chemorefractory metastatic colorectal cancer patients with wild-type KRAS and BRAF. Cancer Genet 205(11):545–551

Mosakhani N, Sarhadi VK, Borze I, Karjalainen-Lindsberg ML, Sundstrom J, Ristamaki R et al (2012b) MicroRNA profiling differentiates colorectal cancer according to KRAS status. Genes Chromosomes Cancer 51(1):1–9

Nagaraja AK, Creighton CJ, Yu Z, Zhu H, Gunaratne PH, Reid JG et al (2010) A link between miR-100 and FRAP1/mTOR in clear cell ovarian cancer. Mol Endocrinol 24(2):447–463

Oneyama C, Ikeda J, Okuzaki D, Suzuki K, Kanou T, Shintani Y et al (2011) MicroRNA-mediated downregulation of mTOR/FGFR3 controls tumor growth induced by SRC-related oncogenic pathways. Oncogene 30(32):3489–3501

Ota T, Doi K, Fujimoto T, Tanaka Y, Ogawa M, Matsuzaki H et al (2012) KRAS up-regulates the expression of miR-181a, miR-200c and miR-210 in a three-dimensional-specific manner in DLD-1 colorectal cancer cells. Anticancer Res 32(6):2271–2275

Pan B, Yi J, Song H (2013) MicroRNA-mediated autophagic signaling networks and cancer chemoresistance. Cancer Biother Radiopharm 28(8):573–578

Patel TA, Colon-Otero G, Bueno Hume C, Copland JA 3rd, Perez EA (2010) Breast cancer in Latinas: gene expression, differential response to treatments, and differential toxicities in Latinas compared with other population groups. Oncologist 15(5):466–475

Perez R, Crombet T, de Leon J, Moreno E (2013) A view on EGFR-targeted therapies from the oncogene-addiction perspective. Front Pharmacol 4:53

Pichler M, Winter E, Stotz M, Eberhard K, Samonigg H, Lax S et al (2012) Down-regulation of KRAS-interacting miRNA-143 predicts poor prognosis but not response to EGFR-targeted agents in colorectal cancer. Br J Cancer 106(11):1826–1832

Politi K, Fan PD, Shen R, Zakowski M, Varmus H (2010) Erlotinib resistance in mouse models of epidermal growth factor receptor-induced lung adenocarcinoma. Dis Models Mech 3(1–2):111–119

Ratner ES, Keane FK, Lindner R, Tassi RA, Paranjape T, Glasgow M et al (2012) A KRAS variant is a biomarker of poor outcome, platinum chemotherapy resistance and a potential target for therapy in ovarian cancer. Oncogene 31(42):4559–4566

Reardon DA, Turner S, Peters KB, Desjardins A, Gururangan S, Sampson JH et al (2011) A review of VEGF/VEGFR-targeted therapeutics for recurrent glioblastoma. J Natl Compr Cancer Netw 9(4):414–427

Roccaro AM, Sacco A, Thompson B, Leleu X, Azab AK, Azab F et al (2009) MicroRNAs 15a and 16 regulate tumor proliferation in multiple myeloma. Blood 113(26):6669–6680

Rossbach M (2012) MicroRNAs in cancer therapy. Expert Opin Ther Targets 16(8):743–745

Roybal JD, Zang Y, Ahn YH, Yang Y, Gibbons DL, Baird BN et al (2011) miR-200 inhibits lung adenocarcinoma cell invasion and metastasis by targeting FLT1/VEGFR1. Mol Cancer Res 9(1):25–35

Ruzzo A, Graziano F, Vincenzi B, Canestrari E, Perrone G, Galluccio N et al (2012) High let-7a microRNA levels in KRAS-mutated colorectal carcinomas may rescue anti-EGFR therapy effects in patients with chemotherapy-refractory metastatic disease. Oncologist 17(6):823–829

Samakoglu S, Deevi DS, Li H, Wang S, Murphy M, Bao C et al (2012) Preclinical rationale for combining an EGFR antibody with cisplatin/gemcitabine for the treatment of NSCLC. Cancer Genomics Proteomics 9(2):77–92

Sasahira T, Kurihara M, Bhawal UK, Ueda N, Shimomoto T, Yamamoto K et al (2012) Downregulation of miR-126 induces angiogenesis and lymphangiogenesis by activation of VEGF-A in oral cancer. Br J Cancer 107(4):700–706

Saydam O, Shen Y, Wurdinger T, Senol O, Boke E, James MF et al (2009) Downregulated microRNA-200a in meningiomas promotes tumor growth by reducing E-cadherin and activating the Wnt/beta-catenin signaling pathway. Mol Cell Biol 29(21):5923–5940

Schepeler T, Holm A, Halvey P, Nordentoft I, Lamy P, Riising EM et al (2012) Attenuation of the beta-catenin/TCF4 complex in colorectal cancer cells induces several growth-suppressive microRNAs that target cancer promoting genes. Oncogene 31(22):2750–2760

Schmitz U, Wolkenhauer O (2013) Web resources for microRNA research. Adv Exp Med Biol 774:225–250

Sebio A, Pare L, Paez D, Salazar J, Gonzalez A, Sala N et al (2013) The LCS6 polymorphism in the binding site of let-7 microRNA to the KRAS 3′-untranslated region: its role in the efficacy of anti-EGFR-based therapy in metastatic colorectal cancer patients. Pharmacogenet Genomics 23(3):142–147

Seike M, Goto A, Okano T, Bowman ED, Schetter AJ, Horikawa I et al (2009) miR-21 is an EGFR-regulated anti-apoptotic factor in lung cancer in never-smokers. Proc Natl Acad Sci U S A 106(29):12085–12090

Smits KM, Paranjape T, Nallur S, Wouters KA, Weijenberg MP, Schouten LJ et al (2011) A let-7 microRNA SNP in the KRAS 3′UTR is prognostic in early-stage colorectal cancer. Clin Cancer Res 17(24):7723–7731

Smits M, Wurdinger T, van het Hof B, Drexhage JA, Geerts D, Wesseling P et al (2012) Myc-associated zinc finger protein (MAZ) is regulated by miR-125b and mediates VEGF-induced angiogenesis in glioblastoma. FASEB J 26(6):2639–2647

Strillacci A, Valerii MC, Sansone P, Caggiano C, Sgromo A, Vittori L et al (2013) Loss of miR-101 expression promotes Wnt/beta-catenin signalling pathway activation and malignancy in colon cancer cells. J Pathol 229(3):379–389

Su J, Zhang A, Shi Z, Ma F, Pu P, Wang T et al (2012) MicroRNA-200a suppresses the Wnt/beta-catenin signaling pathway by interacting with beta-catenin. Int J Oncol 40(4):1162–1170

Sun CY, She XM, Qin Y, Chu ZB, Chen L, Ai LS et al (2013a) miR-15a and miR-16 affect the angiogenesis of multiple myeloma by targeting VEGF. Carcinogenesis 34(2):426–435

Sun J, Chen Z, Tan X, Zhou F, Tan F, Gao Y et al (2013b) MicroRNA-99a/100 promotes apoptosis by targeting mTOR in human esophageal squamous cell carcinoma. Med Oncol 30(1):411

Tanic M, Yanowsky K, Rodriguez-Antona C, Andres R, Marquez-Rodas I, Osorio A et al (2012) Deregulated miRNAs in hereditary breast cancer revealed a role for miR-30c in regulating KRAS oncogene. PLoS One 7(6):e38847

Teixeira AL, Gomes M, Medeiros R (2012) EGFR signaling pathway and related-miRNAs in age-related diseases: the example of miR-221 and miR-222. Front Genet 3:286

Tekirdag KA, Korkmaz G, Ozturk DG, Agami R, Gozuacik D (2013) miR181a regulates starvation- and rapamycin-induced autophagy through targeting of ATG5. Autophagy 9(3):374–385

Uesugi A, Kozaki K, Tsuruta T, Furuta M, Morita K, Imoto I et al (2011) The tumor suppressive microRNA miR-218 targets the mTOR component rictor and inhibits Akt phosphorylation in oral cancer. Cancer Res 71(17):5765–5778

Uhlmann S, Mannsperger H, Zhang JD, Horvat EA, Schmidt C, Kublbeck M et al (2012) Global microRNA level regulation of EGFR-driven cell-cycle protein network in breast cancer. Mol Syst Biol 8:570

Valenta T, Hausmann G, Basler K (2012) The many faces and functions of beta-catenin. EMBO J 31(12):2714–2736

Vangipuram SD, Buck SA, Lyman WD (2012) Wnt pathway activity confers chemoresistance to cancer stem-like cells in a neuroblastoma cell line. Tumour Biol 33(6):2173–2183

Wang S, Aurora AB, Johnson BA, Qi X, McAnally J, Hill JA et al (2008) The endothelial-specific microRNA miR-126 governs vascular integrity and angiogenesis. Dev Cell 15(2):261–271

Wang XC, Du LQ, Tian LL, Wu HL, Jiang XY, Zhang H et al (2011) Expression and function of miRNA in postoperative radiotherapy sensitive and resistant patients of non-small cell lung cancer. Lung Cancer 72(1):92–99
Wang B, Wang H, Yang Z (2012a) miR-122 inhibits cell proliferation and tumorigenesis of breast cancer by targeting IGFLR. PLoS One 7(10):e47053
Wang Y, Wang X, Zhang J, Sun G, Luo H, Kang C et al (2012b) MicroRNAs involved in the EGFR/PTEN/Akt pathway in gliomas. J Neurooncol 106(2):217–224
Wang YS, Wang YH, Xia HP, Zhou SW, Schmid-Bindert G, Zhou CC (2012c) MicroRNA-214 regulates the acquired resistance to gefitinib via the PTEN/Akt pathway in EGFR-mutant cell lines. Asian Pac J Cancer Prev 13(1):255–260
Wang K, Wang X, Zou J, Zhang A, Wan Y, Pu P et al (2013a) miR-96b controls glioma proliferation and invasion through regulating Wnt/beta-catenin signaling via Nemo-like kinase. Neurooncol 15(5):578–588
Wang R, Zhao N, Li S, Fang JH, Chen MX, Yang J et al (2013b) MicroRNA-195 suppresses angiogenesis and metastasis of hepatocellular carcinoma by inhibiting the expression of VEGF, VAV2 and CDC42. Hepatology 58(2):642–653
Wang W, Cheng B, Miao L, Mei Y, Wu M (2013c) Mutant p53-r273h gains new function in sustained activation of EGFR signaling via suppressing miR-27a expression. Cell Death Dis 4:e574
Webster RJ, Giles KM, Price KJ, Zhang PM, Mattick JS, Leedman PJ (2009) Regulation of epidermal growth factor receptor signaling in human cancer cells by microRNA-7. J Biol Chem 284(9):5731–5741
Weiss GJ, Bemis LT, Nakajima E, Sugita M, Birks DK, Robinson WA et al (2008) EGFR regulation by microRNA in lung cancer: correlation with clinical response and survival to gefitinib and EGFR expression in cell lines. Ann Oncol 19(6):1053–1059
Xu B, Niu X, Zhang X, Tao J, Wu D, Wang Z et al (2011) miR-143 decreases prostate cancer cells proliferation and migration and enhances their sensitivity to docetaxel through suppression of KRAS. Mol Cell Biochem 350(1–2):207–213
Xu B, Wang N, Wang X, Tong N, Shao N, Tao J et al (2012a) miR-146a suppresses tumor growth and progression by targeting EGFR pathway and in a p-ERK-dependent manner in castration-resistant prostate cancer. Prostate 72(11):1171–1178
Xu J, Zhu X, Wu L, Yang R, Yang Z, Wang Q et al (2012b) MicroRNA-122 suppresses cell proliferation and induces cell apoptosis in hepatocellular carcinoma by directly targeting Wnt/beta-catenin pathway. Liver Int 32(5):752–760
Xu N, Shen C, Luo Y, Xia L, Xue F, Xia Q et al (2012c) Upregulated miR-130a increases drug resistance by regulating RUNX3 and Wnt signaling in cisplatin-treated HCC cell. Biochem Biophys Res Commun 425(2):468–472
Yamada N, Noguchi S, Mori T, Naoe T, Maruo K, Akao Y (2013) Tumor-suppressive microRNA-145 targets catenin delta-1 to regulate Wnt-beta-catenin signaling in human colon cancer cells. Cancer Lett 335(2):332–342
Yamaguchi G, Takanashi M, Tanaka M, Fujita K, Ohira T, Kuroda M et al (2012) Isolation of miRNAs that target EGFR mRNA in human lung cancer. Biochem Biophys Res Commun 420(2):411–416
Yu S, Lu Z, Liu C, Meng Y, Ma Y, Zhao W et al (2010) miRNA-96 suppresses KRAS and functions as a tumor suppressor gene in pancreatic cancer. Cancer Res 70(14):6015–6025
Yue X, Wang P, Xu J, Zhu Y, Sun G, Pang Q et al (2012) MicroRNA-205 functions as a tumor suppressor in human glioblastoma cells by targeting VEGF-A. Oncol Rep 27(4):1200–1206
Zhang T, Liu M, Wang C, Lin C, Sun Y, Jin D (2011a) Down-regulation of miR-206 promotes proliferation and invasion of laryngeal cancer by regulating VEGF expression. Anticancer Res 31(11):3859–3863
Zhang W, Labonte MJ, Lenz HJ (2011b) KRAS let-7 LCS6 SNP predicts cetuximab efficacy in KRASWT metastatic colorectal cancer patients: does treatment combination partner matter? Ann Oncol 22(2):484–485

Zhang W, Winder T, Ning Y, Pohl A, Yang D, Kahn M et al (2011c) A let-7 microRNA-binding site polymorphism in 3′-untranslated region of KRAS gene predicts response in wild-type KRAS patients with metastatic colorectal cancer treated with cetuximab monotherapy. Ann Oncol 22(1):104–109

Zhang Z, Liu S, Shi R, Zhao G (2011d) miR-27 promotes human gastric cancer cell metastasis by inducing epithelial-to-mesenchymal transition. Cancer Genet 204(9):486–491

Zhang Z, Liu ZB, Ren WM, Ye XG, Zhang YY (2012) The miR-200 family regulates the epithelial-mesenchymal transition induced by EGF/EGFR in anaplastic thyroid cancer cells. Int J Mol Med 30(4):856–862

Zhang F, Liu H, Liu F, Peng Y, Chen M, Liu Y et al (2013) New insights into the pathogenesis and treatment of peritoneal fibrosis: a potential role of Wnt/beta-catenin induced epithelial to mesenchymal transition and stem cells for therapy. Med Hypotheses 81(1):97–100

Zhao WG, Yu SN, Lu ZH, Ma YH, Gu YM, Chen J (2010) The miR-217 microRNA functions as a potential tumor suppressor in pancreatic ductal adenocarcinoma by targeting KRAS. Carcinogenesis 31(10):1726–1733

Zhong M, Bian Z, Wu Z (2013) miR-30a suppresses cell migration and invasion through downregulation of PIK3CD in colorectal carcinoma. Cell Physiol Biochem 31(2–3):209–218

Zhou X, Ren Y, Moore L, Mei M, You Y, Xu P et al (2010) Downregulation of miR-21 inhibits EGFR pathway and suppresses the growth of human glioblastoma cells independent of PTEN status. Lab Invest 90(2):144–155

Zhu N, Zhang D, Xie H, Zhou Z, Chen H, Hu T et al (2011) Endothelial-specific intron-derived miR-126 is down-regulated in human breast cancer and targets both VEGFA and PIK3R2. Mol Cell Biochem 351(1–2):157–164

Zhu X, Li H, Long L, Hui L, Chen H, Wang X et al (2012) miR-126 enhances the sensitivity of non-small cell lung cancer cells to anticancer agents by targeting vascular endothelial growth factor A. Acta Biochim Biophys Sin 44(6):519–526

Zou C, Xu Q, Mao F, Li D, Bian C, Liu LZ et al (2012) miR-145 inhibits tumor angiogenesis and growth by N-Ras and VEGF. Cell Cycle 11(11):2137–2145

Chapter 10
MicroRNA-Modulated Apoptotic and Autophagic Signaling Networks in Cancer

Bo Liu and Yu-quan Wei

Contents

Abstract MicroRNAs (miRNAs), small and non-coding RNAs ~ 22 nucleotides (nt) in length, are estimated to regulate about 30 % human gene expression at the post-transcriptional and the translational levels. MiRNAs are also involved in a series of important cellular processes, such as proliferation, differentiation, apoptosis and autophagy. Of note, apoptosis can invariably contribute to cell death, whereas autophagy can play either a pro-survival or a pro-death role in cancer. Recent evidence has shown that miRNAs can function as oncogenes or tumor suppressive genes in

B. Liu (✉) • Y.-q. Wei
State Key Laboratory of Biotherapy, West China Hospital, Sichuan University, Chengdu 610041, China
e-mail: liubo2400@163.com

S. Babashah (ed.), *MicroRNAs: Key Regulators of Oncogenesis*,
DOI 10.1007/978-3-319-03725-7_10,

diverse types of human cancers. Also, miRNAs are well-characterized to be crucial in tumorigenesis as either oncogenes or tumor suppressors by targeting apoptosis and autophagy. However, the intricate mechanisms of miRNA-modulated apoptosis and autophagy still remain unclear. In this chapter, we focus on summarizing the dual function of miRNAs in regulation of apoptosis and autophagy; thereby revealing the regulatory mechanisms of miRNA-regulated apoptosis and autophagy, which may shed light on developing novel RNA therapeutic strategy in the near future.

Keywords MicroRNA • Cancer • Apoptosis • Autophagy • Therapy

1 Introduction

MicroRNAs (miRNAs), small RNA molecules of approximately 22 nucleotides, are a novel class of endogenously non-coding RNAs that control about 30 % gene expressions by targeting specific mRNAs which bear partially complementary target sequences for degradation and translational repression (Bartel 2004). The first miRNA lin-4 was discovered in 1993, when a small RNA encoded by the lin-4 locus was associated to the developmental timing of the nematode by modulating the protein lin-14, but it was thought to be an idiosyncrasy at that time (Lee et al. 1993). Only after let-7 was characterized and identified to be conserved in many model organisms, the importance of miRNAs in physiology and pathology are beginning to emerge (Hutvágner et al. 2001). Since then, miRNAs have been found to be involved in a series of homeostatic processes, such as cellular proliferation and cell death.

Dysregulation of miRNAs is an important trait in cancer progression since the about 50 % of annotated human miRNAs are located in areas of the genome, known as fragile sites that are associated with cancer. The links between miRNAs and cancer were firstly highlighted when two miRNAs, miR-15 and miR-16, were discovered to locate in a region on chromosome 13 and can be deleted in over 65 % of chronic lymphocytic leukemia (CLL) (Calin et al. 2002). After that, aberrant expression patterns of miRNAs have been found to be associated with some examples of tumorigenesis, and miRNAs correlating with various types of cancer are thought to function as either oncogenes or tumor suppressors (Babashah and Soleimani 2011). The miRNA-17-92 cluster, one of the first identified oncogenic miRNAs, has been also shown to be over-expressed in lung cancers and thus enhancing cell proliferation (Hayashita et al. 2005).

Hitherto, miRNAs have been estimated to account for about 1 % of predicted genes in higher eukaryotic genome and thus being regarded to be involved in modulation of cancer cell apoptosis and autophagy. Apoptosis is characterized by the double stranded cleavage at the linker regions between nucleosomes, resulting in the formation of multiple DNA fragments and phosphatidylserine externalization, accompanied by genes and protein expressions. Currently, there are two signaling pathways mediating apoptosis. In the extrinsic pathway, apoptosis is mediated by death receptors on the cell surface; while in the intrinsic pathway, mitochondrion plays a key role (Wen et al. 2012). Different from apoptosis, autophagy is an evolutionarily conserved lysosomal degradation process in which a cell degrades long-lived proteins and damaged organelles. The complete autophagic flow is a highly regulated, multi-step process that, in general, can be divided into the following five stages: induction, vesicle nucleation, vesicle elongation and completion, docking and fusion, as well as degradation and recycling (Yang and Klionsky 2010). Although both apoptosis and autophagy may decide the fate of cancer cell, apoptosis invariably leads to cancer cell death whereas autophagy plays the Janus role in determining the life or death of cancer cell. The involvement of miRNAs in apoptosis was firstly reported when miRNA-14 was reported to regulate cell death in Drosophila, whereas in 2008, the oncogenic miRNA-221/222 was initially identified to lead to autophagic cell death in human breast carcinoma (Stark et al. 2003; Miller et al. 2008).

Amazingly, miRNAs were firstly described 20 years ago; however, the precise molecular mechanisms of miRNAs still remain obscure. In this chapter, we focus on presenting the relations between miRNAs and apoptosis as well as autophagy in diverse types of cancer, which may spark off better understanding of miRNA-modulated apoptotic and autophagic signaling networks in cancer therapy.

2 Anti-Apoptotic MicroRNAs

Accumulating evidence has recently revealed that a group of up-regulated miRNAs are regarded as oncogenes, including let-7a, miRNA-17-92 cluster, miRNA-21, miR-155 and miRNA-221/222. For example, miR-21, can target the programmed cell death 4 gene (PDCD4) and tropomyosin 1 (TPM1), and also phosphatase PTEN in breast cancer; and miR-17-92 cluster can be regarded as a family of oncogenes, directly targeting many genes involved in apoptosis. Thus, these miRNAs can negatively inhibit tumor suppressors that can control apoptosis, and thus leading to carcinogenesis (Fig. 10.1).

2.1 Let-7a

Let-7a is an important member in the let-7 family and its down-regulation can be associated with the over-expression of oncogenes such as RAS and HMGA2. Further, caspase-3 has been confirmed to be the target of let-7a as the ectopic expression of let-7a can decrease the luciferase activity of a reporter construct containing the 3′-untranslated region (3′-UTR) of caspase-3, and repress the enzyme expression in human squamous carcinoma A431 cells and hepatocellular carcinoma HepG2 cells. Moreover, let-7a is over-expressed while caspase-3 is down-regulated in A10A cells, a doxorubicin-resistant A431 subline. On the other hand, down-regulation

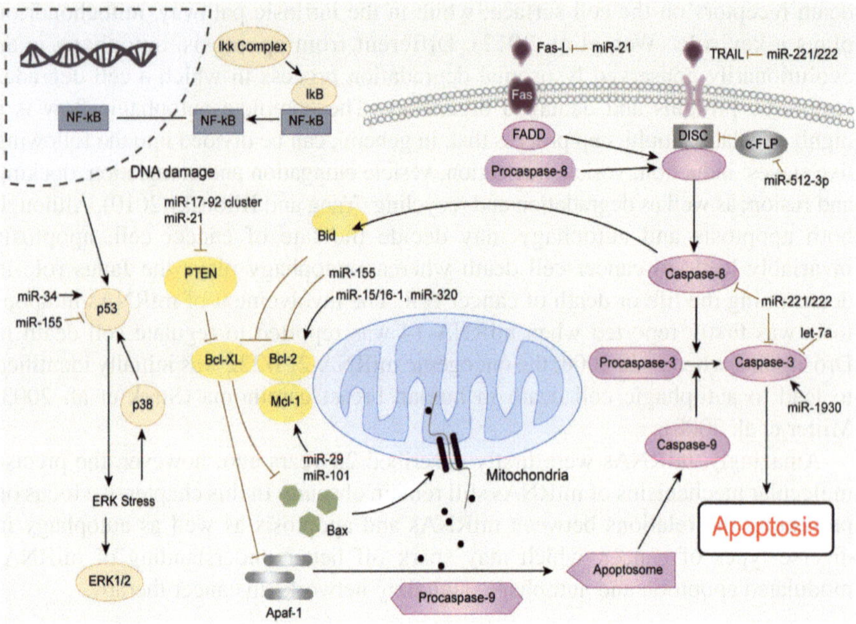

Fig. 10.1 MicroRNAs and apoptosis in cancer

of let-7a has been demonstrated to increase the doxorubicin-induced apoptosis in A431 cells and HepG2 cells while the increase is suppressed by the caspase-3 inhibitor. Therefore, targeting caspase-3, let-7a may play a functional role for modulating apoptosis in cancer (Kannan et al. 2009).

2.2 *MiR-17-92 Cluster*

MiR-17-92 cluster (miR-17-3p, miR-17-5p, miR-18a, miR-20a, miR-19a, miR-19b-1 and miR-92a-1) can be up-regulated in a variety of tumors, including breast, lung, colon, stomach, and pancreas cancers. Moreover, the cluster located on chromosome 13 seems to be frequently over-expressed in a range of hematopoietic malignancies, particularly B-cell lymphomas (Ota et al. 2004).

Because the miR-17-92 cluster can target many genes involved in apoptotic pathways, it seems that the combination of suppressing target mRNAs is responsible for the anti-apoptotic effect. The miR-17-92 cluster-induced cancer is closely associated with the gene c-Myc, which encodes a basic helix-loop-helix transcription factor, promotes proliferation, inhibits apoptosis, induces tumor angiogenesis and cooperates with Myc for accelerating the development of lymphomas (Xiang and Wu 2010). Interaction between the cluster and c-Myc may modulate expression of the E2F transcription factor family (Aguda et al. 2008). The E2F1, a transcription factor promoting cell cycle progression, is induced by c-Myc and creates a reciprocal positive feedback loop by inducing c-Myc expression. MiR-17-5p and miR-20a can directly target the 3′-UTR of the E2F1 in a negative feedback loop of transcriptional regulation, and it seems that c-Myc simultaneously promotes E2F1 transcription and represses following translation, indicating a tightly controlled cell cycle progressive signal (Aguda et al. 2008).

The possible role of miR-17-92 cluster for evading normal apoptotic responses has been strengthened by validation of the pro-apoptotic gene Bim as a direct target. The pro-apoptotic gene Bim is a crucial regulator of B-cell survival and a tumor suppressor in the Eu-Myc model of B-cell lymphoma. Negative regulation of Bim by the miR-17-92 cluster may provide a mechanism for evasion of apoptosis. Further, miR-17-92 cluster can directly target PTEN to increase the level of anti-apoptotic Bcl-2, contributing to the higher rate of proliferation and the lower rate of activation-induced cell death of lymphocytes in tumor organisms (Xiao et al. 2008). MiR-17-92 cluster can bind to the complementary site in the 3′-UTR of the cyclin D1 mRNA, which leads to the inhibition of proliferation of breast cancer cells.

2.3 MiR-21

MiR-21, as an anti-apoptotic factor, has been found to be the most consistently up-regulated miRNA in many types of cancer, such as breast, lung, pancreatic, liver, stomach, colon and prostate cancers. MiR-21 can regulate apoptosis by targeting PTEN, PDCD4 and TPM1 (Wang and Lee 2009).

Down-regulation of PTEN can release its inhibition of protein kinase B (PKB), leading to significantly reduced cancer cell apoptosis (Sayed et al. 2010). In normal hepatocyte cells, the down-regulation of PTEN is accompanied with the induction of phosphorylated Focal Adhesion Kinase (FAK), a major target of PTEN and an inhibitor of apoptosis. In addition, miR-21 regulation of PTEN may increase expressions of two proteins MMP-2 and MMP-9, and expressions of these two proteins can be reduced when the generation of miR-21 can be inhibited (Meng et al. 2007).

Bringing several matched normal and tumor cells into comparison, miR-21 seems to be up-regulated significantly in tumor tissues, while the expression of PDCD4 is indicated to be higher in normal tissues. The similar regulation of PDCD4 is observed in MCF-7 breast cancer cells and embryonic kidney HEK292T cells (Frankel et al. 2008). The regulation on PDCD4 lights the way that miR-21 can positively function on cancer via being up-regulated to inhibit the expression of anti-apoptotic PDCD4 in cancer.

Moreover, miRNA-21 can target TPM1, a member of TPM protein family which plays an important role in stabilizing the microfilament of cytoskeleton and destabilizing in cancer cells. In breast cancer, inhibition of miR-21 may cause the increase of TPM1 protein expression, involved in the 3′-UTR of TPM1 (Zhu et al. 2007). The knockdown of miR-21 in cultured glioblastoma cells triggers activation of caspases and leads to increased apoptotic cell death, indicating that miR-21 is an anti-apoptotic factor in human glioblastoma. Also, its aberrant expression may contribute to the malignant phenotype through blocking expression of critical apoptosis-related genes (Chan et al. 2005).

2.4 *MiR-155*

The anti-apoptotic miR-155 has been found to be dysregulated in both solid and hematological malignancies such as lung cancer, colorectal cancer, pancreatic cancer and CLL. Additionally, its expression has been found to be correlated with chemo-resistance, tumor progression and survival.

CD40 signaling can promote proliferation and rescue B-cells from apoptosis partly through induction of BCL-2A1/BCL-2 L1 and repression of Bcl-2. And, a stromal cell/CD40 ligand (CD154) culture system reproduces this switch in survival protein expression in primary human leukemic B-cells. MiR-155 can require CD154 for further induction to exert its full repressive effects, and anti-miR-155 can prevent CD154-mediated repression of Bcl-2 and thus reducing CD154-mediated proliferation in human leukemic B-cells. Therefore, miR-155 can be induced by CD154 and stromal cell signals, suggesting that Bcl-2 is one of their target mRNAs that can contribute to apoptosis (Willimott and Wagner 2012).

Acting as anti-apoptotic miRNA, miR-155 targets FADD mRNA, and exerts an inhibitory activity on caspase-3 (Tili et al. 2007). Moreover, miR-155 can target the tumor protein p53 inducible nuclear protein 1 (TP53INP1) (Gironella et al. 2007). TP53INP1 is a positive regulator of p53-dependent apoptosis by enhancing Ser46 phosphorylation of p53 which, in turn, induces p53-regulated the expression of apoptosis-inducing protein 1 (p53AIP1) and subsequent apoptotic cell death (Okamura et al. 2001). Hence, the over-expression of miR-155 can inhibit TP53INP1 expression and attenuate apoptosis induced by TP53INP1.

2.5 *MiR-221/222*

MiR-221 and miR-222 have been reported to be significantly up-regulated in primary glioblastomas, papillary thyroid carcinoma and prostate cancer. The miRNA analysis has indicated different expressions of miRNAs in tumor necrosis factor-related apoptosis-inducing ligand (TRAIL)-resistant cells compared to TRAIL-sensitive H460 cells. Over-expression of miR-221 and miR-222, in TRAIL-sensitive cells can increase resistance to TRAIL-induced cell death and reduce activation of caspase-3 and caspase-8 (Garofalo et al. 2008). By contrast, inhibition of the two miRNAs in TRAIL-resistant cells may result in a TRAIL-sensitive phenotype, indicating their roles in determining cell sensitivity to TRAIL. The knockdown of both miR-221 and miR-222 through LNA anti-sense oligonucleotides increases the levels of CDK inhibitor $p27^{kip1}$ in aggressive prostate carcinoma PC3 cells, and considerably reduces *in vitro* clonogenicity feature of the cells. *In vitro* inhibition of miR-221 and miR-222 can modulate expression of both Kit and $p27^{kip1}$ (Galardi et al. 2007). Thus, the role of miR-221/miR-222 has been identified in determining cellular sensitivity to TRAIL-induced apoptosis by targeting caspase-3 and caspase-8 as well as through regulation of Kit and $p27^{kip1}$.

3 Pro-Apoptotic MicroRNAs

MiRNAs as tumor suppressors may function by negatively inhibiting oncogenes and those genes that inhibit apoptosis, and these miRNAs include let-7, which negatively regulates Ras and high mobility group A2 (HMGA2); miR-15a and miR-16-1, which negatively regulate Bcl-2; miR-29, which can target Mcl-1, and miR-34, which is induced by DNA damage and oncogenic stress in a p53-dependent manner which leads to apoptosis. Other miRNAs such as miR-101, miR-193a, miR-206, miR-320 and miR-512-3p can also play pro-apoptotic roles in cancer (Fig. 10.1).

3.1 Let-7 Family

The let-7 family of miRNAs is a highly conserved group comprising 12 closely related members, and they represent 9 distinct let-7 sequences with identical seed sequences, or probably overlapping sets of targets. The let-7 gene was firstly identified at fragile sites associated with human cancers, suggesting its possible role in human cancer. And, let-7 family members have been demonstrated to be responsible for negatively regulating the expression of multiple oncogenes including Ras and MYC (Calin et al. 2004). The let-7 family is complementary to multiple sites in the 3′-UTR of human Ras genes, allowing let-7 to repress the expression of K-Ras and N-Ras. In lung squamous cell carcinoma, the down-regulation of let-7 miRNA in association with the over-expression of Ras has been reported consistent with let-7 that can negatively regulate Ras (Johnson et al. 2007). The let-7 family has also been identified to be down-regulated in Huh7 hepatoma cells in comparison with primary human hepatocytes, as well as possessing a putative target site in the Bcl-XL mRNA. The over-expression of let-7c or let-7g can result in a clear decrease of Bcl-X_L expression in Huh7 and HepG2 cells, and a direct regulation the post-transcriptional level has been revealed in involving let-7c or let-7g and the 3′-UTR of Bcl-X_L mRNA (Kumar et al. 2008). HCC tissues with low expression of let-7c may display higher expression of Bcl-X_L protein than those with high expression of let-7c, suggesting that low let-7 expression may to the over-expression of Bcl-X_L (Shimizu et al. 2010).

3.2 *MiR-15a/16-1*

MiR-15a and miR-16-1 are clustered on human chromosome 13q14 which is frequently lost or down-regulated in CLL and several solid tumors. It has been proposed that both miR-15a and miR-16-1 can promote the normal apoptotic response by direct targeting the anti-apoptotic gene Bcl-2, indicating the probable tumor suppressive function of these two miRNAs in tumorigenesis (Cimmino et al. 2005). The interaction between miR-15a and miR-16-1 and anti-apoptotic gene Bcl-2 leads to cleavage of procaspase-9 and poly ADP-ribose polymerase (PARP), and activation of the intrinsic apoptotic pathway. Given that Bcl-2 is over-expressed in the majority of CLL malignant lymphocytes, the control of Bcl-2 expression by the miR-15/16 cluster can be considered as one of the main molecular mechanisms of this phenomenon (Xia et al. 2008). Moreover, it has been revealed that in non-small-cell lung cancer (NSCLC) cells, cyclin D1, cyclin D2 and cyclin E1 are directly regulated by physiological concentrations of miR-15a and miR-16-1 (Bandi et al. 2009).

3.3 *MiR-29*

Mcl-1 is a potent multi-domain anti-apoptotic protein of the Bcl-2 family which specifically binds to pro-apoptotic members Bim and Bid preventing TRAIL-induced cell death (Lu et al. 2013). Upon binding, the receptor trimerizes, recruits the adaptor protein FADD, and initiates the caspase cascade, which may result in apoptosis. The correlation between miR-29 and Mcl-1 expression has been demonstrated in malignant KMCH cholangiocarcinoma since miR-29 expression can be reduced in cholangiocarcinoma cells which can express Mcl-1 protein. Owing to KMCH cholangiocarcinoma cells that are resistant to TRAIL-induced apoptosis, the over-expression of miR-29 sensitized the cancer cells to TRAIL cytotoxicity, in which miR-29 can negatively regulate Mcl-1 protein levels (Mott et al. 2007). The analysis of primary acute myeloid leukemia (AML) samples reveals the relation between miR-29b and Mcl-1 expression, in which miR-29 can induce global DNA hypomethylation (Garzon et al. 2009). Moreover, an enforced expression of miR-29b renders tumor cells more sensitive to apoptosis-inducing activity of TRAIL, suggesting that the miR-29b/Mcl-1 connection is functionally important and can be exploited for cancer therapy (Marcucci et al. 2009).

3.4 *MiR-34*

Of note, p53 serves as a trans-activator or trans-repressor for many different downstream genes to trigger apoptotic response. The p53-mediated trans-activation of apoptosis-related genes includes pro-apoptotic Bcl-2 family members leading to the intrinsic apoptotic pathway; apoptotic peptidase activating factor-1 (Apaf-1),

Fas/CD95, DR4 and DR5, and components of the extrinsic apoptotic pathways (Solá et al. 2013). MiR-34 family can be direct and conserved p53 target genes, and thus inducing apoptosis. The miR-34 family of miRNAs, including miR-34a, -b and -c, comprises three processed miRNAs that are encoded by two different genes (Bommer et al. 2007). The highly conserved miR-34 family is involved in p53-mediated cell apoptosis; however, miR-34a is also regulated independent of p53 during oncogene-induced senescence (He et al. 2007). The over-expression of miR-34 leads to G1 cell-cycle arrest and apoptosis in various cancer cells, whereas reduction of miR-34 expression attenuates p53-mediated apoptosis, indicating that miRNAs affect tumor suppressive pathways can suppress cancer cell proliferation (Chang et al. 2007). Moreover, the p53-induced transactivation of miR-34 promotes apoptosis and leads to dramatic reprogramming of gene expression, and particularly the genes regulate apoptosis. A genome-wide screening for p53-regulated miRNAs has shown that the most pronounced increase in miRNA abundance after p53 activation is observed for miR-34a, in which a dramatic induction of the miR-34a primary transcript after p53 activation is detected Moreover, the function of p53-induced miR-34a in apoptosis is determined by transfection of H1299 cells with duplex siRNAs corresponding to processed miR-34a (Raver-Shapira et al. 2007). Interestingly, the over-expression of miR-34 leads to a mild decrease in the Bcl-2 protein level, and thus miR-34a can act together with other miRNAs such as miR-15 and miR-16 to suppress efficiently anti-apoptotic Bcl-2. Further, miR-34a is a potent suppressor of cell proliferation through modulation of E2F signaling pathway. Therefore, miR-34a expression may cause reprogramming of genes involved in p53-mediated cell apoptosis (Concepcion et al. 2012). Moreover, miR-34a and miR-34c can target platelet-derived growth factor receptor alpha and beta (PDGFR-α/β) and cell surface tyrosine kinase receptors that can induce cancer cell proliferation, migration and invasion. Thus, down-regulation of miR-34a/c-dependent PDGFR-α/β can inhibit tumorigenesis and then enhancing TRAIL-induced apoptosis in NSCLC cells (Garofalo et al. 2013).

3.5 *Other miRNAs*

Other miRNAs can act as tumor suppressors to induce apoptosis in different cancers. MiR-101 significantly represses the expression of luciferase carrying the 3′-UTR of Mcl-1 and reduces the endogenous protein level of Mcl-1, whereas the miR-101 inhibitor obviously up-regulates Mcl-1 expression and inhibits apoptosis. Moreover, silencing of Mcl-1 can phenocopy the effect of miR-101, and the forced expression of Mcl-1 may reverse the pro-apoptotic effect of miR-101 (Su et al. 2009). The over-expression of miR-193a can induce an activation of caspase-3, leading to apoptosis in A2780 cells. Another genome-wide gene expression with miR-193a-transfected A2780 cells leads to identification of ARHGAP19, CCND1, ERBB4, K-Ras and Mcl-1 as potential targets, and miR-193a can decrease the amount of Mcl-1 by binding 3′-UTR of its mRNA (Nakano et al. 2013). Interestingly, over-expression of miR-206 can decrease levels of Notch 3 protein and its mRNA, and then the expression of miR-206 may markedly induce apoptosis and thus blocking the anti-apoptotic activity of Notch 3. In addition, the ectopic expression of miR-206 inhibits HeLa cell migration. Thus, miR-206 has been identified as a pro-apoptotic activator of cell death, which is associated with its inhibition of Notch 3 signaling and tumor formation (Chen et al. 2009). Moreover, the exogenous expression of miR-320 has been demonstrated to play a negative role in Mcl-1 or Bcl-2 expression and facilitate chemotherapeutic drug-triggered apoptosis in cholangiocarcinoma cells (Chen et al. 2009). The dysregulation of the anti-apoptotic protein cellular FLICE-like inhibitory protein (c-FLIP) is associated with tumorigenesis and progress of most human cancers. MiR-512-3p can negatively regulate c-FLIP expression via a conserved miRNA-binding site in 3′-UTR of c-FLIP, and additional transfection of miR-512-3p remarkably promotes taxol-induced apoptosis, confirming a regulatory mechanism that down-regulation of c-FLIP by miR-512-3p may contribute to apoptosis (Chen et al. 2010).

4 MicroRNAs and Autophagic Signaling Networks

Autophagy is an evolutionarily conserved cellular catabolic process in which proteins and organelles are eliminated through delivery to lysosomes. In cancer, autophagy is an important physiological mechanism that may be a means of temporary survival; but if cellular stress results in continuous or excessively induced autophagy, cell death would ensue. Therefore, autophagy may act as the Roman God Janus by regulating a limited number of autophagy-related (Atg) genes and ultimately sealing the fate of cancer (Choi et al. 2013).

The induction of autophagy is initiated by the complex composed of ULK1/2, mAtg13, FAK family interacting protein of 200 kDa (FIP200) and Atg101, in which miR-106a and miR-885-3p can directly target ULK2 (Huang et al. 2011). Interestingly, conserved and predicted binding sites for miR-885-3p exist in other Atgs, including the ULK1/2-binding partner mAtg13, Atg9A and Atg2B. Then vesicle nucleation occurs, in which proteins and lipids are recruited for construction of the autophagosomal membrane, and this process is initiated by activation of the class III PI3K/Beclin-1 complex. Numerous binding partners of this complex function as either positive or negative regulators, including BAX-interacting factor-1 (Bif-1), Atg14L, UV irradiation resistance-associated gene (UVRAG), activating molecule in Beclin-1-regulated autophagy protein 1 (Ambra1) and Rubicon (Fu et al. 2013). MiR-30a and miR-519a can directly regulate Beclin-1, and over-expression of miR-30a can reduce rapamycin-induced autophagy. Another miRNA that targets Beclin-1 is miR-376b, which is a direct target of Atg4C. The direct regulation of UVRAG, a Beclin-1 binding partner, is regulated by both miR-374a and miR-630 (Frankel and Lund 2012). Additionally, miR-101 has been identified as a potent inhibitor of autophagy via targeting a small guanosine triphosphatase, RAB5A. Two unique ubiquitin-like conjugation systems are involved

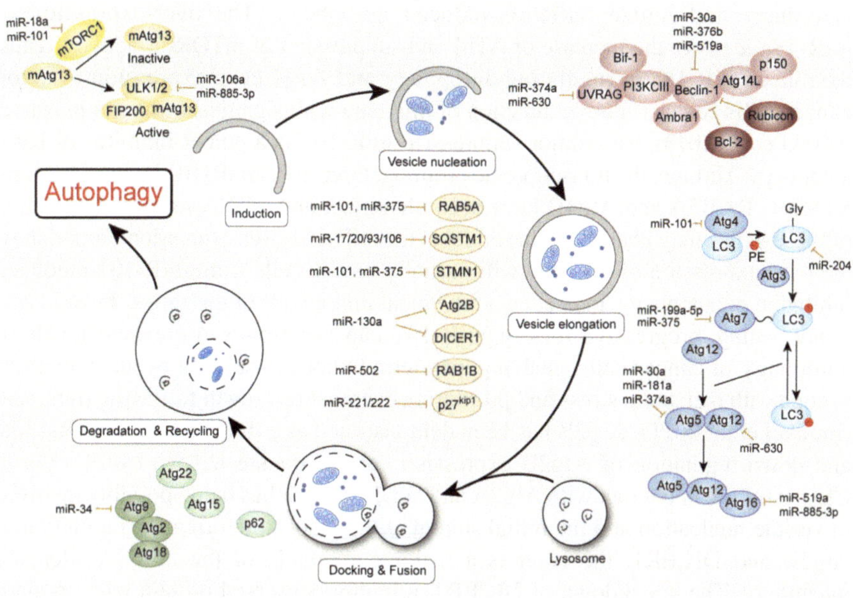

Fig. 10.2 MicroRNAs and autophagy in cancer

in vesicle elongation. One pathway involves the covalent conjugation of Atg12 to Atg5 in a reaction that requires the E1-like and E2-like enzymes, Atg7 and Atg10, respectively (Choi et al. 2013). MiR-204 is a recently identified miRNA regulator of the vesicle elongation process, and its role for autophagic regulation in renal clear cell carcinoma (RCC) via its modulation of LC3B. MiR-375 can directly target Atg7, an E1-like enzyme required for the initial step in both conjugation systems in HCC. MiR-130a may be down-regulated by interfering with Atg9-Atg2-Atg18 complex formation (Xu et al. 2012a). Another regulator of the retrieval step is miR-34a, is identified as a direct regulator of Atg9A. The miR-17/20/93/106 family of miRNAs may share the common seed sequence and are regulators of SQSTM1, which encodes the poly-ubiquitin binding protein p62, and binds to LC3 and acts as a selective autophagy receptor and molecular carrier of cargo to be degraded by autophagy (Frankel and Lund 2012) (Fig. 10.2).

The involvement of miRNA in autophagy was firstly reported in cancer that miR-30a could down-regulate Beclin-1, whose mimics blunted activation of autophagy induced by rapamycin. Recent reports further support a functional significance of miR-30a-mediated autophagy is against CML cells, and miR-30a can also sensitize tumor cells to cisplatin by suppressing Beclin-1-mediated autophagy (Zhu et al. 2009). MiR-18a can potentially regulate Ataxia telangiectasia mutated (ATM) which up-regulates the process of autophagy. The impact of miR-18a on autophagy and ATM expression has recently been revealed in HCT116 colon cancer cells. Then, the over-expression of miR-18a in HCT116 cells is found to potently enhance autophagy and ionize radiation-induced autophagy. The over-expression of miR-18a leads to the increase of ATM and suppression of mTORC1 activity. Thus, the role of miR-18a in regulating autophagy and ATM gene expression in colon cancer cells reveals a novel function of miR-18a with significant impacts in cancer (Qased et al. 2013). The tumor suppressive miR-101 is a potent inhibitor of basal autophagy. Through the transcriptome profiling, three novel miR-101 targets including STMN1, RAB5A and Atg4D have been identified, and siRNA-mediated depletion of these genes may phenocopy the effect of the miR-101 over-expression. Importantly, the over-expression of STMN1 will partially rescue cells from miR-101-mediated inhibition of autophagy, indicating a functional importance of this target. In an *in vivo* tumor setting, progressive loss of miR-101 can contribute to elevated levels of autophagy in cancer cells, enabling long-term tumor cell survival by allowing them to cope with metabolic stress and promoting eventual re-growth following treatment (Frankel et al. 2011). Atg2B has been demonstrated as a direct target of miR-130a, and down-regulation of Atg2B expression can be transfected with miR-130a in CLL. Atg2B can interact with Atg2A and WDR45, and thus being possibly involved in vesicle nucleation and the initial step of autophagosome formation regulation of Atg2B and DICER1, the latter is a major component of the miRNA silencing machinery. The knockdown of DICER1, a highly conserved protein with endonuclease RNase III activity, may lead to reduced autophagic flux in CLL. Thus, miR-130a is involved in autophagy in CLL cells by regulating maturation and activity of many miRNAs, multiplying the amount of indirect effective target genes (Kovaleva et al. 2012). The over-expression of miR-181a may result in the attenuation of starvation-induced and rapamycin-induced autophagy in MCF-7, Huh-7 and K562 cells. The antagomir-mediated inactivation of endogenous miRNA activity can also stimulate autophagy. Indeed, the cellular level of Atg5 can decrease upon miR-181a over-expression and increase following the introduction of antagomirs. Thus, Atg5 has been identified as a miR-181a target (Tekirdag et al. 2013). The level of miR-199a-5p is found to be significantly reduced in HCC patients treated with cisplatin-based chemotherapy. Cisplatin treatment may also lead to decreasing miR-199a-5p levels in HCC cells. The forced expression of miR-199a-5p promotes cisplatin-induced inhibition of cell proliferation, and cisplatin treatment can activate autophagy in Huh7 and HepG2 cells, which may increase cell proliferation. The down-regulation of miR-199a-5p can also enhance autophagy activation by targeting Atg7 (Xu et al. 2012b). The role of tumor suppressive miR-204 in autophagy regulation was initially recognized in cardiomyocytes and further confirmed in the context of RCC via LC3B. The miR-204 over-expression can arrest subcutaneous tumor growth relative to a control miRNA with a mutated seed sequence, and its effects are rescued upon re-expression of LC3B lacking the 3′-UTR. In addition, a negative correlation between LC3B and miR-204 is shown in RCC. Interestingly, the regulation of miR-204 in autophagy and cytotoxity occurs only when the von Hippel-Lindau tumor suppressor gene (VHL) is absent (Mikhaylova et al. 2012). In addition, miR-221/222 can inhibit the cell cycle inhibitor $p27^{Kip1}$, a downstream modulator of PI3KCI/Akt, and thus leading to autophagic cell death in HER2/neu-positive primary human breast carcinoma MCF-7 cells. The ectopic expression of miR-221/222 can render the parental MCF-7 cells resistant to tamoxifen. Thus, a relationship has been demonstrated between miR-221/222 expression and HER2/neu over-expression in breast tumors which are generally resistant to tamoxifen therapy (Miller et al. 2008). MiR-375 plays a predominantly inhibitory role in autophagy activation by attenuating the protective role of autophagy by targeting Atg7, Atg4D, STMN1 and RAB5A in HCC. Further, miR-375, normally down-regulated in HCC when exogenously expressed, was shown to inhibit autophagy in response to hypoxia by targeting Atg7, reducing the conversion of LC3I to LC3II. Xenograft tumors that express miR-375 have fewer autophagic cells, larger areas of necrosis, and grow more slowly than tumors from HCC cells that express lower levels of miR-375 (Chang et al. 2012). Moreover, miR-502 can directly suppress autophagy by decreasing the expression of RAB1B in colon cancer cells. RAB1B is a small GTPase from Ras superfamily to modulate autophagic activity through the regulation of autophagosome formation. RAB1B has been shown to regulate vesicle trafficking at multiple stages and directly impact autophagy in liver cancer. The ectopic expression of miR-502 can interrupt autophagic flux under acute and prolonged nutrient starvation in HCT-116 cells (Zhai et al. 2013) (Fig. 10.2).

Ever since the first autophagy-associated miRNA, miR-30a, was described in 2009, our knowledge of miRNAs in autophagy has rapidly accumulated. Additional tumor-related miRNAs such as miR-18, miR-101, miR-130a, miR-199a-5p, miR-204, miR-221/222, miR-375 and miR-502 have been found to target autophagic pathways in cancer. Moreover, miRNAs also represent an additional player in the intricate interconnection between autophagy and apoptosis. Therefore, multiple miRNAs together with their multiple downstream genes form a complicated regulatory network.

5 Concluding Remarks

During the past two decades, several studies have focused on the understanding of the regulatory role of miRNAs in diverse types of human cancers, acting as either oncogenes or tumor suppressors (Babashah and Soleimani 2011; Babashah et al. 2012). Many lines of evidence support their important roles in apoptosis and autophagy, which may establish a basis for understanding mechanisms linking miRNA deregulation to core apoptotic and autophagic pathways. Several oncogenic or tumor suppressive miRNAs can regulate important apoptotic and autophagic signaling pathways involving Atgs, Bcl-2 family, caspases, the mTORC1 cascade, and p53. Moreover, these oncogenic and tumor suppressive miRNAs regulate apoptosis and autophagy not at the level of single gene or protein product but at the level of the entire network (Fu et al. 2012).

Identification of miRNAs as modulators of gene expression has revealed that they can be used as novel diagnostic and prognostic indicators. Unlike other types of biomarkers, miRNAs have special characteristics, including stability, ease of detection and association with established clinic-pathological prognostic parameters

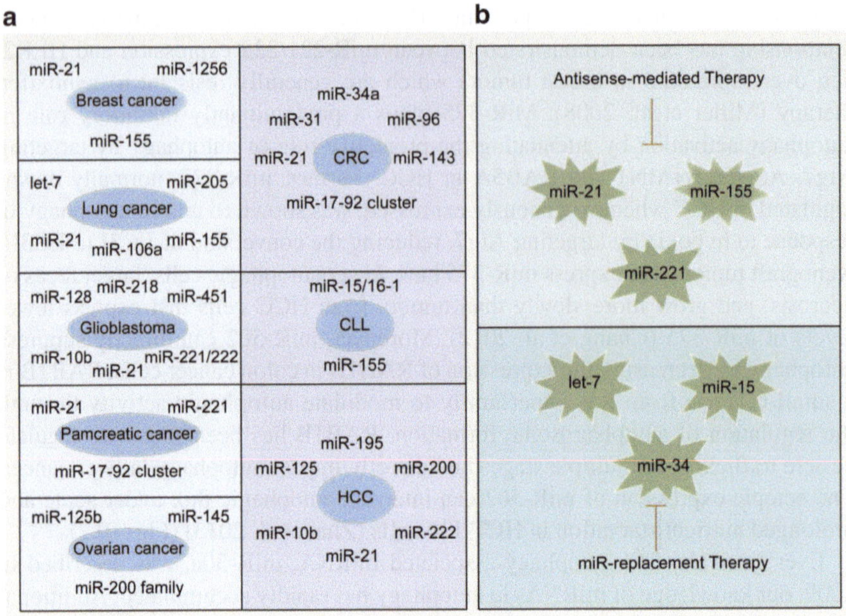

Fig. 10.3 MicroRNAs as (**a**) diagnostic biomarkers and (**b**) therapeutic targets in cancer

which make them robust and reliable biomarkers of cancer (Garzon and Marcucci 2012). Notably, several potential biomarker miRNAs in tumors, such as let-7, miR-21, miR-34, miR-125, miR-155, miR-181 and miR-221/222, are also involved in the apoptotic and autophagic pathways; therefore, additional studies must be conducted to determine whether those biomarkers are involved in apoptosis and autophagy in cancers. Further, the advantage of applying miRNAs for cancer therapy is that they offer an alternative for targeting multiple gene networks controlled by a single miRNA. Through miRNA therapy, personalized cancer treatment can either decrease the activity of oncogenic miRNAs or restore levels of tumor suppressive miRNAs (Nana-Sinkam and Croce 2013). Interestingly, oncogenic miRNAs, including miR-17-29 cluster, miR-21, miR-155 and miR-221/-222, have been found in both the apoptotic and autophagic pathways in cancer, suggesting that targeting the apoptotic and autophagic pathways by miRNAs can be involved in cancer therapy. Similar to oncogenic miRNAs, tumor suppressive miRNAs including let-7 and miR-34 are also involved in the apoptotic and autophagic pathways, further indicating that apoptosis and autophagy may play their important roles in miRNA cancer therapy (Fig. 10.3).

However, despite increasing and encouraging evidence linking miRNAs to apoptosis and autophagy in cancer, many important questions remain to be addressed. Although the identification and validation of miRNA targets has greatly improved during the last few years, we know little regarding the cellular and molecular circuits in which they are involved. It also indicates that there is a complicated regulatory network of multiple miRNAs and multiple downstream genes that are important for future studies. The assessment of the potential for miRNAs as biomarkers is only beginning, because greater attention has been paid to the role of miRNAs in cancer. We may next focus on the expression of miRNAs in different stages of cancer, or we may consider whether some known oncomiRs as biomarkers are involved in the apoptotic and autophagic networks. There are also several limitations of using therapeutic miRNA targets in cancer. Lastly, the ability of miRNAs to target multiple molecules in different cancer cells that belong to apoptotic or autophagic pathways complicates matters further. We can use systems biology to help establish a network of important pathways in apoptosis and autophagy, thereby revealing the role of miRNAs in the network. Therefore, it is conceivable that modulating miRNAs will change cancer cells in response to stress by altering both apoptotic and autophagic processes, which would in turn provide novel therapeutic strategies to fight human cancers.

Acknowledgments We would like to dedicate this chapter to Prof. Xinsheng Yao (Jinan University) who is celebrating his 80th birthday on October 24th, 2013. This work was also supported by grants from the National 973 Basic Research Program of China (No. 2010CB529900).

References

Aguda BD, Kim Y, Piper-Hunter MG, Friedman A, Marsh CB (2008) MicroRNA regulation of a cancer network: consequences of the feedback loops involving miR-17-92, E2F, and Myc. Proc Natl Acad Sci U S A 105(50):19678–19683

Babashah S, Soleimani M (2011) The oncogenic and tumour suppressive roles of microRNAs in cancer and apoptosis. Eur J Cancer 47(8):1127–1137

Babashah S, Sadeghizadeh M, Tavirani MR, Farivar S, Soleimani M (2012) Aberrant microRNA expression and its implications in the pathogenesis of leukemias. Cell Oncol 35(5):317–334

Bandi N, Zbinden S, Gugger M, Arnold M, Kocher V, Hasan L et al (2009) miR-15a and miR-16 are implicated in cell cycle regulation in a Rb-dependent manner and are frequently deleted or down-regulated in non-small cell lung cancer. Cancer Res 69(13):5553–5559

Bartel DP (2004) MicroRNAs: genomics, biogenesis, mechanism, and function. Cell 116(2):281–297

Bommer GT, Gerin I, Feng Y, Kaczorowski AJ, Kuick R, Love RE et al (2007) p53-mediated activation of miRNA34 candidate tumor-suppressor genes. Curr Biol 17(15):1298–1307

Calin GA, Dumitru CD, Shimizu M, Bichi R, Zupo S, Noch E et al (2002) Frequent deletions and down-regulation of micro-RNA genes miR15 and miR16 at 13q14 in chronic lymphocytic leukemia. Proc Natl Acad Sci U S A 99(24):15524–15529

Calin GA, Sevignani C, Dumitru CD, Hyslop T, Noch E, Yendamuri S et al (2004) Human microRNA genes are frequently located at fragile sites and genomic regions involved in cancers. Proc Natl Acad Sci U S A 101(9):2999–3004

Chan JA, Krichevsky AM, Kosik KS (2005) MicroRNA-21 is an antiapoptotic factor in human glioblastoma cells. Cancer Res 65(14):6029–6033

Chang TC, Wentzel EA, Kent OA, Ramachandran K, Mullendore M, Lee KH et al (2007) Transactivation of miR-34a by p53 broadly influences gene expression and promotes apoptosis. Mol Cell 26(5):745–752

Chang Y, Yan W, He X, Zhang L, Li C, Huang H et al (2012) miR-375 inhibits autophagy and reduces viability of hepatocellular carcinoma cells under hypoxic conditions. Gastroenterology 143(1):177–187

Chen L, Yan HX, Yang W, Hu L, Yu LX, Liu Q et al (2009) The role of microRNA expression pattern in human intrahepatic cholangiocarcinoma. J Hepatol 50(2):358–369
Chen F, Zhu HH, Zhou LF, Wu SS, Wang J, Chen Z (2010) Inhibition of c-FLIP expression by miR-512-3p contributes to taxol-induced apoptosis in hepatocellular carcinoma cells. Oncol Rep 23(5):1457–1462
Choi AM, Ryter SW, Levine B (2013) Autophagy in human health and disease. N Engl J Med 368(19):1845–1846
Cimmino A, Calin GA, Fabbri M, Iorio MV, Ferracin M, Shimizu M et al (2005) miR-15 and miR-16 induce apoptosis by targeting BCL2. Proc Natl Acad Sci U S A 102(39):13944–13949
Concepcion CP, Han YC, Mu P, Bonetti C, Yao E, D'Andrea A et al (2012) Intact p53-dependent responses in miR-34-deficient mice. PLoS Genet 8(7):e1002797
Frankel LB, Lund AH (2012) MicroRNA regulation of autophagy. Carcinogenesis 33(11):2018–2025
Frankel LB, Christoffersen NR, Jacobsen A, Lindow M, Krogh A, Lund AH (2008) Programmed cell death 4 (PDCD4) is an important functional target of the microRNA miR-21 in breast cancer cells. J Biol Chem 283(2):1026–1033
Frankel LB, Wen J, Lees M, Høyer-Hansen M, Farkas T, Krogh A et al (2011) microRNA-101 is a potent inhibitor of autophagy. EMBO J 30(22):4628–4641
Fu LL, Wen X, Bao JK, Liu B (2012) MicroRNA-modulated autophagic signaling networks in cancer. Int J Biochem Cell Biol 44(5):733–736
Fu LL, Cheng Y, Liu B (2013) Beclin-1: autophagic regulator and therapeutic target in cancer. Int J Biochem Cell Biol 45(5):921–924
Galardi S, Mercatelli N, Giorda E, Massalini S, Frajese GV, Ciafrè SA et al (2007) miR-221 and miR-222 expression affects the proliferation potential of human prostate carcinoma cell lines by targeting p27Kip1. J Biol Chem 282(32):23716–23724
Garofalo M, Quintavalle C, Di Leva G (2008) MicroRNA signatures of TRAIL resistance in human non-small cell lung cancer. Oncogene 27(27):3845–3855
Garofalo M, Jeon YJ, Nuovo GJ, Middleton J, Secchiero P, Joshi P et al (2013) MiR-34a/c-Dependent PDGFR-α/β Down-regulation Inhibits Tumorigenesis and Enhances TRAIL-Induced Apoptosis in Lung Cancer. PLoS One 8(6):e67581
Garzon R, Marcucci G (2012) Potential of microRNAs for cancer diagnostics, prognostication and therapy. Curr Opin Oncol 24(6):655–659
Garzon R, Heaphy CE, Havelange V, Fabbri M, Volinia S, Tsao T et al (2009) MicroRNA 29b functions in acute myeloid leukemia. Blood 114(26):5331–5341
Gironella M, Seux M, Xie MJ, Cano C, Tomasini R, Gommeaux J et al (2007) Tumor protein 53-induced nuclear protein 1 expression is repressed by miR-155, and its restoration inhibits pancreatic tumor development. Proc Natl Acad Sci U S A 104(41):16170–16175
Hayashita Y, Osada H, Tatematsu Y, Yamada H, Yanagisawa K, Tomida S et al (2005) A polycistronic microRNA cluster, miR-17-92, is overexpressed in human lung cancers and enhances cell proliferation. Cancer Res 65(21):9628–9632
He X, He L, Hannon GJ (2007) The guardian's little helper: microRNAs in the p53 tumor suppressor network. Cancer Res 67(23):11099–11101
Huang Y, Chuang AY, Ratovitski EA (2011) Phospho-ΔNp63α/miR-885-3p axis in tumor cell life and cell death upon cisplatin exposure. Cell Cycle 10(22):3938–3947
Hutvágner G, McLachlan J, Pasquinelli AE, Bálint E, Tuschl T, Zamore PD (2001) A cellular function for the RNA-interference enzyme Dicer in the maturation of the let-7 small temporal RNA. Science 293(5531):834–838
Johnson CD, Esquela-Kerscher A, Stefani G, Byrom M, Kelnar K, Ovcharenko D et al (2007) The let-7 microRNA represses cell proliferation pathways in human cells. Cancer Res 67(16):7713–7722
Kannan M, Mohan KV, Kulkarni S, Atreya C (2009) Membrane array-based differential profiling of platelets during storage for 52 miRNAs associated with apoptosis. Transfusion 49(7):1443–1450
Kovaleva V, Mora R, Park YJ, Plass C, Chiramel AI, Bartenschlager R et al (2012) MiRNA-130a targets ATG2B and DICER1 to inhibit autophagy and trigger killing of chronic lymphocytic leukemia cells. Cancer Res 72(7):1763–1772

Kumar MS, Erkeland SJ, Pester RE, Chen CY, Ebert MS, Sharp PA et al (2008) Suppression of non-small cell lung tumor development by the let-7 microRNA family. Proc Natl Acad Sci U S A 105(10):3903–3908
Lee RC, Feinbaum RL, Ambros V (1993) The C. elegans heterochronic gene lin-4 encodes small RNAs with antisense complementarity to lin-14. Cell 75(5):843–854
Lu T, Shao N, Ji C (2013) Targeting microRNAs to modulate TRAIL-induced apoptosis of cancer cells. Cancer Gene Ther 20(1):33–37
Marcucci G, Mrózek K, Radmacher MD, Bloomfield CD, Croce CM (2009) MicroRNA expression profiling in acute myeloid and chronic lymphocytic leukaemias. Best Pract Res Clin Haematol 22(2):239–248
Meng F, Henson R, Wehbe-Janek H, Ghoshal K, Jacob ST, Patel T (2007) MicroRNA-21 regulates expression of the PTEN tumor suppressor gene in human hepatocellular cancer. Gastroenterology 133(2):647–658
Mikhaylova O, Stratton Y, Hall D (2012) VHL-regulated MiR-204 suppresses tumor growth through inhibition of LC3B-mediated autophagy in renal clear cell carcinoma. Cancer Cell 21(4):532–546
Miller TE, Ghoshal K, Ramaswamy B, Roy S, Datta J, Shapiro CL et al (2008) MicroRNA-221/222 confers tamoxifen resistance in breast cancer by targeting p27Kip1. J Biol Chem 283(44):29897–29903
Mott JL, Kobayashi S, Bronk SF, Gores GJ (2007) Mir-29 regulates Mcl-1 protein expression and apoptosis. Oncogene 26(42):6133–6140
Nakano H, Yamada Y, Miyazawa T, Yoshida T (2013) Gain-of-function microRNA screens identify miR-193a regulating proliferation and apoptosis in epithelial ovarian cancer cells. Int J Oncol 42(6):1875–1882
Nana-Sinkam SP, Croce CM (2013) Clinical applications for microRNAs in cancer. Clin Pharmacol Ther 93(1):98–104
Okamura S, Arakawa H, Tanaka T, Nakanishi H, Ng CC, Taya Y et al (2001) p53DINP1, a p53-inducible gene, regulates p53-dependent apoptosis. Mol Cell 8(1):85–94
Ota A, Tagawa H, Karnan S, Tsuzuki S, Karpas A, Kira S et al (2004) Identification and characterization of a novel gene, C13orf25, as a target for 13q31–q32 amplification in malignant lymphoma. Cancer Res 64(9):3087–3095
Qased AB, Yi H, Liang N, Ma S, Qiao S, Liu X (2013) MicroRNA-18a upregulates autophagy and ataxia telangiectasia mutated gene expression in HCT116 colon cancer cells. Mol Med Rep 7(2):559–564
Raver-Shapira N, Marciano E, Meiri E, Spector Y, Rosenfeld N, Moskovits N et al (2007) Transcriptional activation of miR-34a contributes to p53-mediated apoptosis. Mol Cell 26(5):731–743
Sayed D, He M, Hong C, Gao S, Rane S, Yang Z et al (2010) MicroRNA-21 is a downstream effector of AKT that mediates its antiapoptotic effects via suppression of Fas ligand. J Biol Chem 285(26):20281–20290
Shimizu S, Takehara T, Hikita H, Kodama T, Miyagi T, Hosui A et al (2010) The let-7 family of microRNAs inhibits Bcl-xL expression and potentiates sorafenib-induced apoptosis in human hepatocellular carcinoma. J Hepatol 52(5):698–704
Solá S, Morgado AL, Rodrigues CM (2013) Death receptors and mitochondria: two prime triggers of neural apoptosis and differentiation. Biochim Biophys Acta 1830(1):2160–2166
Stark A, Brennecke J, Russell RB, Cohen SM (2003) Identification of Drosophila MicroRNA targets. PLoS Biol 1(3):E60
Su H, Yang JR, Xu T, Huang J, Xu L, Yuan Y et al (2009) MicroRNA-101, down-regulated in hepatocellular carcinoma, promotes apoptosis and suppresses tumorigenicity. Cancer Res 69(3):1135–1142
Tekirdag KA, Korkmaz G, Ozturk DG, Agami R, Gozuacik D (2013) MIR181A regulates starvation- and rapamycin-induced autophagy through targeting of ATG5. Autophagy 9(3):374–385

Tili E, Michaille JJ, Cimino A, Costinean S, Dumitru CD, Adair B et al (2007) Modulation of miR-155 and miR-125b levels following lipopolysaccharide/TNF-alpha stimulation and their possible roles in regulating the response to endotoxin shock. J Immunol 179(8):5082–5089
Wang Y, Lee CG (2009) MicroRNA and cancer–focus on apoptosis. J Cell Mol Med 13(1):12–23
Wen X, Lin ZQ, Liu B, Wei YQ (2012) Caspase-mediated programmed cell death pathways as potential therapeutic targets in cancer. Cell Prolif 45(3):217–224
Willimott S, Wagner SD (2012) miR-125b and miR-155 contribute to BCL2 repression and proliferation in response to CD40 ligand (CD154) in human leukemic B-cells. J Biol Chem 287(4):2608–2617
Xia L, Zhang D, Du R, Pan Y, Zhao L, Sun S et al (2008) miR-15b and miR-16 modulate multidrug resistance by targeting BCL2 in human gastric cancer cells. Int J Cancer 123(2):372–379
Xiang J, Wu J (2010) Feud or friend? The role of the miR-17-92 cluster in tumorigenesis. Curr Genomics 11(2):129–135
Xiao C, Srinivasan L, Calado DP, Patterson HC, Zhang B, Wang J et al (2008) Lymphoproliferative disease and autoimmunity in mice with increased miR-17-92 expression in lymphocytes. Nat Immunol 9(4):405–414
Xu J, Wang Y, Tan X, Jing H (2012a) MicroRNAs in autophagy and their emerging roles in crosstalk with apoptosis. Autophagy 8(6):873–882
Xu N, Zhang J, Shen C, Luo Y, Xia L, Xue F et al (2012b) Cisplatin-induced downregulation of miR-199a-5p increases drug resistance by activating autophagy in HCC cell. Biochem Biophys Res Commun 423(4):826–831
Yang Z, Klionsky DJ (2010) Eaten alive: a history of macroautophagy. Nat Cell Biol 12(9):814–822
Zhai H, Song B, Xu X, Zhu W, Ju J (2013) Inhibition of autophagy and tumor growth in colon cancer by miR-502. Oncogene 32(12):1570–1579
Zhu S, Si ML, Wu H, Mo YY (2007) MicroRNA-21 targets the tumor suppressor gene tropomyosin 1 (TPM1). J Biol Chem 282(19):14328–14336
Zhu H, Wu H, Liu X, Li B, Chen Y, Ren X et al (2009) Regulation of autophagy by a beclin 1-targeted microRNA, miR-30a, in cancer cells. Autophagy 5(6):816–823

Chapter 11
MicroRNAs in Cancer Drug Resistance and Drug Sensitivity

Hugo Seca, Raquel T. Lima, and M. Helena Vasconcelos

Contents

H. Seca • M.H. Vasconcelos (✉)
Cancer Drug Resistance Group, Instituto de Patologia e Imunologia Molecular da Universidade do Porto (IPATIMUP), Rua Doutor Roberto Frias s/n, 4200 Porto, Portugal

Laboratório de Microbiologia, Departamento de Ciências Biológicas, Faculdade de Farmácia, Universidade do Porto, Rua de Jorge Viterbo Ferreira n.º 228, 4050-313 Porto, Portugal
e-mail: hvasconcelos@ipatimup.pt

R.T. Lima
Cancer Drug Resistance Group, Instituto de Patologia e Imunologia Molecular da Universidade do Porto (IPATIMUP), Rua Doutor Roberto Frias s/n, 4200 Porto, Portugal

Centro de Química Medicinal da Universidade do Porto (CEQUIMED-UP), Faculdade de Farmácia, Universidade do Porto, Rua de Jorge Viterbo Ferreira n.º 228, 4050-313 Porto, Portugal

S. Babashah (ed.), *MicroRNAs: Key Regulators of Oncogenesis*,
DOI 10.1007/978-3-319-03725-7_11,

Abstract MicroRNAs (miRNAs) are important regulators of distinct biological processes and are tightly related with several diseases, including cancer. In the last years, miRNAs have been shown to play an important role in drug resistance/sensitivity to anticancer drugs. This chapter first addresses the phenomenon of drug resistance in cancer cells. It then describes the role of miRNAs in drug resistance or drug sensitivity, including a review of the role of miRNAs and their validated targets in cancer drug resistance. Furthermore, it focuses on miRNAs that target cellular proteins related to drug resistance such as p53 or proteins involved in cellular processes such as apoptosis, autophagy, DNA damage response, epithelial to mesenchymal transition and cancer stem cell phenotype, drug efflux or drug metabolism. In addition, the non-genetic acquisition of drug resistance through the transfer of miRNAs from drug resistant to drug sensitive cells via microvesicles or exosomes is also discussed. Finally, the known miRNA polymorphisms associated with drug resistance are addressed.

Keywords MicroRNA • Drug sensitivity • Drug resistance • Drug efflux • Microvesicles • miRs-polymorphisms

1 Cancer Drug Resistance

Chemotherapy is an important treatment modality for many types of cancer and it is the main treatment for advanced and surgically resectable cancers. Targeted therapies have been more recently developed with the emergence of small molecules which target specific kinases or receptors in the cell that are known to be deregulated in cancer. Despite the benefits in improving overall survival and quality of life for patients, the therapeutic success of both conventional chemotherapy and targeted therapeutics is often limited due to drug resistance [reviewed in Broxterman et al. (2009); Stegmeier et al. (2010); Fojo (2007)].

Cancer drug resistance may be intrinsic, based on genetic characteristics of tumor cells that are insensitive to therapeutic agents even before treatment – intrinsic resistance – or may appear *de novo*, after treatment with a chemotherapeutic agent – acquired resistance. This resistance enables the tumor cells to escape the cytotoxic/cytostatic effect of anticancer drugs and thus allows the tumor to persist and eventually progress.

Cancer drug resistance has been thoroughly studied in cell lines, animal models and cancer patients, in order to elucidate the cellular and molecular mechanisms involved. These mechanisms may comprise genetic alterations, such as a particular gene amplification, translocations and mutations, or epigenetic modifications that alter the function of pivotal genes [reviewed in Fojo (2007); Wilting and Dannenberg (2012)]. In any of the cases, the mechanisms are generally divided into two main groups: those that disturb the delivery and maintenance of the drugs in the cells and those arising from molecular alterations that affect drug sensitivity [reviewed in Gottesman et al. (2002)].

Cancer drug resistance has been shown to be very hard to tackle, mostly due to two factors: (i) the intra-tumor heterogeneity [due to expansion of multiple clones, some being more aggressive than others (Gerlinger et al. 2012)] and (ii) the tumor dynamics [with genetic instability and accumulations of mutations throughout tumor development (Sequist et al. 2011)]. In addition, an evolutive adaptation of cancer cells to drugs has been proposed as a mechanism that promotes drug resistance [reviewed in Gerlinger and Swanton (2010)].

The most common cause for acquisition of cancer drug resistance is the expression of energy-dependent transporters (or drug efflux pumps), that detect some anticancer drugs and carry out the efflux of such drugs from the cancer cells. Since the drug efflux transporters are not specific for a designated substrate, cells that over-express drug efflux pumps such as P-glycoprotein (P-gp) are resistant to various drugs, a phenotype known as multidrug resistance (MDR) [reviewed in Gottesman et al. (2002)]. However, several other mechanisms of drug resistance are known to occur, such as insensitivity to drug-induced apoptosis, enhanced drug metabolism, increased DNA repair or mutations in drug targets [reviewed in Gottesman et al. (2002); Fojo (2007); Borst (2012)]. Table 11.1 summarizes the most common mechanisms of drug resistance. In this chapter, we review the current knowledge on the role of microRNAs (miRNAs), a class of small non-coding regulatory RNAs, in drug resistance.

2 Anticancer Drugs May Affect Cellular MicroRNA Expression

MiRNA expression has been shown to be altered by anticancer drugs. For example, a panel of 60 human cancer cell lines (NCI-60) which has been used to screen more than 100,000 compounds for their anticancer potential, has been profiled for mRNA and protein expression, mutational status, DNA copy number, chromosomal alterations (Boyd and Pauli 1995) and miRNA expression (Gaur et al. 2007; Blower et al. 2007). These studies have shown a correlation between the expression pattern of some miRNAs and the growth inhibitory pattern of certain drugs, which may indicate a role for such miRNAs in drug response (Blower et al. 2007).

Other studies have also shown that anticancer drugs affect miRNAs expression (Rossi et al. 2007; Flamant et al. 2010; Gmeiner et al. 2010). Nevertheless, the results cannot be analyzed straightforward as the observed alterations are due to specific characteristics of the experimental design such as drug concentration, time of exposure to the drug or the type of cells in which the anticancer drugs were tested. Therefore, the precise function of each miRNA might not be easy to assess and still needs extensive work.

Table 11.1 Examples of cancer drug resistance mechanisms

Mechanism	Example	Reviewed in
Over-expression of efflux pumps	Over-expression of P-gp conferring resistance (several drugs)	Fletcher et al. (2010)
Increase drug detoxification	Increased activity of the enzyme CYP3A4 conferring resistance to docetaxel	Akhdar et al. (2012)
Deregulation of apoptosis mechanisms	Over-expression of anti-apoptotic proteins e.g. Bcl-2 (several drugs)	Rebucci and Michiels (2013)
Point mutations in genes coding for drug targets	T315I mutation in BCR-ABL confers resistance to imatinib	Garraway and Janne (2012)
Over-expression of oncogenes causing dysfunction in signaling molecules and kinases	HER2 (ERBB2) amplification (several drugs)	Garraway and Janne (2012)
Activation of alternative mechanisms	MET over-expression with sustained activation of PI3K/Akt signaling confers resistance to TKIs	Kosaka et al. (2011)
Tumor microenvironment	Hypoxia-induced resistance (several drugs)	Rebucci and Michiels (2013)
Epigenetic alterations	Distinct chromatin state mediated by the histone demethylase KDM5A confers resistance to erlotinib	Garraway and Janne (2012)
MiRNAs	MiR-21 over-expression (several drugs)	Chen et al. (2012c)

3 MicroRNAs Involved in Cancer Drug Response: Drug Sensitivity and Drug Resistance

The fact that various cancer drugs affect cellular miRNAs expression indicates that miRNAs may be important regulators of mechanisms involved in treatment outcome. Nevertheless, it is not known if those miRNAs have a function in drug response or in drug resistance, or both. In order to confirm this, functional studies, either by over-expressing or silencing the expression of such miRNAs in cell lines have been performed (Table 11.2).

As can be seen by Table 11.2, a great number of miRNAs have been shown to modulate drug response. Some of them have been described more frequently than others, granting them more recognition in drug resistance and thus will be described here in more detail.

3.1 MiR-21

MiR-21 is over-expressed in almost all types of tumors (Volinia et al. 2006) and it is generally accepted as being an "oncomiR". Its expression is often associated with drug resistance, firstly described when down-regulation of miR-21 sensitized

Table 11.2 Role of some microRNAs and validated targets in cancer drug resistance

MicroRNA	Manipulation of expression carried out[a]	Obtained result	Validated direct target
miR-1	↑	Sensitizes lung cancer cells to DOX (Nasser et al. 2008)	MET, FOXP1
miR-7	↑	Sensitizes breast cancer cells to CDDP (Pogribny et al. 2010)	ABCC1
	↑	Sensitizes head and neck cancer cells to erlotinib (Kalinowski et al. 2012)	EGFR
miR-10a*	↓	Inhibition sensitizes glioblastoma cells to TMZ (Ujifuku et al. 2010)	Unknown
miR-10b	↑	Confers resistance to 5-FU in colon cancer cells (Nishida et al. 2012)	BIM
miR-15a	↑↓	Modulates TAM resistance in breast cancer cells (Cittelly et al. 2010a)	BCL2
miR-15b	↑↓	Modulates CDDP resistance in tongue cancer cells (Sun et al. 2012)	BMI1
	↑↓	Modulates CDDP, DOX, VP-16 and VCR resistance in gastric cancer cells (Xia et al. 2008)	BCL2
miR-16	↑↓	Modulates CDDP, DOX, VP-16 and VCR resistance in gastric cancer cells (Xia et al. 2008)	BCL2
	↑↓	Modulates TAM resistance in breast cancer cells (Cittelly et al. 2010a)	BCL2
miR-17	↑↓	Modulates AZD6244 resistance in lung cancer cells (Dai et al. 2011)	Unknown
miR-17~92	↑↓	Confers resistance to DOX and topotecan and modulates VP-16 resistance in mantle cell lymphoma cells (Rao et al. 2012)	BIM, PTEN, PHLPP2
miR-19	↓	Inhibition sensitizes breast cancer cells to PTX, VP-16 and MX (Liang et al. 2011)	PTEN
miR-19a	↑↓	Modulates 5-FU, CDDP and DOX resistance in gastric cancer cells (Wang et al. 2013a)	PTEN
miR-19b	↑↓	Modulates 5-FU, CDDP and DOX resistance in gastric cancer cells (Wang et al. 2013a)	PTEN
miR-20a	↑↓	Modulates 5-FU, DOX and VM-26 resistance in colon cancer cells (Chai et al. 2011)	BNIP2

(continued)

Table 11.2 (continued)

MicroRNA	Manipulation of expression carried out[a]	Obtained result	Validated direct target
miR-21	↑↓	Modulates DTX resistance in prostate cancer cells (Shi et al. 2010)	PDCD4
	↑	Sensitizes tongue cancer cells to CDDP (Yu et al. 2010)	Unknown
	↑↓	Modulates DNR resistance in leukemia cancer cells (Bai et al. 2011)	PTEN
	↑↓	Modulates CDDP resistance in neuroblastoma cells (Chen et al. 2012d)	PTEN
	↑↓	Modulates gemcitabine resistance in pancreatic cancer cells (Dong et al. 2011)	BCL2
	↑↓	Modulates Trastuzumab resistance in breast cancer cells (Gong et al. 2011)	PTEN
	↓	Inhibition sensitizes prostate cancer cells to STS (Li et al. 2009a)	MARCKS
	↓	Inhibition sensitizes leukemia cells to ATO (Li et al. 2010a)	PDCD4
	↑↓	Modulates CDDP and DTX resistance in lung cancer cells (Liu et al. 2013c)	PTEN
	↑↓	Modulates DOX resistance in bladder cancer cells (Tao et al. 2011)	PTEN
	↑↓	Modulates 5-FU resistance in HCC cells (Tomimaru et al. 2010)	PDCD4, PTEN
	↑	Confers resistance to 5-FU in colon cancer cells (Valeri et al. 2010)	hMSH2
	↑↓	Modulates DOX resistance in breast cancer cells (Wang et al. 2011d)	PTEN
	↑↓	Modulates DEX and DOX resistance in multiple myeloma cells (Wang et al. 2011b)	RHOB
	↓	Inhibition sensitizes lung cancer cells to gefitinib (Garofalo et al. 2012b)	Unknown
	↓	Inhibition sensitizes glioblastoma cells to VM-26 (Li et al. 2009b)	LRRFIP1
	↑	Confers resistance to gemcitabine in pancreatic cancer cells (Giovannetti et al. 2010)	PTEN
	↑↓	Modulates gemcitabine resistance in cholangiocarcinoma cells (Meng et al. 2006)	PTEN
	↓	Inhibition sensitizes CLL cells to fludarabine (Ferracin et al. 2010)	Unknown
	↓	Inhibition sensitizes lung cancer cells to AG1478 (Seike et al. 2009)	Unknown
miR-22	↑	Sensitizes p53-mutated colon cancer cells to PTX (Li et al. 2011b)	Unknown

(continued)

Table 11.2 (continued)

MicroRNA	Manipulation of expression carried out[a]	Obtained result	Validated direct target
miR-23a	↓	Inhibition sensitizes tongue cancer cells to CDDP (Yu et al. 2010)	TOP2B
miR-24	↑	Sensitizes breast and lung cancer cells to TRAIL (Xie et al. 2013)	XIAP
	↑	Confers resistance to MTX in colon cancer cells (Mishra et al. 2007)	DHRF
	↓	Modulates bleomycin and CDDP resistance in CML and HCC cells (Lal et al. 2009)	H2AX
miR-25	↓	Modulates TRAIL resistance in cholangiocarcinoma cells (Razumilava et al. 2012)	DR4
miR-27a	↓	Inhibition sensitizes ovarian cancer cells to vinblastine (Zhu et al. 2008)	Unknown
	↑	Sensitizes leukemia cells to DOX (Feng et al. 2011)	ABCB1
	↑↓	Modulates PTX resistance in ovarian cancer cells (Li et al. 2010b)	HIPK2
	↓	Inhibition sensitizes gastric cancer cells to 5-FU, CDDP, DOX and VCR (Zhao et al. 2011a)	CCND1
miR-29c	↑	Sensitizes nasopharyngeal carcinoma cells to CDDP (Zhang et al. 2013)	BCL2, MCL1
miR-30a	↑↓	Modulates dasatinib, imatinib and nilotinib resistance in CML cells (Yu et al. 2012b)	BECN1
	↑	Sensitizes cervical and gastric cancer cells to CDDP and breast cancer cells to PTX (Zou et al. 2012)	BECN1
	↑	Sensitizes breast cancer cells to PTX (Bockhorn et al. 2013)	Unknown
miR-30a*	↑	Sensitizes breast cancer cells to DOX and PTX (Bockhorn et al. 2013)	Unknown
miR-30b	↑↓	Modulates gefitinib resistance in lung cancer cells (Garofalo et al. 2012b)	BIM
miR-30c	↑↓	Modulates gefitinib resistance in lung cancer cells (Garofalo et al. 2012b)	BIM
	↑	Sensitizes breast cancer cells to DOX and PTX (Bockhorn et al. 2013)	TWF1
miR-31	↑	Sensitizes ovarian cancer cells to PTX (Mitamura et al. 2013)	MET
	↑↓	Modulates DTX resistance in prostate cancer cells and sensitizes prostate cancer cells to CDDP (Bhatnagar et al. 2010)	E2F6
	↑	Sensitizes breast cancer cells to DOX and STS (Korner et al. 2013)	PKCε

(continued)

Table 11.2 (continued)

MicroRNA	Manipulation of expression carried out[a]	Obtained result	Validated direct target
miR-34a	↑↓	Modulates DTX resistance in breast cancer cells (Kastl et al. 2012)	BCL2
	↑	Sensitizes colon cancer cells to 5-FU (Akao et al. 2011)	SIRT1
	↑	Sensitizes prostate cancer cells to DNR, PTX and VP-16 (Kojima et al. 2010)	SIRT1
	↑↓	Modulates DOX resistance in breast cancer cells (Li et al. 2012b)	NOTCH1
	↑	Sensitizes Ewing's sarcoma cells to DOX and VCR (Nakatani et al. 2012)	Unknown
	↑↓	Modulates CDDP resistance in bladder cancer cells (Vinall et al. 2012)	CDK6,SIRT1
	↑	Sensitizes medulloblastoma cells to CDDP and MMC (Weeraratne et al. 2011)	MAGE-A
	↑	Sensitizes prostate cancer cells to CPT (Fujita et al. 2008)	SIRT1
	↑	Sensitizes gastric cancer cells to CDDP, DOX, DTX and gemcitabine (Ji et al. 2008)	BCL2
	↑	Sensitizes retinoblastoma cells to topotecan (Dalgard et al. 2009)	Unknown
	↑↓	Modulates DOX resistance in prostate cancer cells (Rokhlin et al. 2008)	Unknown
miR-34c	↑↓	Modulates DOX resistance in prostate cancer cells (Rokhlin et al. 2008)	Unknown
miR-34c-5p	↑↓	Modulates PTX resistance in lung cancer cells (Catuogno et al. 2013)	BMF
	↑	Sensitizes gastric cancer cells to PTX (Wu et al. 2013b)	MAPT
miR-93	↑↓	Modulates CDDP resistance in ovarian cancer cells (Fu et al. 2012)	PTEN
	↑↓	Modulates TGFβ resistance in gastric cancer cells (Petrocca et al. 2008)	E2F1
miR-96	↑	Sensitizes osteosarcoma, cervical and ovarian cancer cells to CDDP (Wang et al. 2012b)	RAD51, REV1
miR-98	↑	Sensitizes lung cancer and cells to CDDP (Xiang et al. 2013a)	HMGA2
	↑	Confers resistance to CDDP and DOX in HNSCC cells (Hebert et al. 2007)	HMGA2
miR-100	↑	Sensitizes lung cancer cells to DTX (Feng et al. 2012a)	PLK1
	↑	Sensitizes ovarian cancer cells to everolimus (Nagaraja et al. 2010)	mTOR

(continued)

Table 11.2 (continued)

MicroRNA	Manipulation of expression carried out[a]	Obtained result	Validated direct target
miR-101	↑	Sensitizes breast cancer cells to TAM (Frankel et al. 2011)	ATG4D, RAB5A, STMN1
	↑	Sensitizes colon cancer cells to ABT-263 (Lam et al. 2010)	MCL1
	↑	Sensitizes HCC cells to curcumin, DOX and VP-16 (Su et al. 2009)	MCL1
miR-103	↑	Sensitizes lung cancer cells to gefitinib (Garofalo et al. 2012b)	PKCε
miR-106a	↓	Inhibition sensitizes ovarian cancer cells to PTX (Huh et al. 2013)	BCL10, CASP7
miR-106b	↑↓	Modulates TGFβ resistance in gastric cancer cells (Petrocca et al. 2008)	E2F1
	↑	Sensitizes HMEC to DOX (Ivanovska et al. 2008)	CDKN1A
miR-122	↑	Sensitizes HCC cells to DOX and VCR (Xu et al. 2011b)	Unknown
	↑	Sensitizes HCC cells to DOX (Fornari et al. 2009)	CCNG1
miR-125a/b	↑	Sensitizes colon cancer cells to PTX (Chen et al. 2013a)	Unknown
miR-125b	↑↓	Modulates PTX resistance in breast cancer cells (Zhou et al. 2010)	BAK1
	↑↓	Modulates CDDP resistance in ovarian cancer cells (Kong et al. 2011)	BAK1
	↑↓	Modulates gemcitabine and PTX resistance in breast cancer cells (Liu et al. 2013b)	BAK1
	↑↓	Modulates 5-FU resistance in breast cancer cells (Wang et al. 2012a)	E2F3
	↑	Confers resistance to DOX in leukemia cells (Zhang et al. 2011a)	BAK1
miR-125b-2	↓	Inhibition confers resistance to TMZ in glioblastoma cells (Shi et al. 2012a)	Unknown
miR-126	↑↓	Modulates DOX and VCR resistance in lung cancer cells (Zhu et al. 2012c)	VEGFA
	↑	Sensitizes lung cancer cells to gefitinib (Zhong et al. 2010)	Unknown
miR-128	↑	Sensitizes breast cancer cells to DOX (Zhu et al. 2011)	BMI1, ABCC5
miR-128b	↑	Sensitizes ALL cells to DEX (Kotani et al. 2009)	AF4, MLL
miR-128-2	↑	Confers resistance to 5-FU, CDDP and DOX in lung cancer cells (Donzelli et al. 2012)	E2F5
miR-129	↑	Sensitizes colon cancer cells to 5-FU (Karaayvaz et al. 2013)	BCL2

(continued)

Table 11.2 (continued)

MicroRNA	Manipulation of expression carried out[a]	Obtained result	Validated direct target
miR-130a	↑	Sensitizes NSCLC cells to TRAIL (Acunzo et al. 2012)	MET
	↑↓	Modulates CDDP resistance in HCC cells (Xu et al. 2012c)	RUNX3
	↑↓	Modulates CDDP resistance in ovarian cancer cells (Yang et al. 2012c)	Unknown
miR-130b	↑↓	Modulates CDDP and PTX resistance in ovarian cancer cells (Yang et al. 2012b)	CSF1
miR-134	↑↓	Modulates CDDP, DOX and VP-16 resistance in SCLC cells (Guo et al. 2010)	ABCC1
miR-135	↑↓	Modulates PTX resistance in lung cancer cells (Holleman et al. 2011)	APC
	↑↓	Modulates CDDP resistance in lung cancer cells (Zhou et al. 2013)	MCL1
miR-138	↑	Sensitizes leukemia cells to 5-FU, CDDP, DOX and VCR (Zhao et al. 2010)	ABCB1
miR-140	↑↓	Modulates 5-FU resistance and confers resistance to MTX in colon cancer cells (Song et al. 2009a)	HDAC4
miR-141	↑	Confers resistance to CDDP in esophageal cancer cells (Imanaka et al. 2011)	YAP1
	↑↓	Modulates CDDP resistance in ovarian cancer cells (van Jaarsveld et al. 2012)	KEAP1
miR-143	↑	Sensitizes gastric cancer cells to 5-FU (Borralho et al. 2009)	ERK5
	↑	Sensitizes colon cancer cells to DTX (Xu et al. 2011a)	KRAS
	↑	Sensitizes colon cancer cells to L-OHP (Qian et al. 2013)	IGF-IR
miR-144	↑	Sensitizes CML cells to imatinib (Liu et al. 2012b)	Unknown
miR-145	↑	Sensitizes cervical cancer cells to MMC (Shi et al. 2012b)	MUC1, MYC
	↑	Sensitizes glioblastoma cells to CDDP and TMZ (Yang et al. 2012f)	OCT4, SOX2
	↑	Sensitizes lung cancer cells to gefitinib (Zhong et al. 2010)	Unknown
miR148a	↑	Sensitizes prostate cancer cells to PTX (Fujita et al. 2010)	MSK1
	↑	Sensitizes esophageal cancer cells to 5-FU and CDDP (Hummel et al. 2011)	Unknown

(continued)

Table 11.2 (continued)

MicroRNA	Manipulation of expression carried out[a]	Obtained result	Validated direct target
miR-148b	↑	Sensitizes colon cancer cells to ABT-263 (Lam et al. 2010)	MCL1
miR-152	↑	Sensitizes ovarian cancer cells to CDDP (Xiang et al. 2013b)	DNMT1
miR-153	↑	Sensitizes CML cells to ATO (Liu et al. 2012a)	Unknown
	↑	Sensitizes colon cancer cells to ABT-263 (Lam et al. 2010)	MCL1
miR-155	↓	Inhibition sensitizes B-cell lymphoma cells to rituximab (Kim et al. 2012)	Unknown
	↑↓	Modulates CDDP resistance in colon cancer cells (Pu et al. 2012)	Unknown
	↓	Inhibition sensitizes lung cancer cells to CDDP (Zang et al. 2012)	Unknown
	↑↓	Modulates DOX, PTX and VP-16 resistance in breast cancer cells (Kong et al. 2010)	FOXO3A
miR-181a	↑	Sensitizes AML cells to Ara-C (Bai et al. 2012)	BCL2
	↑↓	Modulates DNR resistance in CML cells (Li et al. 2012a)	BCL2
	↓	Inhibition sensitizes B-cell lymphomas to MX and SAHA (Lwin et al. 2010)	BIM
	↑	Sensitizes lung cancer cells to CDDP, L-OHP and carboplatin (Galluzzi et al. 2010)	Unknown
miR-181b	↑	Sensitizes gastric and lung cancer cells to 5-FU, CDDP, DOX, VCR and VP-16 (Zhu et al. 2010)	BCL2
miR-182	↑↓	Modulates DEX resistance in acute T-cell leukemia cells (Yang et al. 2012a)	Unknown
miR-185	↑	Sensitizes ovarian cancer cells to CDDP (Xiang et al. 2013b)	DNMT1
miR-186*	↑↓	Modulates curcumin resistance in lung cancer cells (Zhang et al. 2010b)	Unknown
miR-192	↑	Confers resistance to 5-FU in colon cancer cells (Boni et al. 2010)	TYMS
miR-193a	↑	Sensitizes colon cancer cells to ABT-263 (Lam et al. 2010)	MCL1
miR-193a-3p	↑↓	Modulates 5-FU resistance in HCC cells (Ma et al. 2012)	E2F1, SRSF2
miR-193b*	↑	Confers resistance to carboplatin in ovarian cancer cells (Ziliak et al. 2012)	Unknown

(continued)

Table 11.2 (continued)

MicroRNA	Manipulation of expression carried out[a]	Obtained result	Validated direct target
miR-195	↑↓	Modulates DOX resistance in colon cancer cells (Qu et al. 2013)	BCL-w
	↓	Inhibition sensitizes glioblastoma cells to TMZ (Ujifuku et al. 2010)	Unknown
	↑↓	Modulates 5-FU resistance in HCC cells (Yang et al. 2012e)	BCL-w
miR-199a	↑	Sensitizes ovarian cancer cells to CDDP, DOX and PTX (Cheng et al. 2012)	$CD44^+$
	↓	Inhibition sensitizes cervical cancer cells to CDDP (Lee et al. 2008)	Unknown
miR-199a-3p	↑	Sensitizes HCC cells to DOX (Fornari et al. 2010)	mTOR
miR-199a-5p	↑	Sensitizes HCC cells to CDDP (Xu et al. 2012d)	ATG7
miR-200b	↑	Sensitizes pancreatic cancer cells to gemcitabine (Li et al. 2009c)	Unknown
	↑↓	Modulates CDDP resistance in tongue cancer cells (Sun et al. 2012)	BMI1
	↑	Sensitizes lung cancer cells to DTX (Feng et al. 2012b)	E2F3
	↑↓	Modulates CDDP, DOX, VCR and VP-16 resistance in gastric and lung cancer cells (Zhu et al. 2012a)	BCL2, XIAP
	↑↓	Modulates gemcitabine resistance in cholangiocarcinoma cells (Meng et al. 2006)	PTPN12
	↑	Sensitizes breast cancer cells DOX (Tryndyak et al. 2010)	Unknown
miR-200c	↑	Sensitizes gastric cancer cells to 5-FU, CDDP, DOX and PTX (Chen et al. 2010b)	Unknown
	↑	Sensitizes lung cancer cells to CDDP and cetuximab (Ceppi et al. 2010)	Unknown
	↑	Sensitizes breast cancer cells to epirubicin (Chen et al. 2012b)	ABCB1
	↑	Sensitizes ovarian cancer cells to PTX (Cittelly et al. 2012)	Unknown
	↓	Inhibition sensitizes esophageal cancer cells to CDDP (Hamano et al. 2011)	Unknown
	↑↓	Modulates DOX resistance in breast cancer cells (Kopp et al. 2012)	Unknown
	↑	Sensitizes melanoma cells to CDDP, PLX4720 and U0126 (Liu et al. 2012d)	Unknown
	↑↓	Modulates CDDP, DOX, VCR and VP-16 resistance in gastric and lung cancer cells (Zhu et al. 2012a)	BCL2, XIAP
	↑	Sensitizes breast cancer cells to DOX (Tryndyak et al. 2010)	Unknown
	↑	Sensitizes bladder cancer cells to erlotinib and gefitinib (Adam et al. 2009)	ERRFI-1

(continued)

Table 11.2 (continued)

MicroRNA	Manipulation of expression carried out[a]	Obtained result	Validated direct target
miR-203	↑	Sensitizes p53-mutated colon cancer cells to PTX (Li et al. 2011a)	AKT2
	↓	Inhibition sensitizes breast cancer cells to CDDP (Ru et al. 2011)	SOCS3
	↑	Sensitizes lung cancer cells to gefitinib (Garofalo et al. 2012b)	SRC
miR-205	↑↓	Modulates DTX resistance in prostate cancer cells and sensitizes prostate cancer cells to CDDP (Bhatnagar et al. 2010)	BCL-w
	↑	Sensitizes breast cancer cells to gefitinib and lapatinib (Iorio et al. 2009)	HER3
miR-212	↑↓	Modulates cetuximab resistance in HNSCC cells (Hatakeyama et al. 2010)	HB-EGF
	↑	Sensitizes lung cancer cells to TRAIL (Incoronato et al. 2010)	PED/PEA-15
miR-214	↓	Inhibition sensitizes tongue cancer cells to CDDP (Yu et al. 2010)	Unknown
	↑↓	Modulates CDDP resistance in ovarian cancer cells (Yang et al. 2008a)	PTEN
	↓	Inhibition confers resistance to gefitinib (Wang et al. 2012c)	PTEN
	↑	Sensitizes cervical cancer cells to CDDP (Wang et al. 2013b)	BCL-w
miR-215	↑↓	Modulates MTX resistance in colon cancer cells and confers resistance to TDX in osteosarcoma and colon cancer cells (Song et al. 2010)	DHFR, TYMS
miR-216a	↑	Confers resistance to sorafenib in HCC cells (Xia et al. 2013)	PTEN, SMAD7
miR-217	↑	Confers resistance to sorafenib in HCC cells (Xia et al. 2013)	PTEN, SMAD7
miR-221	↑↓	Modulates TRAIL resistance in NSCLC cells (Garofalo et al. 2008; Garofalo et al. 2009)	CDKN1B/p27^{Kip1}, PTEN, TIMP3
	↑	Confers resistance to TAM in breast cancer cells (Miller et al. 2008)	CDKN1B/p27^{Kip1}
	↑	Confers resistance to fulvestrant in breast cancer cells (Rao et al. 2011)	CDKN1B/p27^{Kip1}, ERα
	↑↓	Modulates gefitinib resistance in lung cancer cells (Garofalo et al. 2012b)	APAF1
	↑↓	Modulates TAM resistance in breast cancer cells (Zhao et al. 2008)	ERα
miR-222	↑↓	Modulates TRAIL resistance in NSCLC cells (Garofalo et al. 2008, 2009)	CDKN1B/p27^{Kip1}, PTEN, TIMP3
	↑	Confers resistance to TAM in breast cancer cells (Miller et al. 2008)	CDKN1B/p27^{Kip1}
	↑	Sensitizes colon cancer cells to L-OHP and VCR (Xu et al. 2012b)	ADAM17

(continued)

Table 11.2 (continued)

MicroRNA	Manipulation of expression carried out[a]	Obtained result	Validated direct target
	↑	Confers resistance to fulvestrant in breast cancer cells (Rao et al. 2011)	CDKN1B/p27^{Kip1}, ERα
	↑↓	Modulates TAM resistance in breast cancer cells (Zhao et al. 2008)	ERα
	↑↓	Modulates gefitinib resistance in lung cancer cells (Garofalo et al. 2012b)	APAF1
	↓	Inhibition sensitizes CLL cells to fludarabine (Ferracin et al. 2010)	Unknown
miR-224	↓	Inhibition confers resistance to MTX in colon cancer cells (Mencia et al. 2011)	CDS2, SLC4A4, HSPC159
miR-296	↓	Inhibition sensitizes esophageal cancer cells to 5-FU, CDDP, DOX and VCR (Hong et al. 2010)	ABCB1
miR-296-3p	↑↓	Modulates imatinib, TMZ and VP-16 resistance in glioblastoma cells (Bai et al. 2013)	EAG1
miR-297	↑↓	Modulates DOX, L-OHP and VCR resistance in colon cancer cells (Xu et al. 2012a)	ABCC2
miR-298	↑↓	Modulates DOX resistance in breast cancer cells (Bao et al. 2012)	ABCB1
miR-301	↓	Inhibition sensitizes breast cancer cells to TAM (Shi et al. 2011)	BBC3, COL2A1, FOXF2, PTEN
miR-302	↓	Inhibition sensitizes HNSCC cells to CDDP (Bourguignon et al. 2012)	Unknown
miR-320c	↑↓	Modulates gemcitabine resistance in pancreatic cancer cells (Iwagami et al. 2013)	SMARCC1
miR-326	↑	Sensitizes breast cancer cells to DOX and VP-16 (Liang et al. 2010)	ABCC1
miR-328	↑↓	Modulates 5-FU and HCPT resistance in colon cancer cells (Xu et al. 2012e)	ABCG2, MMP16
	↑	Sensitizes breast cancer cells to MX (Pan et al. 2009b)	ABCG2
miR-331-5p	↑	Sensitizes leukemia cells to DOX (Feng et al. 2011)	ABCB1
miR-337-3p	↑↓	Modulates PTX resistance and sensitizes NSCLC cells to DTX (Du et al. 2012)	RAP1A, STAT3
miR-342	↑↓	Modulates TAM resistance in breast cancer cells (Cittelly et al. 2010b)	BMP7, GEMIN4
miR-345	↑	Sensitizes breast cancer cells to CDDP (Pogribny et al. 2010)	ABCC1
miR-375	↑↓	Modulates TAM resistance in breast cancer cells (Ward et al. 2013)	MTDH
	↑	Confers resistance to PTX in cervical cancer cells (Shen et al. 2013)	Unknown
miR-376	↑↓	Modulates CDDP resistance in ovarian cancer cells (Ye et al. 2011)	ALK7

(continued)

Table 11.2 (continued)

MicroRNA	Manipulation of expression carried out[a]	Obtained result	Validated direct target
miR-379	↑↓	Modulates CDDP, DOX and VP-16 resistance in SCLC cells (Guo et al. 2010)	Unknown
miR-421	↑↓	Modulates CDDP resistance in nasopharyngeal carcinoma cells (Chen et al. 2013b)	FOXO4
miR-429	↑↓	Modulates CDDP, DOX, VCR and VP-16 resistance in gastric and lung cancer cells (Zhu et al. 2012a)	BCL2, XIAP
miR-451	↑↓	Modulates DOX resistance in breast cancer (Kovalchuk et al. 2008)	ABCB1
	↑	Sensitizes CML cells to imatinib (Liu et al. 2012b)	Unknown
	↑↓	Modulates TAM resistance and sensitizes breast cancer cells to raloxifene and fulvestrant (Bergamaschi and Katzenellenbogen 2012)	14-3-3ζ
	↓	Inhibition sensitizes ovarian cancer cells to vinblastine (Zhu et al. 2008)	Unknown
	↑	Sensitizes lung cancer cells to CDDP (Bian et al. 2011)	Unknown
	↑	Sensitizes colon cancer stem cells to irinotecan (Bitarte et al. 2011)	MIF
	↑	Sensitizes glioblastoma cells to imatinib (Gal et al. 2008)	Unknown
miR-455-3p	↓	Inhibition sensitizes glioblastoma cells to TMZ (Ujifuku et al. 2010)	Unknown
miR-494	↑↓	Modulates TRAIL resistance in lung cancer cells (Romano et al. 2012)	BIM
miR-495	↑↓	Modulates CDDP, DOX and VP-16 resistance in SCLC cells (Guo et al. 2010)	Unknown
miR-497	↑↓	Modulates CDDP, DOX, VCR and VP-16 resistance in gastric and lung cancer cells (Zhu et al. 2012b)	BCL2
miR-504	↑	Confers resistance to VP-16 in osteosarcoma and lung cancer cells (Hu et al. 2010b)	P53
miR-505	↑	Sensitizes breast cancer cells to DOX (Yamamoto et al. 2011)	Unknown
miR-512-3p	↑	Sensitizes HCC cells to PTX (Chen et al. 2010a)	c-FLIP
miR-513a-3p	↑	Sensitizes lung cancer cells to CDDP (Zhang et al. 2012)	GSTP1
miR-519c	↓	Inhibition confers resistance to MX in breast and colon cancer cells (To et al. 2009)	ABCG2
miR-520h	↓	Inhibition confers resistance to MX in breast and colon cancer cells (To et al. 2009)	ABCG2

(continued)

Table 11.2 (continued)

MicroRNA	Manipulation of expression carried out[a]	Obtained result	Validated direct target
miR-558a-5p	↑	Sensitizes colon cancer cells to ABT-263 (Lam et al. 2010)	MCL1
miR-582-5p	↑	Sensitizes colon cancer cells to ABT-263 (Lam et al. 2010)	MCL1
miR-591	↑	Sensitizes ovarian cancer cells to PTX (Huh et al. 2013)	ZEB1
miR-605	↑	Sensitizes colon cancer cells to ABT-263 (Lam et al. 2010)	Unknown
miR-630	↑	Confers resistance to CDDP and carboplatin in lung cancer cells (Galluzzi et al. 2010)	Unknown
miR-661	↑	Sensitizes colon cancer cells to ABT-263 (Lam et al. 2010)	MCL1
miR-663	↑↓	Modulates CTX, DTX and DOX resistance in breast cancer cells (Hu et al. 2013)	HSPG2
miR-744*	↑	Sensitizes colon cancer cells to ABT-263 (Lam et al. 2010)	Unknown
miR-876-3p	↑	Sensitizes colon cancer cells to ABT-263 (Lam et al. 2010)	MCL1
miR-886-3p	↑	Sensitizes colon cancer cells to ABT-263 (Lam et al. 2010)	MCL1
miR-892b	↑	Sensitizes colon cancer cells to ABT-263 (Lam et al. 2010)	MCL1
miR-1915	↑↓	Modulates DOX, L-OHP, MMC and VCR resistance in colon cancer cells (Xu et al. 2013)	BCL2
let-7	↑	Sensitizes pancreatic cancer cells to gemcitabine (Bhutia et al. 2013)	RRM2
let-7a	↑	Sensitizes NSCLC cells to gefitinib (Zhong et al. 2010)	Unknown
	↑	Sensitizes breast cancer cells to PTX (Lv et al. 2012)	Unknown
	↑↓	Modulates DOX, interferon-γ and PTX resistance in HCC cells (Tsang and Kwok 2008)	CASP3
let-7b	↑	Sensitizes breast cancer cells to TAM (Zhao et al. 2011b)	ER-α36
let-7c	↑	Sensitizes HCC cells to sorafenib (Shimizu et al. 2010)	BCL-xL
let-7d	↑↓	Modulates 5-FU, CDDP and PTX resistance in OSCC cells (Chang et al. 2011)	Unknown
let-7g	↑	Sensitizes ovarian cancer cells to PTX and vinblastine (Boyerinas et al. 2012)	IMP-1

(continued)

Table 11.2 (continued)

MicroRNA	Manipulation of expression carried out[a]	Obtained result	Validated direct target
let-7i	↑↓	Modulates CDDP resistance in ovarian cancer cells (Yang et al. 2008b)	Unknown
	↑	Sensitizes ovarian cancer cells to PTX (Liu et al. 2012c)	Unknown
	↑	Sensitizes breast cancer cells to TAM (Zhao et al. 2011b)	ER-α36

[a]Up- (↑) or down-regulation (↓)

miR-10a*, miR-30a*, miR-186*, miR-193b*, miR-744* are the names of a certain microRNA

Abbreviations: *5-FU* 5-fluorouracil, *ABCB1* ATP-binding cassette, sub-family B (MDR/TAP), member 1, *ABCC1* multidrug resistance-associated protein 1, *ABCC2* multidrug resistance-associated protein 2, *ABCC5* multidrug resistance-associated protein 5, *ABCG2* ATP-binding cassette sub-family G member 2, *ABT-263* navitoclax (first-in-class BCL2 family inhibitor), *ADAM17* A disintegrin and metalloprotease-17, *AG1478* EGFR inhibitor, *ALDH1A3* aldehyde dehydrogenase 1 family, member A3, *ALK7* activin receptor-like kinase 7, *AML* acute myeloid leukemia, *APAF1* apoptotic peptidase activating factor 1, *Ara-C* cytarabine, *ASO* antisense oligonucleotides, *ATO* arsenic trioxide, *AZD6244* small molecule inhibitor of the MEK (MAP/ERK kinase), *BAK1* Bcl-2 homologous antagonist/killer, *BBC3* BCL2 binding component 3, *BIM* BCL2-like 11, *BMF* Bcl-2-modifying factor, *BMP7* bone morphogenetic protein 7, *BNIP2* BCL2/adenovirus E1B 19 kDa protein-interacting protein 2, *CCND1* cyclin D1, *CCNG1* cyclin G1, *CDDP* cisplatin, *CDK6* cyclin-dependent kinase 6, *CDKN1A* cyclin-dependent kinase inhibitor 1A, *CDKN1B* cyclin-dependent kinase inhibitor 1B ($p27^{Kip1}$), *CDS2* CDP-diacylglycerol synthase (phosphatidate cytidylyltransferase) 2, *CLL* chronic lymphocytic leukemia, *CML* chronic myeloid leukemia, *COL2A1* collagen, type II, alpha 1, *COX-2* cyclooxygenase-2, *CPT* camptothecin, *CTNNB1* β-catenin, *CTX* cyclophosphamide, *DHFR* dihydrofolate reductase, *DNMT1* DNA methyltransferase 1, *DNR* daunorubicin, *DEX* dexamethasone, *DHFR* dihydrofolate reductase, *DOX* doxorubicin, *DR4* death receptor-4, *DTX* docetaxel, *EAG1* ether-à-go-go 1, *EGFR* epidermal growth factor receptor, *ER* estrogen receptor, *ERRFI-1* ErbB receptor inhibitor-1, *FOXF2* forkhead box F2, *FOXO3A* forkhead box O3, *FOXO4* forkhead box O4, *GEMIN4* component of gems 4, *GSTP1* glutathione S-transferase P1, *HB-EGF* heparin-binding EGF-like growth factor, *HCC* hepatocellular carcinoma, *HCPT* hydroxycamptothecin, *HDAC4* histone deacetylase 4, *HIPK2* homeodomain-interacting protein kinase-2, *HMEC* human mammary epithelial cells, *HMGA2* high-mobility group AT-hook 2, *HNSCC* head and neck squamous cell carcinoma, *HSPC159* galectin-related protein, *HSPG2* heparin sulfate proteoglycan 2, *IGF-IR* insulin-like growth factor-I receptor, *IMP-1* insulin-like growth factor 2 mRNA binding protein 1, *KEAP1* kelch-like ECH-associated protein 1, *LRRFIP1* leucine rich repeat (in FLII) interacting protein 1, *L-OHP* oxaliplatin, *MAGE-A* melanoma antigen family A, *MAPT* microtubule-associated protein tau, *MARCKS* myristoylated alanine-rich protein kinase c substrate, *MIF* macrophage inhibitory factor, *MMC* mitomycin C, *MMP16* matrix metalloproteinase-16, *MSK1* mitogen and stress-activated protein kinase, *MTDH* metadherin, *MTX* methotrexate, *MX* mitoxantrone, *MYBL2* v-Myb myeloblastosis viral oncogene homolog-like2, *NSCLC* non-small cell lung cancer, *OSCC* oral squamous cell carcinoma, *PDCD4* programmed cell death protein 4, *PHLPP2* PH domain and leucine rich repeat protein phosphatase 2, *PKCε* protein kinase C epsilon, *PLK1* serine/threonine-protein kinase, *PLX4720* specific inhibitor of $B\text{-}RAF^{V600E}$, *PPP2R1B* protein phosphatase 2, regulatory subunit A, beta, *PTEN* phosphatase and tensin homolog, *PTPN12* protein tyrosine phosphatase, nonreceptor type 12, *PTX* paclitaxel, *RAB5A* Ras-related protein Rab-5A, *RAP1A* Ras-related protein Rap-1A, *RRM2* ribonucleotide reductase M2, *SAHA* suberoylanilide hydroxamic acid, *SCLC* small cell lung cancer, *SRC* sarcoma viral oncogene homolog, *SFPQ* splicing factor proline and glutamate-rich, *SIRT1* silent mating type information regulation 2 homolog 1, *SLC4A4* Na/bicarbonate cotransporter 1, *SMARCC1* SWI/SNF related matrix associated actin dependent regulator of chromatin subfamily C member 1, *SOCS3* suppressor of cytokine signaling 3; *STAT3* signal transducer and activator of transcription 3, *STMN1* stathmin 1/oncoprotein 18, *STS* staurosporine, *TAM* tamoxifen, *TDX* Tomudex, *TMZ* temozolomide, *TOP2B* DNA topoisomerase 2-beta, *TRAIL* TNF-related apoptosis-inducing ligand, *TWF1* actin-binding protein twinfilin 1, *TYMS* thymidylate synthetase, *U0126* selective inhibitor of MEK1/2 kinase, *VCR* vincristine, *VM-26* teniposide, *VP-16* etoposide, *YAP1* Yes-associated protein 1

cholangiocarcinoma cell lines to gemcitabine (Meng et al. 2006). MiR-21 has been shown to regulate the translation of several mRNAs such as PTEN, Bcl-2 or PDCD4 (programmed cell death 4), all of which play important roles in cellular mechanisms such as cellular growth, apoptosis, invasion and cell cycle. The miR-21 involvement in drug resistance has been demonstrated in several cancer cell types such as breast (Wang et al. 2011d), colon (Valeri et al. 2010), gastric (Xiang et al. 2013a), prostate (Li et al. 2009a), among others, and it has also been shown to modulate resistance to several drugs, from the conventional chemotherapies such as 5-fluorouracil (Tomimaru et al. 2010), cisplatin (Chen et al. 2012d) or doxorubicin (Tao et al. 2011), to targeted compounds/drugs such as AG1478 (Seike et al. 2009), gefitinib (Garofalo et al. 2012b) or trastuzumab (Gong et al. 2011). Although miR-21 expression is most often associated with cancer drug resistance, its expression can modulate the sensitivity of cytotoxic drugs in opposite directions depending on the compound class, indicating that different mechanisms determine toxic and protective effects (Blower et al. 2008). In fact, miR-21 ectopic expression has been associated with chemosensitivity to cisplatin in tongue squamous cell carcinoma cell lines (Yu et al. 2010).

Another study has shown a non-canonical mechanism of mRNA regulation by miR-21. Indeed, it has been demonstrated that miR-21 binds the 3′-UTR region of Bcl-2 mRNA causing an increase in the expression of Bcl-2 protein and consequent gemcitabine resistance in pancreatic cancer cells (Dong et al. 2011). MiR-21 has also been shown to be involved in resistance to doxorubicin and etoposide by modulation of autophagy and inhibition of miR-21 expression increased the sensitivity of leukemia cells to the above mentioned drugs (Seca et al. 2013).

3.2 MiR-34a

The family of miR-34, which includes miR-34a, miR-34b and miR-34c, is transcriptionally activated by p53 (He et al. 2007). This discovery prompted the evaluation of miR-34 expression in cancer cells since p53 is a pivotal tumor suppressor protein often deregulated in cancer. In fact, miR-34a has been shown to be down-regulated in many p53 deficient cell lines and this down-regulation was associated with drug resistance (Fujita et al. 2008; Kojima et al. 2010; Ji et al. 2008). Furthermore, ectopic expression of miR-34a was shown to sensitize prostate cancer cells to camptothecin (Fujita et al. 2008), bladder cancer cells to cisplatin (Vinall et al. 2012) or breast cancer cells to doxorubicin (Li et al. 2012b). MiR-34a has been shown to target several proteins known to be regulated by p53 (Wong et al. 2011).

SIRT1 (a protein deacetylase which is involved in apoptosis), Bcl-2 (an anti-apoptotic protein), cyclin D1 and Cdk6 (involved in cell cycle control) have all been shown to be direct or indirect targets of miR-34a (Vinall et al. 2012; Fujita et al. 2008; Ji et al. 2008). The fact that miR-34a inhibits the expression of these proteins underlies its effect in sensitizing different cancer cell types to various cancer drugs. However, a different study has shown that miR-34a targeted Bcl-2 mRNA, thus decreasing its expression, but that its ectopic expression was associated with increased resistance to docetaxel in breast cancer cells through an unidentified mechanism (Kastl et al. 2012). Two recent papers have described two other targets for miR-34a. Notch1 signaling may contribute to cancer chemoresistance (Gu et al. 2010) and miR-34a over-expression sensitized breast cancer cells to doxorubicin by targeting Notch1 (Li et al. 2012b). MAGE-A, which is aberrantly expressed in many cancers (Kasuga et al. 2008), was also shown to be a target of miR-34a. In addition, miR-34a over-expression sensitized medulloblastoma cells to both cisplatin and mitomycin C through modulation of MAGE-A (Weeraratne et al. 2011).

3.3 *MiR-221/222*

MiR-221/222 are encoded in the X chromosome and share the same seed sequence between them (Garofalo et al. 2012a). They are commonly regarded as "oncomiRs" as they target several tumor suppressor proteins such as Apaf-1, $p27^{Kip1}$, PTEN or TIMP3 (Garofalo et al. 2012a). In fact, it has been shown that by targeting these mRNAs, miR-221/222 could increase resistance to TRAIL and gefitinib in non-small cell lung cancer (NSCLC) cells (Garofalo et al. 2008, 2009, 2012b). MiR-221/222 has also been shown to target and down-regulate ERα expression, thus being partially responsible for resistance to tamoxifen (Zhao et al. 2008). Therefore, down-regulation of miR-221/222 sensitized breast cancer cells to tamoxifen-induced cell growth arrest and apoptosis (Zhao et al. 2008). Another study has shown that $p27^{Kip1}$ is also decreased by miR-221/222 in tamoxifen resistant cells (Miller et al. 2008). Other studies showed that ectopic expression of miR-221/222 conferred resistance to fulvestrant in breast cancer cells through the deregulation of multiple oncogenic signaling pathways that involved β-catenin activation and repression of TGF-β (Rao et al. 2011). Interestingly, miR-222 ectopic expression has been also shown to sensitize HCT-8 and HCT116 resistant gastric cancer cells to oxaliplatin. MiR-222 over-expression decreased ADAM-17 expression, thus decreasing EGFR activation and TGF-α shedding, which in turn reduces P-gp and MRP1 efflux pump expression (Xu et al. 2012b).

3.4 *MiR-451*

MiR-451 is involved in erythropoiesis, being essential for maintenance and/or late-stage maturation of committed erythroid precursors (Dore et al. 2008). Its association with cancer drug resistance was first described in breast cancer cells resistant to doxorubicin which presented a clear up-regulation of P-gp protein, a known drug efflux transporter. Ectopic expression of miR-451 targeted MDR1 mRNA and thus sensitized these cells to doxorubicin (Kovalchuk et al. 2008). MiR-451 has been also shown to sensitize both chronic myeloid leukemia (CML) and glioblastoma cells to imatinib, although without an identified target (Liu et al. 2012b; Gal et al. 2008). Moreover, miR-451 has been shown to be involved in the inhibition of colon cancer stem cells (CSCs) (Bitarte et al. 2011). By binding the migration inhibitory factor (MIF) mRNA and hampering its translation, as well as indirectly controlling COX-2 expression, miR-451 regulates the Wnt signaling activity [which defines colon CSCs (Vermeulen et al. 2010)] sensitizing colon CSCs to irinotecan (Bitarte et al. 2011). Another study described 14-3-3ζ [factor that binds and stabilizes key proteins involved in signal transduction and cell cycle regulation (Zannis-Hadjopoulos et al. 2008)] as a direct target of miR-451 that is involved in tamoxifen resistance in breast cancer cells (Bergamaschi and Katzenellenbogen 2012). Additionally, it has been shown that miR-451 was down-regulated in NSCLC tissue, compared to non-cancerous adjacent tissue, and that ectopic expression of miR-451 sensitized lung cancer cells to cisplatin (Bian et al. 2011).

4 Regulation of Drug Response Pathways by MicroRNAs

4.1 Regulation of Apoptosis

MiRNAs have been shown to target one or simultaneously various mRNAs involved in the core apoptotic signaling, acting both as anti-apoptotic and pro-apoptotic molecules [reviewed in Lima et al. (2011)]. Moreover, various miRNAs which target apoptosis related proteins, particularly of the intrinsic apoptotic pathway, have already been described to be associated with drug response. The anti-apoptotic Bcl-2 is targeted by several miRNAs and this has been shown to modulate the response of different cancer cellular models to various drugs. For example, miR-15a, -15b, -16, -34a, -129, -181b, -200bc/429cluster, -497 and -1915, all of which down-regulate Bcl-2 expression, have been related to the decrease of resistance to tamoxifen in breast cancer cells (Cittelly et al. 2010a) or to other drugs such as cisplatin, doxorubicin, etoposide, docetaxel, vincristine, gemcitabine and 5-fluorouracil in colon, gastric, breast or lung cancer cells (Xia et al. 2008; Ji et al. 2008; Kastl et al. 2012; Karaayvaz et al. 2013; Zhu et al. 2010, 2012a, b; Xu et al. 2013). In addition, miR-181a down-regulation and consequent up-regulation of Bcl-2 was shown to confer resistance to cytarabine in acute myeloid leukemia cells (Bai et al. 2012) while its over-expression sensitized CML cells to daunorubicin by targeting Bcl-2 (Li et al. 2012a). Moreover, miR-21 which has been shown to directly bind to the 3′-UTR of Bcl-2 mRNA, increasing its expression, caused gemcitabine resistance in pancreatic cancer cells (Dong et al. 2011).

Another anti-apoptotic Bcl-2 family member is Mcl-1, which has been found to be directly targeted by miR-101, -148b, -153, -193a, -558a-5p, -582-5p, -661, -876-3p, -886-3p and -892b in a study aiming at identifying miRNAs which are modulators of colon cancer cells sensitivity to ABT-263 (Lam et al. 2010). Another study has also shown the ability of miR-101, found down-regulated in hepatocellular carcinoma (HCC), to sensitize these tumor cells to drug-induced apoptosis (Su et al. 2009). Moreover, and at least in part through the modulation of apoptosis via targeting Mcl-1, miR-135a/b have been suggested to play a role in the development of cisplatin resistance in lung cancer cells (Zhou et al. 2013).

In addition, the down-regulation of Bcl-w (anti-apoptotic Bcl-2 family protein) by miR-195, -205 and -214 has been described to modulate resistance to cisplatin, doxorubicin, docetaxel or 5-fluorouracil in cervical, colon, hepatocellular or prostate cancer cells (Qu et al. 2013; Yang et al. 2012e; Bhatnagar et al. 2010; Wang et al. 2013b). Moreover, inhibition of Bcl-xL (anti-apoptotic Bcl-2 family member) by let-7c has been found to sensitize HCC cells to apoptosis induced by sorafenib (Shimizu et al. 2010).

On the other hand, the pro-apoptotic Bcl-2 antagonist killer 1 (Bak1) is targeted by miR-125b. This has been shown to increase resistance of breast cancer cells to paclitaxel and gemcitabine (Zhou et al. 2010; Liu et al. 2013b), of ovarian cancer cells to cisplatin (Kong et al. 2011) and of leukemia cells to doxorubicin (Zhang et al. 2011a).

Likewise, the regulation of the BH3-only protein, Bim, by miRNAs also plays a role in drug response. Indeed, the blockage of Bim expression by miR-17 in lung cancer cells caused resistance to AZD6244 (a small molecule inhibitor of the MEK pathway) (Dai et al. 2011). Moreover, miR-17-92 cluster, which targets Bim and PTEN, has been shown to confer resistance to doxorubicin and to etoposide in mantle cell lymphoma (Rao et al. 2012). In addition, miR-30b and -30c were shown to modulate the resistance of lung cancer cells to gefitinib (Garofalo et al. 2012b) while miR-494 induced TRAIL resistance in NSCLC, all through a down-regulation of Bim (Romano et al. 2012). Moreover, miR-10b expression (which also down-regulates Bim) was described as a potential indicator of chemosensitivity to 5-fluorouracil-based chemotherapy regimens in colorectal cancer (Nishida et al. 2012).

Other apoptotic proteins are known to be regulated by miRNAs, altering drug response of cancer cells. For example, by targeting caspase-3, let-7a suppressed HCC death induced by doxorubicin, IFN-γ and paclitaxel (Tsang and Kwok 2008). Also, the expression of miR-106a, which targets caspase-7 (and also BCL10) has been shown to be associated with paclitaxel resistance in ovarian cancer cells (Huh et al. 2013). Furthermore, by down-regulating the apoptotic peptidase activating factor-1 (Apaf-1), miR-221 and -222 increased lung cancer cells resistance to gefitinib (Garofalo et al. 2012b).

Finally, the most potent member of the inhibitor of apoptosis family, XIAP, is targeted by miR-200bc/429 cluster (all of which, interestingly, also target Bcl-2) and this seems to be associated with increased sensitivity of gastric and lung cancer cells to cisplatin, doxorubicin, vincristine and etoposide (Zhu et al. 2012a).

4.2 Regulation of p53 Levels

The p53 pathway is one of the major players in the regulation of cellular response to drugs. The p53, considered the guardian of the genome, may initiate DNA repair, cell-cycle arrest, senescence and, importantly, apoptosis [reviewed in Vazquez et al. (2008)]. Several miRNAs have been described to alter p53 expression levels.

In particular, miR-25, -30d and -504 have been shown to directly target p53. Their over-expression reduces p53 protein levels impairing p53 functions, such as p53-mediated transcriptional activation, apoptosis and cell-cycle arrest as well as conferring resistance to etoposide induced cell death (Hu et al. 2010b; Kumar et al. 2011). In addition, miR-375 has also been shown to target and down-regulate p53 expression, affecting the response to ionizing radiation and etoposide treatment. In fact, the expression of miR-375 desensitized AGS cells to ionizing radiation and etoposide, abrogating the cell cycle arrest and apoptosis after DNA damage (Liu et al. 2013a).

There are other miRNAs, such as miR-122 and miR-630, that have been shown to affect p53 stability and activity, leading to alterations in drug sensitivity (Fornari et al. 2009; Galluzzi et al. 2010). Besides the miRNAs that directly or indirectly regulate p53 expression, there are also miRNAs that are regulated by p53 and are involved in drug response, such as miR-34 (Bommer et al. 2007) and miR-519 (Fornari et al. 2012).

4.3 Regulation of Autophagy

The role of autophagy in drug response is complex and not fully understood but several studies have shown that autophagy is involved in chemoresistance.

MiR-30a is one of the miRNAs which is frequently found associated with autophagy. It targets Beclin-1, resulting in decreased autophagic activity. Through the inhibition of Beclin-1 expression, miR-30a regulated autophagy induced by rapamycin in breast and lung cancer cells (Zhu et al. 2009). In addition, in a CML cell line and in CML stem/progenitor cells, over-expression of miR-30a decreased Beclin-1 and ATG5 expression (and subsequently autophagy), enhancing imatinib-induced apoptosis (Yu et al. 2012a, b). Moreover, in a study using cell lines and animal models, cisplatin or taxol treatment were found to increase autophagy but to decrease the levels of miR-30a. Importantly, the inhibition of Beclin-1-mediated tumor cell autophagy which was possible following the increase in miR-30a levels, significantly promoted cisplatin-induced tumor cell apoptosis (Zou et al. 2012).

MiR-101 was identified as potent inhibitor of autophagic flux in MCF-7 breast cancer cells. This miRNA is described to directly target STMN1, RAB5A and ATG4D, blocking autophagy, and to increase the cytotoxicity of 4-hydroxytamoxifen in breast cancer cells (Frankel et al. 2011).

MiR-199a-5p was found to be significantly reduced in HCC patients treated with cisplatin-based chemotherapy and in human HCC cell lines following cisplatin treatment. This cisplatin-induced down-regulation of miR-199a-5p resulted in drug resistance. In addition, the decrease in miR-199a-5p has resulted in enhanced autophagy activation by targeting ATG7 (Xu et al. 2012d).

Finally, through miRNA profiling, miR-375 was found to be up-regulated in DOX-induced senescent K562 (CML) cells. This up-regulated expression was found to be associated with decreased expression of 14-3-3ζ and SP1 genes and with initiation of autophagy (Yang et al. 2012d).

4.4 Regulation of DNA Damage Response

The mechanisms involved in DNA damage response (DDR) are mediated through several proteins including those acting as DNA damage sensors (such as H2AX, BRCA1), as signal transducers [such as ATM (ataxia telangiectasia mutated) and ATR (ATM-Rad3-related)] and as effectors [including DNA repair, cell-cycle checkpoint (G1/S, intra-S and G2/M) and apoptosis mechanisms] [reviewed in Green and Lin (2012)]. Several reports have already described miRNAs which regulate DDR and DNA repair (Hu et al. 2010a; Lal et al. 2009; Pothof et al. 2009; Bhatnagar et al. 2010).

MiR-24 was found to down-regulate H2AX, suppressing DNA repair in terminally differentiated blood cells. MiR-24 up-regulation, in post-replicative cells, reduced H2AX and rendered cells vulnerable to DNA damage (Lal et al. 2009). Another study has shown that miR-24 over-expression was associated with decreased expression of H2AX and decreased DDR upon etoposide treatment in highly differentiated $CD8^+$ T cells (Brunner et al. 2012). Over-expressing miR-24 in breast cancer cells increased sensitivity to cisplatin by direct targeting of H2AX (Srivastava et al. 2011). In addition, in an osteosarcoma cell line, miR-138 was also found to enhance cellular sensitivity to cisplatin and camptothecin by directly targeting H2AX mRNA (Wang et al. 2011c). In another study with cisplatin-resistant human NSCLC cells (A549/DDP), it was shown that the up-regulation of miR-138 increased the sensitivity of these cells to cisplatin. In that same study, the excision repair cross-complementation group 1 (ERCC1) was also found to be negatively regulated by miR-138, with its levels being inversely correlated with the levels of miR-138 in A549/DDP cells (Wang et al. 2011a).

MiR-182 has been described to target BRCA1. In breast cancer cells, miR-182 was shown to mediate BRCA1 down-regulation, affecting DNA repair (Moskwa et al. 2011).

ATM was shown to be directly targeted by miR-18a, their expression being inversely correlated in rectal tumor tissues. The repair of DNA damage induced by etoposide was inhibited by miR-18a, leading to increased apoptosis and accumulation of DNA damage (Wu et al. 2013a). MiR-181a/b was also shown to directly target ATM and to sensitize breast cancer cells to PARP inhibitors (Bisso et al. 2013).

The mismatch repair (MMR) system is involved in DNA damage recognition and repair. MiR-21 was shown to target the core MMR complex, human mutS homolog 2 (hMSH2) and 6 (hMSH6). Over-expression of this miRNA in colorectal cancer cells led to a significantly reduced sensitivity to 5-fluorouracil (Valeri et al. 2010). DNA repair protein RAD51 and the trans-lesion synthesis DNA polymerase REV1 are transcriptionally regulated by miR-96, which directly targets the 3′-UTR of REV1 and the coding region of RAD51. Over-expression of miR-96 increased sensitivity to PARP inhibitor AZD2281 *in vitro* and to cisplatin both *in vitro* and *in vivo* (Wang et al. 2012b). Matrix extracellular phosphoglycoprotein (MEPE) is a cofactor of CHK1 that protects cells from DNA damage-induced killing (Zhang et al. 2010c). MiR-376a was shown to target MEPE, reducing G2 arrest and sensitizing cervical and liver cancer cells to camptothecin and etoposide (Sheng et al. 2013).

It has been shown that c-Myc over-expression causes increased replication origin activity with subsequent DNA damage (Dominguez-Sola et al. 2007). MiR-34c has been shown to target and repress c-Myc upon etoposide induced DNA damage, inhibiting DNA synthesis as well as blocking cells in S-phase, thus preventing replication of damaged DNA (Cannell et al. 2010).

4.5 Regulation of Cancer Stem-Like Cells and Epithelial to Mesenchymal Transition

CSCs are thought to have a major role in tumor proliferation, invasion or metastasis. Epithelial to mesenchymal transition (EMT) is a process that encompasses a transition from a cobblestone phenotype to a fibroblastic phenotype, together with acquisition of increased cell motility and invasion [reviewed in Thiery et al. (2009)]. These cellular alterations are accompanied by molecular adjustments including the decreased expression of E-cadherin and β-catenin and increased expression of vimentin, fibronectin or smooth-muscle actin [reviewed in Thiery and Sleeman (2006)]. Current data suggests that the attainment of the EMT phenotype and the appearance of CSCs share biological alterations and cooperate in the development of cancer metastasis, recurrence, and chemoresistance (Hollier et al. 2009; Mani et al. 2008).

Analysis of miRNAs that regulate stem cell function and their association with the response to chemotherapy in esophageal cancers pointed to miR-200c as being involved in chemoresistance to cisplatin through directly interaction with PPP2R1B, a protein that inhibits tumor invasiveness and is involved in the Akt pathway (Hamano et al. 2011). In addition, miR-145 was shown to directly target the 3′-UTR of Oct4 and Sox2, two known stem cell markers. Over-expression of miR-145 in glioblastoma (GBM) CSCs reduced the expression of these factors as well as of other stemness genes, including Nanog, c-Myc, and of the oncogene Bmi-1. It also reduced MDR1 and ABCG2 proteins, sensitizing these cells to both cisplatin and temozolomide. All these results demonstrate that miR-145 is a key hub in the stemness network of GBM CSCs (Yang et al. 2012f). MiR-125a/b has been shown to directly target and decrease both Mcl-1 and ALDH1A3 (Aldehyde dehydrogenase 1 family member A3), a cancer stem cell marker (Douville et al. 2009). ALDH1-positive cells from the HT29 colon cancer cell line were resistant to paclitaxel and this resistance could be overcome by over-expression of miR-125a/b (Chen et al. 2013a). Intriguingly, miR-125b over-expression was shown to increase CSC population and to confer resistance to taxol and gemcitabine in human breast epithelial cells by targeting Bak1 (Liu et al. 2013b). In another study, docetaxel-resistant pancreatic cells presented various stem cell markers as well as EMT phenotype. Transfection of either miR-200c or miR-205 could revert the phenotype and sensitize these cells through directly targeting ZEB1 and ZEB2 (zinc finger E-box–binding homeobox 1 and 2) (Puhr et al. 2012). ZEB1 was also shown to be directly targeted by miR-591. Moreover, over-expression of miR-591 decreased cell proliferation and migration and sensitized ovarian cancer resistant cells to paclitaxel (Huh et al. 2013). Over-expression of miR-216a/217 was shown to directly target PTEN and SMAD7 and thus activate both TGF-β and PI3K/Akt pathways. This over-expression induced EMT and increased the stem-like cell population conferring resistance to sorafenib in HCC cells (Xia et al. 2013).

E-cadherin, known to be involved in EMT, is directly targeted by miR-23a. In NSCLC cell lines, over-expression of miR-23a led to increased resistance to gefitinib and increased EMT phenotype (Cao et al. 2012). Bmi1 is another protein reported to be involved in EMT and stem cell self-renewal (Song et al. 2009b; Yang et al. 2010). MiR-15b and -200b have been shown to directly target Bmi1, revert EMT phenotype, suppress motility and migration and sensitize tongue squamous cell carcinoma resistant cells to cisplatin (Sun et al. 2012). Other studies have shown the involvement of miR-200 family in the regulation of EMT and drug resistance. In bladder cancer cells, over-expression of miR-200b and -200c inhibited cell migration and sensitized cells to cetuximab by directly binding ZEB1, ZEB2 and ERRFI-1 (Adam et al. 2009). In another study, re-expression of miR-200b in gemcitabine resistant pancreatic cancer cells reverted the EMT phenotype and sensitized cells to gemcitabine (Li et al. 2009c). Additionally, miR-200c was shown to directly target other mesenchymal genes such as class III β-tubulin (TUBB3). Ectopic expression of miR-200c sensitized endometrial Hec50 cancer cells to paclitaxel and vincristine but not to doxorubicin or cisplatin, indicating that targeting TUBB3 is a major mechanism through which miR-200c restores sensitivity to microtubule-binding chemotherapeutic agents (Cochrane et al. 2009). In tamoxifen resistant cells, re-expression of miR-375 sensitized cells to tamoxifen, inhibiting invasion and reverting EMT-like properties. miR-375 was shown to directly target metadherin (MTDH), increasing E-cadherin and ZO-1 expression and decreasing the mesenchymal markers ZEB1 and SNAI2, indicating that the involvement of miR-375 in EMT could, at least in part, be via targeting MTDH (Ward et al. 2013).

4.6 Regulation of Drug Efflux

By regulating genes involved in drug efflux, several miRNAs have been shown to play a role in drug resistance. In particular, several of these miRNAs are directly or indirectly related with MDR1/P-gp expression, one of the main contributors to the MDR phenotype [reviewed in Lopes-Rodrigues et al. (2013)].

Various miRNAs have been described to directly target the 3′-UTR of the MDR1 mRNA. An inverse correlation between the expression of such miRNAs and MDR1 mRNA and P-gp is observed. Indeed, over-expression of miR-451, -331-5p or -27a, which target MDR1/P-gp, has increased sensitivity of breast cancer and leukemia cells to doxorubicin (Feng et al. 2011; Kovalchuk et al. 2008). On the other hand, the down-regulation of miR-298 in breast cancer cells increased P-gp expression and resistance to doxorubicin. Importantly, miR-298 expression has been proposed as potential predictor of doxorubicin chemoresistance in human breast cancer (Bao et al. 2012).

Other miRNAs indirectly regulate MDR1/P-gp expression, by targeting mRNAs involved in MDR1 gene activation. The expression of such miRNAs has also been associated with drug response. Indeed, increased miR-21 caused a decrease in the levels of the tumor suppressor protein PDCD4, which was found to be associated with increased translation of several transcripts, including MDR1. MiR-21 down-regulation enhanced multidrug sensitivity in MCF-7 cells (Bourguignon et al. 2009). Moreover, the down-regulation of miR-130a in cisplatin resistant SKOV3 ovarian cancer cells caused an indirect inhibition of MDR1 mRNA and of P-gp expression and reverted resistance to cisplatin (Yang et al. 2012c). In addition, let-7 affected acquired resistance of ovarian cancer cells to taxanes by targeting IMP-1, resulting in MDR1 mRNA destabilization (Boyerinas et al. 2012).

Interestingly, in ovarian and cervical cancer cells, miR-27a and -451 (previously referred as direct MDR1/P-gp regulators) also seem to be activators of P-gp expression (Zhu et al. 2008). Indeed, antagomirs of miR-27a or -451 decreased MDR1 mRNA and P-gp expression, increasing sensitivity to drugs which are P-gp substrates (Zhu et al. 2008).

In addition to the above referred miRNAs, regulating P-gp expression at a post-transcriptional level, some miRNAs have been proposed to regulate MDR1 expression at a transcriptional level by interfering with its promoter region [reviewed in Toscano-Garibay and Aquino-Jarquin 2012; Lopes-Rodrigues et al. (2013)]. MiR-138 up-regulation in promyelocytic leukemia cells reverted the MDR phenotype by down-regulating P-gp (Zhao et al. 2010). In addition, down-regulation of miR-27a decreased P-gp expression, possibly reverting drug resistance in esophageal squamous cells and gastric cancer cells (Zhang et al. 2010a; Dong et al. 2011).

Furthermore, some miRNAs regulate MDR1 by unknown mechanisms. MiR-200c was down-regulated in breast cancer patients which were non-responders to neoadjuvant chemotherapy (compared to responders) and in doxorubicin-resistant human breast cancer cells MCF-7/ADR (compared to parental MCF-7 cells). MiR-200c restoration in MCF-7 cells reduced MDR1 mRNA and P-gp and increased sensitivity to epirubicin (Chen et al. 2012b). The over-expression of miR-122, a liver specific miRNA frequently down-regulated in HCC, could modulate the sensitivity of HCC cells to drugs through the down-regulation of MDR related genes including MDR1 and MRP (Xu et al. 2011b).

In addition to MDR1/P-gp, the regulation of other efflux proteins by miRNAs has also been associated with drug response/resistance. For example, miR-181a, -328, -519c and -520h target BCRP/ABCG2. This has been shown to modulate colon and breast cancer cells sensitivity to mitoxantrone, 5-fluorouracil or hydroxycamptothecin (Xu et al. 2012e; Jiao et al. 2013; Pan et al. 2009b; To et al. 2009).

MiR-326, which was down-regulated in etoposide resistant MCF-7 cells (MCF-7/VP) and in a panel of advanced breast cancer tissues, negatively correlated with MRP1/ABCC1 levels. MiR-326 over-expression in MCF-7/VP cells sensitized cells to etoposide and doxorubicin (Liang et al. 2010).

Finally, miR-297 targets MRP2/ABCC2 and was down-regulated in a panel of human colorectal carcinoma tissues, as well as in the oxaliplatin resistant cell line HCT116/L-OHP (comparing with its parental cells). The ectopic expression of miR-297 in MDR colorectal carcinoma cells sensitized cells to anticancer drugs *in vitro* and *in vivo* (Xu et al. 2012a).

4.7 Regulation of Drug Metabolism

Drugs are metabolized mainly by enzymes of the cytochrome P450 (CYP) family, which may be post-transcriptionally regulated by miRNAs. CYP members are over-expressed in a wide range of cancers, metabolizing cancer drugs and probably conferring drug resistance (Murray et al. 1997).

The first miRNA found to be associated with a CYP member was miR-27b, which was shown to directly target CYP1B1 in breast cancer cells (Tsuchiya et al. 2006). MiR-27b has also been shown to directly bind and regulate CYP3A4 expression (Pan et al. 2009a). Other CYP members have since been found to be targets of miRNAs. CYP24 is targeted by miR-125b (Komagata et al. 2009), CYP2E1 by miR-378 (Mohri et al. 2010), CYP1A1 by miR-892a (Choi et al. 2012) and CYP2J2 by let-7b (Chen et al. 2012a). Besides directly targeting CYP proteins, miRNAs have been shown to target transcription factors that regulate the expression of a variety of drug-metabolizing enzymes, such as Pregnane X receptor (PXR). For example, miR-148a directly targets PXR affecting the expression of CYP3A4 (Takagi et al. 2008).

5 Non-genetic Acquisition of MicroRNAs via Microvesicles and Exosomes

A possible association between the number of exosomes produced by cancer cells and cancer drug resistance has been described [reviewed in Muralidharan-Chari et al. (2010); Camussi et al. (2010)]. In fact, chemoresistant cancer cells were shown to release more microvesicles than chemosensitive cancer cells (Safaei et al. 2005). In addition, it is known that miRNAs may be included in exosomes and transfer drug resistance to other cells (Jaiswal et al. 2012a). A recent study has shown that P-gp over-expressing drug resistant cells can selectively package some miRNAs into microvesicles, which upon release on the recipient drug sensitive cells can "retemplate" those cells conferring them a MDR phenotype (Jaiswal et al. 2012a, b).

Interestingly, the transfer of miR-21-mediated chemoresistance in pancreatic adenocarcinoma cells has very recently been shown to occur via cell-to-cell contact involving the Systemic RNA Interference-defective-1 Transmembrane Family Member 1 (SIDT1) (Elhassan et al. 2012).

6 MicroRNA Polymorphisms Associated with Drug Resistance

MicroRNA polymorphisms (miRs-polymorphisms or miRSNPs) have been described as a novel class of functional polymorphisms, usually present at (or near) a miRNA binding site of functional genes and in genes involved in miRNA biogenesis. These polymorphisms may affect gene expression by interfering with a miRNA function. Therefore, this can lead to drug resistance/sensitivity depending on the function of the targeted mRNA that presents the polymorphism [reviewed in Mishra et al. (2008); Zhang and Dolan (2010)].

In particular, a miR-200b/200c/429-binding site polymorphism in the 3′-UTR region of AP-2α gene (a transcription factor which regulates several genes involved in cell proliferation and apoptosis) was found to associate with resistance to cisplatin in endometrial cancer cells (Wu et al. 2011). MiR-24 expression targeting the common drug target DHFR (dihydrofolate reductase) has been shown to regulate cellular proliferation. Indeed, miR-24 SNP has been shown to interfere with miR-24 function, to increase DHFR mRNA and protein expression and resistance to methotrexate (Mishra et al. 2007, 2008, 2009; Zhang and Dolan 2010). Moreover, for the let-7 family of miRNAs, which binds to and regulates the expression of KRAS genes, a polymorphism in the let-7 complementary-binding site (lcs6) of the 3′-UTR of KRAS gene has been described. The presence of this polymorphism has been shown to increase KRAS expression *in vitro* (Chin et al. 2008). Interestingly, let-7 lcs6 polymorphism was shown to predict response in wild-type KRAS patients with metastatic colorectal cancer treated with cetuximab monotherapy, and may therefore be considered as a predictive marker of cetuximab efficacy in this context (Zhang et al. 2011b).

MiRNA polymorphic variants have also been hypothesized as useful predictors of clinical outcome in metastatic colorectal cancer patients treated with the combination of 5-fluorouracil and irinotecan (Boni et al. 2011). In addition, studies in cervical cancer have shown that, in the absence of information on human papillomavirus infection, the tumor necrosis factor-alpha-induced protein 8 (TNFAIP8)-rs11064 SNP may affect the affinity of miR-22 binding to the 3′-UTR of TNFAIP8, regulating its expression. This may contribute not only for the risk of developing cervical cancer but also, the increased in TNFAIP8 protein expression may help predict platinum resistance and clinical outcomes in cervical cancer patients (Shi et al. 2013).

7 Conclusion and Perspectives

Although research into the miRNAs area is still quite recent, the evidence for their relevance in cancer drug sensitivity and drug resistance is overwhelming. Care will need to be taken when trying to understand all the evidence that is being put together concerning the role of miRNAs in drug resistance, since this evidence is drawn from various tumor models, different experimental conditions and testing various drugs and different concentrations. Moreover, the fact that miRNAs are multi-target and may regulate the expression of various proteins, either directly or indirectly, adds another layer of complexity into this drug resistance mechanism. Furthermore, miRNAs may operate through the canonical or the non-canonical mechanism of regulation of gene expression, i.e. they may down-regulate or up-regulate the target mRNAs expression, which makes this complex system very difficult to unravel. Nevertheless, some miRNAs such as miR-21 have been shown to be responsible for drug resistance in most of the studies performed to date. This indicates that the pathways targeted by such miRNAs very often lead to a drug resistant phenotype, irrespective of the particularly targeted mRNA in any specific study. This opens up new avenues to the discovery of miRNAs which may be considered good biomarkers for drug resistance and/or therapeutic targets to overcome drug resistance. MiRNAs are a cutting edge area of research and hopefully in the near future further multidisciplinary and interdisciplinary studies will emerge to help develop this particular area of cancer drug resistance research.

References

Acunzo M, Visone R, Romano G, Veronese A, Lovat F, Palmieri D et al (2012) miR-130a targets MET and induces TRAIL-sensitivity in NSCLC by downregulating miR-221 and 222. Oncogene 31(5):634–642

Adam L, Zhong M, Choi W, Qi W, Nicoloso M, Arora A et al (2009) miR-200 expression regulates epithelial-to-mesenchymal transition in bladder cancer cells and reverses resistance to epidermal growth factor receptor therapy. Clin Cancer Res 15(16):5060–5072

Akao Y, Noguchi S, Iio A, Kojima K, Takagi T, Naoe T (2011) Dysregulation of microRNA-34a expression causes drug-resistance to 5-FU in human colon cancer DLD-1 cells. Cancer Lett 300(2):197–204

Akhdar H, Legendre C, Aninat C, More F (2012) Anticancer drug metabolism: chemotherapy resistance and new therapeutic approaches. In: Paxton DJ (ed) Topics on drug metabolism. InTech, Rijeka, pp 137–171

Bai H, Xu R, Cao Z, Wei D, Wang C (2011) Involvement of miR-21 in resistance to daunorubicin by regulating PTEN expression in the leukaemia K562 cell line. FEBS Lett 585(2):402–408

Bai H, Cao Z, Deng C, Zhou L, Wang C (2012) miR-181a sensitizes resistant leukaemia HL-60/Ara-C cells to Ara-C by inducing apoptosis. J Cancer Res Clin Oncol 138(4):595–602

Bai Y, Liao H, Liu T, Zeng X, Xiao F, Luo L et al (2013) MiR-296-3p regulates cell growth and multi-drug resistance of human glioblastoma by targeting ether-a-go-go (EAG1). Eur J Cancer 49(3):710–724

Bao L, Hazari S, Mehra S, Kaushal D, Moroz K, Dash S (2012) Increased expression of P-glycoprotein and doxorubicin chemoresistance of metastatic breast cancer is regulated by miR-298. Am J Pathol 180(6):2490–2503

Bergamaschi A, Katzenellenbogen BS (2012) Tamoxifen downregulation of miR-451 increases 14-3-3zeta and promotes breast cancer cell survival and endocrine resistance. Oncogene 31(1):39–47

Bhatnagar N, Li X, Padi SK, Zhang Q, Tang MS, Guo B (2010) Downregulation of miR-205 and miR-31 confers resistance to chemotherapy-induced apoptosis in prostate cancer cells. Cell Death Dis 1:e105

Bhutia YD, Hung SW, Krentz M, Patel D, Lovin D, Manoharan R et al (2013) Differential processing of let-7a precursors influences RRM2 expression and chemosensitivity in pancreatic cancer: role of LIN-28 and SET oncoprotein. PloS One 8(1):e53436

Bian HB, Pan X, Yang JS, Wang ZX, De W (2011) Upregulation of microRNA-451 increases cisplatin sensitivity of non-small cell lung cancer cell line (A549). J Exp Clin Cancer Res 30:20

Bisso A, Faleschini M, Zampa F, Capaci V, De Santa J, Santarpia L et al (2013) Oncogenic miR-181a/b affect the DNA damage response in aggressive breast cancer. Cell Cycle 12(11):1679–1687

Bitarte N, Bandres E, Boni V, Zarate R, Rodriguez J, Gonzalez-Huarriz M et al (2011) MicroRNA-451 is involved in the self-renewal, tumorigenicity, and chemoresistance of colorectal cancer stem cells. Stem Cells 29(11):1661–1671

Blower PE, Verducci JS, Lin SL, Zhou J, Chung JH, Dai ZY et al (2007) MicroRNA expression profiles for the NCI-60 cancer cell panel. Mol Cancer Ther 6(5):1483–1491

Blower PE, Chung JH, Verducci JS, Lin S, Park JK, Dai Z et al (2008) MicroRNAs modulate the chemosensitivity of tumor cells. Mol Cancer Ther 7(1):1–9

Bockhorn J, Dalton R, Nwachukwu C, Huang S, Prat A, Yee K et al (2013) MicroRNA-30c inhibits human breast tumour chemotherapy resistance by regulating TWF1 and IL-11. Nat Commun 4:1393

Bommer GT, Gerin I, Feng Y, Kaczorowski AJ, Kuick R, Love RE et al (2007) p53-mediated activation of miRNA34 candidate tumor-suppressor genes. Curr Biol 17(15):1298–1307

Boni V, Bitarte N, Cristobal I, Zarate R, Rodriguez J, Maiello E et al (2010) miR-192/miR-215 influence 5-fluorouracil resistance through cell cycle-mediated mechanisms complementary to its post-transcriptional thymidilate synthase regulation. Mol Cancer Ther 9(8):2265–2275

Boni V, Zarate R, Villa JC, Bandres E, Gomez MA, Maiello E et al (2011) Role of primary miRNA polymorphic variants in metastatic colon cancer patients treated with 5-fluorouracil and irinotecan. Pharmacogenomics J 11(6):429–436

Borralho PM, Kren BT, Castro RE, da Silva IB, Steer CJ, Rodrigues CM (2009) MicroRNA-143 reduces viability and increases sensitivity to 5-fluorouracil in HCT116 human colorectal cancer cells. FEBS J 276(22):6689–6700

Borst P (2012) Cancer drug pan-resistance: pumps, cancer stem cells, quiescence, epithelial to mesenchymal transition, blocked cell death pathways, persisters or what? Open Biol 2(5):120066

Bourguignon LY, Spevak CC, Wong G, Xia W, Gilad E (2009) Hyaluronan-CD44 interaction with protein kinase C(epsilon) promotes oncogenic signaling by the stem cell marker Nanog and the Production of microRNA-21, leading to down-regulation of the tumor suppressor protein PDCD4, anti-apoptosis, and chemotherapy resistance in breast tumor cells. J Biol Chem 284(39):26533–26546

Bourguignon LY, Wong G, Earle C, Chen L (2012) Hyaluronan-CD44v3 interaction with Oct4-Sox2-Nanog promotes miR-302 expression leading to self-renewal, clonal formation, and cisplatin resistance in cancer stem cells from head and neck squamous cell carcinoma. J Biol Chem 287(39):32800–32824

Boyd MR, Pauli KD (1995) Some practical considerations and applications of the National-Cancer-Institute in-vitro anticancer drug discovery screen. Drug Dev Res 34(2):91–109

Boyerinas B, Park SM, Murmann AE, Gwin K, Montag AG, Zillhardt M et al (2012) Let-7 modulates acquired resistance of ovarian cancer to Taxanes via IMP-1-mediated stabilization of multidrug resistance 1. Int J Cancer/Journal International du Cancer 130(8):1787–1797

Broxterman HJ, Gotink KJ, Verheul HM (2009) Understanding the causes of multidrug resistance in cancer: a comparison of doxorubicin and sunitinib. Drug Resist Update 12(4–5):114–126

Brunner S, Herndler-Brandstetter D, Arnold CR, Wiegers GJ, Villunger A, Hackl M et al (2012) Upregulation of miR-24 is associated with a decreased DNA damage response upon etoposide treatment in highly differentiated CD8(+) T cells sensitizing them to apoptotic cell death. Aging Cell 11(4):579–587

Camussi G, Deregibus MC, Bruno S, Cantaluppi V, Biancone L (2010) Exosomes/microvesicles as a mechanism of cell-to-cell communication. Kidney Int 78(9):838–848

Cannell IG, Kong YW, Johnston SJ, Chen ML, Collins HM, Dobbyn HC et al (2010) p38 MAPK/MK2-mediated induction of miR-34c following DNA damage prevents Myc-dependent DNA replication. Proc Natl Acad Sci U S A 107(12):5375–5380
Cao M, Seike M, Soeno C, Mizutani H, Kitamura K, Minegishi Y et al (2012) MiR-23a regulates TGF-beta-induced epithelial-mesenchymal transition by targeting E-cadherin in lung cancer cells. Int J Oncol 41(3):869–875
Catuogno S, Cerchia L, Romano G, Pognonec P, Condorelli G, de Franciscis V (2013) miR-34c may protect lung cancer cells from paclitaxel-induced apoptosis. Oncogene 32(3):341–351
Ceppi P, Mudduluru G, Kumarswamy R, Rapa I, Scagliotti GV, Papotti M et al (2010) Loss of miR-200c expression induces an aggressive, invasive, and chemoresistant phenotype in non-small cell lung cancer. Mol Cancer Res 8(9):1207–1216
Chai H, Liu M, Tian R, Li X, Tang H (2011) miR-20a targets BNIP2 and contributes chemotherapeutic resistance in colorectal adenocarcinoma SW480 and SW620 cell lines. Acta Biochim Biophys Sin 43(3):217–225
Chang CJ, Hsu CC, Chang CH, Tsai LL, Chang YC, Lu SW et al (2011) Let-7d functions as novel regulator of epithelial-mesenchymal transition and chemoresistant property in oral cancer. Oncol Rep 26(4):1003–1010
Chen F, Zhu HH, Zhou LF, Wu SS, Wang J, Chen Z (2010a) Inhibition of c-FLIP expression by miR-512-3p contributes to taxol-induced apoptosis in hepatocellular carcinoma cells. Oncol Rep 23(5):1457–1462
Chen Y, Zuo J, Liu Y, Gao H, Liu W (2010b) Inhibitory effects of miRNA-200c on chemotherapy-resistance and cell proliferation of gastric cancer SGC7901/DDP cells. Chin J Cancer 29(12):1006–1011
Chen F, Chen C, Yang S, Gong W, Wang Y, Cianflone K et al (2012a) Let-7b inhibits human cancer phenotype by targeting cytochrome P450 epoxygenase 2J2. PloS One 7(6):e39197
Chen J, Tian W, Cai H, He H, Deng Y (2012b) Down-regulation of microRNA-200c is associated with drug resistance in human breast cancer. Med Oncol 29(4):2527–2534
Chen PS, Su JL, Hung MC (2012c) Dysregulation of microRNAs in cancer. J Biomed Sci 19:90
Chen Y, Tsai YH, Fang Y, Tseng SH (2012d) Micro-RNA-21 regulates the sensitivity to cisplatin in human neuroblastoma cells. J Pediatr Surg 47(10):1797–1805
Chen J, Chen Y, Chen Z (2013a) MiR-125a/b regulates the activation of cancer stem cells in paclitaxel-resistant colon cancer. Cancer Invest 31(1):17–23
Chen L, Tang Y, Wang J, Yan Z, Xu R (2013b) miR-421 induces cell proliferation and apoptosis resistance in human nasopharyngeal carcinoma via downregulation of FOXO4. Biochem Biophys Res Commun 435(4):745–750
Cheng W, Liu T, Wan X, Gao Y, Wang H (2012) MicroRNA-199a targets CD44 to suppress the tumorigenicity and multidrug resistance of ovarian cancer-initiating cells. FEBS J 279(11): 2047–2059
Chin LJ, Ratner E, Leng S, Zhai R, Nallur S, Babar I et al (2008) A SNP in a let-7 microRNA complementary site in the KRAS 3′ untranslated region increases non-small cell lung cancer risk. Cancer Res 68(20):8535–8540
Choi YM, An S, Lee EM, Kim K, Choi SJ, Kim JS et al (2012) CYP1A1 is a target of miR-892a-mediated post-transcriptional repression. Int J Oncol 41(1):331–336
Cittelly DM, Das PM, Salvo VA, Fonseca JP, Burow ME, Jones FE (2010a) Oncogenic HER2{Delta}16 suppresses miR-15a/16 and deregulates BCL-2 to promote endocrine resistance of breast tumors. Carcinogenesis 31(12):2049–2057
Cittelly DM, Das PM, Spoelstra NS, Edgerton SM, Richer JK, Thor AD et al (2010b) Downregulation of miR-342 is associated with tamoxifen resistant breast tumors. Mol Cancer 9:317
Cittelly DM, Dimitrova I, Howe EN, Cochrane DR, Jean A, Spoelstra NS et al (2012) Restoration of miR-200c to ovarian cancer reduces tumor burden and increases sensitivity to paclitaxel. Mol Cancer Ther 11(12):2556–2565
Cochrane DR, Spoelstra NS, Howe EN, Nordeen SK, Richer JK (2009) MicroRNA-200c mitigates invasiveness and restores sensitivity to microtubule-targeting chemotherapeutic agents. Mol Cancer Ther 8(5):1055–1066

Dai B, Meng J, Peyton M, Girard L, Bornmann WG, Ji L et al (2011) STAT3 mediates resistance to MEK inhibitor through microRNA miR-17. Cancer Res 71(10):3658–3668

Dalgard CL, Gonzalez M, deNiro JE, O'Brien JM (2009) Differential microRNA-34a expression and tumor suppressor function in retinoblastoma cells. Invest Ophthalmol Vis Sci 50(10):4542–4551

Dominguez-Sola D, Ying CY, Grandori C, Ruggiero L, Chen B, Li M et al (2007) Non-transcriptional control of DNA replication by c-Myc. Nature 448(7152):445–451

Dong J, Zhao YP, Zhou L, Zhang TP, Chen G (2011) Bcl-2 upregulation induced by miR-21 via a direct interaction is associated with apoptosis and chemoresistance in MIA PaCa-2 pancreatic cancer cells. Arch Med Res 42(1):8–14

Donzelli S, Fontemaggi G, Fazi F, Di Agostino S, Padula F, Biagioni F et al (2012) MicroRNA-128-2 targets the transcriptional repressor E2F5 enhancing mutant p53 gain of function. Cell Death Differ 19(6):1038–1048

Dore LC, Amigo JD, Dos Santos CO, Zhang Z, Gai X, Tobias JW et al (2008) A GATA-1-regulated microRNA locus essential for erythropoiesis. Proc Natl Acad Sci U S A 105(9):3333–3338

Douville J, Beaulieu R, Balicki D (2009) ALDH1 as a functional marker of cancer stem and progenitor cells. Stem Cells Dev 18(1):17–25

Du L, Subauste MC, DeSevo C, Zhao Z, Baker M, Borkowski R et al (2012) miR-337-3p and its targets STAT3 and RAP1A modulate taxane sensitivity in non-small cell lung cancers. PloS One 7(6):e39167

Elhassan MO, Christie J, Duxbury MS (2012) Homo sapiens systemic RNA interference-defective-1 transmembrane family member 1 (SIDT1) protein mediates contact-dependent small RNA transfer and microRNA-21-driven chemoresistance. J Biol Chem 287(8): 5267–5277

Feng DD, Zhang H, Zhang P, Zheng YS, Zhang XJ, Han BW et al (2011) Down-regulated miR-331-5p and miR-27a are associated with chemotherapy resistance and relapse in leukaemia. J Cell Mol Med 15(10):2164–2175

Feng B, Wang R, Chen LB (2012a) MiR-100 resensitizes docetaxel-resistant human lung adenocarcinoma cells (SPC-A1) to docetaxel by targeting Plk1. Cancer Lett 317(2):184–191

Feng B, Wang R, Song HZ, Chen LB (2012b) MicroRNA-200b reverses chemoresistance of docetaxel-resistant human lung adenocarcinoma cells by targeting E2F3. Cancer 118(13):3365–3376

Ferracin M, Zagatti B, Rizzotto L, Cavazzini F, Veronese A, Ciccone M et al (2010) MicroRNAs involvement in fludarabine refractory chronic lymphocytic leukemia. Mol Cancer 9:123

Flamant S, Ritchie W, Guilhot J, Holst J, Bonnet ML, Chomel JC et al (2010) Micro-RNA response to imatinib mesylate in patients with chronic myeloid leukemia. Haematologica 95(8):1325–1333

Fletcher JI, Haber M, Henderson MJ, Norris MD (2010) ABC transporters in cancer: more than just drug efflux pumps. Nat Rev Cancer 10(2):147–156

Fojo T (2007) Multiple paths to a drug resistance phenotype: mutations, translocations, deletions and amplification of coding genes or promoter regions, epigenetic changes and microRNAs. Drug Resist Update 10(1–2):59–67

Fornari F, Gramantieri L, Giovannini C, Veronese A, Ferracin M, Sabbioni S et al (2009) MiR-122/cyclin G1 interaction modulates p53 activity and affects doxorubicin sensitivity of human hepatocarcinoma cells. Cancer Res 69(14):5761–5767

Fornari F, Milazzo M, Chieco P, Negrini M, Calin GA, Grazi GL et al (2010) MiR-199a-3p regulates mTOR and c-Met to influence the doxorubicin sensitivity of human hepatocarcinoma cells. Cancer Res 70(12):5184–5193

Fornari F, Milazzo M, Chieco P, Negrini M, Marasco E, Capranico G et al (2012) In hepatocellular carcinoma miR-519d is up-regulated by p53 and DNA hypomethylation and targets CDKN1A/p21, PTEN, AKT3 and TIMP2. J Pathol 227(3):275–285

Frankel LB, Wen J, Lees M, Hoyer-Hansen M, Farkas T, Krogh A et al (2011) microRNA-101 is a potent inhibitor of autophagy. EMBO J 30(22):4628–4641

Fu X, Tian J, Zhang L, Chen Y, Hao Q (2012) Involvement of microRNA-93, a new regulator of PTEN/Akt signaling pathway, in regulation of chemotherapeutic drug cisplatin chemosensitivity in ovarian cancer cells. FEBS Lett 586(9):1279–1286

Fujita Y, Kojima K, Hamada N, Ohhashi R, Akao Y, Nozawa Y et al (2008) Effects of miR-34a on cell growth and chemoresistance in prostate cancer PC3 cells. Biochem Biophys Res Commun 377(1):114–119

Fujita Y, Kojima K, Ohhashi R, Hamada N, Nozawa Y, Kitamoto A et al (2010) MiR-148a attenuates paclitaxel resistance of hormone-refractory, drug-resistant prostate cancer PC3 cells by regulating MSK1 expression. J Biol Chem 285(25):19076–19084

Gal H, Pandi G, Kanner AA, Ram Z, Lithwick-Yanai G, Amariglio N et al (2008) MIR-451 and Imatinib mesylate inhibit tumor growth of Glioblastoma stem cells. Biochem Biophys Res Commun 376(1):86–90

Galluzzi L, Morselli E, Vitale I, Kepp O, Senovilla L, Criollo A et al (2010) miR-181a and miR-630 regulate cisplatin-induced cancer cell death. Cancer Res 70(5):1793–1803

Garofalo M, Quintavalle C, Di Leva G, Zanca C, Romano G, Taccioli C et al (2008) MicroRNA signatures of TRAIL resistance in human non-small cell lung cancer. Oncogene 27(27): 3845–3855

Garofalo M, Di Leva G, Romano G, Nuovo G, Suh SS, Ngankeu A et al (2009) miR-221&222 regulate TRAIL resistance and enhance tumorigenicity through PTEN and TIMP3 downregulation. Cancer Cell 16(6):498–509

Garofalo M, Quintavalle C, Romano G, Croce CM, Condorelli G (2012a) miR221/222 in cancer: their role in tumor progression and response to therapy. Curr Mol Med 12(1):27–33

Garofalo M, Romano G, Di Leva G, Nuovo G, Jeon YJ, Ngankeu A et al (2012b) EGFR and MET receptor tyrosine kinase-altered microRNA expression induces tumorigenesis and gefitinib resistance in lung cancers. Nat Med 18(1):74–82

Garraway LA, Janne PA (2012) Circumventing cancer drug resistance in the era of personalized medicine. Cancer Discov 2(3):214–226

Gaur A, Jewell DA, Liang Y, Ridzon D, Moore JH, Chen CF et al (2007) Characterization of microRNA expression levels and their biological correlates in human cancer cell lines. Cancer Res 67(6):2456–2468

Gerlinger M, Swanton C (2010) How Darwinian models inform therapeutic failure initiated by clonal heterogeneity in cancer medicine. Br J Cancer 103(8):1139–1143

Gerlinger M, Rowan AJ, Horswell S, Larkin J, Endesfelder D, Gronroos E et al (2012) Intratumor heterogeneity and branched evolution revealed by multiregion sequencing. New Engl J Med 366(10):883–892

Giovannetti E, Funel N, Peters GJ, Del Chiaro M, Erozenci LA, Vasile E et al (2010) MicroRNA-21 in pancreatic cancer: correlation with clinical outcome and pharmacologic aspects underlying its role in the modulation of gemcitabine activity. Cancer Res 70(11):4528–4538

Gmeiner WH, Reinhold WC, Pommier Y (2010) Genome-wide mRNA and microRNA profiling of the NCI 60 cell-line screen and comparison of FdUMP[10] with fluorouracil, floxuridine, and topoisomerase 1 poisons. Mol Cancer Ther 9(12):3105–3114

Gong C, Yao Y, Wang Y, Liu B, Wu W, Chen J et al (2011) Up-regulation of miR-21 mediates resistance to trastuzumab therapy for breast cancer. J Biol Chem 286(21):19127–19137

Gottesman MM, Fojo T, Bates SE (2002) Multidrug resistance in cancer: role of ATP-dependent transporters. Nat Rev Cancer 2(1):48–58

Green LJ, Lin S-Y (2012) DNA damage response and breast cancer: an overview. In: Aft RL (ed) Targeting new pathways and cell death in breast cancer. InTech, Rijeka, pp 97–113

Gu F, Ma Y, Zhang Z, Zhao J, Kobayashi H, Zhang L et al (2010) Expression of Stat3 and Notch1 is associated with cisplatin resistance in head and neck squamous cell carcinoma. Oncol Rep 23(3):671–676

Guo L, Liu Y, Bai Y, Sun Y, Xiao F, Guo Y (2010) Gene expression profiling of drug-resistant small cell lung cancer cells by combining microRNA and cDNA expression analysis. Eur J Cancer 46(9):1692–1702

Hamano R, Miyata H, Yamasaki M, Kurokawa Y, Hara J, Moon JH et al (2011) Overexpression of miR-200c induces chemoresistance in esophageal cancers mediated through activation of the Akt signaling pathway. Clin Cancer Res 17(9):3029–3038

Hatakeyama H, Cheng H, Wirth P, Counsell A, Marcrom SR, Wood CB et al (2010) Regulation of heparin-binding EGF-like growth factor by miR-212 and acquired cetuximab-resistance in head and neck squamous cell carcinoma. PloS One 5(9):e12702

He X, He L, Hannon GJ (2007) The guardian's little helper: microRNAs in the p53 tumor suppressor network. Cancer Res 67(23):11099–11101

Hebert C, Norris K, Scheper MA, Nikitakis N, Sauk JJ (2007) High mobility group A2 is a target for miRNA-98 in head and neck squamous cell carcinoma. Mol Cancer 6:5

Holleman A, Chung I, Olsen RR, Kwak B, Mizokami A, Saijo N et al (2011) miR-135a contributes to paclitaxel resistance in tumor cells both in vitro and in vivo. Oncogene 30(43):4386–4398

Hollier BG, Evans K, Mani SA (2009) The epithelial-to-mesenchymal transition and cancer stem cells: a coalition against cancer therapies. J Mammary Gland Biol Neoplasia 14(1):29–43

Hong L, Han Y, Zhang H, Li M, Gong T, Sun L et al (2010) The prognostic and chemotherapeutic value of miR-296 in esophageal squamous cell carcinoma. Ann Surg 251(6):1056–1063

Hu H, Du L, Nagabayashi G, Seeger RC, Gatti RA (2010a) ATM is down-regulated by N-Myc-regulated microRNA-421. Proc Natl Acad Sci U S A 107(4):1506–1511

Hu W, Chan CS, Wu R, Zhang C, Sun Y, Song JS et al (2010b) Negative regulation of tumor suppressor p53 by microRNA miR-504. Mol Cell 38(5):689–699

Hu H, Li S, Cui X, Lv X, Jiao Y, Yu F et al (2013) The overexpression of hypomethylated miR-663 induces chemotherapy resistance in human breast cancer cells by targeting heparin sulfate proteoglycan 2 (HSPG2). J Biol Chem 288(16):10973–10985

Huh JH, Kim TH, Kim K, Song JA, Jung YJ, Jeong JY et al (2013) Dysregulation of miR-106a and miR-591 confers paclitaxel resistance to ovarian cancer. Br J Cancer 109(2):452–461

Hummel R, Watson DI, Smith C, Kist J, Michael MZ, Haier J et al (2011) Mir-148a improves response to chemotherapy in sensitive and resistant oesophageal adenocarcinoma and squamous cell carcinoma cells. J Gastrointest Surg 15(3):429–438

Imanaka Y, Tsuchiya S, Sato F, Shimada Y, Shimizu K, Tsujimoto G (2011) MicroRNA-141 confers resistance to cisplatin-induced apoptosis by targeting YAP1 in human esophageal squamous cell carcinoma. J Human Genetics 56(4):270–276

Incoronato M, Garofalo M, Urso L, Romano G, Quintavalle C, Zanca C et al (2010) miR-212 increases tumor necrosis factor-related apoptosis-inducing ligand sensitivity in non-small cell lung cancer by targeting the antiapoptotic protein PED. Cancer Res 70(9):3638–3646

Iorio MV, Casalini P, Piovan C, Di Leva G, Merlo A, Triulzi T et al (2009) microRNA-205 regulates HER3 in human breast cancer. Cancer Res 69(6):2195–2200

Ivanovska I, Ball AS, Diaz RL, Magnus JF, Kibukawa M, Schelter JM et al (2008) MicroRNAs in the miR-106b family regulate p21/CDKN1A and promote cell cycle progression. Mol Cell Biol 28(7):2167–2174

Iwagami Y, Eguchi H, Nagano H, Akita H, Hama N, Wada H et al (2013) miR-320c regulates gemcitabine-resistance in pancreatic cancer via SMARCC1. Br J Cancer 109(2):502–511

Jaiswal R, Gong J, Sambasivam S, Combes V, Mathys JM, Davey R et al (2012a) Microparticle-associated nucleic acids mediate trait dominance in cancer. FASEB J 26(1):420–429

Jaiswal R, Luk F, Gong J, Mathys JM, Grau GE, Bebawy M (2012b) Microparticle conferred microRNA profiles–implications in the transfer and dominance of cancer traits. Mol Cancer 11:37

Ji Q, Hao X, Meng Y, Zhang M, Desano J, Fan D et al (2008) Restoration of tumor suppressor miR-34 inhibits human p53-mutant gastric cancer tumorspheres. BMC Cancer 8:266

Jiao X, Zhao L, Ma M, Bai X, He M, Yan Y et al (2013) MiR-181a enhances drug sensitivity in mitoxantone-resistant breast cancer cells by targeting breast cancer resistance protein (BCRP/ABCG2). Breast Cancer Res Treat 139(3):717–730

Kalinowski FC, Giles KM, Candy PA, Ali A, Ganda C, Epis MR et al (2012) Regulation of epidermal growth factor receptor signaling and erlotinib sensitivity in head and neck cancer cells by miR-7. PloS One 7(10):e47067

Karaayvaz M, Zhai H, Ju J (2013) miR-129 promotes apoptosis and enhances chemosensitivity to 5-fluorouracil in colorectal cancer. Cell Death Dis 4:e659

Kastl L, Brown I, Schofield AC (2012) miRNA-34a is associated with docetaxel resistance in human breast cancer cells. Breast Cancer Res Treat 131(2):445–454

Kasuga C, Nakahara Y, Ueda S, Hawkins C, Taylor MD, Smith CA et al (2008) Expression of MAGE and GAGE genes in medulloblastoma and modulation of resistance to chemotherapy. Laboratory investigation. J Neurosurg Pediatr 1(4):305–313

Kim JH, Kim WS, Park C (2012) Epstein-Barr virus latent membrane protein-1 protects B-cell lymphoma from rituximab-induced apoptosis through miR-155-mediated Akt activation and up-regulation of Mcl-1. Leuk Lymphoma 53(8):1586–1591

Kojima K, Fujita Y, Nozawa Y, Deguchi T, Ito M (2010) MiR-34a attenuates paclitaxel-resistance of hormone-refractory prostate cancer PC3 cells through direct and indirect mechanisms. Prostate 70(14):1501–1512

Komagata S, Nakajima M, Takagi S, Mohri T, Taniya T, Yokoi T (2009) Human CYP24 catalyzing the inactivation of calcitriol is post-transcriptionally regulated by miR-125b. Mol Pharmacol 76(4):702–709

Kong W, He L, Coppola M, Guo J, Esposito NN, Coppola D et al (2010) MicroRNA-155 regulates cell survival, growth, and chemosensitivity by targeting FOXO3a in breast cancer. J Biol Chem 285(23):17869–17879

Kong F, Sun C, Wang Z, Han L, Weng D, Lu Y et al (2011) miR-125b confers resistance of ovarian cancer cells to cisplatin by targeting pro-apoptotic Bcl-2 antagonist killer 1. J Huazhong Univ Sci Technol Med Sci 31(4):543–549

Kopp F, Oak PS, Wagner E, Roidl A (2012) miR-200c sensitizes breast cancer cells to doxorubicin treatment by decreasing TrkB and Bmi1 expression. PloS One 7(11):e50469

Korner C, Keklikoglou I, Bender C, Worner A, Munstermann E, Wiemann S (2013) MicroRNA-31 sensitizes human breast cells to apoptosis by direct targeting of protein kinase C epsilon (PKCepsilon). J Biol Chem 288(12):8750–8761

Kosaka T, Yamaki E, Mogi A, Kuwano H (2011) Mechanisms of resistance to EGFR TKIs and development of a new generation of drugs in non-small-cell lung cancer. J Biomed Biotechnol 2011:165214

Kotani A, Ha D, Hsieh J, Rao PK, Schotte D, den Boer ML et al (2009) miR-128b is a potent glucocorticoid sensitizer in MLL-AF4 acute lymphocytic leukemia cells and exerts cooperative effects with miR-221. Blood 114(19):4169–4178

Kovalchuk O, Filkowski J, Meservy J, Ilnytskyy Y, Tryndyak VP, Chekhun VF et al (2008) Involvement of microRNA-451 in resistance of the MCF-7 breast cancer cells to chemotherapeutic drug doxorubicin. Mol Cancer Ther 7(7):2152–2159

Kumar M, Lu Z, Takwi AA, Chen W, Callander NS, Ramos KS et al (2011) Negative regulation of the tumor suppressor p53 gene by microRNAs. Oncogene 30(7):843–853

Lal A, Pan Y, Navarro F, Dykxhoorn DM, Moreau L, Meire E et al (2009) miR-24-mediated downregulation of H2AX suppresses DNA repair in terminally differentiated blood cells. Nat Struct Mol Biol 16(5):492–498

Lam LT, Lu X, Zhang H, Lesniewski R, Rosenberg S, Semizarov D (2010) A microRNA screen to identify modulators of sensitivity to BCL2 inhibitor ABT-263 (navitoclax). Mol Cancer Ther 9(11):2943–2950

Lee JW, Choi CH, Choi JJ, Park YA, Kim SJ, Hwang SY et al (2008) Altered MicroRNA expression in cervical carcinomas. Clin Cancer Res 14(9):2535–2542

Li T, Li D, Sha J, Sun P, Huang Y (2009a) MicroRNA-21 directly targets MARCKS and promotes apoptosis resistance and invasion in prostate cancer cells. Biochem Biophys Res Commun 383(3):280–285

Li Y, Li W, Yang Y, Lu Y, He C, Hu G et al (2009b) MicroRNA-21 targets LRRFIP1 and contributes to VM-26 resistance in glioblastoma multiforme. Brain Res 1286:13–18

Li Y, VandenBoom TG 2nd, Kong D, Wang Z, Ali S, Philip PA et al (2009c) Up-regulation of miR-200 and let-7 by natural agents leads to the reversal of epithelial-to-mesenchymal transition in gemcitabine-resistant pancreatic cancer cells. Cancer Res 69(16):6704–6712

Li Y, Zhu X, Gu J, Dong D, Yao J, Lin C et al (2010a) Anti-miR-21 oligonucleotide sensitizes leukemic K562 cells to arsenic trioxide by inducing apoptosis. Cancer Sci 101(4):948–954

Li Z, Hu S, Wang J, Cai J, Xiao L, Yu L et al (2010b) MiR-27a modulates MDR1/P-glycoprotein expression by targeting HIPK2 in human ovarian cancer cells. Gynecol Oncol 119(1):125–130

Li J, Chen Y, Zhao J, Kong F, Zhang Y (2011a) miR-203 reverses chemoresistance in p53-mutated colon cancer cells through downregulation of Akt2 expression. Cancer Lett 304(1):52–59

Li J, Zhang Y, Zhao J, Kong F, Chen Y (2011b) Overexpression of miR-22 reverses paclitaxel-induced chemoresistance through activation of PTEN signaling in p53-mutated colon cancer cells. Mol Cell Biochem 357(1–2):31–38

Li H, Hui L, Xu W (2012a) miR-181a sensitizes a multidrug-resistant leukemia cell line K562/A02 to daunorubicin by targeting BCL-2. Acta Biochim Biophys Sin 44(3):269–277

Li XJ, Ji MH, Zhong SL, Zha QB, Xu JJ, Zhao JH et al (2012b) MicroRNA-34a modulates chemosensitivity of breast cancer cells to adriamycin by targeting Notch1. Arch Med Res 43(7):514–521

Liang Z, Wu H, Xia J, Li Y, Zhang Y, Huang K et al (2010) Involvement of miR-326 in chemotherapy resistance of breast cancer through modulating expression of multidrug resistance-associated protein 1. Biochem Pharm 79(6):817–824

Liang Z, Li Y, Huang K, Wagar N, Shim H (2011) Regulation of miR-19 to breast cancer chemoresistance through targeting PTEN. Pharm Res 28(12):3091–3100

Lima RT, Busacca S, Almeida GM, Gaudino G, Fennell DA, Vasconcelos MH (2011) MicroRNA regulation of core apoptosis pathways in cancer. Eur J Cancer 47(2):163–174

Liu L, Chen R, Huang S, Wu Y, Li G, Zhang B et al (2012a) miR-153 sensitized the K562 cells to As2O3-induced apoptosis. Med Oncol 29(1):243–247

Liu L, Wang S, Chen R, Wu Y, Zhang B, Huang S et al (2012b) Myc induced miR-144/451 contributes to the acquired imatinib resistance in chronic myelogenous leukemia cell K562. Biochem Biophys Res Commun 425(2):368–373

Liu N, Zhou C, Zhao J, Chen Y (2012c) Reversal of paclitaxel resistance in epithelial ovarian carcinoma cells by a MUC1 aptamer-let-7i chimera. Cancer Invest 30(8):577–582

Liu S, Tetzlaff MT, Cui R, Xu X (2012d) miR-200c inhibits melanoma progression and drug resistance through down-regulation of BMI-1. Am J Pathol 181(5):1823–1835

Liu Y, Xing R, Zhang X, Dong W, Zhang J, Yan Z et al (2013a) miR-375 targets the p53 gene to regulate cellular response to ionizing radiation and etoposide in gastric cancer cells. DNA Repair 12(9):741–750

Liu Z, Liu H, Desai S, Schmitt DC, Zhou M, Khong HT et al (2013b) miR-125b functions as a key mediator for snail-induced stem cell propagation and chemoresistance. J Biol Chem 288(6):4334–4345

Liu ZL, Wang H, Liu J, Wang ZX (2013c) MicroRNA-21 (miR-21) expression promotes growth, metastasis, and chemo- or radioresistance in non-small cell lung cancer cells by targeting PTEN. Mol Cell Biochem 372(1–2):35–45

Lopes-Rodrigues V, Seca H, Sousa D, Sousa E, Lima RT, Vasconcelos MH (2013) The network of P-glycoprotein and microRNAs interactions. Int J Cancer. doi:10.1002/ijc.28500. [Epub ahead of print]

Lv K, Liu L, Wang L, Yu J, Liu X, Cheng Y et al (2012) Lin28 mediates paclitaxel resistance by modulating p21, Rb and Let-7a miRNA in breast cancer cells. PloS One 7(7):e40008

Lwin T, Lin J, Choi YS, Zhang X, Moscinski LC, Wright KL et al (2010) Follicular dendritic cell-dependent drug resistance of non-Hodgkin lymphoma involves cell adhesion-mediated Bim down-regulation through induction of microRNA-181a. Blood 116(24):5228–5236

Ma K, He Y, Zhang H, Fei Q, Niu D, Wang D et al (2012) DNA methylation-regulated miR-193a-3p dictates resistance of hepatocellular carcinoma to 5-fluorouracil via repression of SRSF2 expression. J Biol Chem 287(8):5639–5649

Mani SA, Guo W, Liao MJ, Eaton EN, Ayyanan A, Zhou AY et al (2008) The epithelial-mesenchymal transition generates cells with properties of stem cells. Cell 133(4):704–715

Mencia N, Selga E, Noe V, Ciudad CJ (2011) Underexpression of miR-224 in methotrexate resistant human colon cancer cells. Biochem Pharm 82(11):1572–1582

Meng F, Henson R, Lang M, Wehbe H, Maheshwari S, Mendell JT et al (2006) Involvement of human micro-RNA in growth and response to chemotherapy in human cholangiocarcinoma cell lines. Gastroenterology 130(7):2113–2129

Miller TE, Ghoshal K, Ramaswamy B, Roy S, Datta J, Shapiro CL et al (2008) MicroRNA-221/222 confers tamoxifen resistance in breast cancer by targeting p27Kip1. J Biol Chem 283(44):29897–29903

Mishra PJ, Humeniuk R, Longo-Sorbello GS, Banerjee D, Bertino JR (2007) A miR-24 microRNA binding-site polymorphism in dihydrofolate reductase gene leads to methotrexate resistance. Proc Natl Acad Sci U S A 104(33):13513–13518

Mishra PJ, Banerjee D, Bertino JR (2008) MiRSNPs or MiR-polymorphisms, new players in microRNA mediated regulation of the cell: introducing microRNA pharmacogenomics. Cell Cycle 7(7):853–858

Mishra PJ, Song B, Wang Y, Humeniuk R, Banerjee D, Merlino G et al (2009) MiR-24 tumor suppressor activity is regulated independent of p53 and through a target site polymorphism. PloS One 4(12):e8445

Mitamura T, Watari H, Wang L, Kanno H, Hassan MK, Miyazaki M et al (2013) Downregulation of miRNA-31 induces taxane resistance in ovarian cancer cells through increase of receptor tyrosine kinase MET. Oncogenesis 2:e40

Mohri T, Nakajima M, Fukami T, Takamiya M, Aoki Y, Yokoi T (2010) Human CYP2E1 is regulated by miR-378. Biochem Pharmacol 79(7):1045–1052

Moskwa P, Buffa FM, Pan Y, Panchakshari R, Gottipati P, Muschel RJ et al (2011) miR-182-mediated downregulation of BRCA1 impacts DNA repair and sensitivity to PARP inhibitors. Mol Cell 41(2):210–220

Muralidharan-Chari V, Clancy JW, Sedgwick A, D'Souza-Schorey C (2010) Microvesicles: mediators of extracellular communication during cancer progression. J Cell Sci 123(Pt 10):1603–1611

Murray GI, Taylor MC, McFadyen MC, McKay JA, Greenlee WF, Burke MD et al (1997) Tumor-specific expression of cytochrome P450 CYP1B1. Cancer Res 57(14):3026–3031

Nagaraja AK, Creighton CJ, Yu Z, Zhu H, Gunaratne PH, Reid JG et al (2010) A link between mir-100 and FRAP1/mTOR in clear cell ovarian cancer. Mol Endocrinol 24(2):447–463

Nakatani F, Ferracin M, Manara MC, Ventura S, Del Monaco V, Ferrari S et al (2012) miR-34a predicts survival of Ewing's sarcoma patients and directly influences cell chemo-sensitivity and malignancy. J Pathol 226(5):796–805

Nasser MW, Datta J, Nuovo G, Kutay H, Motiwala T, Majumder S et al (2008) Down-regulation of micro-RNA-1 (miR-1) in lung cancer. Suppression of tumorigenic property of lung cancer cells and their sensitization to doxorubicin-induced apoptosis by miR-1. J Biol Chem 283(48):33394–33405

Nishida N, Yamashita S, Mimori K, Sudo T, Tanaka F, Shibata K et al (2012) MicroRNA-10b is a prognostic indicator in colorectal cancer and confers resistance to the chemotherapeutic agent 5-fluorouracil in colorectal cancer cells. Ann Surg Oncol 19(9):3065–3071

Pan YZ, Gao W, Yu AM (2009a) MicroRNAs regulate CYP3A4 expression via direct and indirect targeting. Drug Metab Dispos 37(10):2112–2117

Pan YZ, Morris ME, Yu AM (2009b) MicroRNA-328 negatively regulates the expression of breast cancer resistance protein (BCRP/ABCG2) in human cancer cells. Mol Pharmacol 75(6): 1374–1379

Petrocca F, Visone R, Onelli MR, Shah MH, Nicoloso MS, de Martino I et al (2008) E2F1-regulated microRNAs impair TGFbeta-dependent cell-cycle arrest and apoptosis in gastric cancer. Cancer Cell 13(3):272–286

Pogribny IP, Filkowski JN, Tryndyak VP, Golubov A, Shpyleva SI, Kovalchuk O (2010) Alterations of microRNAs and their targets are associated with acquired resistance of MCF-7 breast cancer cells to cisplatin. Int J Cancer (Journal International du Cancer) 127(8):1785–1794

Pothof J, Verkaik NS, van IJcken W, Wiemer EA, Ta VT, van der Horst GT et al (2009) MicroRNA-mediated gene silencing modulates the UV-induced DNA-damage response. EMBO J 28(14):2090–2099

Pu J, Bai D, Yang X, Lu X, Xu L, Lu J (2012) Adrenaline promotes cell proliferation and increases chemoresistance in colon cancer HT29 cells through induction of miR-155. Biochem Biophys Res Commun 428(2):210–215

Puhr M, Hoefer J, Schafer G, Erb HH, Oh SJ, Klocker H et al (2012) Epithelial-to-mesenchymal transition leads to docetaxel resistance in prostate cancer and is mediated by reduced expression of miR-200c and miR-205. Am J Pathol 181(6):2188–2201

Qian X, Yu J, Yin Y, He J, Wang L, Li Q et al (2013) MicroRNA-143 inhibits tumor growth and angiogenesis and sensitizes chemosensitivity to oxaliplatin in colorectal cancers. Cell Cycle 12(9):1385–1394

Qu J, Zhao L, Zhang P, Wang J, Xu N, Mi W et al (2013) MicroRNA-195 chemosensitizes colon cancer cells to the chemotherapeutic drug doxorubicin by targeting the first binding site of BCL2L2 mRNA. J Cell Physiol. doi:10.1002/jcp.24366. [Epub ahead of print]

Rao X, Di Leva G, Li M, Fang F, Devlin C, Hartman-Frey C et al (2011) MicroRNA-221/222 confers breast cancer fulvestrant resistance by regulating multiple signaling pathways. Oncogene 30(9):1082–1097

Rao E, Jiang C, Ji M, Huang X, Iqbal J, Lenz G et al (2012) The miRNA-17 approximately 92 cluster mediates chemoresistance and enhances tumor growth in mantle cell lymphoma via PI3K/AKT pathway activation. Leukemia 26(5):1064–1072

Razumilava N, Bronk SF, Smoot RL, Fingas CD, Werneburg NW, Roberts LR et al (2012) miR-25 targets TNF-related apoptosis inducing ligand (TRAIL) death receptor-4 and promotes apoptosis resistance in cholangiocarcinoma. Hepatology 55(2):465–475

Rebucci M, Michiels C (2013) Molecular aspects of cancer cell resistance to chemotherapy. Biochem Pharmacol 85(9):1219–1226

Rokhlin OW, Scheinker VS, Taghiyev AF, Bumcrot D, Glover RA, Cohen MB (2008) MicroRNA-34 mediates AR-dependent p53-induced apoptosis in prostate cancer. Cancer Biol Ther 7(8): 1288–1296

Romano G, Acunzo M, Garofalo M, Di Leva G, Cascione L, Zanca C et al (2012) MiR-494 is regulated by ERK1/2 and modulates TRAIL-induced apoptosis in non-small-cell lung cancer through BIM down-regulation. Proc Natl Acad Sci U S A 109(41):16570–16575

Rossi L, Bonmassar E, Faraoni I (2007) Modification of miR gene expression pattern in human colon cancer cells following exposure to 5-fluorouracil in vitro. Pharmacol Res 56(3): 248–253

Ru P, Steele R, Hsueh EC, Ray RB (2011) Anti-miR-203 upregulates SOCS3 expression in breast cancer cells and enhances cisplatin chemosensitivity. Genes Cancer 2(7):720–727

Safaei R, Larson BJ, Cheng TC, Gibson MA, Otani S, Naerdemann W et al (2005) Abnormal lysosomal trafficking and enhanced exosomal export of cisplatin in drug-resistant human ovarian carcinoma cells. Mol Cancer Ther 4(10):1595–1604

Seca H, Lima RT, Lopes-Rodrigues V, Guimaraes JE, Almeida GM, Vasconcelos MH (2013) Targeting miR-21 induces autophagy and chemosensitivity of leukemia cells. Curr Drug Targets 14(10):1135–1143

Seike M, Goto A, Okano T, Bowman ED, Schetter AJ, Horikawa I et al (2009) MiR-21 is an EGFR-regulated anti-apoptotic factor in lung cancer in never-smokers. Proc Natl Acad Sci U S A 106(29):12085–12090

Sequist LV, Waltman BA, Dias-Santagata D, Digumarthy S, Turke AB, Fidias P et al (2011) Genotypic and histological evolution of lung cancers acquiring resistance to EGFR inhibitors. Sci Transl Med 3(75):75ra26

Shen Y, Wang P, Li Y, Ye F, Wang F, Wan X et al (2013) miR-375 is upregulated in acquired paclitaxel resistance in cervical cancer. Br J Cancer 109(1):92–99

Sheng J, Luo W, Yu F, Gao N, Hu B (2013) MicroRNA-376a sensitizes cells following DNA damage by downregulating MEPE expression. Cancer Biother Radiopharm 28(7):523–529

Shi GH, Ye DW, Yao XD, Zhang SL, Dai B, Zhang HL et al (2010) Involvement of microRNA-21 in mediating chemo-resistance to docetaxel in androgen-independent prostate cancer PC3 cells. Acta Pharmacol Sin 31(7):867–873

Shi W, Gerster K, Alajez NM, Tsang J, Waldron L, Pintilie M et al (2011) MicroRNA-301 mediates proliferation and invasion in human breast cancer. Cancer Res 71(8):2926–2937

Shi L, Zhang S, Feng K, Wu F, Wan Y, Wang Z et al (2012a) MicroRNA-125b-2 confers human glioblastoma stem cells resistance to temozolomide through the mitochondrial pathway of apoptosis. Int J Oncol 40(1):119–129

Shi M, Du L, Liu D, Qian L, Hu M, Yu M et al (2012b) Glucocorticoid regulation of a novel HPV-E6-p53-miR-145 pathway modulates invasion and therapy resistance of cervical cancer cells. J Pathol 228(2):148–157

Shi TY, Cheng X, Yu KD, Sun MH, Shao ZM, Wang MY et al (2013) Functional variants in TNFAIP8 associated with cervical cancer susceptibility and clinical outcomes. Carcinogenesis 34(4):770–778

Shimizu S, Takehara T, Hikita H, Kodama T, Miyagi T, Hosui A et al (2010) The let-7 family of microRNAs inhibits Bcl-xL expression and potentiates sorafenib-induced apoptosis in human hepatocellular carcinoma. J Hepatol 52(5):698–704

Song B, Wang Y, Xi Y, Kudo K, Bruheim S, Botchkina GI et al (2009a) Mechanism of chemoresistance mediated by miR-140 in human osteosarcoma and colon cancer cells. Oncogene 28(46):4065–4074

Song LB, Li J, Liao WT, Feng Y, Yu CP, Hu LJ et al (2009b) The polycomb group protein Bmi-1 represses the tumor suppressor PTEN and induces epithelial-mesenchymal transition in human nasopharyngeal epithelial cells. J Clin Invest 119(12):3626–3636

Song B, Wang Y, Titmus MA, Botchkina G, Formentini A, Kornmann M et al (2010) Molecular mechanism of chemoresistance by miR-215 in osteosarcoma and colon cancer cells. Mol Cancer 9:96

Srivastava N, Manvati S, Srivastava A, Pal R, Kalaiarasan P, Chattopadhyay S et al (2011) miR-24-2 controls H2AFX expression regardless of gene copy number alteration and induces apoptosis by targeting antiapoptotic gene BCL-2: a potential for therapeutic intervention. Breast Cancer Res 13(2):R39

Stegmeier F, Warmuth M, Sellers WR, Dorsch M (2010) Targeted cancer therapies in the twenty-first century: lessons from imatinib. Clin Pharmacol Ther 87(5):543–552

Su H, Yang JR, Xu T, Huang J, Xu L, Yuan Y et al (2009) MicroRNA-101, down-regulated in hepatocellular carcinoma, promotes apoptosis and suppresses tumorigenicity. Cancer Res 69(3):1135–1142

Sun L, Yao Y, Liu B, Lin Z, Lin L, Yang M et al (2012) MiR-200b and miR-15b regulate chemotherapy-induced epithelial-mesenchymal transition in human tongue cancer cells by targeting BMI1. Oncogene 31(4):432–445

Takagi S, Nakajima M, Mohri T, Yokoi T (2008) Post-transcriptional regulation of human pregnane X receptor by micro-RNA affects the expression of cytochrome P450 3A4. J Biol Chem 283(15):9674–9680

Tao J, Lu Q, Wu D, Li P, Xu B, Qing W et al (2011) microRNA-21 modulates cell proliferation and sensitivity to doxorubicin in bladder cancer cells. Oncol Rep 25(6):1721–1729

Thiery JP, Sleeman JP (2006) Complex networks orchestrate epithelial-mesenchymal transitions. Nat Rev Mol Cell Biol 7(2):131–142

Thiery JP, Acloque H, Huang RY, Nieto MA (2009) Epithelial-mesenchymal transitions in development and disease. Cell 139(5):871–890

To KK, Robey RW, Knutsen T, Zhan Z, Ried T, Bates SE (2009) Escape from hsa-miR-519c enables drug-resistant cells to maintain high expression of ABCG2. Mol Cancer Ther 8(10):2959–2968

Tomimaru Y, Eguchi H, Nagano H, Wada H, Tomokuni A, Kobayashi S et al (2010) MicroRNA-21 induces resistance to the anti-tumour effect of interferon-alpha/5-fluorouracil in hepatocellular carcinoma cells. Br J Cancer 103(10):1617–1626

Toscano-Garibay JD, Aquino-Jarquin G (2012) Regulation exerted by miRNAs in the promoter and UTR sequences: MDR1/P-gp expression as a particular case. DNA Cell Biol 31(8):1358–1364

Tryndyak VP, Beland FA, Pogribny IP (2010) E-cadherin transcriptional down-regulation by epigenetic and microRNA-200 family alterations is related to mesenchymal and drug-resistant phenotypes in human breast cancer cells. Int J Cancer 126(11):2575–2583

Tsang WP, Kwok TT (2008) Let-7a microRNA suppresses therapeutics-induced cancer cell death by targeting caspase-3. Apoptosis 13(10):1215–1222

Tsuchiya Y, Nakajima M, Takagi S, Taniya T, Yokoi T (2006) MicroRNA regulates the expression of human cytochrome P450 1B1. Cancer Res 66(18):9090–9098

Ujifuku K, Mitsutake N, Takakura S, Matsuse M, Saenko V, Suzuki K et al (2010) miR-195, miR-455-3p and miR-10a(*) are implicated in acquired temozolomide resistance in glioblastoma multiforme cells. Cancer Lett 296(2):241–248

Valeri N, Gasparini P, Braconi C, Paone A, Lovat F, Fabbri M et al (2010) MicroRNA-21 induces resistance to 5-fluorouracil by down-regulating human DNA MutS homolog 2 (hMSH2). Proc Natl Acad Sci U S A 107(49):21098–21103

van Jaarsveld MT, Helleman J, Boersma AW, van Kuijk PF, van Ijcken WF, Despierre E et al (2012) miR-141 regulates KEAP1 and modulates cisplatin sensitivity in ovarian cancer cells. Oncogene 32(36):4284–4293

Vazquez A, Bond EE, Levine AJ, Bond GL (2008) The genetics of the p53 pathway, apoptosis and cancer therapy. Nat Rev Drug Discov 7(12):979–987

Vermeulen L, De Sousa EMF, van der Heijden M, Cameron K, de Jong JH, Borovski T et al (2010) Wnt activity defines colon cancer stem cells and is regulated by the microenvironment. Nat Cell Biol 12(5):468–476

Vinall RL, Ripoll AZ, Wang S, Pan CX, deVere White RW (2012) MiR-34a chemosensitizes bladder cancer cells to cisplatin treatment regardless of p53-Rb pathway status. Int J Cancer 130(11):2526–2538

Volinia S, Calin GA, Liu CG, Ambs S, Cimmino A, Petrocca F et al (2006) A microRNA expression signature of human solid tumors defines cancer gene targets. Proc Natl Acad Sci U S A 103(7):2257–2261

Wang Q, Zhong M, Liu W, Li J, Huang J, Zheng L (2011a) Alterations of microRNAs in cisplatin-resistant human non-small cell lung cancer cells (A549/DDP). Exp Lung Res 37(7):427–434

Wang X, Li C, Ju S, Wang Y, Wang H, Zhong R (2011b) Myeloma cell adhesion to bone marrow stromal cells confers drug resistance by microRNA-21 up-regulation. Leuk Lymphoma 52(10):1991–1998

Wang Y, Huang JW, Li M, Cavenee WK, Mitchell PS, Zhou X et al (2011c) MicroRNA-138 modulates DNA damage response by repressing histone H2AX expression. Mol Cancer Res 9(8):1100–1111

Wang ZX, Lu BB, Wang H, Cheng ZX, Yin YM (2011d) MicroRNA-21 modulates chemosensitivity of breast cancer cells to doxorubicin by targeting PTEN. Arch Med Res 42(4):281–290

Wang H, Tan G, Dong L, Cheng L, Li K, Wang Z et al (2012a) Circulating MiR-125b as a marker predicting chemoresistance in breast cancer. PloS One 7(4):e34210

Wang Y, Huang JW, Calses P, Kemp CJ, Taniguchi T (2012b) MiR-96 downregulates REV1 and RAD51 to promote cellular sensitivity to cisplatin and PARP inhibition. Cancer Res 72(16):4037–4046

Wang YS, Wang YH, Xia HP, Zhou SW, Schmid-Bindert G, Zhou CC (2012c) MicroRNA-214 regulates the acquired resistance to gefitinib via the PTEN/AKT pathway in EGFR-mutant cell lines. Asian Pac J Cancer Prev 13(1):255–260

Wang F, Li T, Zhang B, Li H, Wu Q, Yang L et al (2013a) MicroRNA-19a/b regulates multidrug resistance in human gastric cancer cells by targeting PTEN. Biochem Biophys Res Commun 434(3):688–694

Wang F, Liu M, Li X, Tang H (2013b) MiR-214 reduces cell survival and enhances cisplatin-induced cytotoxicity via down-regulation of Bcl2l2 in cervical cancer cells. FEBS Lett 587(5):488–495

Ward A, Balwierz A, Zhang JD, Kublbeck M, Pawitan Y, Hielscher T et al (2013) Re-expression of microRNA-375 reverses both tamoxifen resistance and accompanying EMT-like properties in breast cancer. Oncogene 32(9):1173–1182

Weeraratne SD, Amani V, Neiss A, Teider N, Scott DK, Pomeroy SL et al (2011) miR-34a confers chemosensitivity through modulation of MAGE-A and p53 in medulloblastoma. Neuro Oncol 13(2):165–175

Wilting RH, Dannenberg JH (2012) Epigenetic mechanisms in tumorigenesis, tumor cell heterogeneity and drug resistance. Drug Resist Update 15(1–2):21–38

Wong MY, Yu Y, Walsh WR, Yang JL (2011) microRNA-34 family and treatment of cancers with mutant or wild-type p53 (Review). Int J Oncol 38(5):1189–1195

Wu Y, Xiao Y, Ding X, Zhuo Y, Ren P, Zhou C et al (2011) A miR-200b/200c/429-binding site polymorphism in the 3′ untranslated region of the AP-2alpha gene is associated with cisplatin resistance. PloS One 6(12):e29043

Wu CW, Dong YJ, Liang QY, He XQ, Ng SS, Chan FK et al (2013a) MicroRNA-18a attenuates DNA damage repair through suppressing the expression of ataxia telangiectasia mutated in colorectal cancer. PloS One 8(2):e57036

Wu H, Huang M, Lu M, Zhu W, Shu Y, Cao P et al (2013b) Regulation of microtubule-associated protein tau (MAPT) by miR-34c-5p determines the chemosensitivity of gastric cancer to paclitaxel. Cancer Chemother Pharmacol 71(5):1159–1171

Xia L, Zhang D, Du R, Pan Y, Zhao L, Sun S et al (2008) miR-15b and miR-16 modulate multidrug resistance by targeting BCL2 in human gastric cancer cells. Int J Cancer 123(2):372–379

Xia H, Ooi LL, Hui KM (2013) MicroRNA-216a/217-induced epithelial-mesenchymal transition targets PTEN and SMAD7 to promote drug resistance and recurrence of liver cancer. Hepatology 58(2):629–641

Xiang Q, Tang H, Yu J, Yin J, Yang X, Lei X (2013a) MicroRNA-98 sensitizes cisplatin-resistant human lung adenocarcinoma cells by up-regulation of HMGA2. Pharmazie 68(4):274–281

Xiang Y, Ma N, Wang D, Zhang Y, Zhou J, Wu G et al (2013b) MiR-152 and miR-185 co-contribute to ovarian cancer cells cisplatin sensitivity by targeting DNMT1 directly: a novel epigenetic therapy independent of decitabine. Oncogene. doi:10.1038/onc.2012.575. [Epub ahead of print]

Xie Y, Tobin LA, Camps J, Wangsa D, Yang J, Rao M et al (2013) MicroRNA-24 regulates XIAP to reduce the apoptosis threshold in cancer cells. Oncogene 32(19):2442–2451

Xu B, Niu X, Zhang X, Tao J, Wu D, Wang Z et al (2011a) miR-143 decreases prostate cancer cells proliferation and migration and enhances their sensitivity to docetaxel through suppression of KRAS. Mol Cell Biochem 350(1–2):207–213

Xu Y, Xia F, Ma L, Shan J, Shen J, Yang Z et al (2011b) MicroRNA-122 sensitizes HCC cancer cells to adriamycin and vincristine through modulating expression of MDR and inducing cell cycle arrest. Cancer Lett 310(2):160–169

Xu K, Liang X, Shen K, Cui D, Zheng Y, Xu J et al (2012a) miR-297 modulates multidrug resistance in human colorectal carcinoma by down-regulating MRP-2. Biochem J 446(2):291–300

Xu K, Liang X, Shen K, Sun L, Cui D, Zhao Y et al (2012b) MiR-222 modulates multidrug resistance in human colorectal carcinoma by down-regulating ADAM-17. Exp Cell Res 318(17):2168–2177

Xu N, Shen C, Luo Y, Xia L, Xue F, Xia Q et al (2012c) Upregulated miR-130a increases drug resistance by regulating RUNX3 and Wnt signaling in cisplatin-treated HCC cell. Biochem Biophys Res Commun 425(2):468–472

Xu N, Zhang J, Shen C, Luo Y, Xia L, Xue F et al (2012d) Cisplatin-induced downregulation of miR-199a-5p increases drug resistance by activating autophagy in HCC cell. Biochem Biophys Res Commun 423(4):826–831

Xu XT, Xu Q, Tong JL, Zhu MM, Nie F, Chen X et al (2012e) MicroRNA expression profiling identifies miR-328 regulates cancer stem cell-like SP cells in colorectal cancer. Br J Cancer 106(7):1320–1330

Xu K, Liang X, Cui D, Wu Y, Shi W, Liu J (2013) miR-1915 inhibits Bcl-2 to modulate multidrug resistance by increasing drug-sensitivity in human colorectal carcinoma cells. Mol Carcinog 52(1):70–78

Yamamoto Y, Yoshioka Y, Minoura K, Takahashi RU, Takeshita F, Taya T et al (2011) An integrative genomic analysis revealed the relevance of microRNA and gene expression for drug-resistance in human breast cancer cells. Mol Cancer 10:135

Yang H, Kong W, He L, Zhao JJ, O'Donnell JD, Wang J et al (2008a) MicroRNA expression profiling in human ovarian cancer: miR-214 induces cell survival and cisplatin resistance by targeting PTEN. Cancer Res 68(2):425–433

Yang N, Kaur S, Volinia S, Greshock J, Lassus H, Hasegawa K et al (2008b) MicroRNA microarray identifies Let-7i as a novel biomarker and therapeutic target in human epithelial ovarian cancer. Cancer Res 68(24):10307–10314

Yang MH, Hsu DS, Wang HW, Wang HJ, Lan HY, Yang WH et al (2010) Bmi1 is essential in Twist1-induced epithelial-mesenchymal transition. Nat Cell Biol 12(10):982–992

Yang A, Ma J, Wu M, Qin W, Zhao B, Shi Y et al (2012a) Aberrant microRNA-182 expression is associated with glucocorticoid resistance in lymphoblastic malignancies. Leuk Lymphoma 53(12):2465–2473

Yang C, Cai J, Wang Q, Tang H, Cao J, Wu L et al (2012b) Epigenetic silencing of miR-130b in ovarian cancer promotes the development of multidrug resistance by targeting colony-stimulating factor 1. Gynecol Oncol 124(2):325–334

Yang L, Li N, Wang H, Jia X, Wang X, Luo J (2012c) Altered microRNA expression in cisplatin-resistant ovarian cancer cells and upregulation of miR-130a associated with MDR1/P-glycoprotein-mediated drug resistance. Oncol Rep 28(2):592–600

Yang MY, Lin PM, Liu YC, Hsiao HH, Yang WC, Hsu JF et al (2012d) Induction of cellular senescence by doxorubicin is associated with upregulated miR-375 and induction of autophagy in K562 cells. PloS One 7(5):e37205

Yang X, Yin J, Yu J, Xiang Q, Liu Y, Tang S et al (2012e) miRNA-195 sensitizes human hepatocellular carcinoma cells to 5-FU by targeting BCL-w. Oncol Rep 27(1):250–257

Yang YP, Chien Y, Chiou GY, Cherng JY, Wang ML, Lo WL et al (2012f) Inhibition of cancer stem cell-like properties and reduced chemoradioresistance of glioblastoma using microRNA145 with cationic polyurethane-short branch PEI. Biomaterials 33(5):1462–1476

Ye G, Fu G, Cui S, Zhao S, Bernaudo S, Bai Y et al (2011) MicroRNA 376c enhances ovarian cancer cell survival by targeting activin receptor-like kinase 7: implications for chemoresistance. J Cell Sci 124(Pt 3):359–368

Yu ZW, Zhong LP, Ji T, Zhang P, Chen WT, Zhang CP (2010) MicroRNAs contribute to the chemoresistance of cisplatin in tongue squamous cell carcinoma lines. Oral Oncol 46(4):317–322

Yu Y, Cao L, Yang L, Kang R, Lotze M, Tang D (2012a) microRNA 30A promotes autophagy in response to cancer therapy. Autophagy 8(5):853–855

Yu Y, Yang L, Zhao M, Zhu S, Kang R, Vernon P et al (2012b) Targeting microRNA-30a-mediated autophagy enhances imatinib activity against human chronic myeloid leukemia cells. Leukemia 26(8):1752–1760

Zang YS, Zhong YF, Fang Z, Li B, An J (2012) MiR-155 inhibits the sensitivity of lung cancer cells to cisplatin via negative regulation of Apaf-1 expression. Cancer Gene Ther 19(11):773–778

Zannis-Hadjopoulos M, Yahyaoui W, Callejo M (2008) 14-3-3 cruciform-binding proteins as regulators of eukaryotic DNA replication. Trends Biochem Sci 33(1):44–50

Zhang W, Dolan ME (2010) The emerging role of microRNAs in drug responses. Curr Opin Mol Ther 12(6):695–702

Zhang H, Li M, Han Y, Hong L, Gong T, Sun L et al (2010a) Down-regulation of miR-27a might reverse multidrug resistance of esophageal squamous cell carcinoma. Dig Dis Sci 55(9): 2545–2551

Zhang J, Zhang T, Ti X, Shi J, Wu C, Ren X et al (2010b) Curcumin promotes apoptosis in A549/DDP multidrug-resistant human lung adenocarcinoma cells through an miRNA signaling pathway. Biochem Biophys Res Commun 399(1):1–6

Zhang P, Wang H, Rowe PS, Hu B, Wang Y (2010c) MEPE/OF45 as a new target for sensitizing human tumour cells to DNA damage inducers. Br J Cancer 102(5):862–866

Zhang H, Luo XQ, Feng DD, Zhang XJ, Wu J, Zheng YS et al (2011a) Upregulation of microRNA-125b contributes to leukemogenesis and increases drug resistance in pediatric acute promyelocytic leukemia. Mol Cancer 10:108

Zhang W, Winder T, Ning Y, Pohl A, Yang D, Kahn M et al (2011b) A let-7 microRNA-binding site polymorphism in 3′-untranslated region of KRAS gene predicts response in wild-type KRAS patients with metastatic colorectal cancer treated with cetuximab monotherapy. Ann Oncol 22(1):104–109
Zhang X, Zhu J, Xing R, Tie Y, Fu H, Zheng X et al (2012) miR-513a-3p sensitizes human lung adenocarcinoma cells to chemotherapy by targeting GSTP1. Lung Cancer 77(3):488–494
Zhang JX, Qian D, Wang FW, Liao DZ, Wei JH, Tong ZT et al (2013) MicroRNA-29c enhances the sensitivities of human nasopharyngeal carcinoma to cisplatin-based chemotherapy and radiotherapy. Cancer Lett 329(1):91–98
Zhao JJ, Lin J, Yang H, Kong W, He L, Ma X et al (2008) MicroRNA-221/222 negatively regulates estrogen receptor alpha and is associated with tamoxifen resistance in breast cancer. J Biol Chem 283(45):31079–31086
Zhao X, Yang L, Hu J, Ruan J (2010) miR-138 might reverse multidrug resistance of leukemia cells. Leuk Res 34(8):1078–1082
Zhao X, Yang L, Hu J (2011a) Down-regulation of miR-27a might inhibit proliferation and drug resistance of gastric cancer cells. J Exp Clin Cancer Res 30:55
Zhao Y, Deng C, Lu W, Xiao J, Ma D, Guo M et al (2011b) let-7 microRNAs induce tamoxifen sensitivity by downregulation of estrogen receptor alpha signaling in breast cancer. Mol Med 17(11–12):1233–1241
Zhong M, Ma X, Sun C, Chen L (2010) MicroRNAs reduce tumor growth and contribute to enhance cytotoxicity induced by gefitinib in non-small cell lung cancer. Chem Biol Interact 184(3):431–438
Zhou M, Liu Z, Zhao Y, Ding Y, Liu H, Xi Y et al (2010) MicroRNA-125b confers the resistance of breast cancer cells to paclitaxel through suppression of pro-apoptotic Bcl-2 antagonist killer 1 (Bak1) expression. J Biol Chem 285(28):21496–21507
Zhou L, Qiu T, Xu J, Wang T, Wang J, Zhou X et al (2013) miR-135a/b modulate cisplatin resistance of human lung cancer cell line by targeting MCL1. Pathol Oncol Res 19(4):677–683
Zhu H, Wu H, Liu X, Evans BR, Medina DJ, Liu CG et al (2008) Role of MicroRNA miR-27a and miR-451 in the regulation of MDR1/P-glycoprotein expression in human cancer cells. Biochem Pharmacol 76(5):582–588
Zhu H, Wu H, Liu X, Li B, Chen Y, Ren X et al (2009) Regulation of autophagy by a beclin 1-targeted microRNA, miR-30a, in cancer cells. Autophagy 5(6):816–823
Zhu W, Shan X, Wang T, Shu Y, Liu P (2010) miR-181b modulates multidrug resistance by targeting BCL2 in human cancer cell lines. Int J Cancer 127(11):2520–2529
Zhu Y, Yu F, Jiao Y, Feng J, Tang W, Yao H et al (2011) Reduced miR-128 in breast tumor-initiating cells induces chemotherapeutic resistance via Bmi-1 and ABCC5. Clin Cancer Res 17(22):7105–7115
Zhu W, Xu H, Zhu D, Zhi H, Wang T, Wang J et al (2012a) miR-200bc/429 cluster modulates multidrug resistance of human cancer cell lines by targeting BCL2 and XIAP. Cancer Chemother Pharmacol 69(3):723–731
Zhu W, Zhu D, Lu S, Wang T, Wang J, Jiang B et al (2012b) miR-497 modulates multidrug resistance of human cancer cell lines by targeting BCL2. Med Oncol 29(1):384–391
Zhu X, Li H, Long L, Hui L, Chen H, Wang X et al (2012c) miR-126 enhances the sensitivity of non-small cell lung cancer cells to anticancer agents by targeting vascular endothelial growth factor A. Acta Biochim Biophys Sin 44(6):519–526
Ziliak D, Gamazon ER, Lacroix B, Kyung Im H, Wen Y, Huang RS (2012) Genetic variation that predicts platinum sensitivity reveals the role of miR-193b* in chemotherapeutic susceptibility. Mol Cancer Ther 11(9):2054–2061
Zou Z, Wu L, Ding H, Wang Y, Zhang Y, Chen X et al (2012) MicroRNA-30a sensitizes tumor cells to cis-platinum via suppressing beclin 1-mediated autophagy. J Biol Chem 287(6):4148–4156

Zhang W, Winder T, Ning Y, Pohl A, Yang D, Kahn M et al (2011) A let-7 microRNA-binding site polymorphism in 3′-untranslated region of KRAS gene predicts response in wild-type KRAS patients with metastatic colorectal cancer treated with cetuximab monotherapy. [illegible]

Chapter 12
The Biological Roles of MicroRNAs in Cancer Stem Cells

Bin Bao, Asfar S. Azmi, Aamir Ahmad, Yiwei Li, Sanjeev Banerjee, Dejuan Kong, Shadan Ali, and Fazlul H. Sarkar

Contents

B. Bao • A.S. Azmi • A. Ahmad • Y. Li • S. Banerjee • D. Kong
Department of Pathology, Karmanos Cancer Institute, Wayne State University, Detroit, MI, USA

S. Ali
Department of Oncology, Karmanos Cancer Institute, Wayne State University, Detroit, MI, USA

F.H. Sarkar(✉)
Department of Pathology, Karmanos Cancer Institute, Wayne State University, Detroit, MI, USA

Department of Oncology, Karmanos Cancer Institute, Wayne State University, Detroit, MI, USA
e-mail: fsarkar@med.wayne.edu

S. Babashah (ed.), *MicroRNAs: Key Regulators of Oncogenesis*,
DOI 10.1007/978-3-319-03725-7_12,

Abstract The concept of cancer stem cells (CSCs) has great clinical implications because small sub-populations of CSCs have been identified in many different tumors that are associated with poor clinical outcome. Sufficient evidence supports central functions of CSCs in tumorigenesis, due to its distinct high potentials of self-renewal, pluripotent differentiation and apoptosis-resistance, contributing to tumor aggressiveness. Therefore, inhibiting/eliminating CSCs will provide a new effective therapeutic approach for the treatment of aggressive tumors. However, the mechanistic roles of CSCs in tumorigenesis are not well understood. MicroRNAs (miRNAs) have been discovered to act as key regulators of gene expression in tumorigenesis. Aberrant expression of miRNAs has been discovered to be related to worse clinical outcome of many different tumors. Evidence shows that these tumor-related miRNAs have key functions in the regulation of cell cycle/proliferation, migration/invasion, chemo-radiation resistance, and metastasis. Moreover, miRNAs may also exert important functions in modulating CSC characteristics; however, its detailed mechanism(s) has not been fully elucidated. Here, we will summarize the potential role of CSC-related miRNAs in CSC function, and will further define the role of genistein in targeting these CSC-related miRNAs.

Keywords MicroRNA • Cancer stem cells • Therapeutic resistance • Genistein

1 Introduction

The past decade witnessed significant efforts in the areas of cancer stem cells (CSCs) research in the field of cancer. The CSCs documents distinct properties such as greater self-renewal capacity, prolonging life-span, un-limited cell differentiation of multiple cell lineages and proliferation of daughter tumor cells, all of which can contribute to tumor aggressiveness such as therapeutic resistance, tumor recurrence, and metastasis. This strongly suggests that CSCs exhibits critical functional role in the development of cancer and tumor progression. Therefore, targeting molecular signatures of CSCs is believed to be a new and effective therapeutic approach for the treatment and/or prevention of tumor aggressiveness, which would likely lead to a better clinical prognosis of cancer patients. However, the molecular pathogenesis of CSCs and its signaling networks are still in its infancy. Here, we will summarize the potential function of microRNAs (miRNAs) as endogenous mediators of gene transcription in the regulation of CSC characteristics during tumor development and progression. We will also summarize the potential role of genistein, a natural agent rich in soy-based food, in targeting CSC-related miRNAs in cancers.

2 The Concept of Cancer Stem Cell

Although the concept of CSCs or tumor initiating cells (TICs) as special small sub-populations of tumor cells contributing to the initiation and development of cancer has been proposed more than a decade, it still in its infancy. Significant progress in the concept and molecular biology of CSC has not been elucidated until the small sub-population of CSCs was first isolated and characterized from acute myeloid leukemia (AML) patients in 1997 (Bonnet and Dick 1997). Since then, these small sub-populations of CSCs have been identified and substantially characterized from a variety of different malignant diseases such as breast, gastric, prostate, pancreatic, lung, colon, and brain tumors as well as leukemia and melanoma. Similar to normal stem cells, for example, embryonic and adult stem cells, CSCs share common features such as dormancy, long life-span, self-renewal capacity, and the potential of its differentiation into multiple cell lineages, contributing to tumor development and progression. It has been believed that normal stem cells such as mesenchymal stem cells are able of reprogramming into CSC cells because of the aberrations of certain micro-environments such as hypoxia, chronic inflammation, and defective DNA repair systems in the body, leading to tumorigenesis (Bao et al. 2012b).

CSCs has been identified to account for a very small percentage (0.05–1 %) of tumor cells in a tumor tissue, or in a tumor micro-environment, and have the greatest ability of self-renewal and the highest potential of unlimited differentiation capacity into heterogeneous tumor cell populations, consistent with over-expression of stem cell genes, leading to tumor development and progression (Hermann et al. 2010; Ischenko et al. 2010; Lee et al. 2008; Sarkar et al. 2009; Yu et al. 2012). The concept of CSCs has great clinical implications because the small sub-populations of CSCs have been identified in a lot of different tumor tissues and highly associated with poorer clinical outcome such as short disease-free survival time, increased tumor recurrent rate, and remarkable resistance to chemo-radiation therapy (Creighton et al. 2010; Hermann et al. 2010; Lee et al. 2008; Ischenko et al. 2010). This concept also provides a reasonable explanation for clinical observations of why tumor shrinkage alone may not be connected to the disease-free survival rate of cancer patients (Creighton et al. 2010), which is due in part to the presence of these small sub-populations of CSCs after conventional treatment. It has been noted from experimental studies that inhibition of CSC characteristics results in the suppression of tumor development and progression. These findings support that CSC exhibits a critical functional role in therapeutic resistance, tumor recurrence/relapse and metastasis after conventional therapy. Therefore, inhibition or elimination of CSCs would likely become a new and targeted therapeutic approach for the treatment of tumor aggressiveness.

3 The Clinical Implications of Cancer Stem Cells in Tumors

Over the past decade, great numbers of clinical reports have been documented elucidating the role of small sub-populations of CSCs in the involvement of chemo-radiation therapy resistance and tumor metastasis, eventually leading to poorer clinical outcome of patients diagnosed with different malignant diseases such as breast, lung, prostate, pancreatic, liver, colon, gastric, and brain tumors. For example, one early clinical study indicates that the over-expression of CSC surface marker CD133 in colorectal cancer patients are highly associated with poorer overall survival rate (Kojima et al. 2008). A recent clinical report indicates that CD44+/CD24− CSC ratio in breast tumor tissues collected from 1,350 of patients who had breast cancer, accompanied by histological grades, molecular types, and clinical stages is independent factors of clinical outcomes in breast cancer patients, and it was significantly associated with tumor aggressive phenotypes including estrogen receptor, progesterone receptor, and Ki67, a known proliferation index (Liu et al. 2012). Moreover, the sub-populations of CD44+/CD24− CSC cells are highly related to a 5-year disease-free survival rate of breast cancer. The breast cancer patients who had high values of CD44+/CD24− CSC ratio have poor clinical outcome such as higher distant tumor recurrence rate (Liu et al. 2012), consistent with the findings from other investigators (Lee et al. 2011). These data strongly suggest an important implication of CSC sub-populations in tumor development and progression.

4 The Role of Cancer Stem Cells in Treatment Resistance

The contribution of small subpopulations of CSCs in tumor tissues has been associated with poorer clinical outcome in different tumors and the presence of CSCs appears to be associated with treatment resistance, which in part due to its intrinsic properties such as self-renewal capacity, dormancy, and DNA damage repair mechanism. Clinical data have revealed that the sub-populations of CSCs identified by either CD44+/CD24− or ALDH1+ markers are significantly increased in breast cancer patients after chemotherapy. The patients with increased sub-populations of CD44+/CD24− CSCs after chemotherapy have the high values of ki67 proliferation index in post-chemotherapy tumor tissues. The patients with increased sub-populations of ALDH1-positive CSCs after chemotherapy are also associated with estrogen receptor negativity and p53 over-expression in the post-chemotherapy tumor samples. Furthermore, the patients with such an increased CSC sub-population after chemotherapy has shown to have a significantly poor clinical outcome such as shorter disease-free survival period (Lee et al. 2011). The clinical data also documented the identification of small sub-populations of CD133-positive CSCs in tumor tissues of pancreatic cancer patients having high potential of tumorigenesis and chemotherapy resistance. Inhibitions of these CSC sub-populations decreased metastasis of pancreatic tumors without alternation of its potential of tumorigenesis (Hermann et al. 2007). Similarly, the sub-populations of CSCs in human colon tumors have also been identified which attributed to increased resistance to cancer drugs (Dylla et al. 2008; Todaro et al. 2007). These data clearly support that these CSC sub-populations are strongly associated with the resistance to chemotherapy and tumor metastasis, contributing to poorer clinic outcome of cancer patients.

In addition, the sub-populations of CSCs also render tumors the resistance to radiation therapy, potentially through preferential activation of DNA damage repair system, leading to poorer clinical outcome. An early study revealed that the sub-populations of CSC-like (CD133+) cells in glioma are enhanced after radiation exposure in human patient (Bao et al. 2006). One experimental report showed that the CSC-like cells of glioma are significantly resistant to radiation exposure, compared to non-CSC-like (CD133−) tumor cells of glioma, which was consistent with increased activation of radiation-induced DNA damage repair systems in CSC-like cells (Bao et al 2006). A recent study confirms that the sub-populations of CSCs are more radiation-resistant than normal stem cells, potentially due to increased expression of nicotinamide N-methyltransferase (NNMT, which participates in the modulation of DNA repair mechanisms) and it is associated with radiation-resistant marker such as poly (ADP-ribose) polymerase 1 (PARP1), a DNA single strand repair protein, which is considered as a radiation-resistant marker (D'Andrea et al. 2011; D'Andrea 2012). Moreover, radiation has been documented to induce "stemness" by enhancement of the CSC self-renewal capacity and the expression of CSC signature genes such as Sox2 and Oct3/4 in liver cancer cells (Ghisolfi et al. 2012). These findings strongly support the critical role of CSCs in regulating radiation resistance, leading to tumor aggressive phenotype.

5 The Contribution of Cancer Stem Cells in Tumor Aggressive Phenotypes *In Vitro* and *In Vivo*

Although the molecular pathogenesis of CSCs in tumor biology has not been fully elucidated, it is clear though that CSCs possess more aggressive behavior such as pluripotent differentiation, unlimited proliferation, higher angionenic potential, higher migration/invasion potential, and preferential resistance to malicious micro-environments such as conventional therapeutics, low O2, inflammation, and radiation exposure.

Great numbers of *in vitro* and *in vivo* studies have provided solid evidence in support of the critical role of CSCs in tumor development and progression. In 1997, Dick and his group first identified and characterized the small sub-populations of CD34+/CD38− CSCs isolated from the bone marrow of AML patients by flow cytometry-based technique (Bonnet and Dick 1997; Bhatia et al. 1997). The injections of 5,000 of these leukemic CSCs showing high self-renewal capacity produced human leukemia in immunological defective mice (Bonnet and Dick 1997). Another early experimental study revealed that the transplantation of as few as 2,000 CSC-like (CD133+) cells of hepatocellular carcinoma SMMC-7721 cells generates tumor formation in SCID mice. However, transplantation of the same numbers of non-CSC-like (CD133−) tumor cells fails to generate tumor formation in SCID mice. These CSC-like cells also displayed higher clonogenic potential (Yin et al. 2007). Similar findings have been reported in many other different tumors by many investigators (Al-Hajj et al. 2003; Klonisch et al. 2008; Patrawala et al. 2005).

Our published and unpublished data show that cancer cell sphere-enriched or CD44+/CD133+/EpCAM+ sorted CSC-like cells of pancreatic cancer MiaPaCa-2 cells exhibit a significantly increased potentials of cell growth, clonogenicity, migration, and sphere formation, along with increased expression of CSC-related molecular signatures. These cells also show a greater potential of tumor formation in a tumor xenograft model by the transplantation of as few as 5,000 sphere-enriched or CSC triple positive (CD44+/CD133+/EpCAM+) cells of MiaPaCa-2 cells, compared with the injection of 10^6–10^7 of its parental cells within the same time period. These findings clearly suggest that such sub-populations of CSC cells display greater tumor-forming ability. Several animal experimental reports revealed that the sub-populations of CSCs also contribute to resistance to conventional chemotherapeutic drugs and radiation exposure, potentially due to preferential activation of DNA damage repair systems, leading to an increased DNA repair capacity (Dylla et al. 2008; Hermann et al. 2007; Shafee et al. 2008), suggesting a very important role of CSCs in treatment resistance. Therefore, targeting these CSCs would likely provide an effective newer therapeutic strategy for the successful treatment of cancer patients.

6 The Role of MicroRNAs in the Regulation of Cancer Stem Cell Characteristics

The discovery of microRNAs (miRNAs), a large family of non-protein-coding RNA molecules, has provided a newer perspective on cancer research. It has been identified that miRNAs could function as potent endogenous mediators of post-transcriptional regulators of gene transcription by its site-specific binding in the 3′ un-translated region (3′-UTR) of its target genes, leading to either mRNA degradation or protein synthesis inhibition (Garzon et al. 2007; Liu and Tang 2011). Currently, more than 2,000 miRNAs have been discovered in humans and animals to participate in the regulation of at least 1/3 of genes, which are the regulators of many different biological processes including cell cycle/proliferation, differentiation, cell survival/apoptosis, and metabolism (DeSano and Xu 2009; Perera and Ray 2007; Babashah and Soleimani 2011). Clinically, aberrant expression of miRNAs has been discovered to be highly related to poorer clinical outcome, treatment resistance, and tumor recurrence/relapse. Sufficient evidence has suggested that miRNAs exert important functions in tumor development and progression. More importantly, miRNAs have been recognized as potential mediators of CSC characteristics mediated through the modulation of multiple pro-oncogenic signaling pathways, contributing to tumor aggressive phenotypes. In the following paragraphs, we will discuss some well-characterized tumor-related miRNAs, which are intertwined with the regulation of CSC characteristics and tumor aggressive phenotypes.

6.1 Let-7

Sufficient numbers of clinical and experimental studies have provided clear evidence to support that let-7 seems to exert an important function in tumor development and progression by regulating several pro-oncogenic signaling networks. The aberrant expressions of let-7 family are found to be highly related to poorer clinical outcome such as short 5 year disease-free survival and high tumor recurrence rate. Let-7 has also been considered as a potential tumor suppressor molecule. It was noted that the expression of let-7a, b, and c served as negative mediators of important cellular characteristics of epithelial-to-mesenchymal transition (EMT) phenotype of cancer cells, a biological process that is believed to be reminiscent of CSCs, leading to treatment resistance, tumor recurrence/relapse and metastasis, which is in part mediated through the regulation in the expression of the modulation of phosphatase and tensin homolog deleted on chromosome 10 (PTEN), a well-established tumor suppressor that suppresses the activation of Akt/mTOR signaling network, and the regulation of stem cell signature gene Lin-28 which control cell differentiation and dormancy of CSCs, as documented in several different tumors including breast, lung, pancreatic, colon and prostate cancer cells (Kong et al. 2010; Chang et al. 2011a; Li et al. 2009; Peter 2009; McCarty 2012). It has been observed that the level of let-7 is remarkably decreased in embryonic stem cells and in CSCs of several different tumor cells (Golestaneh et al 2012; Yu et al. 2007; Zhong et al. 2010), and is inversely related to the expression of Lin-28, a known stem cell signature gene. The data further revealed that let-7 expression is bi-directionally repressed by Lin-28 (Zhong et al. 2010), consistent with other findings in which let-7 target several pro-oncogenic genes including Lin-28 (De et al. 2011; Hua et al. 2012; Li et al. 2012; Huang 2012). It has also been noted that over-expression of let-7a by its mimics in breast CSCs results in the suppression of cell proliferation, CSC self-renewal capacity, tumor formation ability, and metastasis in mouse xenograft tumor model, which was in agreement with the reduction of the populations of undifferentiated tumor cells *in vitro* (Yu et al. 2007). Moreover, our recent report demonstrates that let-7 inhibits the expression of enhancer of zeste homolog 2 (EZH2), a key epigenetic mediator of gene expression which shows important functions in embryonic and adult stem cells to regulate the expression of the genes that are preferentially and selectively activated during development and differentiation (Kong et al. 2012b). EZH2 is believed to be a potential mediator for maintaining the CSC characteristics through the regulation of multiple pro-oncogenic signaling pathways (Chang and Hung 2012; Chou et al. 2011; Crea et al. 2012). Therefore, let-7 appears to exhibit an important role in the modulation of CSC characteristics by targeting multiple cell signaling pathways.

6.2 MiR-21

MiR-21 is one of the most studied miRNAs and is shown to be a potential pro-oncogenic molecule, which is in part due to deregulation of multiple signaling pathways. Many clinical reports revealed higher expression of miR-21 in many different tumors including pancreatic, gastric, prostate, colon, liver, breast and brain tumors, and was found to be highly related to worse clinical outcome of patients (Dillhoff et al. 2008; Moriyama et al. 2009). Our recent data revealed that higher levels of miR-21 expression are related to lower levels of PTEN expression in pancreatic cancer cells and its drug-resistant isogenic cells (Ali et al. 2010; Bao et al. 2012d). Furthermore, miR-21 has been identified to suppress the expression of PTEN in liver cancer cells, consistent with the promotion of tumor growth and progression by the up-regulation of Akt/ERK signaling networks (Bao et al. 2013). It has also been known that miR-21 has pro-oncogenic properties through attenuation of cell death and enrichment of cell growth, migration, invasion, angiogenesis, and metastasis (Moriyama et al. 2009; Olson et al. 2009; Zhang et al. 2007, 2009). Moreover, increased numbers of experimental studies have shown that the levels of miR-21 are remarkably increased in CSCs, compared to non-CSC tumor cells (Golestaneh et al. 2012; Han et al. 2012). The functional gain of miR-21 by transfection of its mimics can prolong the hypoxia-mediated survival of the mesenchymal stem cells of bone marrow. The functional loss of miR-21 by its inhibitor induces apoptosis of these stem cells (Nie et al. 2011). In addition, the loss of miR-21 by its antisense inhibitor has also shown to enrich the cell differentiation, and attenuate the sphere formation of drug-resistant colon cancer cells, which is in direct agreement with the inhibition of CD44 (a CSC cell surface marker) and TCF/LEF, a positive mediator of CSC characteristics (Yu et al. 2013b) Our recent data shows that the loss of miR-21 by transfection of its inhibitor decreased CSC self-renewal capacity, which is in direct agreement with the inhibition in the expression of CSC cell surface markers such as CD44 and EpCAM in CSC-like sphere cells of prostate and pancreatic cancers under hypoxia (Bao et al. 2012a, c). These findings strongly suggest that miR-21 exhibit key role in the regulation of CSC characteristics mediated through the regulation of multiple signaling pathways.

6.3 MiR-22

Increased numbers of clinical and experimental studies have provided convincing evidence in support of the notion that miR-22 may have an important function during tumor development and progression in a cell-context manner. It was noted that the expression of miR-22 is remarkably increased in human senescent fibroblasts and in epithelial cells, but its expression is significantly reduced in many different tumors such as colon, liver, ovarian, colon, gastric, prostate, and breast tumor cells (Li et al. 2010a, 2011a; Tsuchiya et al. 2011; Zhang et al. 2010; Pandey and Picard 2009; Xiong et al. 2010). Reduced levels of miR-22 in liver tumor patients have been identified to be highly associated with poorer clinical outcome such as shorter disease-free survival and increased tumor recurrence rate (Zhang et al. 2010). Forced over-expression of miR-22 by its mimics results in the suppression of cell proliferation, migration, invasion, and metastasis in several different tumor cells mediated through activation of PTEN, p21, and p53 (Li et al. 2010a, 2011a; Tsuchiya et al. 2011; Zhang et al. 2010), which imply that miR-22 may function as a tumor suppressor molecule. It has been found that the expression of miR-22 is lost during the induction of TGF-β-mediated EMT phenotype in colon cancer HT-29 cells (Cai et al. 2013), implying a potential function of miR-22 in the regulation of EMT phenotype. It has also been noted that miR-22 can regulate hypoxia signaling pathway by targeting HIF-1α, a potential regulator of CSC characteristics (Yamakuchi et al. 2011); however, the detailed functional role of miR-22 in the regulation of CSC characteristic is still not clear, suggesting that more investigations are required for molecular characterization of the biological function of miR-22 in the regulation of CSCs.

6.4 MiR-26a

Emerging evidence from clinical and experimental studies clearly suggests a potential function of miR-26a as a suppressor which is mediated through targeting multiple signaling pathways such as IL-6/STAT, Myc, and MCL-1. Loss or reduced expression of miR-26a has been documented in a lot of different tumors including bladder, pancreatic, prostate, gastric, and breast cancers as well as melanoma and nasopharyngeal carcinoma, and it has been found to be associated with poorer clinical outcome (Lu et al. 2012; Pang et al. 2010). Several recent reports have suggested that miR-26a can modulate cancer epigenome through repression in the expression of EZH2, a potential positive mediator of CSC characteristics (Banerjee et al. 2011; Bracken et al. 2003; Chen et al. 2010; Friedman et al. 2009; Fujii et al. 2008; Lu et al. 2011).

Over-expression of miR-26a through transfection of mimics in tumor cells can decrease the expression of EZH2, attributing to the attenuation of tumor invasion and metastasis (Banerjee et al. 2011; Friedman et al. 2009; Lu et al. 2011). Therefore, targeting of miR-26a expression provides a new therapeutic approach for the treatment of aggressive tumors. Our recent report showed that functional gain of miR-26a expression by transfection of its mimics results in the inhibition of the expression of CSC markers such as EZH2, Oct4, Notch-1, and EpCAM in pancreatic cancer cells (Bao et al. 2012e), which strongly suggest that miR-26a plays key roles in maintaining CSC characteristic through regulation of CSC signature genes.

6.5 MiR-30

The evidence from limited numbers of experimental studies suggests that miR-30 appears to exert an important function in the modulation of CSC characteristics, and other tumor aggressive phenotypes. The miR-30 has been considered to act as a potential suppressor of tumorigenesis by regulating several signaling networks such as EGF and EMT during the tumor development and progression (Kao et al. 2013). It has been observed that the expression of miR-30 is lost during TGF-β-mediated acquisition of EMT phenotype of colon cancer HT-29 cells (Cai et al. 2013). The miR-30 expression is inversely connected with the expression of Lin-28 during differentiation and proliferation of embryonic stem cells, and CSC-like cells (Zhong et al. 2010). The levels of miR-30 are significantly suppressed in Lin-28+ tumor (CSC-like) cells, compared to Lin-28– tumor (non-CSC-like) cells (Zhong et al. 2010). Forced over-expression of miR-30 by transfection of its mimics in breast CSC-like cells inhibits CSC self-renewal capacity, and increases cell death by targeting Ubc9 and ITGB3, respectively. The blockage of miR-30 by its anti-sense inhibitor revived the self-renewal capacity of CSCs (Yu et al. 2010). The over-expression of miR-30 in CSCs inhibits the growth of xenograft tumor *in vivo* and also lung metastasis *in vivo*. In contrast, the blockage of miR-30 by its anti-sense inhibitor promotes tumorigenesis and lung metastasis *in vivo* (Yu et al. 2010). These findings clearly imply that miR-30 appears to exert a key function in the regulation of CSC characteristics and other tumor aggressive phenotypes. Its detailed role in tumorigenesis would likely be further investigated in the future.

6.6 MiR-34a

Emerging evidence from clinical and experimental studies has shown that miR-34a functions as a potential suppressor of tumorigenesis by targeting multiple pro-oncogenic signaling pathways. The miR-34a expression has been identified to be lost or attenuated in many different tumors such as prostate cancer, breast cancer, lung cancer, gastric cancer, oral squamous cell carcinoma, colon cancer and pancreatic cancer. Such an aberrant expression of miR-34a is highly connected with worse clinical outcome of cancer patients (Kent et al. 2009; Kong et al. 2012a), suggesting that the expression of miR-34a plays a key role in tumor development and progression.

The loss of miR-34a expression in tumors has been identified to be related to DNA methylation of the promoter/enhancer region of the miR-34a gene (Kent et al. 2009; Kong et al. 2012a). It is clear that miR-34a exerts anti-tumor activity in both cell culture system and in animal models through repression of cell growth, proliferation, migration, invasion, and metastasis, which are in part due to activation of p53 and inactivation of Cyclin D1, E2F1/2, and CDK6 in tumors (Aranha et al. 2011; Guo et al. 2011; Lodygin et al. 2008; Sun et al. 2008; Wang et al. 2011). Moreover, several recent experimental reports have shown that miR-34a can inhibit the expression of CSC signature genes such as CD44, CD133, and Notch-1, which is in direct agreement with the inhibition of the CSC self-renewal capacity of different tumor cells (Chang et al. 2011b; Kong et al. 2012a; Liu and Tang 2011; Nalls et al. 2011).

It has also been observed that the functional gain of miR-34a by transfection of mimics could inhibit EMT phenotype mediated through the inhibition of ZEB1, Snail and Slug, known mesenchymal markers of EMT phenotype (Siemens et al. 2011). Additionally, it has been shown that miR-34a and Snail could generate a bi-directionally negative feedback loop to regulate EMT characteristics (Siemens et al. 2011). Moreover, p53 over-expression has been found to increase the expression of miR-34a, suggesting that miR-34a may be a downstream target of p53, which activates the miR-34a expression (Chang et al. 2007). One recent experimental report has shown that the functional gain of expression of miR-34a in CSC-like cells of pancreatic cancer inhibits cell proliferation, migration/invasion, and the self-renewal capacity, which is in direct agreement with the up-regulation in the expression of p53, p21, and p27 (Nalls et al. 2011). Our unpublished data showed that over-expression of miR-34a by transfection of its mimics inhibits cell growth, migration, and the formation of pancreatospheres, which was consistent with the down-regulation in the expression of CSC cell surface marker proteins CD44 and EpCAM in the CSC-like sphere-forming cells of pancreatic cancer. These findings clearly suggest that miR-34a plays a key role as a potential suppressor of tumorigenesis mediated through the inhibition of CSC characteristics.

6.7 MiR-101

Increased numbers of clinical and experimental studies have demonstrated that miR-101 appears to function as a suppressor of tumorigenesis mediated through the regulation of multiple signaling pathways. The loss of expression or decreased expression of miR-101 has been found in many different tumors including lung, prostate, bladder, breast, colon, pancreatic, and liver cancers, and was found to be related to poorer clinical outcome of cancer patients (Lu et al. 2012; Pang et al. 2010). The functional gain of miR-101 is shown to inhibit cell growth, migration/invasion, and *in vivo* tumor growth by targeting several signaling pathways such as c-Met and COX2 (Hu et al. 2013; Hao et al. 2011; He et al. 2012; Zhang et al. 2013). Several *in vitro* and *in vivo* study reports have provided clear evidence in support of the anti-tumor function of miR-101 in different tumors, which is in art mediated through the deregulation of CSC characteristics by negatively targeting EZH2 (Bao et al. 2012d; Sparmann and van 2006), an epigenetic mediator of gene expression (Chang and Hung 2012). The inhibition of miR-101 by transfection of its anti-sense inhibitor caused up-regulation in the expression of EZH2 and Wnt/β-catenin signaling pathways, which was consistent with the up-regulation of CSC self-renewal capacity and down-regulation of E-cadherin, an epithelial marker of EMT phenotype, leading to tumor aggressive phenotype (Strillacci et al. 2013; Bao et al. 2012e; Cao et al. 2010; Cho et al. 2011; Sakurai et al. 2012; Semaan et al. 2011; Zhang et al. 2011).

Our recent report revealed that the functional gain of miR-101 expression leads to the down-regulation of EZH2, which is in direct agreement with the attenuation of CSC self-renewal capacity of human pancreatic cancer cells and its sphere forming (CSC-like) cells (Bao et al. 2012d), consistent with the similar results reported by other authors (Alajez et al. 2010; Au et al. 2012; Smits et al. 2010; Zhang et al. 2011). These findings strongly suggest a pivotal function of miR-101 in the modulation of CSC characteristics. However, the detailed mechanistic role of miR-101 as a potential suppressor of tumorigenesis mediated through the modulation of CSC phenotype and function requires further in-depth examination.

6.8 MiR-128

Emerging evidence suggest that miR-128 appears to exert a very important function in the modulation of CSC and tumor aggressive phenotypes within the tumor micro-environment. The clinical data have shown that aberrant expression of miR-128 is related to worse clinical outcome such as decreased disease-free survival and increased tumor recurrence rate in many different tumors including breast cancer and prostate cancer (Zhu et al. 2011). Several experimental studies have shown that the functional loss of expression of miR-128 by its oligonucleotide inhibitor or siRNA increased cell proliferation and invasion in different tumor cells. Forced over-expression of miR-128 was able to reverse tumor aggressive phenotypes of different tumor cells such as breast cancer and prostate cancer (Peruzzi et al. 2013). Furthermore, accumulated evidence suggests that miR-128 serves as a potential tumor suppressor by targeting several oncogenic signaling networks such as Bmil, SUZ12, and ABCC5, and mitogenic kinases (Godlewski et al. 2008; Peruzzi et al. 2013; Zhu et al. 2011).

Recently, the miR-128 expression has been found to be significantly down-regulated in stem cells including CSC-like cells. Forced over-expression of miR-128 resulted in the inhibition of CSC self-renewal capacity, which was in direct agreement with the inactivation of Bmil, a major epigenetic mediator which positively control CSC characteristics in CSC-like cells of human gliomas (Godlewski et al. 2008). It has also been noted that miR-128 can induce apoptosis by targeting Bax, a known apoptotic mediator in human embryonic kidney cells. Additionally, the expression of miR-128 has been shown to be remarkably decreased in drug-resistant CSC-like cells of breast cancer and breast tumor tissues (Zhu et al. 2011). Moreover, low levels of miR-128 in tumor tissues are related to resistance to chemotherapy, which was associated with poor clinical outcome. These data obtained to-date clearly imply the potential role of miR-128 in the regulation of CSC characteristics.

6.9 MiR-145

The miR-145 clearly participates in the modulation of CSC and EMT characteristics during tumor development and progression. Increased numbers of clinical and experimental studies have provided convincing evidence showing that miR-145 acts as a potential tumor suppressor molecule in a variety of tumors including prostate, breast, and colon cancers. Loss of miR-145 expression or lower levels of miR-145 expression has been identified to be highly connected with poor clinical outcome such as decreased disease-free survival and increased incidence of tumor recurrence in several different tumors including head and neck, pancreatic, prostate and bladder cancers (Avgeris et al. 2013; Yoshino et al. 2013; Yu et al. 2013a), suggesting the potential function of miR-145 in tumor development and progression.

The expression of miR-145 has been shown to have anti-tumor activity in many different tumor cells by targeting Myc, IRS-1, CDK6, catenin-δ-1 (Shao et al. 2013; Yin et al. 2013; Takaoka et al. 2012; Yamada et al. 2013). One recent experimental study has revealed that p53, a known tumor suppressor, inhibits EMT phenotype and CSC characteristics including colony formation, sphere formation, and CSC signature genes (CD44, Myc, and Klf4) in prostate cancer PC3 cells through modulation of miR-145 (Ren et al. 2013). Moreover, miR-145 has been shown to directly suppress the expressions of Oct4, Sox9/ADAM17 (a known positive mediator of tumor metastasis) in different tumors such as renal cell cancer, and head and neck cancer (Doberstein et al. 2013; Yu et al. 2013a). Overexpression of miR-145 by transfection of its mimics resulted in the suppression of cell migration, invasion, and CSC characteristics which was in part mediated by targeting Sox9 in melanoma cells (Dynoodt et al. 2013). It has also been noted that the expression of miR-145 was down-regulated in CSC-like (ALDH1+/CD44+) cells of human head and neck cancer in the comparison to the non-CSC-like (ALDH1−/CD44−) tumor cells. The functional gain of miR-145 expression by infection of cells with lentiviral-miR-145 inhibited tumor progression in mouse tumor xenografts model derived from CSC-like cells of human head and neck cancer (Yu et al. 2013a). These results strongly support that miR-145 elicits an important function in the regulation of CSC characteristics and tumor aggressive phenotypes by targeting multiple pro-oncogenic signaling networks including CSC signature genes.

6.10 MiR-146a

The loss of expression or reduced expression of miR-146a has been found to be correlated with worse clinical outcome of different tumors (Li et al. 2010b; Pang et al. 2010). The data from many experimental studies indicate that miR-146a may function as a potent tumor suppressor molecule via the inhibition of cell cycle, migration, invasiveness and metastasis ability of many different tumors including pancreatic, breast, liver, gastric, colon, and prostate tumors by regulating multiple signaling networks such as NF-κB, K-ras, and Notch-1 (Ali et al. 2011; Li et al. 2010b; Mei et al. 2011). It has been noted that miR-146a inhibits the activation of NF-κB, leading to the suppression of gene expression of NF-κB-mediated interleukin-1β (IL-1β), IL-6, IL-8, and TNF-α by down-regulation of IL-1 receptor associated kinase 1 (IRAK1) and TNF receptor associated factor 6 (TRAF6) (Bhaumik et al. 2008). It has been documented that the activation of NF-κB signaling pathway is engaged in the enhancement of CSC characteristics by the regulation of CSC signature genes such as Nanog, Sox2, and Lin-28 (Liu et al. 2010). Our recent report confirms that the expression of miR-146a is down-regulated in pancreatic cancer cells and K-ras-transgenic mice of pancreatic cancer (Ali et al. 2011; Li et al. 2010b). Forced over-expression of miR-146a resulted in the inhibition of EGFR and NF-κB activity, leading to the suppression of NF-κB-mediated targets, accompanied by the inhibition of tumor cell invasive potential (Li et al. 2010b), which suggest that miR-146a may function as a tumor suppressor in part mediated by targeting NF-κB signaling pathway.

Several experimental studies have shown that miR-146a appears to have a key role in maintaining the hemostasis of normal stem cells such as embryonic, mesenchymal, and hematopoietic stem cells. Moreover, miR-146a appears to inhibit the development of glioma by down-regulation of Notch-1 (Mei et al. 2011). However, one recent study showed high expression levels of miR-146a in CD133+ sphere forming cells of ovarian cancer (Nam et al. 2012). These findings suggest that miR-146a might have an important role in regulating CSC phenotype and function; however, the detailed mechanistic studies are required delineating the precise role of miR-146a in human tumors.

6.11 MiR-200

A great number of clinical and experimental reports have provided clear evidence in favor of the role of miR-200 in human tumor development and progression. Aberrant expression of miR-200 have been documented in many different tumors, and it is highly related to poorer clinical outcome such as decreased disease-free survival and increased incidence of tumor recurrence associated with increased therapeutic resistance in many different tumors including breast, prostate, gastric, pancreatic, colon, lung, liver, and brain tumors. Currently, miR-200 has been widely considered to function as potential tumor suppressor molecule, which is in part mediated through the inhibition of EMT phenotype by primarily targeting ZEB1/2, a well-known marker of EMT phenotype. Early experimental studies have reported that miR-200b, c are able to bind to the 3′UTR of ZEB1/2 genes, attributing to the inhibition in the expression of these genes (Brabletz and Brabletz 2010).

Our recent reports have shown that drug-resistant human pancreatic cancer cells have decreased expression of miR-200a,b,c, and display more mesenchymal phenotype consistent with EMT characteristics, compared to its drug-sensitive tumor cells (Ali et al. 2010). Forced over-expression of miR-200a,b,c by transfection of its mimics in these drug-resistant tumor cells or PDGF-D-induced EMT phenotypic prostate cancer PC3 cells inhibited the expression of ZEB1, ZEB2, Slug, which was consistent with the up-regulation in the expression of E-cadherin, an epithelial marker (Ali et al. 2010; Kong et al. 2009). These results are in direct agreement with results published by other investigators (Peter 2009; Kent et al. 2009; Li et al. 2009). Additionally miR-200 has been shown to decrease the expression of Bmil, Suz12, and Notch-1, well-established mediators of CSC and EMT characteristics in many different tumor cells, which is in direct agreement with the inhibition of CSC self-renewal capacity (Bao et al. 2011b; Iliopoulos et al. 2010; Leal and Lleonart 2012). Importantly, lower levels of miR-200a, b, and c expression have been identified in CSC-like (CD44+/CD24−) cells of breast cancer (Shimono et al. 2009). Our unpublished data from micro-array analysis reveals that CD44+/CD133+/EpCAM+ CSC cells of pancreatic cancer exhibit deregulations in the expression of miR-200 family members. These data clearly suggest that miR-200 may exert an important function in the regulation of CSCs characteristics potentially by regulating multiple signaling networks.

6.12 MiR-210

A great number of clinical and experimental reports have produced solid evidence in favor of the fact that miR-210 may play a key function in tumor development and progression. High levels of miRN-210 expression has been found in serum/plasma and tumor tissues and it was correlated with poor clinical outcome in many different tumors including breast, prostate, pancreatic tumors as well as, lymphoma (Huang et al. 2009; Ivan et al. 2008). It has been shown that miR-210 functions as a hypoxia-responsive molecule in many different cells including tumor cells (Camps et al. 2008; Chan and Loscalzo 2010; Devlin et al. 2011; Favaro et al. 2010; Gee et al. 2010; Hebert et al. 2007; Ho et al. 2010; Kulshreshtha et al. 2007; Puissegur et al. 2011; Quero et al. 2011), and it can up-regulate VEGF expression in pancreatic tumor cells in HIF-1α-control manner. The expression of miR-210 has also been found to participate in the modulation of DNA-defective repair system by targeting the gene expression of RAD52, an important mediator of repair machinery. Over-expression of miR-210 by transfection of its mimics resulted in defective DNA repair, leading to genetic instability (Crosby et al. 2009), implying a critical function of miR-210 in the homeostasis of DNA repair systems in tumors. Our unpublished data reveals that CD44+/CD133+/EpCAM+ MiaPaCa-2 cells of human pancreatic cancer are characteristic of CSCs with greater self-renewal capacity, clonogenicity, and migration/invasion, consistent with higher expression of CSC markers. We found that these CSC cells have increased levels of miR-210, compared to CD44-/CD133-/EpCAM- MiaPaCa-2 cells, suggesting that miR-210 might exert a pivotal functional role in the regulation of CSC characteristics leading to tumor aggressiveness.

6.13 MiR-335

Increased numbers of clinical and experimental reports have clearly shown that miR-335 exhibits a key role in tumor development and progression. For example, a recent report has demonstrated that lower level of miR-335 is associated with poor clinical outcome such as advanced clinical stage, increased metastasis, higher frequency of recurrence and poor disease-free survival in different tumors including breast, gastric and prostate cancers (Tavazoie et al. 2008; Xiong et al. 2013; Xu et al. 2012). The expression of miR-335 has been found to be serve as a prognostic marker/signature in several tumors including gastric cancer and prostate cancer, and has been considered as a potential tumor suppressor molecule in prostate cancer (Xiong et al. 2013; Xu et al. 2012; Yan et al. 2012). The signaling pathway analysis reveals that miR-335 may participate in the modulation of different target genes such as several pro-oncogenic signaling networks including p53, MAPK, TGF-β, Wnt, Rb, and mTOR (Yan et al. 2012), leading to the inhibition of tumor cell aggressiveness. However, limited number of experimental studies suggest that miR-335 promotes tumor cell growth in certain types of tumors, supporting that the regulatory function of miR-335 in tumorigenesis may be tumor site-specific. Our unpublished data reveals that CSC-like cells sorted from pancreatic cancer MiaPaCa-2 cells have increased levels of miR-335, in comparison to its MiaPaCa-2 parental cells or non-CSC tumor cells, implying a potential role of miR-335 in regulating CSC characteristics during tumor development and progression. However, further mechanistic investigations are required to clarify the functional role of miR-335 in the regulation of CSCs.

6.14 MiR-451

A significant number of clinical and experimental studies have provided clear evidence showing that miR-451 may exert a key functional role in the regulation of tumor development and progression. Low levels of miR-451 expression have been identified in different tumors such as bladder cancer, lung cancer, urothelial carcinoma, gliomas, and renal cell carcinoma (Redova et al. 2012; Solomides et al. 2012; Xie et al. 2012; Tian et al. 2012). The functional gain of miR-451 by transfection of its mimics causes suppression of cell survival, proliferation and migration, consistent with inhibition in the expression of Akt and Bcl-2 in esophageal carcinoma cells. Injection of miR-451 inhibits tumor growth in tumor xenograft model of human esophageal carcinoma (Wang et al. 2013). It has also been found that miR-451 displays an anti-tumor function in cell culture and mouse xenograft tumor models, consistent with inhibition of PI3K/Akt pathway targeting calcium binding protein 39 (CAB39) in glioma (Tian et al. 2012). Therefore, miR-451 has been recognized as potential tumor suppressor (Li et al. 2011b).

One recent experimental study has shown that the expression of miR-451 was significantly decreased in CSC-like sphere cells of colon cancer, in comparison to its parental cells (Bitarte et al. 2011). The functional gain of miR-451 resulted in the suppression of CSC self-renewal ability, tumorigenicity and chemo-radiation resistance in the CSC sphere cells, which is in direct agreement with the attenuation of COX2 by miR-451 (Bitarte et al. 2011). Notch-1, a known mediator of CSC and EMT characteristics, has been shown to be decreased in the expression of miR-451, which can directly suppress the expression of Myc. Moreover, the suppression of miR-451 expression by Notch-1 is required for oncogenesis in ALL (acute lymphoblastic leukemia) (Li et al. 2011b). Such findings clearly imply that miR-451 appears to have a key function in the modulation of CSC characteristics during tumor development and progression.

7 The Role of Genistein as an Anti-tumor Agent by Targeting Tumor-Related microRNAs

It has been well known that genistein is one the major class of isoflavones primarily existed in the Leguminosae family of plants such as soybean. Genistein has a similar structure like estrogen, displaying a weak estrogenic activity by its binding to estrogen receptor, thereby inhibiting estrogen receptor signaling pathway. Therefore, genistein has also been known as phytoestrogen. The epidemiological and clinical reports have clearly shown that genistein could exert an anti-tumor activity in prostate cancer in Japan and USA (Adlercreutz et al. 1991, 1993; Hebert et al. 1998; Jacobsen et al. 1998).

A great number of experimental reports including reports from our group have demonstrated that genistein displays potent anti-tumor activity via down-regulation of several pro-oncogenic signaling networks including NF-κB, Wnt, FoxM1, Notch-1, and Akt/mTOR in many different tumors (Kuang et al. 2009; Sarkar and Li 2009; Sarkar et al. 2010; Su et al. 2007; Wang et al. 2008; Su and Simmen 2009; Wagner and Lehmann 2006). Our recent reports also showed that genistein can up-regulate let-7, miR-26a, miR-101, miR-146a, and miR-200, and down-regulate miR-21, CSC cell surface markers (CD44 and EpCAM), and CSC self-renewal capacity in pancreatic cancer cells, and its CSC-like sphere forming cells, which is consistent with its anti-tumor activity against pancreatic tumor in mouse xenograft models (Bao et al. 2011a, b; Li et al. 2009). The above findings clearly imply that genistein appears to be a potential anti-tumor agent, which may be partially mediated through deregulation of CSC-related miRNAs that may be critical during tumor development and progression.

8 Conclusions and Perspectives

We made efforts to summarize the "state-of-our-knowledge" on the potential roles of tumor-related miRNAs in the regulation of CSC characteristics and other aggressive tumor phenotypes during tumor development and progression. In summary, a great amount of persuasive evidence clearly supports that a small sub-population of CSCs have a great clinical implications in tumor progression because the presence of CSCs in tumor tissues is highly associated with poorer clinical outcomes. The poor clinical outcome is in part associated with distinct properties of CSCs including their ability of self-renewal, potential of differentiation into multiple cell lineages and unlimited proliferation capacity, consistent with high propensity of tumorigenesis *in vivo*, and acquisition of treatment resistance phenotype and tumor metastasis, contributing to tumor aggressiveness. Eradication or inhibition of these small sub-populations of cells has been designated as a new and effective therapeutic approach for the treatment of aggressive tumors although the pathogenesis of CSCs in cancer biology is still not completely understood. Emerging evidence clearly support that tumor-related miRNAs appears to have a very important role in the modulation of CSC phenotype and function during tumor development and progression. However, more investigations are required to ascertain detailed role of these tumor-related miRNAs in the regulation of CSC characteristics.

Acknowledgements We presented evidence as concisely as possible and during such effort we could not cite all the published data, and thus we sincerely apologize to those whose work could not be cited here.

References

Adlercreutz H, Honjo H, Higashi A, Fotsis T, Hamalainen E, Hasegawa T et al (1991) Urinary excretion of lignans and isoflavonoid phytoestrogens in Japanese men and women consuming a traditional Japanese diet. Am J Clin Nutr 54(6):1093–1100

Adlercreutz H, Markkanen H, Watanabe S (1993) Plasma concentrations of phyto-oestrogens in Japanese men. Lancet 342(8881):1209–1210

Alajez NM, Shi W, Hui AB, Bruce J, Lenarduzzi M, Ito E et al (2010) Enhancer of Zeste homolog 2 (EZH2) is overexpressed in recurrent nasopharyngeal carcinoma and is regulated by miR-26a, miR-101, and miR-98. Cell Death Dis 1:e85

Al-Hajj M, Wicha MS, Benito-Hernandez A, Morrison SJ, Clarke MF (2003) Prospective identification of tumorigenic breast cancer cells. Proc Natl Acad Sci U S A 100(7): 3983–3988

Ali S, Ahmad A, Banerjee S, Padhye S, Dominiak K, Schaffert JM et al (2010) Gemcitabine sensitivity can be induced in pancreatic cancer cells through modulation of miR-200 and miR-21 expression by curcumin or its analogue CDF. Cancer Res 70(9):3606–3617

Ali S, Banerjee S, Logna F, Bao B, Philip PA, Korc M et al (2011) Inactivation of Ink4a/Arf leads to deregulated expression of miRNAs in K-Ras transgenic mouse model of pancreatic cancer. J Cell Physiol 227(10):3373–3380

Aranha MM, Santos DM, Sola S, Steer CJ, Rodrigues CM (2011) miR-34a regulates mouse neural stem cell differentiation. PLoS One 6(8):e21396

Au SL, Wong CC, Lee JM, Fan DN, Tsang FH, Ng IO et al (2012) Enhancer of zeste homolog 2 epigenetically silences multiple tumor suppressor microRNAs to promote liver cancer metastasis. Hepatology 56(2):622–631

Avgeris M, Stravodimos K, Fragoulis EG, Scorilas A (2013) The loss of the tumour-suppressor miR-145 results in the shorter disease-free survival of prostate cancer patients. Br J Cancer 108(12):2573–2581

Babashah S, Soleimani M (2011) The oncogenic and tumour suppressive roles of microRNAs in cancer and apoptosis. Eur J Cancer 47(8):1127–1137

Banerjee R, Mani RS, Russo N, Scanlon CS, Tsodikov A, Jing X et al (2011) The tumor suppressor gene rap1GAP is silenced by miR-101-mediated EZH2 overexpression in invasive squamous cell carcinoma. Oncogene 30(42):4339–4349

Bao S, Wu Q, McLendon RE, Hao Y, Shi Q, Hjelmeland AB et al (2006) Glioma stem cells promote radioresistance by preferential activation of the DNA damage response. Nature 444(7120):756–760

Bao B, Wang Z, Ali S, Kong D, Banerjee S, Ahmad A et al (2011a) Over-expression of FoxM1 leads to epithelial-mesenchymal transition and cancer stem cell phenotype in pancreatic cancer cells. J Cell Biochem 112(9):2296–2306

Bao B, Wang Z, Ali S, Kong D, Li Y, Ahmad A et al (2011b) Notch-1 induces epithelial-mesenchymal transition consistent with cancer stem cell phenotype in pancreatic cancer cells. Cancer Lett 307(1):26–36

Bao B, Ahmad A, Kong D, Ali S, Azmi AS, Li Y et al (2012a) Hypoxia induced aggressiveness of prostate cancer cells is linked with deregulated expression of VEGF, IL-6 and miRNAs that are attenuated by CDF. PLoS One 7(8):e43726

Bao B, Ahmad A, Li Y, Azmi AS, Ali S, Banerjee S et al (2012b) Targeting CSCs within the tumor microenvironment for cancer therapy: a potential role of mesenchymal stem cells. Expert Opin Ther Targets 16(10):1041–1054

Bao B, Ali S, Ahmad A, Azmi AS, Li Y, Banerjee S et al (2012c) Hypoxia-induced aggressiveness of pancreatic cancer cells is due to increased expression of VEGF, IL-6 and miR-21, which can be attenuated by CDF treatment. PLoS One 7(12):e50165

Bao B, Ali S, Banerjee S, Wang Z, Logna F, Azmi AS et al (2012d) Curcumin analogue CDF inhibits pancreatic tumor growth by switching on suppressor microRNAs and attenuating EZH2 expression. Cancer Res 72(1):335–345

Bao B, Wang Z, Ali S, Ahmad A, Azmi AS, Sarkar SH et al (2012e) Metformin inhibits cell proliferation, migration and invasion by attenuating CSC function mediated by deregulating miRNAs in pancreatic cancer cells. Cancer Prev Res (Phila) 5(3):355–364

Bao L, Yan Y, Xu C, Ji W, Shen S, Xu G et al (2013) MicroRNA-21 suppresses PTEN and hSulf-1 expression and promotes hepatocellular carcinoma progression through AKT/ERK pathways. Cancer Lett 337(2):226–236

Bhatia M, Bonnet D, Kapp U, Wang JC, Murdoch B, Dick JE (1997) Quantitative analysis reveals expansion of human hematopoietic repopulating cells after short-term ex vivo culture. J Exp Med 186(4):619–624

Bhaumik D, Scott GK, Schokrpur S, Patil CK, Campisi J, Benz CC (2008) Expression of microRNA-146 suppresses NF-kappaB activity with reduction of metastatic potential in breast cancer cells. Oncogene 27(42):5643–5647

Bitarte N, Bandres E, Boni V, Zarate R, Rodriguez J, Gonzalez-Huarriz M et al (2011) MicroRNA-451 is involved in the self-renewal, tumorigenicity, and chemoresistance of colorectal cancer stem cells. Stem Cells 29(11):1661–1671

Bonnet D, Dick JE (1997) Human acute myeloid leukemia is organized as a hierarchy that originates from a primitive hematopoietic cell. Nat Med 3(7):730–737

Brabletz S, Brabletz T (2010) The ZEB/miR-200 feedback loop–a motor of cellular plasticity in development and cancer? EMBO Rep 11(9):670–677

Bracken AP, Pasini D, Capra M, Prosperini E, Colli E, Helin K (2003) EZH2 is downstream of the pRB-E2F pathway, essential for proliferation and amplified in cancer. EMBO J 22(20): 5323–5335

Cai ZG, Zhang SM, Zhang H, Zhou YY, Wu HB, Xu XP (2013) Aberrant expression of microRNAs involved in epithelial-mesenchymal transition of HT-29 cell line. Cell Biol Int 37(7):669–674

Camps C, Buffa FM, Colella S, Moore J, Sotiriou C, Sheldon H et al (2008) hsa-miR-210 is induced by hypoxia and is an independent prognostic factor in breast cancer. Clin Cancer Res 14(5):1340–1348

Cao P, Deng Z, Wan M, Huang W, Cramer SD, Xu J et al (2010) MicroRNA-101 negatively regulates Ezh2 and its expression is modulated by androgen receptor and HIF-1alpha/HIF-1beta. Mol Cancer 9:108

Chan SY, Loscalzo J (2010) MicroRNA-210: a unique and pleiotropic hypoxamir. Cell Cycle 9(6):1072–1083

Chang CJ, Hung MC (2012) The role of EZH2 in tumour progression. Br J Cancer 106(2):243–247

Chang TC, Wentzel EA, Kent OA, Ramachandran K, Mullendore M, Lee KH et al (2007) Transactivation of miR-34a by p53 broadly influences gene expression and promotes apoptosis. Mol Cell 26(5):745–752

Chang CJ, Hsu CC, Chang CH, Tsai LL, Chang YC, Lu SW et al (2011a) Let-7d functions as novel regulator of epithelial-mesenchymal transition and chemoresistant property in oral cancer. Oncol Rep 26(4):1003–1010

Chang SJ, Weng SL, Hsieh JY, Wang TY, Chang MD, Wang HW (2011b) MicroRNA-34a modulates genes involved in cellular motility and oxidative phosphorylation in neural precursors derived from human umbilical cord mesenchymal stem cells. BMC Med Genomics 4:65

Chen Y, Xie D, Yin LW, Man CC, Yao H, Chan CY et al (2010) RNAi targeting EZH2 inhibits tumor growth and liver metastasis of pancreatic cancer in vivo. Cancer Lett 297(1):109–116

Cho HM, Jeon HS, Lee SY, Jeong KJ, Park SY, Lee HY et al (2011) MicroRNA-101 inhibits lung cancer invasion through the regulation of enhancer of zeste homolog 2. Exp Ther Med 2(5):963–967

Chou RH, Yu YL, Hung MC (2011) The roles of EZH2 in cell lineage commitment. Am J Transl Res 3(3):243–250

Crea F, Fornaro L, Bocci G, Sun L, Farrar WL, Falcone A et al (2012) EZH2 inhibition: targeting the crossroad of tumor invasion and angiogenesis. Cancer Metastasis Rev 31(3–4):753–761

Creighton CJ, Chang JC, Rosen JM (2010) Epithelial-mesenchymal transition (EMT) in tumor-initiating cells and its clinical implications in breast cancer. J Mammary Gland Biol Neoplasia 15(2):253–260

Crosby ME, Kulshreshtha R, Ivan M, Glazer PM (2009) MicroRNA regulation of DNA repair gene expression in hypoxic stress. Cancer Res 69(3):1221–1229

D'Andrea FP (2012) Intrinsic radiation resistance of mesenchymal cancer stem cells and implications for treatment response in a murine sarcoma model. Dan Med J 59(2):B4388

D'Andrea FP, Safwat A, Kassem M, Gautier L, Overgaard J, Horsman MR (2011) Cancer stem cell overexpression of nicotinamide N-methyltransferase enhances cellular radiation resistance. Radiother Oncol 99(3):373–378

De VC, Riggi N, Suva ML, Janiszewska M, Horlbeck J, Baumer K et al (2011) Let-7a is a direct EWS-FLI-1 target implicated in Ewing's sarcoma development. PLoS One 6(8):e23592

DeSano JT, Xu L (2009) MicroRNA regulation of cancer stem cells and therapeutic implications. AAPS J 11(4):682–692

Devlin C, Greco S, Martelli F, Ivan M (2011) miR-210: more than a silent player in hypoxia. IUBMB Life 63(2):94–100

Dillhoff M, Liu J, Frankel W, Croce C, Bloomston M (2008) MicroRNA-21 is overexpressed in pancreatic cancer and a potential predictor of survival. J Gastrointest Surg 12(12):2171–2176

Doberstein K, Steinmeyer N, Hartmetz AK, Eberhardt W, Mittelbronn M, Harter PN et al (2013) MicroRNA-145 targets the metalloprotease ADAM17 and is suppressed in renal cell carcinoma patients. Neoplasia 15(2):218–230

Dylla SJ, Beviglia L, Park IK, Chartier C, Raval J, Ngan L et al (2008) Colorectal cancer stem cells are enriched in xenogeneic tumors following chemotherapy. PLoS One 3(6):e2428

Dynoodt P, Speeckaert R, De WO, Chevolet I, Brochez L, Lambert J et al (2013) miR-145 overexpression suppresses the migration and invasion of metastatic melanoma cells. Int J Oncol 42(4):1443–1451

Favaro E, Ramachandran A, McCormick R, Gee H, Blancher C, Crosby M et al (2010) MicroRNA-210 regulates mitochondrial free radical response to hypoxia and krebs cycle in cancer cells by targeting iron sulfur cluster protein ISCU. PLoS One 5(4):e10345

Friedman JM, Liang G, Liu CC, Wolff EM, Tsai YC, Ye W et al (2009) The putative tumor suppressor microRNA-101 modulates the cancer epigenome by repressing the polycomb group protein EZH2. Cancer Res 69(6):2623–2629

Fujii S, Ito K, Ito Y, Ochiai A (2008) Enhancer of zeste homologue 2 (EZH2) down-regulates RUNX3 by increasing histone H3 methylation. J Biol Chem 283(25):17324–17332

Garzon R, Pichiorri F, Palumbo T, Visentini M, Aqeilan R, Cimmino A et al (2007) MicroRNA gene expression during retinoic acid-induced differentiation of human acute promyelocytic leukemia. Oncogene 26(28):4148–4157

Gee HE, Camps C, Buffa FM, Patiar S, Winter SC, Betts G et al (2010) hsa-mir-210 is a marker of tumor hypoxia and a prognostic factor in head and neck cancer. Cancer 116(9):2148–2158

Ghisolfi L, Keates AC, Hu X, Lee DK, Li CJ (2012) Ionizing radiation induces stemness in cancer cells. PLoS One 7(8):e43628

Godlewski J, Nowicki MO, Bronisz A, Williams S, Otsuki A, Nuovo G et al (2008) Targeting of the Bmi-1 oncogene/stem cell renewal factor by microRNA-128 inhibits glioma proliferation and self-renewal. Cancer Res 68(22):9125–9130

Golestaneh AF, Atashi A, Langroudi L, Shafiee A, Ghaemi N, Soleimani M (2012) miRNAs expressed differently in cancer stem cells and cancer cells of human gastric cancer cell line MKN-45. Cell Biochem Funct 30(5):411–418

Guo Y, Li S, Qu J, Wang S, Dang Y, Fan J et al (2011) MiR-34a inhibits lymphatic metastasis potential of mouse hepatoma cells. Mol Cell Biochem 354(1–2):275–282

Han M, Wang Y, Liu M, Bi X, Bao J, Zeng N et al (2012) MiR-21 regulates epithelial-mesenchymal transition phenotype and hypoxia-inducible factor-1alpha expression in third-sphere forming breast cancer stem cell-like cells. Cancer Sci 103(6):1058–1064

Hao Y, Gu X, Zhao Y, Greene S, Sha W, Smoot DT et al (2011) Enforced expression of miR-101 inhibits prostate cancer cell growth by modulating the COX-2 pathway in vivo. Cancer Prev Res (Phila) 4(7):1073–1083

He XP, Shao Y, Li XL, Xu W, Chen GS, Sun HH et al (2012) Downregulation of miR-101 in gastric cancer correlates with cyclooxygenase-2 overexpression and tumor growth. FEBS J 279(22):4201–4212

Hebert JR, Hurley TG, Olendzki BC, Teas J, Ma Y, Hampl JS (1998) Nutritional and socioeconomic factors in relation to prostate cancer mortality: a cross-national study. J Natl Cancer Inst 90(21):1637–1647

Hebert C, Norris K, Scheper MA, Nikitakis N, Sauk JJ (2007) High mobility group A2 is a target for miRNA-98 in head and neck squamous cell carcinoma. Mol Cancer 6:5

Hermann PC, Huber SL, Herrler T, Aicher A, Ellwart JW, Guba M et al (2007) Distinct populations of cancer stem cells determine tumor growth and metastatic activity in human pancreatic cancer. Cell Stem Cell 1(3):313–323

Hermann PC, Bhaskar S, Cioffi M, Heeschen C (2010) Cancer stem cells in solid tumors. Semin Cancer Biol 20(2):77–84

Ho AS, Huang X, Cao H, Christman-Skieller C, Bennewith K, Le QT et al (2010) Circulating miR-210 as a novel hypoxia marker in pancreatic cancer. Transl Oncol 3(2):109–113

Hu Z, Lin Y, Chen H, Mao Y, Wu J, Zhu Y et al (2013) MicroRNA-101 suppresses motility of bladder cancer cells by targeting c-Met. Biochem Biophys Res Commun 435(1):82–87

Hua S, Xiaotao X, Renhua G, Yongmei Y, Lianke L, Wen G et al (2012) Reduced miR-31 and let-7 maintain the balance between differentiation and quiescence in lung cancer stem-like side population cells. Biomed Pharmacother 66(2):89–97

Huang Y (2012) A mirror of two faces: Lin28 as a master regulator of both miRNA and mRNA. Wiley Interdiscip Rev RNA 3(4):483–494
Huang X, Ding L, Bennewith KL, Tong RT, Welford SM, Ang KK et al (2009) Hypoxia-inducible mir-210 regulates normoxic gene expression involved in tumor initiation. Mol Cell 35(6):856–867
Iliopoulos D, Lindahl-Allen M, Polytarchou C, Hirsch HA, Tsichlis PN, Struhl K (2010) Loss of miR-200 inhibition of Suz12 leads to polycomb-mediated repression required for the formation and maintenance of cancer stem cells. Mol Cell 39(5):761–772
Ischenko I, Seeliger H, Kleespies A, Angele MK, Eichhorn ME, Jauch KW et al (2010) Pancreatic cancer stem cells: new understanding of tumorigenesis, clinical implications. Langenbecks Arch Surg 395(1):1–10
Ivan M, Harris AL, Martelli F, Kulshreshtha R (2008) Hypoxia response and microRNAs: no longer two separate worlds. J Cell Mol Med 12(5A):1426–1431
Jacobsen BK, Knutsen SF, Fraser GE (1998) Does high soy milk intake reduce prostate cancer incidence? The Adventist Health Study (United States). Cancer Causes Control 9(6):553–557
Kao CJ, Martiniez A, Shi XB, Yang J, Evans CP, Dobi A et al (2013) miR-30 as a tumor suppressor connects EGF/Src signal to ERG and EMT. Oncogene. doi: 10.1038/onc.2013.200. [Epub ahead of print]
Kent OA, Mullendore M, Wentzel EA, Lopez-Romero P, Tan AC, Alvarez H et al (2009) A resource for analysis of microRNA expression and function in pancreatic ductal adenocarcinoma cells. Cancer Biol Ther 8(21):2013–2024
Klonisch T, Wiechec E, Hombach-Klonisch S, Ande SR, Wesselborg S, Schulze-Osthoff K et al (2008) Cancer stem cell markers in common cancers – therapeutic implications. Trends Mol Med 14(10):450–460
Kojima M, Ishii G, Atsumi N, Fujii S, Saito N, Ochiai A (2008) Immunohistochemical detection of CD133 expression in colorectal cancer: a clinicopathological study. Cancer Sci 99(8):1578–1583
Kong D, Li Y, Wang Z, Banerjee S, Ahmad A, Kim HR et al (2009) miR-200 regulates PDGF-D-mediated epithelial-mesenchymal transition, adhesion, and invasion of prostate cancer cells. Stem Cells 27(8):1712–1721
Kong D, Banerjee S, Ahmad A, Li Y, Wang Z, Sethi S et al (2010) Epithelial to mesenchymal transition is mechanistically linked with stem cell signatures in prostate cancer cells. PLoS One 5(8):e12445
Kong D, Heath E, Chen W, Cher M, Powell I, Heilbrun L et al (2012a) Epigenetic silencing of miR-34a in human prostate cancer cells and tumor tissue specimens can be reversed by BR-DIM treatment. Am J Transl Res 4(1):14–23
Kong D, Heath E, Chen W, Cher ML, Powell I, Heilbrun L et al (2012b) Loss of let-7 up-regulates EZH2 in prostate cancer consistent with the acquisition of cancer stem cell signatures that are attenuated by BR-DIM. PLoS One 7(3):e33729
Kuang HB, Miao CL, Guo WX, Peng S, Cao YJ, Duan EK (2009) Dickkopf-1 enhances migration of HEK293 cell by beta-catenin/E-cadherin degradation. Front Biosci 14:2212–2220
Kulshreshtha R, Ferracin M, Negrini M, Calin GA, Davuluri RV, Ivan M (2007) Regulation of microRNA expression: the hypoxic component. Cell Cycle 6(12):1426–1431
Leal JA, Lleonart ME (2012) MicroRNAs and cancer stem cells: therapeutic approaches and future perspectives. Cancer Lett 338(1):174–183
Lee CJ, Dosch J, Simeone DM (2008) Pancreatic cancer stem cells. J Clin Oncol 26(17):2806–2812
Lee HE, Kim JH, Kim YJ, Choi SY, Kim SW, Kang E et al (2011) An increase in cancer stem cell population after primary systemic therapy is a poor prognostic factor in breast cancer. Br J Cancer 104(11):1730–1738
Li Y, VandenBoom TG, Kong D, Wang Z, Ali S, Philip PA et al (2009) Up-regulation of miR-200 and let-7 by natural agents leads to the reversal of epithelial-to-mesenchymal transition in gemcitabine-resistant pancreatic cancer cells. Cancer Res 69(16):6704–6712

Li J, Liang S, Yu H, Zhang J, Ma D, Lu X (2010a) An inhibitory effect of miR-22 on cell migration and invasion in ovarian cancer. Gynecol Oncol 119(3):543–548

Li Y, VandenBoom TG, Wang Z, Kong D, Ali S, Philip PA et al (2010b) miR-146a suppresses invasion of pancreatic cancer cells. Cancer Res 70(4):1486–1495

Li J, Zhang Y, Zhao J, Kong F, Chen Y (2011a) Overexpression of miR-22 reverses paclitaxel-induced chemoresistance through activation of PTEN signaling in p53-mutated colon cancer cells. Mol Cell Biochem 357(1–2):31–38

Li X, Sanda T, Look AT, Novina CD, von Boehmer H (2011b) Repression of tumor suppressor miR-451 is essential for NOTCH1-induced oncogenesis in T-ALL. J Exp Med 208(4):663–675

Li N, Zhong X, Lin X, Guo J, Zou L, Tanyi JL et al (2012) Lin-28 homologue A (LIN28A) promotes cell cycle progression via regulation of cyclin-dependent kinase 2 (CDK2), cyclin D1 (CCND1), and cell division cycle 25 homolog A (CDC25A) expression in cancer. J Biol Chem 287(21):17386–17397

Liu C, Tang DG (2011) MicroRNA regulation of cancer stem cells. Cancer Res 71(18): 5950–5954

Liu M, Sakamaki T, Casimiro MC, Willmarth NE, Quong AA, Ju X et al (2010) The canonical NF-kappaB pathway governs mammary tumorigenesis in transgenic mice and tumor stem cell expansion. Cancer Res 70(24):10464–10473

Liu C, Luo Y, Liu X, Lu P, Zhao Z (2012) Clinical implications of CD44+/. Cancer Biother Radiopharm 27(5):324–328

Lodygin D, Tarasov V, Epanchintsev A, Berking C, Knyazeva T, Korner H et al (2008) Inactivation of miR-34a by aberrant CpG methylation in multiple types of cancer. Cell Cycle 7(16):2591–2600

Lu J, He ML, Wang L, Chen Y, Liu X, Dong Q et al (2011) MiR-26a inhibits cell growth and tumorigenesis of nasopharyngeal carcinoma through repression of EZH2. Cancer Res 71(1):225–233

Lu L, Tang D, Wang L, Huang LQ, Jiang GS, Xiao XY et al (2012) Gambogic acid inhibits TNF-alpha-induced invasion of human prostate cancer PC3 cells in vitro through PI3K/Akt and NF-kappaB signaling pathways. Acta Pharmacol Sin 33(4):531–541

McCarty MF (2012) Metformin may antagonize Lin28 and/or Lin28B activity, thereby boosting let-7 levels and antagonizing cancer progression. Med Hypotheses 78(2):262–269

Mei J, Bachoo R, Zhang CL (2011) MicroRNA-146a inhibits glioma development by targeting Notch1. Mol Cell Biol 31(17):3584–3592

Moriyama T, Ohuchida K, Mizumoto K, Yu J, Sato N, Nabae T et al (2009) MicroRNA-21 modulates biological functions of pancreatic cancer cells including their proliferation, invasion, and chemoresistance. Mol Cancer Ther 8(5):1067–1074

Nalls D, Tang SN, Rodova M, Srivastava RK, Shankar S (2011) Targeting epigenetic regulation of miR-34a for treatment of pancreatic cancer by inhibition of pancreatic cancer stem cells. PLoS One 6(8):e24099

Nam EJ, Lee M, Yim GW, Kim JH, Kim S, Kim SW et al (2012) MicroRNA profiling of a CD133(+) spheroid-forming subpopulation of the OVCAR3 human ovarian cancer cell line. BMC Med Genomics 5:18

Nie Y, Han BM, Liu XB, Yang JJ, Wang F, Cong XF et al (2011) Identification of MicroRNAs involved in hypoxia- and serum deprivation-induced apoptosis in mesenchymal stem cells. Int J Biol Sci 7(6):762–768

Olson P, Lu J, Zhang H, Shai A, Chun MG, Wang Y et al (2009) MicroRNA dynamics in the stages of tumorigenesis correlate with hallmark capabilities of cancer. Genes Dev 23(18):2152–2165

Pandey DP, Picard D (2009) miR-22 inhibits estrogen signaling by directly targeting the estrogen receptor alpha mRNA. Mol Cell Biol 29(13):3783–3790

Pang Y, Young CY, Yuan H (2010) MicroRNAs and prostate cancer. Acta Biochim Biophys Sin (Shanghai) 42(6):363–369

Patrawala L, Calhoun T, Schneider-Broussard R, Zhou J, Claypool K, Tang DG (2005) Side population is enriched in tumorigenic, stem-like cancer cells, whereas ABCG2+ and A. Cancer Res 65(14):6207–6219

Perera RJ, Ray A (2007) MicroRNAs in the search for understanding human diseases. BioDrugs 21(2):97–104

Peruzzi P, Bronisz A, Nowicki MO, Wang Y, Ogawa D, Price R et al (2013) MicroRNA-128 coordinately targets Polycomb Repressor Complexes in glioma stem cells. Neuro Oncol 15(9):1212–1224

Peter ME (2009) Let-7 and miR-200 microRNAs: guardians against pluripotency and cancer progression. Cell Cycle 8(6):843–852

Puissegur MP, Mazure NM, Bertero T, Pradelli L, Grosso S, Robbe-Sermesant K et al (2011) miR-210 is overexpressed in late stages of lung cancer and mediates mitochondrial alterations associated with modulation of HIF-1 activity. Cell Death Differ 18(3):465–478

Quero L, Dubois L, Lieuwes NG, Hennequin C, Lambin P (2011) miR-210 as a marker of chronic hypoxia, but not a therapeutic target in prostate cancer. Radiother Oncol 101(1):203–208

Redova M, Poprach A, Nekvindova J, Iliev R, Radova L, Lakomy R et al (2012) Circulating miR-378 and miR-451 in serum are potential biomarkers for renal cell carcinoma. J Transl Med 10:55

Ren D, Wang M, Guo W, Zhao X, Tu X, Huang S et al (2013) Wild-type p53 suppresses the epithelial-mesenchymal transition and stemness in PC-3 prostate cancer cells by modulating miR145. Int J Oncol 42(4):1473–1481

Sakurai T, Bilim VN, Ugolkov AV, Yuuki K, Tsukigi M, Motoyama T et al (2012) The enhancer of zeste homolog 2 (EZH2), a potential therapeutic target, is regulated by miR-101 in renal cancer cells. Biochem Biophys Res Commun 422(4):607–614

Sarkar FH, Li Y (2009) Harnessing the fruits of nature for the development of multi-targeted cancer therapeutics. Cancer Treat Rev 35(7):597–607

Sarkar FH, Li Y, Wang Z, Kong D (2009) Pancreatic cancer stem cells and EMT in drug resistance and metastasis. Minerva Chir 64(5):489–500

Sarkar FH, Li Y, Wang Z, Padhye S (2010) Lesson learned from nature for the development of novel anti-cancer agents: implication of isoflavone, curcumin, and their synthetic analogs. Curr Pharm Des 16(16):1801–1812

Semaan A, Qazi AM, Seward S, Chamala S, Bryant CS, Kumar S et al (2011) MicroRNA-101 inhibits growth of epithelial ovarian cancer by relieving chromatin-mediated transcriptional repression of p21(waf(1)/cip(1)). Pharm Res 28(12):3079–3090

Shafee N, Smith CR, Wei S, Kim Y, Mills GB, Hortobagyi GN et al (2008) Cancer stem cells contribute to cisplatin resistance in Brca1/p53-mediated mouse mammary tumors. Cancer Res 68(9):3243–3250

Shao Y, Qu Y, Dang S, Yao B, Ji M (2013) MiR-145 inhibits oral squamous cell carcinoma (OSCC) cell growth by targeting c-Myc and Cdk6. Cancer Cell Int 13(1):51

Shimono Y, Zabala M, Cho RW, Lobo N, Dalerba P, Qian D et al (2009) Downregulation of miRNA-200c links breast cancer stem cells with normal stem cells. Cell 138(3):592–603

Siemens H, Jackstadt R, Hunten S, Kaller M, Menssen A, Gotz U et al (2011) miR-34 and SNAIL form a double-negative feedback loop to regulate epithelial-mesenchymal transitions. Cell Cycle 10(24):4256–4271

Smits M, Nilsson J, Mir SE, van der Stoop PM, Hulleman E, Niers JM et al (2010) miR-101 is down-regulated in glioblastoma resulting in EZH2-induced proliferation, migration, and angiogenesis. Oncotarget 1(8):710–720

Solomides CC, Evans BJ, Navenot JM, Vadigepalli R, Peiper SC, Wang ZX (2012) MicroRNA profiling in lung cancer reveals new molecular markers for diagnosis. Acta Cytol 56(6):645–654

Sparmann A, van Lohuizen M (2006) Polycomb silencers control cell fate, development and cancer. Nat Rev Cancer 6(11):846–856

Strillacci A, Valerii MC, Sansone P, Caggiano C, Sgromo A, Vittori L et al (2013) Loss of miR-101 expression promotes Wnt/beta-catenin signalling pathway activation and malignancy in colon cancer cells. J Pathol 229(3):379–389

Su Y, Simmen RC (2009) Soy isoflavone genistein upregulates epithelial adhesion molecule E-cadherin expression and attenuates beta-catenin signaling in mammary epithelial cells. Carcinogenesis 30(2):331–339

Su Y, Simmen FA, Xiao R, Simmen RC (2007) Expression profiling of rat mammary epithelial cells reveals candidate signaling pathways in dietary protection from mammary tumors. Physiol Genomics 30(1):8–16
Sun F, Fu H, Liu Q, Tie Y, Zhu J, Xing R et al (2008) Downregulation of CCND1 and CDK6 by miR-34a induces cell cycle arrest. FEBS Lett 582(10):1564–1568
Takaoka Y, Shimizu Y, Hasegawa H, Ouchi Y, Qiao S, Nagahara M et al (2012) Forced expression of miR-143 represses ERK5/c-Myc and p68/p72 signaling in concert with miR-145 in gut tumors of Apc(Min) mice. PLoS One 7(8):e42137
Tavazoie SF, Alarcon C, Oskarsson T, Padua D, Wang Q, Bos PD et al (2008) Endogenous human microRNAs that suppress breast cancer metastasis. Nature 451(7175):147–152
Tian Y, Nan Y, Han L, Zhang A, Wang G, Jia Z et al (2012) MicroRNA miR-451 downregulates the PI3K/AKT pathway through CAB39 in human glioma. Int J Oncol 40(4):1105–1112
Todaro M, Alea MP, Di Stefano AB, Cammareri P, Vermeulen L, Iovino F et al (2007) Colon cancer stem cells dictate tumor growth and resist cell death by production of interleukin-4. Cell Stem Cell 1(4):389–402
Tsuchiya N, Izumiya M, Ogata-Kawata H, Okamoto K, Fujiwara Y, Nakai M et al (2011) Tumor suppressor miR-22 determines p53-dependent cellular fate through post-transcriptional regulation of p21. Cancer Res 71(13):4628–4639
Wagner J, Lehmann L (2006) Estrogens modulate the gene expression of Wnt-7a in cultured endometrial adenocarcinoma cells. Mol Nutr Food Res 50(4–5):368–372
Wang Z, Desmoulin S, Banerjee S, Kong D, Li Y, Deraniyagala RL et al (2008) Synergistic effects of multiple natural products in pancreatic cancer cells. Life Sci 83(7–8):293–300
Wang X, Meyers C, Guo M, Zheng ZM (2011) Upregulation of p18Ink4c expression by oncogenic HPV E6 via p53-miR-34a pathway. Int J Cancer 129(6):1362–1372
Wang T, Zang WQ, Li M, Wang N, Zheng YL, Zhao GQ (2013) Effect of miR-451 on the biological behavior of the esophageal carcinoma cell line EC9706. Dig Dis Sci 58(3):706–714
Xie P, Xu F, Cheng W, Gao J, Zhang Z, Ge J et al (2012) Infiltration related miRNAs in bladder urothelial carcinoma. J Huazhong Univ Sci Technol Med Sci 32(4):576–580
Xiong J, Yu D, Wei N, Fu H, Cai T, Huang Y et al (2010) An estrogen receptor alpha suppressor, microRNA-22, is downregulated in estrogen receptor alpha-positive human breast cancer cell lines and clinical samples. FEBS J 277(7):1684–1694
Xiong SW, Lin TX, Xu KW, Dong W, Ling XH, Jiang FN et al (2013) MicroRNA-335 acts as a candidate tumor suppressor in prostate cancer. Pathol Oncol Res 19(3):529–537
Xu Y, Zhao F, Wang Z, Song Y, Luo Y, Zhang X et al (2012) MicroRNA-335 acts as a metastasis suppressor in gastric cancer by targeting Bcl-w and specificity protein 1. Oncogene 31(11):1398–1407
Yamada N, Noguchi S, Mori T, Naoe T, Maruo K, Akao Y (2013) Tumor-suppressive microRNA-145 targets catenin delta-1 to regulate Wnt/beta-catenin signaling in human colon cancer cells. Cancer Lett 335(2):332–342
Yamakuchi M, Yagi S, Ito T, Lowenstein CJ (2011) MicroRNA-22 regulates hypoxia signaling in colon cancer cells. PLoS One 6(5):e20291
Yan Z, Xiong Y, Xu W, Gao J, Cheng Y, Wang Z et al (2012) Identification of hsa-miR-335 as a prognostic signature in gastric cancer. PLoS One 7(7):e40037
Yin S, Li J, Hu C, Chen X, Yao M, Yan M et al (2007) CD133 positive hepatocellular carcinoma cells possess high capacity for tumorigenicity. Int J Cancer 120(7):1444–1450
Yin Y, Yan ZP, Lu NN, Xu Q, He J, Qian X et al (2013) Downregulation of miR-145 associated with cancer progression and VEGF transcriptional activation by targeting N-RAS and IRS1. Biochim Biophys Acta 1829(2):239–247
Yoshino H, Seki N, Itesako T, Chiyomaru T, Nakagawa M, Enokida H (2013) Aberrant expression of microRNAs in bladder cancer. Nat Rev Urol 10(7):396–404
Yu F, Yao H, Zhu P, Zhang X, Pan Q, Gong C et al (2007) let-7 regulates self renewal and tumorigenicity of breast cancer cells. Cell 131(6):1109–1123

Yu F, Deng H, Yao H, Liu Q, Su F, Song E (2010) Mir-30 reduction maintains self-renewal and inhibits apoptosis in breast tumor-initiating cells. Oncogene 29(29):4194–4204

Yu C, Yao Z, Jiang Y, Keller ET (2012) Prostate cancer stem cell biology. Minerva Urol Nefrol 64(1):19–33

Yu CC, Tsai LL, Wang ML, Yu CH, Lo WL, Chang YC et al (2013a) miR145 targets the SOX9/ADAM17 axis to inhibit tumor-initiating cells and IL-6-mediated paracrine effects in head and neck cancer. Cancer Res 73(11):3425–3440

Yu Y, Sarkar FH, Majumdar AP (2013b) Down-regulation of miR-21 induces differentiation of chemoresistant colon cancer cells and enhances susceptibility to therapeutic regimens. Transl Oncol 6(2):180–186

Zhang B, Pan X, Cobb GP, Anderson TA (2007) microRNAs as oncogenes and tumor suppressors. Dev Biol 302(1):1–12

Zhang Z, Sun H, Dai H, Walsh RM, Imakura M, Schelter J et al (2009) MicroRNA miR-210 modulates cellular response to hypoxia through the MYC antagonist MNT. Cell Cycle 8(17):2756–2768

Zhang J, Yang Y, Yang T, Liu Y, Li A, Fu S et al (2010) microRNA-22, downregulated in hepatocellular carcinoma and correlated with prognosis, suppresses cell proliferation and tumourigenicity. Br J Cancer 103(8):1215–1220

Zhang JG, Guo JF, Liu DL, Liu Q, Wang JJ (2011) MicroRNA-101 exerts tumor-suppressive functions in non-small cell lung cancer through directly targeting enhancer of zeste homolog 2. J Thorac Oncol 6(4):671–678

Zhang J, Han C, Zhu H, Song K, Wu T (2013) miR-101 inhibits cholangiocarcinoma angiogenesis through targeting vascular endothelial growth factor (VEGF). Am J Pathol 182(5):1629–1639

Zhong X, Li N, Liang S, Huang Q, Coukos G, Zhang L (2010) Identification of microRNAs regulating reprogramming factor LIN28 in embryonic stem cells and cancer cells. J Biol Chem 285(53):41961–41971

Zhu Y, Yu F, Jiao Y, Feng J, Tang W, Yao H et al (2011) Reduced miR-128 in breast tumor-initiating cells induces chemotherapeutic resistance via Bmi-1 and ABCC5. Clin Cancer Res 17(22):7105–7115

Part II
MicroRNAs and Their Clinical Implications

Chapter 13
MicroRNAs and Clinical Implications in Cancer

Jürgen Wittmann

Contents

J. Wittmann (✉)
Division of Molecular Immunology, Department of Internal Medicine III,
Nikolaus-Fiebiger Center of Molecular Medicine, University of Erlangen-Nürnberg,
Glückstraße 6, D-91054 Erlangen, Germany
e-mail: jwittman@molmed.uni-erlangen.de

S. Babashah (ed.), *MicroRNAs: Key Regulators of Oncogenesis*,
DOI 10.1007/978-3-319-03725-7_13,

Abstract MicroRNAs (miRNAs) are 18–25 nucleotides long, non-coding RNA molecules that regulate gene expression at the posttranscriptional level. Due to their fundamental roles in development and differentiation, corruption of their expression can lead to the development of disease. In fact, numerous studies have revealed that miRNAs play important regulatory roles in virtually every cancer type studied so far. In this book chapter, I review the current literature to pinpoint the role of miRNAs and their clinical implications in cancer. After presenting some examples to illustrate the roles of miRNAs in several types of cancer, I discuss miRNAs as potential cancer biomarkers for diagnosis, prognosis and therapy and also explain their advantages and limitations. I describe what is currently known about the use of miRNAs as potential predictors of therapeutic outcome and discuss the potential of integrating miRNA data into clinical trials and if and how miRNA could be implemented into "personalized medicine" approaches. Finally, I outline the perspectives and challenges of miRNAs and their clinical implications in cancer and discuss criteria that must be fulfilled before miRNAs can be used as reliable tools in diagnostic, prognostic and therapeutic settings.

Keywords MicroRNA • Cancer • Biomarker • Prognosis • Diagnosis • Therapy

1 The Potential Roles of MicroRNAs in Cancer

The high conservation of microRNA (miRNA) genes, their expression in every tissue and cell type in multicellular eukaryotes as well as their demonstrated involvement in fundamental processes like cellular proliferation, programmed cell death, differentiation, cell motility and invasiveness early suggested their important roles in diseases and cancers [reviewed in Esquela-Kerscher and Slack (2006)]. In fact, numerous associations of miRNAs with virtually all cancer types have meanwhile been demonstrated. The potential relevance of miRNAs in cancer and their clinical implications will be addressed in the following chapters for some of the most widespread cancer types, where miRNAs have not only shed light on the molecular biology of the respective cancer type, but also promise to improve cancer diagnosis, prognosis, and treatment.

1.1 Breast Cancer

The oldest description of breast cancer (BC) dates back to an approximately 5,000-year-old papyrus inscription discovered in Egypt ("The History of Cancer", American Cancer society). Although, of course, diagnosis and treatment options for BC have strongly improved from back then, BC still remains the most frequent cancer among American women, with an estimated 232,000 new cases diagnosed and about 40,000 predicted deaths in 2013, according to data of the American Cancer Society (http://www.cancer.org). Early BC detection, as with many cancers, is important and is still inadequate. Invasive procedures like tissue biopsies for histological examination are still the methods of choice for determining diagnosis and prognosis of BC.

The breast is made up of glandular, fibrous, and adipose tissue. Most commonly, BC originates from the inner lining of the milk ducts (ductal carcinomas) or the lobules supplying the ducts with milk (lobular carcinomas). In the mammary gland, multipotent epithelial stem cells and progenitors are thought to generate and maintain both myo-epithelial and luminal epithelial lineages. What makes BC a complicated disease to understand and to control is its' complex development. Deregulation can already take place in mammary stem cells, leading to the development of BC stem cells or metastatic cancer stem cells. In its' final form, therapy-resistant cancer stem cells can arise.

Many different *in vivo* and *in vitro* models to study BC development exist. It is therefore not surprising that using these models, several conducted miRNA profiling studies reported the involvement of a multitude of miRNAs during steps of cancer initiation, metastasis and therapy resistance (Sempere et al. 2007; Lowery et al. 2009; Van der Auwera et al. 2010; Janssen et al. 2010; Radojicic et al. 2011; Sieuwerts et al. 2011; Farazi et al. 2011; Rothé et al. 2011; Tjensvoll et al. 2012; Schrauder et al. 2012; Guo et al. 2013). Although evidently every model has its' limitations, the following three exemplary studies have helped to advance our understanding of BC at the molecular level and to increase our understanding of the biology of this disease. The final goal must be to identify BC biomarkers that contribute to more favorable patient outcomes by helping in early diagnosis, the determination of prognosis and possibly the prediction of treatment response, ideally also in patient subgroups.

The study of Iorio and colleagues in the year 2005 was one of the first researches to investigate different miRNA expression between normal breast- and tumor tissue (Iorio et al. 2005). They used miRNA microarray profiling to analyze 76 BC and 10 normal breast samples and identified in total 29 miRNAs whose expression was significantly deregulated. Some miRNAs like miR-21 and miR-155 were up-regulated in BC, while others like miR-10b and miR-145 were down-regulated. These findings could suggest that these miRNAs may act as oncogenes and tumor suppressor genes, respectively, and this could also point the way towards utilizing these discoveries therapeutically by sequestering or over-expressing these miRNAs. Of interest, a set of 15 miRNAs was able to correctly predict the nature of the analyzed sample, i.e., normal breast tissue or BC tissue. Learning about the mRNA targets of these miRNAs may reveal additional information at the molecular level on the biological function of these miRNAs in BC. The authors interestingly also found miRNAs that were differentially expressed in BC with various distinctive biopathologic features like estrogen receptor (ER) or progesterone receptor (PR) status, proliferation index, human epidermal growth factor receptor 2 (HER2) status, lymph node status or the presence or absence of vascular invasion. Although the number of differentially expressed miRNAs distinguishing these conditions was rather small, this could also reflect technical limitations, because at that time, only 245 miRNA probes were spotted onto the microarray slide.

To classify molecular BC tumor subtypes such as luminal A, luminal B, basal-like, normal-like or HER2-positive BC, Blenkiron et al. (2007) used a bead-based miRNA expression profiling method to show that a number of individual miRNAs were associated with these clinicopathological factors. The predictive potential of the identified miRNAs was tested by classifying basal versus luminal subtypes in an independent, although small data set. The authors noted that they only found little agreement between the miRNAs they identified as associated with clinicopathological factors when compared to the data of Iorio and colleagues (2005). Probably, mixed cell populations consisting of tumor and stromal cells to a different degree were used for profiling analyses, pointing towards the important pre-requisite to use highly purified, uniform cell populations for profiling. This of course also applies to profiling BC subpopulations like tumor initiating cells or cancer stem cells. Interestingly, the authors also discovered that deregulation of miRNA biogenesis pathway genes like DICER1 could also be involved in the etiology of BC, as DICER1 expression is significantly down-regulated in basal-like, HER2-positive and luminal B type tumors.

A recent study by Buffa et al. (2011) was the first large study to integrate global miRNA- and mRNA profiles in BC to identify miRNAs associated with distant relapse-free survival (DRFS) that provide independent prognostic information. They identified four miRNAs independently associated with DRFS in ER-positive and six in ER-negative cases. Detailed analyses revealed that these prognostic miRNAs are associated with key biological processes in BC such as proliferation (miR-135a), hypoxia (miR-210, miR-342), invasion (miR-27b), or immune responses (miR-150). When validated in independent cohorts, these results could elucidate potential novel therapeutic targets in BC.

1.2 Ovarian Cancer

Ovarian cancer (OC) comprises approximately 25 % of all gynecological cancers, making it the sixth leading tumor of women in the Western world. Due to a lack of effective screening programs for early detection and absent or only subtle early signs and symptoms of OC, more than 75 % of all patients present with an already advanced stage disease. The significant heterogeneity of OC comprising histologically defined subgroups such as serous, mucinous, endometrioid and clear cell carcinomas, as well as the variable clinical manifestations and underlying molecular genetic events further complicate diagnosis and treatment of OC. Inactivation of tumor suppressor genes or the activation of oncogenes are as in other human cancer types the major contributors to OC. The complex biological process during OC initiation and progression involves cell de-differentiation and proliferation, angiogenesis, invasion and metastasis. Treatment usually consists of a combination of surgery and chemotherapy or radiotherapy; for advanced stages, an antibody-based therapy targeting VEGF is available.

Improved outcomes in OC could be achieved by a more successful early detection and an enhanced understanding of the molecular basis of disease, which would both lead to improved therapies. Various groups have meanwhile studied the role of miRNAs in OC. In analogy to the heterogeneity of OC, investigation of many different specimen (OC samples vs. normal ovarian tissues and ovarian epithelia cell lines) with many different methods [comparative genomic hybridization, microarray analysis, next-generation sequencing (NGS), quantitative reverse transcriptase polymerase chain reaction (qRT-PCR) analysis], also lead to heterogeneous results in the discovered up- and down-regulated miRNA patterns (Iorio et al. 2007; Zhang et al. 2008; Lee et al. 2009). However, some miRNAs were nevertheless identified as aberrantly expressed in more than one study, suggesting their likely involvement in OC.

One of the first NGS studies in OC reported an inventory of miRNAs known to be expressed in primary cultures of normal human ovarian surface epithelium and in tissue from three of the most common histotypes of OC (Wyman et al. 2009). This group discovered 124 miRNAs that were differentially expressed in normal versus OC samples and, in addition to some miRNAs that were conserved between OC subtypes, 38 miRNAs were differentially expressed across histologic OC subtypes. This is of special interest, as alterations in OC subtype-specific miRNA patterns could indicate their potential as new biomarkers for OC subtypes. For example, miR-449a was shown to be a serous-specific miRNA, while miRNAs miR-499-5p, miR-375, miR-196a, miR-196b and miR-182 are endometrioid-specific and miR-486-5p, miR-144, miR-30a, and miR-199a-5p are clear cell-specific.

Heterogeneity of OC also applies for the many pathways involved in OC cell proliferation and apoptosis. Several miRNAs fine-tuning important pathway master regulator proteins like p53, BRAF, PTEN or cell cycle genes have been reported [reviewed in Mezzanzanica et al. (2011); Lech et al. (2013)]. To get a global understanding of miRNAs involved in cell proliferation and cell death in OC, Nakana and colleagues performed a gain-of-function miRNA screen by transfecting 319 miRNA species into OC cells (Nakano et al. 2013). In addition to discovering miRNAs that increased or decreased cell viability or miRNAs that asserted pro-proliferative and anti-proliferative effects, the authors also revealed miRNAs affecting cell cycle. miR-193a was identified as a strong anti-proliferative miRNA, as it induced the inhibition of DNA synthesis as measured by BrdU incorporation and induced activation of caspase 3/7, resulting in apoptotic cell death. The possible tumor suppressive role of this miRNA in OC cells could be attributed to mRNA targets like ARHGAP19, CCND1, ERBB4, KRAS, and MCL1, which were identified in a genome-wide gene expression analysis in OC cells over-expressing this miRNA.

Ideally, miRNAs could be used for prognostic purposes to predict outcome of treatment, but at the same time, they could also serve as a potential therapeutic tool. Depending on whether a miRNA has oncogenic or tumor suppressive properties, miRNA silencing by antisense oligonucleotides or antagomirs or miRNA restoration by application of chemically modified synthetic miRNA mimics could be employed. In advanced OC, chemotherapy is the preferred therapeutical approach. However, chemoresistance to platinum-based drugs coupled with the mitotic inhibitor paclitaxel is the main limitation to successful long-term treatment. Several studies therefore investigated if selected miRNAs were involved in the development of drug-resistance in OC and if this drug resistance was associated with a distinct miRNA fingerprint: while a panel of seven miRNAs (higher expression in platinum-based chemotherapy resistant OC: miR-27a, miR-23a, miR-30c, let-7g, miR-199a-3p; higher expression in platinum-based chemotherapy sensitive OC: miR-378, miR-625) could indicate platinum-resistance (Eitan et al. 2009), down-regulation of the three miRNAs miR-30c, miR-130a, and miR-335 suggested chemoresistance to platinum-based drugs and paclitaxel (Sorrentino et al. 2008). These findings could imply that miRNAs might not only be used as a prognostic tool to predict and monitor the outcome of chemotherapy, but that miRNAs might also serve as potential therapeutic tools for modulating the response to chemotherapy in OC.

1.3 Prostate Cancer

Prostate cancer (PCa) is a malign tumor disease originating from the prostate, a gland in the male reproductive system. Besides lung and colorectal cancer, PCa is the third most deadly cancer disease in men and responsible for approximately 10 % of male cancer deaths. In its early phase, the disease is without signs or symptoms, while in the advanced stage, one or several of symptoms like frequent urination, nocturia, hematuria or dysuria can occur. Diagnosis of PCa involves prostate imaging by ultrasound and magnetic resonance imaging (MRI). The prostate-specific antigen (PSA) test measures the blood level of PSA, a glycoprotein enzyme that is produced and secreted by epithelial cells of the prostate gland. The initial assumption that the higher a man's PSA level, the more likely it is that he has PCa has changed in recent years, as there are additional reasons for having an elevated PSA level, and also some men with PCa who do not have elevated PSA levels. Therefore, the U.S. Preventive Services Task Force recommends against PSA-based screening for PCa: due to a high number of false-positive results, overdiagnosis and overtreatment, the potential benefit does not outweigh the expected harms (http://www.uspreventiveservicestaskforce.org/prostatecancerscreening/prostatecancerscript.pdf; retrieved August 05, 2013). Only an invasive biopsy can reveal if the patient really suffers from PCa. Staging the tumor will indicate if the tumor already metastasized or if it is still restricted to the prostate and will also help to prognose and to select appropriate therapies. PCa most commonly metastasizes to the bones and lymph nodes. Pillars of therapy are surgical intervention, radiation therapy or suppression of androgen production by surgical or chemical castration. Even after decades of research, the exact causes of PCa remain elusive. Risk factors like obesity, age, genetic background or dietary factors are imprecisely defined, necessitating new research avenues like studying miRNAs involved in PCa biology, diagnosis or prognosis.

The androgen ablation therapy (AAT) of PCa dates back to a publication in 1941, when Charles Huggins and colleagues reported that AAT causes regression of primary and metastatic androgen-dependent prostate cancer (Huggins et al. 1941). While more than 80 % of PCa patients undergo remission after AAT for a median progression-free survival of 1–3 years [androgen-dependent (AnD) state], an androgen-independent (AnI) phenotype inevitably emerges, for which no effective treatment has been developed yet. To shed light on the mechanisms for the development and progression of PCa, Shi and co-workers studied miRNA patterns in prostate cell lines, in benign and malignant cells, in androgen receptor (AR)- positive and -negative cells, and in AnD and AnI cells (Shi et al. 2007). They discovered differential expression of miR-125b in AnD and AnI PCa cells, as well as in benign and malignant prostate tissues. Interestingly, androgen signaling leads to an up-regulation of miR-125b expression, possibly by androgen responsive elements in the miR-125b-2 promoter region. Mechanistically, transfection of miR-125b stimulated AnI growth of PCa cells and down-regulated the expression of Bak1, which enhances apoptotic cell death. Therefore, miR-125b might contribute to the pathogenesis of PCa by acting as an oncogene.

Follow-up studies of the same group investigated the precise mechanism of miR-125b on the pathogenesis of PCa (Shi et al. 2011). They found that enforced expression of miR-125b promoted xenograft tumor growth in intact and castrated male athymic mice when subcutaneously injected with PCa cells that stably over-expressed miR-125b. The authors further discovered that miR-125b directly targets eight mRNA transcripts and attenuates apoptosis by targeting three key pro-apoptotic genes BAK1, BBC3, and p53. Repression of miR-125b activity enabled sensitization of PCa cells to different therapeutic interventions, possibly making it an attractive therapeutic target in PCa.

In the clinical context, improved approaches for noninvasive and reliable PCa detection are urgently needed. Measurement of tumor-derived miRNAs in the blood (serum or plasma) could be a promising avenue, as shown by Mitchell and co-authors (2008). In a mouse xenograft model and in human cancer patients, Mitchell et al. showed that serum levels of miR-141 can distinguish patients with PCa from healthy controls. The finding of miR-141 as PCa serum biomarker has meanwhile been confirmed by several studies. Brase and co-workers also discovered increased levels of miR-141 (together with miR-375) in serum samples from patients with metastatic and localized PCa (Brase et al. 2011). Yaman and colleagues reported that miR-141 levels in serum were higher in patients diagnosed with metastatic PCa than in patients with localized/local advanced disease (Yaman et al. 2011). Selth and co-workers demonstrated that miR-141 was commonly increased in serum of PCa patients as well as in a PCa mouse model (Selth et al. 2012). A report by Gonzales and collaborators investigating the role of miR-141 as a potential biomarker of therapeutic response in PCa patients came to the conclusion that miR-141 demonstrated a similar ability to predict clinical progression when compared with other clinically validated biomarkers like PSA, circulating tumor cells or lactate dehydrogenase levels (Gonzales et al. 2011).

1.4 Melanoma

The skin that protects our body from heat, light irradiation, infections and injuries consists of two main layers, the epidermis and the dermis. The epidermis contains melanocytes, which contribute as pigment-producing cells by the production of melanin to the individual skin color and to the protection against UV irradiation. Malignant melanoma (MM) is a tumor which arises from these cells. Although MM only makes up 3 % of all skin neoplasias, it accounts for 65 % of skin cancer deaths. One of the hallmarks of MM is it's high potential to metastasize and it's high resistance to apoptosis and chemotherapy. Early diagnosis is therefore of crucial relevance, as the excision of very thin tumors can only occur successfully when not already micrometastases have developed, which will later form macrometastases.

Possibly because of changes in spare time activities and increased exposure to UV irradiation, incidences in MM have increased in the last decades all over the world, making enforced innovative research on the biology and treatment of MM mandatory. A pivotal requirement will be the availability of biomarkers, which could allow the selection of patients and tumors which respond to the planned therapy. Such biomarkers are lacking until now. MicroRNAs could show new ways for diagnostic and, in the long term, also for therapeutic approaches.

Several steps have to occur before melanocytes turn into melanoma cells. In the last years, several studies have been performed to examine the role of miRNAs during these processes. While many of them yielded interesting insights into melanoma biology and identified potential mRNA targets of miRNAs deregulated in MM, one has to be aware of the limitations of these studies, like comparison of melanoma cell lines with cultured melanocytes or tumor tissues, which often still contain fibroblasts or lymphocytes. Also, most of the studies performed until now were done *in vitro*, which sometimes questions their relevance to human disease. Finally, one of the major aspects of miRNA biology and function is sometimes not fully appreciated: it is only in very rare cases that one miRNA along with only one mRNA target is responsible for the investigated biological effects. Normally, one miRNA molecule regulates many mRNA targets and also, one mRNA can be regulated by many miRNAs. Therefore, oversimplification of models might underestimate the role of miRNAs in modulating global gene expression and protein abundance.

One of the first essential steps in MM is uncontrolled proliferation, which is caused by dysregulation of cell cycle proteins or regulators of proliferation. Cyclins and cyclin-dependent kinases have been shown to be targets for several miRNAs. Schultz and colleagues analyzed miRNA expression in laser-microdissected tissues from benign melanocytic nevi and primary MMs by qRT-PCR analysis and discovered differential expression of several miRNAs (Schultz et al. 2008). In primary melanomas as compared to benign nevi, several members of the let-7 miRNA family were down-regulated, acting possibly as tumor suppressors in MM. The authors could show that when let-7 was ectopically expressed in melanoma cell lines, the expression of several cyclins as well as cyclin-dependent kinase 4 was repressed, which lead to inhibition of cell cycle progression and anchorage-independent growth of melanoma cells.

For progression of tumor cells to metastases, several properties like migratory capacity or ability to invade tissues to enter the blood or the lymphatic circulation need to be acquired. Many studies have revealed that miRNAs indeed are able to modulate these properties in melanoma cells. Cellular targets for miRNAs are adhesion molecules like integrins (Penna et al. 2011), the receptor c-MET important for invasive growth (Migliore et al. 2008) or MITF, the microphthalmia-associated transcription factor important for MM plasticity [reviewed by Bell and Levy (2011)]. One group of enzymes important for remodeling the extracellular matrix are matrix metalloproteinases (MMPs). Studying the regulation of Basigin, a protein which stimulates adjacent fibroblasts or tumor cells to produce MMPs, Fu and colleagues identified let-7b as a tumor suppressor that represses MM proliferation and migration as well as tumor metastasis in melanoma cells (Fu et al. 2011). let-7b suppresses the expression of Basigin in melanoma cells, which could result in the indirect suppression of MMP-9 and the ability to invade or migrate to neighboring tissues.

Especially in MM, early detection is very important for successful treatment. Possibly, miRNA might be used as valuable diagnostic and prognostic biomarkers if they faithfully can follow disease progression or predict response to therapy. First analyses of miRNAs as noninvasive biomarkers for MM have been reported, e.g. from Leidinger and collaborators by analyzing miRNA expression signatures of blood cells (Leidinger et al. 2010). miRNA microarray analysis was used to detect deregulated miRNAs in blood cells of 35 MM patients when compared to blood samples from 20 healthy individuals. The authors could show that analyzing a subset of 16 deregulated miRNAs enabled them to reach a classification accuracy of 97.4 %, a specificity of 95 % and a sensitivity of 98.9 % by supervised analysis.

Analyzing miRNAs in formalin-fixed paraffin embedded (FFPE) tissue samples is a good way to conduct retrospective studies, as there is a very good correlation to the miRNome in frozen material (Glud et al. 2009). This technique also enables the definition of a miRNA expression signature that is predictive of disease outcome. Segura and co-workers determined the miRNA profile of 59 FFPE melanoma metastases and correlated the miRNA expression with post-recurrence survival and other clinicopathologic criteria (Segura et al. 2010). They found that higher expression of 18 miRNAs was significantly correlated with longer survival, defined as more than 18 months post-recurrence survival. A 6 miRNA signature was able to stratify stage III MM patients into "better" and "worse" prognostic categories, predicting post-recurrence survival in patients with an estimated accuracy of 80.2 %.

1.5 Leukemia

Leukemias are hematological neoplasms of the blood or the lymphoid system, which are characterized by a strong increase in white blood cell (leukocyte) counts and especially dysfunctional leukocyte precursors. They can spread to the bone marrow, compromise normal hematopoiesis and lead to an altered composition of normal blood components. Due to a lack of oxygen-carrying erythrocytes, blood clotting thrombocytes and of functional leukocytes, anemia can develop. Infiltration of organs like spleen, liver or lymph nodes can affect their functions.

Leukemias are usually diagnosed by blood count analysis and/or a bone marrow aspiration. Classification is based on morphological, immunological as well as cytogenetic and molecular parameters of the leukemic cells. Depending on the cell type involved, one distinguishes myeloid from lymphatic leukemias. While myeloid leukemias arise from precursor cells of granulocytes, lymphatic leukemias affect lymphocytes and their precursor cells. Depending on the course of disease, acute and chronic leukemias can be distinguished: acute leukemias are live-threatening and can lead to death in a few weeks or months if not treated. Chronic leukemias progress mostly over years and often show no signs or symptoms. Aberrant expression of miRNAs is a common characteristic of leukemias (Babashah et al. 2012). In the following passages, the four most important leukemia diseases will be described along with the role some miRNAs have been found to play in these diseases.

1.5.1 Chronic Lymphocytic Leukemia

Chronic lymphocytic leukemia (CLL) is the most often occurring leukemic disease in the Western world and most often occurs with advanced age rather than in teenagers or children (the median age at first diagnosis is between 70 and 75 years). Depending on the status of the markers zeta-chain (TCR)-associated protein kinase 70 kDa (ZAP-70) and CD38, two major types of CLL can be distinguished: CD38- and ZAP-70-positive CLL cases have an average survival of 8 years, while CD38- and ZAP-70-negative CLL cases have an average survival of more than 25 years. In contrast to therapies of acute leukemias, therapeutic approaches of chronic forms mostly rely on less aggressive, but therefore continuous treatments. At first, the "watch and wait" strategy is applied for CLL patients; current standard therapies are chemotherapy or a therapy with cortisol analogs. Some patients are also treated with the monoclonal antibodies Alemtuzumab, which targets CD52 on B- and T cells, or Rituximab, which targets CD20 on B cells, in combination with chemotherapy. A complete cure with eradication of all leukemic cells with these treatments is not possible until now, but the symptoms can be alleviated.

A very high impact paper by the group of Carlo Croce in 2002 (cited more than 1,750 times until August 2013) was the first to draw a link between the observation that a part of chromosome 13 (13q14.3) was frequently deleted or altered in many patients with CLL and the deregulation of miRNAs in CLL (Calin et al. 2002). The clustered miRNAs miR-15a/16-1 were discovered to reside in this deleted region and were subsequently shown to be essential for CLL pathogenesis by the generation of several transgenic CLL mouse models by Klein and colleagues (2010; Lia et al. 2012). Among the identified target genes of miR-15a/16-1 are the anti-apoptotic protein Bcl2 (Cimmino et al. 2005) or the cell cycle regulators Cyclin D1, Cyclin D3, Cyclin E1, CDK6 (Liu et al. 2008) and Cyclin T2 (Teng et al. 2011). Recently, a novel feedback loop was identified in which the miR-15a/miR-16-1 cluster directly targets p53 and its downstream effectors, while p53 stimulates the transcription of miR-15a/miR-16-1. Interestingly, p53 also stimulates the miR-34b/miR-34c cluster, which directly targets the ZAP-70 kinase (Fabbri et al. 2011).

1.5.2 Chronic Myeloid Leukemia

The transformation and subsequent unrestricted proliferation of a single pluripotent hematopoietic precursor cell is mostly the reason for chronic myeloid leukemia (CML). In almost all CML cases, a reciprocal translocation between the chromosomes 9 and 22 [t(9;22)] is involved. The break point involves the ABL1 gene on chromosome 9 and the BCR ("breakpoint cluster region") on chromosome 22. Cytogenetically, this translocation is visible as a shortened chromosome 22, the so-called "Philadelphia chromosome". The ABL1 gene, which encodes for a tyrosine-kinase, is critically mutated by the generation of the BCR-ABL1 fusion gene, as its kinase activity is continuously active, thereby operating as an oncogene and leading to increased and unrestricted proliferation of the concerned cells.

The only cure for CML is bone marrow transplantation. As this is a rather risky surgical procedure, it is only feasible for a small number of CML patients. The new first-line treatment for CML is the tyrosine-kinase inhibitor Imatinib, which targets the BCR-ABL1 kinase fusion protein. Further treatment options include chemotherapy and interferon therapy, which both by far are not as successful as Imatinib treatment.

An interesting publication by Bueno and co-workers investigated the role of miR-203 in several hematological malignancies and reported that the miR-203 gene is hypermethylated in CML (Bueno et al. 2008). The ABL1 gene was a predicted miR-203 target gene and is highly expressed as BCR-ABL1 transcript in CML patients. The authors found out that miR-203 has a tumor suppressive role in CML, as its re-expression reduced the levels of ABL1 and BCR-ABL1 fusion proteins and inhibited tumor cell proliferation in an ABL1-dependent manner. These findings open up the exciting prospect of re-expressing miR-203 in CML with a therapeutic benefit.

1.5.3 Acute Lymphoblastic Leukemia

Acute lymphoblastic leukemia (ALL) is the most common neoplasm in children. Different subtypes of ALL can be identified by immunological, cytogenetical and molecular analyses and they are categorized by the French-American-British (FAB) classification system. It originates from uncontrolled clonal proliferation of lymphoid progenitor cells in the bone marrow and is often characterized by cytogenetic abnormalities like changes in the chromosome ploidy or chromosomal translocations. Treatment options include chemotherapy, steroids, radiation therapy or a combination thereof and bone marrow/stem cell transplantations. In children, more than 90 % of ALL cases can be cured.

Probably due to the biological heterogeneity and distinct lineage origins of ALL, only very few studies dealing with the role of miRNAs in ALL have been performed (Zanette et al. 2007; Schotte et al. 2009). In order to identify novel prognostic and therapeutic predictors in ALL, Yan and collaborators performed miRNA microarray analyses on diagnostic bone marrow samples (Yan et al. 2013). They discovered expression of miR-335 as the most significantly down-regulated miRNA associated with poor outcome. When re-introduced into ALL cells, an increased sensitization to prednisolone-mediated apoptosis could be observed, which could fit to the observation that glucocorticoid resistance is one of the major reasons contributing to poor ALL outcome. On the molecular level, they could show that MAPK1 is a target gene of miR-335, and that MEK/ERK inhibitor treatment enhanced prednisolone-induced cell death through activation of the proapoptotic protein Bim. Reconstitution of miR-335 expression may represent a promising therapeutic strategy for the treatment of ALL.

1.5.4 Acute Myeloid Leukemia

Acute myeloid leukemia (AML) is the most common acute leukemia affecting adults and derives from the myeloid line of blood cells. It is a heterogeneous group of malignancies with variable response to treatment. Characteristic is the rapid growth of abnormal white blood cells that have lost the ability to mature and which accumulate in the bone marrow, where they interfere with the production of normal blood cells. First-line treatment consists primarily of chemotherapy; under certain circumstances, also hematopoietic stem cell transplantation can be considered. Many chromosomal aberrations have been characterized in AML. In translocations, often genes important for normal cell regulation are involved, like AML1/ETO t(8;21) or PML/RARα t(15;17).

The MLL (mixed lineage leukemia) gene on chromosome 11 encodes for a histone methyltransferase and is frequently involved in chromosome translocations with >60 different partner genes. AMLs with MLL rearrangements are associated with intermediate or poor survival. By miRNA expression profiling, Jiang et al. identified miR-495 as down-regulated in MLL-rearranged AML samples when compared to other AML or normal control samples (Jiang et al. 2012). Using several *in vitro* and *in vivo* approaches, Jiang and colleagues could demonstrate that cellular transformation by MLL-fusion proteins as well as leukemogenesis could be efficiently inhibited by over-expression of miR-495. In leukemic human AML/MLL cells, ectopic miR-495 expression decreased cell viability and increased apoptosis. The opposite effect could be observed when two verified target genes of miR-495, PBX3 and MEIS1, were over-expressed in these cells, implying a tumor suppressive role for miR-495 in AML/MLL cells by targeting important leukemia-related genes.

1.6 Colorectal Cancer

Colorectal cancer (CRC) is the second leading cause of cancer-related deaths of men and women, which approximately 10 % of the world population are diagnosed with during their lives. CRC almost always develops from benign polyps to a malignant adenocarcinoma and at first, there are hardly any signs and symptoms of this disease. The prospects of being cured by an operation and chemotherapy with a mean 5-year survival rate of 40–60 % are critically dependent on the stage at which CRC is discovered. Approximately 35 % of all CRC cases can be attributed to inherited genetic factors (Lichtenstein et al. 2000). A high intake of fat or red meat, obesity, smoking, lack of physical exercise as well as high alcohol consumption or simply older age can also be risk factors for CRC.

Colonoscopy for removal of potentially precancerous polyps is considered to be the best screening tool for CRC and provides the best chance for successful treatment. In addition, the fecal occult blood test is frequently used, as it is the only noninvasive procedure, although less sensitive and specific than colonoscopy. Therefore, great interest in discovering novel, noninvasive screening methodologies for increased accuracy and a higher screening rate for CRC in the population is desired. Complicating the diagnosis is the existence of different phenotypic subgroups. These include CRC with microsatellite instability or the status of the oncogene KRAS or the tumor suppressor gene p53. As these CRC subgroups also differ in therapeutic response, biomarker information for precise subtype prediction will in the end also have a direct impact on the projected patient survival. Potentially, miRNA expression patterns might help to classify CRC tumors into different phenotypic groups and these expression patterns may also explain some of the biological differences between each subtype.

More than 300 studies have meanwhile investigated the role of miRNAs in the molecular biology and development of CRC. Interestingly, while in most cancers, miRNAs seem to be rather down-regulated, i.e., have a tumor suppressive function; two thirds of deregulated miRNAs are elevated in CRC (Luo et al. 2011). The following section will deal with two miRNAs consistently altered in CRC, the tumor suppressive miR-143/miR-145 and the oncogenic miR-21. One of the first studies examining the role of miRNAs in CRC identified miR-143 and miR-145 as potential tumor suppressors, as their expression was reduced in CRC (Michael et al. 2003).

MiR-143 and miR-145 are two clustered miRNAs which are located at the long arm of chromosome 5 (5q32), a region commonly deleted in leukemias (Ohyashiki et al. 1987). Many studies of the last 5 years considerably extended our understanding of the precise molecular function of miR-143 and miR-145 in CRC (Akao et al. 2010; Pagliuca et al. 2012; Qian et al. 2013), demonstrating that both miRNAs negatively regulate a complex network of genes implicated in cell cycle control, proliferation or apoptosis by targeting factors like c-Myc (Sachdeva et al. 2009), KRAS and BRAF (Chen et al. 2009; Pagliuca et al. 2012) or Bcl2 (Zhang et al. 2010). A promising therapeutic approach could be the restoration of miR-143/miR-145 expression in CRC by synthetic miRNA mimics to target highly proliferating tumors.

The oncogenic miR-21 is another highly relevant miRNA in CRC and is the most generally up-regulated miRNA in cancer (Volinia et al. 2006) with important roles in cancer initiation, progression, and metastasis. At least 40 studies have meanwhile investigated or identified a role for miR-21 in CRC. One of the reasons why miR-21 expression is altered in so many different cancer diseases could be that it is regulated via pathways such as Ras/MAPK or NF-κB, which are themselves frequently deregulated in cancer.

The multiple-hit hypothesis of cancer postulates that multiple "hits" to DNA are necessary to cause cancer (Knudson 1971). The stepwise progression of CRC from normal colonic mucosa to colon adenomas, and finally to colon carcinomas could also involve the increasing deregulation of miRNAs. In adenomas, which are benign precursor lesions of colon adenocarcinoma, and colon carcinomas, miR-21 levels are elevated (Schetter et al. 2008). As higher expression levels of miR-21 correlate with more advanced stages of CRC, this hints at a potential role for miR-21 in the initiation and progression of CRC. *In situ* hybridization using locked nucleic acid (LNA)-modified probes revealed that the frequency and amount of miR-21 expression increases during the transition from adenoma to advanced CRC (Yamamichi et al. 2009). Numerous cancer-related miR-21 target genes have been identified that provide molecular explanations for the strong effect of miR-21 deregulation, among them the tumor suppressors PTEN and PDCD4. Down-regulation of PTEN by miR-21 leads to increased activation of the PI3-kinase/Akt pathway and thereby promotion of sustained proliferation and survival of tumor cells (Meng et al. 2006). PDCD4 is involved in apoptosis, inhibition of migration and invasion and modulation of immune responses, and it's deregulation by miR-21 therefore promotes tumor development (Asangani et al. 2008). From a therapeutic perspective, modulation of miR-21 levels by an anti-miR-21-based approach may result in reduced proliferation, invasion and increased cell death in CRC and other cancer diseases.

A recent publication reported the potential use of serum miR-21 levels as biomarker for the early detection and prognosis of CRC (Toiyama et al. 2013). In an initial study of sera from 12 CRC patients and 12 control subjects, the authors found an up-regulation of miR-21 in serum of patients with adenomas and CRC, which could be validated in a larger independent cohort. Interestingly, Toiyama et al. (2013) observed that miR-21 levels significantly dropped in post-operative serum from patients who underwent curative surgery. Analysis of miR-21 expression in matched primary CRC tissues and serum revealed that high miR-21 expression was statistically significantly associated with tumor size, distant metastasis, and poor survival. The authors concluded from their data that serum miR-21 was an independent prognostic marker for CRC.

1.7 Lung Cancer

Lung cancer (LC) is the most common cause of cancer-related deaths in men and women and can be divided into the main types small cell lung carcinoma (SCLC), which is the most aggressive type, and non-small cell lung carcinoma (NSCLC).

NSCLC accounts for about 85 % of LCs and its three major histological subtypes adenocarcinoma (AD), squamous cell carcinoma (SCC), and large cell carcinoma (LCC) show heterogeneity in pathogenesis, prognosis, and response to treatment. Inhaled tobacco smoke is by far the greatest risk factor for developing LC, which increases with the amount and duration of smoking. In approximately 90 % of all LC cases of men and 80 % of all women, smoking is responsible for LC death. Other LC cases can be attributed to passive smoking, long-term exposure to high amounts of radon gas, a combination of genetic risk factors, or being frequently exposed to tumor promoting substances like asbestos, chromium, or air pollution. Due to a lack of symptoms, LC is usually diagnosed only at advanced stages when the prognosis is poor. Treatment strategies depend on the cancer's specific cell type, how far the tumor has spread and the person's general health status, and include surgery followed by radiation and/or chemotherapy or palliative care.

An early study to investigate the miRNA expression profile in LC and to determine the ability to use miRNAs as diagnostic and/or prognostic markers for LC was conducted by Yanaihara and colleagues (2006). By comparing miRNA microarray expression profiles of LC tissues with corresponding noncancerous lung tissues, they discovered 43 miRNAs with significant differences in expression between the groups. The microarray data, which was confirmed by qRT-PCR analysis and cross-validated in an independent set of ADs, also revealed a specific molecular miRNA signature for subsets of LC that differ in tumor histology and clinical behavior. Also, a correlation with the survival of lung ADs, including those classified as early disease stage I, could be shown. High miR-155 and low let-7a-2 expression correlated with poor survival by univariate analysis as well as multivariate analysis for miR-155.

The detection of the presence of a disease is a key goal of tumor biomarker research and the identification of ideally noninvasive markers for diagnostic applications an impetus for cancer research. Due to the noninvasive procedure, miRNAs in blood represent an interesting possibility for biomarkers. One of the first comprehensive analyses of miRNAs in serum and plasma of humans and other animals was reported by Chen and colleagues (2008). By employing Solexa NGS technology, they reported the miRNA repertoire in serum of patients with NSCLC and of healthy Chinese subjects and identified a specific serum miRNA expression pattern for LC. Two miRNA were further examined by qRT-PCR analysis in an independent sample set of 75 healthy donors and 152 LC patients, confirming that both miRNAs were present at significantly higher levels in sera of LC patients than controls. Several other studies have meanwhile also identified prognostic and diagnostic miRNA as circulating biomarker for LC (e.g. Franchina et al. 2013; Lin et al. 2013; Kaduthanam et al. 2013; Chen et al. 2013). However, results of miRNA biomarker studies in LC are sometimes not overlapping or even contradictory, which implies the need for standardization of the utilized analytical methodologies, like the use of internal and external controls in each assay, and a consensus normalization method of the results. Many improvements still need to be made before the establishment of miRNAs as routine biomarkers in the clinical laboratory.

As deregulated miRNA expression has been meanwhile shown for LC, certain miRNAs might be employed as therapeutic targets. The so-called "miRNA replacement therapy" makes use of tumor suppressor miRNAs. Re-introduction of them into tumor cells is expected to lead to reactivation or correction of cellular pathways, eventually leading to a therapeutic response. Difficulties in executing this strategy are until now e.g. obstacles associated with miRNA delivery, either in terms of safety (e.g. virus-based delivery) or efficiency [e.g. complexing nucleic acids with polyethylenimine (PEI)]. Wiggins and colleagues studied a cohort of NSCLC tumor samples and observed reduced miR-34a expression in all histotypes of NSCLC (Wiggins et al. 2010). As miR-34a inhibited the growth of cultured LC cells, the authors devised a therapeutic formulation using chemically synthesized miR-34a and a lipid-based delivery vehicle for intratumoral delivery. They observed that administration of this formulation blocked tumor growth in mouse models of NSCLC when administered locally or systemically, accompanied by an accumulation of miR-34a in the tumor tissue and down-regulation of miR-34a target genes. The safety profile of systemically delivered miR-34a showed no elevation of cytokines or liver and kidney enzymes in serum, which suggests that the "miRNA replacement therapy" in LC is well tolerated and does not induce an immune response.

1.8 Pancreatic Cancer

Pancreatic cancer (PC) is a devastating disease with very poor prognosis and is the fourth most common carcinoma in men and the fifth most common in women in Western countries. Like other tumors of the gastrointestinal tract, PC is not a homogenous disease, but can be divided into tumors arising from the exocrine pancreas and those arising from endocrine cells. The pancreatic ductal adenocarcinoma (cancer of the pancreas) is the most frequent tumor of the pancreas (approximately 80 %); in addition, there are a number of other tumors (pancreatic cystic neoplasms, pancreatic neuroendocrine tumors, acinic cell carcinoma), which differ considerably in their aggressiveness and the ability to metastasize. Smoking and diets low in vegetables and fruits and high in red meat seem to be the most important controllable risk factors. Patients with chronic pancreatitis have been shown to have an increased risk of PC. Although PC can also occur in young people, the risk increases with age. PC is an almost incurable disease, as there are hardly any symptoms in early stage disease and in later stages, symptoms are usually nonspecific and varied. The late time point of diagnosis at an already advanced stage hinders successful treatment, resulting in a 5-year survival rate of approximately only 5 %. Treatment options depend on the stage of the cancer and consist of surgery with adjuvant chemotherapy, chemotherapy or radiation therapy or palliative chemotherapy for patients not suitable for resection with curative intent.

MiRNAs have been show to play a role in different stages of PC; from degeneration of pancreatic cells to metastasis or finally to resistance against chemo- or radiotherapy. Several studies have investigated the involvement of miRNAs in metastasis. A recent publication by Wang and co-workers demonstrated that miRNAs can suppress metastasis (Wang et al. 2013). The authors discovered that all three miR-124 genes are highly methylated in PC tissue compared with noncancerous tissue, leading to silencing of miR-124. On the molecular level, Wang and colleagues could demonstrate that miR-124 is a tumor suppressor, as miR-124 inhibited cell proliferation, invasion and metastasis. Rac1, a GTPase belonging to the RAS superfamily of small GTP-binding proteins and a putative tumor promoter in PC, was shown to be a direct target of miR-124, leading to inactivation of the MKK4-JNK-c-Jun pathway in PC.

The Ras pathway performs an essential function in the transmission of growth-promoting signals from cell surface receptors, and mutation of the KRAS gene, which encodes for a GTPase, is an essential step in the development of many cancers. Mutations of KRAS occur in over 90 % of PCs, which is a remarkably high mutation frequency. Zhao and colleagues investigated the expression and possible role of miR-217 in PC by *in situ* hybridization and qRT-PCR analysis and showed that miR-217 was down-regulated in 16/21 of PC tissues when compared with the corresponding normal pancreatic tissue (Zhao et al. 2010). Ectopic expression of miR-217 in PC cells inhibited tumor cell growth and anchorage-independent colony formation. *In silico* and *in vitro* analyses revealed that KRAS was a direct target of miR-217. While miR-217 up-regulation reduced the constitutive phosphorylation of the downstream AKT signaling pathway, miR-217 down-regulation in PC cells increased cell anchorage-independent colony formation, implying a tumor suppressive role for miR-217 in PC and the prospect of utilizing miR-217 as a therapeutic agent for miRNA-based PC therapy. A similar study was performed by Yu and collaborators (2010). They found that miR-96 directly targets KRAS in PC cells and that ectopic expression of miR-96 lead to diminished KRAS levels, dampened the AKT signaling pathway, triggered apoptosis in PC cell lines, decreased cancer cell invasion and migration and slowed tumor growth.

One of the reasons for the poor survival of PC patients is the (too) late discovery of the tumor. Detection of aberrant miRNA expression may offer a diagnostic, minimally invasive biomarker assay for the early detection of PC and for effective clinical management. A first step in this direction was the identification of "signature miRNAs" that allow discrimination between benign and malignant pancreas tissues as well as between different types of malignancies. Bloomston et al. (2007) investigated differentially expressed miRNAs in patients with ductal adenocarcinoma of the pancreas and chronic pancreatitis by miRNA microarray analysis. They identified 21 miRNAs with increased and 4 with decreased expression in PC that were able to correctly differentiate PC from benign pancreatic tissue in 90 % of the samples. Using a miRNA signature of 15 increased and 8 decreased miRNAs, PC could be differentiated from chronic pancreatitis with 93 % accuracy. Also, a subgroup of 6 miRNAs was able to distinguish long-term survivors with node-positive PC from those dying within 2 years. The high expression of only one miRNA, miR-196a-2, was already found to predict poor survival (median survival 14.3 months vs. 26.5 months).

Another study confirming the applicability of miRNAs to distinguish benign from malign tissues in PC was reported by Lee and co-workers (2007). This group studied more than 200 miRNA precursors in specimens of human pancreatic adenocarcinoma, paired benign tissue, normal pancreas, chronic pancreatitis and several PC cell lines by qRT-PCR analysis. The expression of the biologically active mature miRNA was validated by using qRT-PCR- and Northern blot analysis. By hierarchical clustering, 28 of 28 tumors, 6 of 6 normal pancreas and 11 of 15 adjacent benign tissues could be correctly classified.

2 MicroRNAs as Potential Cancer Biomarkers

Biomarkers in cancer medicine would ideally fulfill several requirements:

- they should predict who will develop disease, so precautionary measures might be taken
- they should detect disease at an early stage, so treatment can be started in time to increase the likelihood of complete cure
- they should correctly classify related or similar diseases
- they should distinguish between different subtypes of tumors in order to choose the most efficient therapy
- they should detect disease relapse, so countermeasures can be applied early
- they should guide therapeutic decision-making to apply suitable therapeutic regimen, i.e., to discontinue treatment with ineffective agents early to decrease morbidity from drug toxicity, to help reducing costs and to allow the patient to be treated by an alternative therapy

Proteins, RNA, DNA, or also metabolites can be used as biomarkers and plenty of them for many diseases are presently in use. These include e.g. the mutation status of the BRCA1/BRCA2 genes in breast or ovarian cancer, serum alpha-fetoprotein (AFP) levels in liver cancer, or the occurrence of the BCR-ABL1 chromosomal translocation in CML. Although many biomarkers are currently in use, the sensitivity and specificity of detection are often not satisfactory. Histological evaluation of tumor material obtained from tissue biopsies remains the gold standard, as tumor markers greatly improve diagnosis. Due to the invasive, unpleasant, and inconvenient nature of current diagnostic procedures, their clinical application is often limited and the identification of noninvasive tumor markers is highly desirable.

Large-scale mRNA expression profiling techniques such as microarray analysis or NGS technologies have allowed the identification of key differences in mRNA gene expression between tumor tissues and nonaffected normal tissues. This holds also true for miRNAs, as different types of cancers have differing miRNA profiles that allow distinction from adjacent normal tissue [reviewed by Munker and Calin (2011)]. In fact, a study of 20 different leukemias and solid cancers revealed that each cancer had a specific miRNA profile and that most poorly differentiated tumors could be assigned to their tissues of origin based on their miRNA expression levels (Lu et al. 2005). This research team also determined that the majority of miRNAs are down-regulated in cancer specimens and that miRNA expression profiles could classify human cancers by their developmental lineage and differentiation state even better than mRNA expression profiling. Therefore, unique patterns of altered miRNA expression could potentially provide complex fingerprints that may serve as molecular biomarkers, resulting in an important advance in cancer diagnosis and prognosis.

MiRNAs have been found associated with circulating exosomes, which are small (30–100 nm) membrane vesicles of endocytic origin. They are released into the extracellular environment upon fusion of multivesicular bodies with the plasma membrane [reviewed by Ajit (2012)]. As tumor cells have been shown to release miRNAs into the circulation and miRNA profiles are altered in plasma/serum of cancer patients, this finding makes them very appealing as potential novel biomarkers (Taylor and Gercel-Taylor 2008).

Further innate properties of miRNAs make them highly interesting from a clinical biomarker perspective. Due to their small size, they are highly stable and relatively resistant to RNase degradation (Mitchell et al. 2008) and have been isolated from most body fluids, including serum, plasma, urine, semen, milk, tears and saliva [reviewed by Cortez et al. (2011)]. They can be reliably detected in small volume samples using highly developed means of detection like qRT-PCR analysis, and they are highly conserved between species, enabling the use of animal disease models for preclinical studies. The lack of post-processing modifications and their low complexity when compared to proteins make miRNAs a potentially rich source of diagnostic, prognostic, and predictive information in cancer medicine.

In the following sections, I will use hepatocellular carcinoma (HCC) as an example to demonstrate how miRNAs could serve as diagnostic or prognostic markers, predictors of therapeutic outcome, how they can be used as therapeutic agents, their potential use in clinical trials or personalized medicine and finally, I will discuss the perspectives and challenges of miRNAs as potential cancer biomarkers.

2.1 MiRNAs in Diagnosis and Prognosis

Development of HCC is a complex process which generally affects patients already suffering from chronic liver diseases, like hepatitis B or C virus (HBV/HCV) infection, liver cirrhosis, which can occur after alcohol abuse, or type 2 diabetes. For detection and diagnosis of HCC, liver imaging studies play a key role. Therefore, common diagnostic methods include techniques such as triphasic computed tomography scanning, MRI, and abdominal ultrasound (Befeler and Di Bisceglie 2002), which also help in identifying patients still at an asymptomatic HCC stage. Imaging techniques are also frequently combined with the measurement of serum AFP levels, which is generally considered to be a significant marker for HCC (Tong et al. 2001). A review of several studies using AFP as biomarker in HCC revealed that approximately 30 % of HCC cases are not related to AFP production, calling into question the sensitivity and specificity of AFP and showing the need for additional tests for detecting HCC (Gupta et al. 2003). Profiling of circulating miRNAs could represent such an additional test, and numerous studies have been performed to assess the usefulness of miRNAs for such purposes (Table 13.1).

Table 13.1 Circulating microRNAs in hepatocellular carcinoma diagnosis and prognosis

MicroRNA	Functions	Reference
miR-21	miR-21 serum levels are elevated in patients with chronic hepatitis C virus infection (CHC) compared to healthy controls No difference between miR-21 serum levels in patients with CHC and CHC-associated hepatocellular carcinoma (HCC) miR-21 serum levels can discriminate between minimal and mild-severe necroinflammation	Bihrer et al. (2011)
miR-885-5p	miR-885-5p serum levels are significantly higher in patients with HCC, liver cirrhosis and chronic hepatitis B virus infection (CHB) than in healthy controls	Gui et al. (2011)
miR-1, miR-122	HCC patients with higher miR-1 and miR-122 serum levels show longer overall survival than individuals with lower miR-1 and miR-122 serum concentrations miR-1 serum levels are independently associated with overall survival miR-122 serum levels correlate with clinical chemistry parameters of hepatic necroinflammation, liver function and synthetic capacity	Köberle et al. (2013)
miR-25, miR-375, let-7f	Description of a 13 serum miRNA signature for detection of CHB infection miR-25, miR-375, and let-7f serum levels can separate HCC cases from healthy controls	Li et al. (2010)
miR-221	High miR-221 serum levels correlate with HCC tumor size, liver cirrhosis and tumor stage Overall survival rate of the high miR-221 expression HCC group is significantly lower than that of the low miR-221 expression HCC group	Li et al. (2011)
miR-18a, miR-378	Serum miR-18a levels are significantly higher in CHB patients with HCC than healthy controls Serum miR-378 levels are significantly lower in CHB patients with HCC compared to healthy controls	Li et al. (2012)
miR-15b, miR-130b	miR-15b, miR-21, miR-130b and miR-183 serum levels are markedly reduced after surgery miR-15b and miR-130b are highly expressed in serum of HCC patients	Liu et al. (2012)
miR-122	miR-122 serum levels are significantly higher in HCC patients than in healthy controls miR-122 serum levels are significantly reduced in HCC post-operative serum samples when compared to pre-operative samples	Qi et al. (2011)
miR-16, miR-199a	miR-16 and miR-199a serum levels are significantly lower in HCC than in healthy controls	Qu et al. (2011)
miR-92a	miR-92a in plasma from HCC patients is decreased when compared to that of healthy controls Amount of plasma miR-92a is elevated after surgical treatment	Shigoka et al. (2010)

(continued)

Table 13.1 (continued)

MicroRNA	Functions	Reference
miR-21	miR-21 plasma levels significantly diminish after HCC surgery compared with pre-operative values miR-21 plasma levels in HCC patients are significantly higher than in patients with chronic hepatitis and healthy controls	Tomimaru et al. (2012)
miR-21, miR-122, miR-223	miR-21, miR-122, and miR-223 serum levels are significantly higher in patients with HCC and chronic hepatitis than those in healthy controls miR-21 and miR-122 serum levels are higher in patients with chronic hepatitis than in patients with HCC	Xu et al. (2011)
miR-500	miR-500 serum levels are increased in HCC patients compared to healthy controls	Yamamoto et al. (2009)
miR-17-5p	HCC patients with high miR-17-5p serum levels have a significantly shortened overall survival miR-17-5p serum levels are an independent risk factor for overall survival miR-17-5p serum levels are significantly down-regulated in the HCC post-operative group and get up-regulated in the relapsed group	Zheng et al. (2012)

Abbreviations: *CHB* chronic hepatitis B virus infection, *CHC* chronic hepatitis C virus infection, *HCC* hepatocellular carcinoma, *miRNA* microRNA

Ideally, miRNA profiling should contribute to the identification of putative HCC subtypes (etiology, genotype, and phenotype) as well as aid in diagnosis and prognosis. Table 13.1 shows that, in total, 18 circulating miRNAs have been described in HCC diagnosis and prognosis until now. These heterogeneous studies are rather small and restricted to special cases like HCC patients with or without HBV and/or HCV infection or with or without liver cirrhosis. Before the feasibility of serum miRNA in diagnosis and prognosis can be determined (exemplary for HCC), larger scale studies involving hundreds of well-defined HCC patients and controls studying the 18 until now reported circulating miRNAs in HCC should be conducted. Only then the suitability of miRNAs as additional tests for detecting HCC at a still asymptomatic stage can be estimated.

2.2 *MiRNAs as Predictors of Therapeutic Outcome*

The ability of cells to evade engagement of apoptosis plays a significant role in tumor development. Therefore, cancer cells must bypass this major barrier on the way towards transformation and tumor progression and eventually, to survive in the tumor environment. One therapeutic approach is to drive cancer cells into apoptosis by (again) sensitizing them to certain stimuli, enabling them (again) to regulate the apoptotic cascades. A frequent side-effect of anti-cancer drug treatment is the development of drug resistance. For example, the regained response of cancer cells to apoptotic stimuli upon drug treatment is lost again. The prediction of who will develop anti-cancer drug resistance upon treatment is therefore a crucial question in need to be answered. As miRNAs have been shown to regulate mRNAs important for programmed cell death, an interesting aspect has been the investigation of them during apoptosis. In HCC tumors and HCC cell lines, miR-21 was noted to be highly over-expressed (Meng et al. 2007). PTEN was shown to be a direct target of miR-21 and to contribute to the promotion of HCC growth and spread as well as augmentation of cancer cell survival, leading to enhanced cancer cell survival by loss of PI3 kinase pathway inhibition. One of the combination therapies in advanced HCC is the administration of interferon (IFN)-α and 5-fluorouracil (5-FU). As the prediction of the response to this therapy was unsatisfactory, Tomimaru and colleagues studied the anti-tumor effect of miR-21 on the sensitivity of HCC cells to IFN-α/5-FU (Tomimaru et al. 2010). Modulation of miR-21 levels by over-expression or down-modulation lead to resistance or sensitivity of HCC cells to IFN-α/5-FU treatment. Importantly, miR-21 expression in clinical HCC specimens was significantly associated with clinical response to IFN-α/5-FU combination therapy and survival rate. The authors concluded that miR-21 could be used as a predictor of the clinical response to the IFN-α/5-FU combination therapy in HCC.

2.3 *Modulating miRNA Levels as Therapeutic Approach*

Like transcription factors, miRNAs can regulate gene expression in a complex manner, as one miRNA can fine-tune the expression of hundreds of mRNAs. This feature of miRNA biology can be utilized for the development of a novel cancer therapy approach. Depending on whether a miRNA has oncogenic or tumor suppressive properties (Babashah and Soleimani 2011), manipulating gene networks by miRNA silencing with antisense oligonucleotides or antagomirs or miRNA restoration by application of chemically modified synthetic miRNA mimics could be used. The miRNA with the furthest advancement of this therapeutic approach is miR-122, which is the most highly expressed miRNA in the liver.

One of the reasons for liver damage and subsequent development of cirrhosis and HCC is infection with HCV, which persists in approximately 170 million people worldwide. Although the recent introduction of HCV protease inhibitors improved therapy, inhibitors are active against only the dominant viral genomes. Effective therapeutic approaches against the diverse strains of HCV and with a low likelihood of creating resistance would be a welcome addition to the drug arsenal. Building upon the finding that the stability and propagation of HCV is dependent on a functional interaction between the HCV genome and miR-122 (Jopling et al. 2005), down-modulation of miR-122 by a LNA-modified oligonucleotide complementary to miR-122 (“miravirsen”) in chimpanzees lead to long-lasting suppression of HCV viremia, with no evidence of side effects or viral resistance in the treated animals (Lanford et al. 2010). The Phase 2a dose-finding study of subcutaneous weekly miravirsen administration in patients with chronic HCV infection now showed a prolonged, dose-dependent reduction in HCV RNA levels without evidence of viral resistance (Janssen et al. 2013). Miravirsen enters liver cells and binds tightly and stably to miR-122, preventing the latter from binding to HCV RNA. Sequencing the targeted viral sequence revealed no sign of drug resistance. These exciting experiments benefitted from the fact that the hurdle for nucleic acid delivery to the liver is lowest, as this organ is designed to remove toxic substances from the body. A side-effect of miravirsen treatment was the long-lasting decrease in serum cholesterol levels of approximately 25 %, which miR-122 normally also controls independently of its effect on HCV. The sustained biologic effect, which lasted 14 weeks after the final injection, suggests that such drugs could be administered infrequently. An issue that needs careful study design and safety monitoring for long-term use will be the finding that miR-122 is a tumor suppressor gene for HCC, as mice lacking miR-122 are viable, but have a high risk of fatty liver, fibrosis, and HCC (Tsai et al. 2012; Hsu et al. 2012).

2.4 MiRNAs in Clinical Trials and Personalized Medicine

A more detailed understanding of the impact of genetics in disease lead to the proposal of personalized medicine, i.e., that patients are not only diagnosed and managed according to their clinical signs and symptoms or based on the mechanisms of disease, but also to consider results of molecular testing methods for global gene or protein expression or somatic mutations in cancer cells. This includes looking for markers associated with prognosis and likely treatment responses, to better define prognosis in these patients and to suggest treatment options that are most likely to succeed.

That personalized cancer therapy can show beneficial effects and could one day become the standard of care is implied by results of recent clinical trials performed at the MD Anderson Cancer Center (Tsimberidou et al. 2012). Matching targeted therapies with certain gene mutations in patients in a Phase 1 clinical trials program showed a higher overall response rate, a longer time-to-treatment failure and longer survival when compared with non-matched patients. Although the study was not randomized and patients with diverse tumor types and several prior therapies were included, this indicates that genetic testing of newly diagnosed patients at early stages in their disease is beneficial. As we have seen that miRNA profiling can contribute many important information to prognosis, diagnosis and therapy prediction in many cancer types, this implicates the potential of integrating miRNA data into clinical trials as well.

In May 2013, Mirna Therapeutics, Inc. reported that they have initiated a Phase 1 clinical study of the first miRNA to advance into a human clinical trial for cancer (http://clinicaltrials.gov/ct2/show/NCT01829971). MRX34, which is a miRNA mimic of the tumor suppressor miR-34, is given intravenously in patients with unresectable primary liver cancer or advanced or metastatic cancer with liver involvement.

Up to 48 patients are enrolled in this Phase 1 clinical study, which consists of an initial dose-escalation phase followed by an enrichment phase. It will be very interesting to follow this first miRNA clinical trial possibly progressing through several clinical phases and to learn about the challenges and opportunities of miRNAs in personalized medicine and clinical trials.

3 Perspectives and Challenges

A lot of hopes and expectations are set on miRNAs as biomarkers in cancer medicine. They should fulfil not less than to improve disease diagnosis by distinguishing healthy from malignant tissues, to provide a significant and independent prognostic value, to identify the tissue of origin in poorly differentiated tumors or tumors of unknown origin, to distinguish different subtypes of the same tumor, to predict therapeutic efficacy, to identify population subgroups that are responsive to drugs, to forecast disease recurrence and to allow maintenance of surveillance following surgery.

Now, it is obvious that miRNAs cannot live up to all biomarker expectations. When thinking about the perspectives and challenges of miRNAs and their clinical implications in cancer, the prospects for these tiny molecules do not look bad:

- The expression of miRNAs is frequently dysregulated in cancer
- The expression patterns of miRNAs in cancer seem to be tissue-specific
- MiRNAs show an unexpectedly high stability in tissues, serum and plasma

Like with other biomarkers, it is unlikely that any single miRNA will achieve the desired level of diagnostic or prognostic accuracy. Increased levels of particular miRNAs like miR-21 can be associated with several different types of tumors. It is therefore rather likely that a more accurate assessment would be if a miRNA biomarker signature for a certain cancer type can be established. Maybe, sensitivity and specificity of diagnosis could be even more improved by combining cancer miRNA biomarkers with other available screening tests. However, until now, the comparability between results of different publications is rather low. Therefore, measures should be taken to standardize protocols to increase the translatability of studies. Some of the potential pre- and post analytical variations could include:

- Precise classification of patients and the history of their disease by generating comprehensive databases. This is also important in regard to personalized medicine approaches.
- Use of consistent study designs. This includes an adequate sample size of training and test sets, multiple independent patient populations, and appropriately designed and powered blinded and randomized clinical trials. To assess the obtained results, subsequent validations in follow-up trials should be considered.
- Choice of samples. Depending on what type of specimen should be analyzed (primary tumor material, plasma, serum, whole blood, etc.), it has to be ascertained

that the material is free of "contamination" with other tissues, cell types, etc. Ideally, biomarker measurement should not consume tissue needed for other tests, particularly for histopathological evaluation.

- Sample processing. Standardization of specimen storage, blood sample collection, sample transportation, will help avoiding sources of variability.
- Use of common miRNA extraction methodology. The vast amount of methods available (use of Phenol/Chloroform, Trizol, spin columns, etc.) makes development of standard protocols and adherence to them mandatory to exclude variation from this important step.
- Reliable quantification methods. Development of protocols for standard technology platforms like microarray, qRT-PCR analysis, NGS approaches, etc. and the type of miRNA to be analyzed (pri-miRNA, pre-miRNA, mature miRNA). The determination of a miRNA signature must also be feasible for non-specialists, must be reproducible and include quality controls.
- Appropriate and generally applicable normalization controls. A lot of variation is introduced by the use of inappropriate normalization controls and methods. Should miRNAs be compared to other classes of small RNAs, such as the snoRNA RNU6B, should a mix of several, apparently unchanged miRNAs be used as normalization controls, should miRNA levels be compared to the level of a mix of several mRNAs, etc. This need for a housekeeping miRNA is especially true for miRNAs in body fluids, as most other classes of RNA are unstable in them.
- Consistent data processing and use of consistent statistical methods. The results should be readily interpretable by a clinician.

In some cases, miRNA biomarkers could at the same time also be used for therapeutic purposes, which would allow the development of exciting therapeutic strategies to intervene with miRNA levels by targeting oncogenic miRNAs with antagomirs or by restoring suppressor miRNAs with degradation-resistant chemically synthesized miRNA molecules. First use of miRNAs as therapeutics will necessitate exact investigations, as miRNAs with their pleiotropic mode of regulation could give rise to a complex phenotype that may not be readily predictable. This could mean that other pathways are also targeted, which might be unwanted, but it might also mean that with one reagent, you could synergistically target several pathways involved in cancer at the same time.

The first decade of intensive research on miRNAs has already brought us many insights into the molecular biology of many cancer types and into cancer medicine. MiRNAs are increasingly studied as molecular markers for cancer prognosis and diagnosis, and their potential role as possible therapeutic target is also starting to be appreciated. Although some aspects of miRNAs in cancer medicine are still in their infancy, if miRNA research in cancer medicine continuous to develop at its current pace, the future promises to hold many encouraging findings on miRNAs in cancer medicine.

Acknowledgements I would like to thank Prof. Hans-Martin Jäck for his continuous support.

References

Ajit SK (2012) Circulating microRNAs as biomarkers, therapeutic targets, and signaling molecules. Sensors (Basel) 12(3):3359–3369

Akao Y, Nakagawa Y, Hirata I, Iio A, Itoh T, Kojima K et al (2010) Role of anti-oncomirs miR-143 and –145 in human colorectal tumors. Cancer Gene Ther 17(6):398–408

Asangani IA, Rasheed SA, Nikolova DA, Leupold JH, Colburn NH, Post S et al (2008) MicroRNA-21 (miR-21) post-transcriptionally downregulates tumor suppressor Pdcd4 and stimulates invasion, intravasation and metastasis in colorectal cancer. Oncogene 27(15):2128–2136

Babashah S, Soleimani M (2011) The oncogenic and tumour suppressive roles of microRNAs in cancer and apoptosis. Eur J Cancer 47(8):1127–1137

Babashah S, Sadeghizadeh M, Tavirani MR, Farivar S, Soleimani M (2012) Aberrant microRNA expression and its implications in the pathogenesis of leukemias. Cell Oncol (Dordr) 35(5):317–334

Befeler AS, Di Bisceglie AM (2002) Hepatocellular carcinoma: diagnosis and treatment. Gastroenterology 122(6):1609–1619

Bell RE, Levy C (2011) The three M's: melanoma, microphthalmia-associated transcription factor and microRNA. Pigment Cell Melanoma Res 24(6):1088–1106

Bihrer V, Waidmann O, Friedrich-Rust M, Forestier N, Susser S, Haupenthal J et al (2011) Serum microRNA-21 as marker for necroinflammation in hepatitis C patients with and without hepatocellular carcinoma. PLoS One 6(10):e26971

Blenkiron C, Goldstein LD, Thorne NP, Spiteri I, Chin SF, Dunning MJ et al (2007) MicroRNA expression profiling of human breast cancer identifies new markers of tumor subtype. Genome Biol 8(10):R214

Bloomston M, Frankel WL, Petrocca F, Volinia S, Alder H, Hagan JP et al (2007) MicroRNA expression patterns to differentiate pancreatic adenocarcinoma from normal pancreas and chronic pancreatitis. JAMA 297(17):1901–1908

Brase JC, Johannes M, Schlomm T, Fälth M, Haese A, Steuber T et al (2011) Circulating miRNAs are correlated with tumor progression in prostate cancer. Int J Cancer 128(3):608–616

Bueno MJ, Pérez de Castro I, Gómez de Cedrón M, Santos J, Calin GA, Cigudosa JC et al (2008) Genetic and epigenetic silencing of microRNA-203 enhances ABL1 and BCR-ABL1 oncogene expression. Cancer Cell 13(6):496–506

Buffa FM, Camps C, Winchester L, Snell CE, Gee HE, Sheldon H et al (2011) microRNA-associated progression pathways and potential therapeutic targets identified by integrated mRNA and microRNA expression profiling in breast cancer. Cancer Res 71(17):5635–5645

Calin GA, Dumitru CD, Shimizu M, Bichi R, Zupo S, Noch E et al (2002) Frequent deletions and down-regulation of micro- RNA genes miR15 and miR16 at 13q14 in chronic lymphocytic leukemia. Proc Natl Acad Sci U S A 99(24):15524–15529

Chen X, Ba Y, Ma L, Cai X, Yin Y, Wang K et al (2008) Characterization of microRNAs in serum: a novel class of biomarkers for diagnosis of cancer and other diseases. Cell Res 18(10):997–1006

Chen X, Guo X, Zhang H, Xiang Y, Chen J, Yin Y et al (2009) Role of miR-143 targeting KRAS in colorectal tumorigenesis. Oncogene 28(10):1385–1392

Chen Q, Si Q, Xiao S, Xie Q, Lin J, Wang C et al (2013) Prognostic significance of serum miR-17-5p in lung cancer. Med Oncol 30(1):353

Cimmino A, Calin GA, Fabbri M, Iorio MV, Ferracin M, Shimizu M et al (2005) miR-15 and miR-16 induce apoptosis by targeting BCL2. Proc Natl Acad Sci U S A 102(39):13944–13949

Cortez MA, Bueso-Ramos C, Ferdin J, Lopez-Berestein G, Sood AK, Calin GA (2011) MicroRNAs in body fluids–the mix of hormones and biomarkers. Nat Rev Clin Oncol 8(8):467–477

Eitan R, Kushnir M, Lithwick-Yanai G, David MB, Hoshen M, Glezerman M et al (2009) Tumor microRNA expression patterns associated with resistance to platinum based chemotherapy and survival in ovarian cancer patients. Gynecol Oncol 114(2):253–259

Esquela-Kerscher A, Slack FJ (2006) Oncomirs – microRNAs with a role in cancer. Nat Rev Cancer 6(4):259–269

Fabbri M, Bottoni A, Shimizu M, Spizzo R, Nicoloso MS, Rossi S et al (2011) Association of a microRNA/TP53 feedback circuitry with pathogenesis and outcome of B-cell chronic lymphocytic leukemia. JAMA 305(1):59–67

Farazi TA, Horlings HM, Ten Hoeve JJ, Mihailovic A, Halfwerk H, Morozov P et al (2011) MicroRNA sequence and expression analysis in breast tumors by deep sequencing. Cancer Res 71(13):4443–4453

Franchina T, Amodeo V, Bronte G, Savio G, Ricciardi GR, Picciotto M et al (2013) Circulating miR-22, miR-24 and miR-34a as novel predictive biomarkers to pemetrexed-based chemotherapy in advanced non small cell lung cancer. J Cell Physiol 229(1):97–99

Fu TY, Chang CC, Lin CT, Lai CH, Peng SY, Ko YJ et al (2011) Let-7b-mediated suppression of basigin expression and metastasis in mouse melanoma cells. Exp Cell Res 317(4):445–451

Glud M, Klausen M, Gniadecki R, Rossing M, Hastrup N, Nielsen FC et al (2009) MicroRNA expression in melanocytic nevi: the usefulness of formalin-fixed, paraffin-embedded material for miRNA microarray profiling. J Invest Dermatol 129(5):1219–1224

Gonzales JC, Fink LM, Goodman OB Jr, Symanowski JT, Vogelzang NJ, Ward DC (2011) Comparison of circulating MicroRNA 141 to circulating tumor cells, lactate dehydrogenase, and prostate-specific antigen for determining treatment response in patients with metastatic prostate cancer. Clin Genitourin Cancer 9(1):39–45

Gui J, Tian Y, Wen X, Zhang W, Zhang P, Gao J et al (2011) Serum microRNA characterization identifies miR-885-5p as a potential marker for detecting liver pathologies. Clin Sci (Lond) 120(5):183–193

Guo L, Zhao Y, Yang S, Cai M, Wu Q, Chen F (2013) Genome-wide screen for aberrantly expressed miRNAs reveals miRNA profile signature in breast cancer. Mol Biol Rep 40(3):2175–2186

Gupta S, Bent S, Kohlwes J (2003) Test characteristics of alpha-fetoprotein for detecting hepatocellular carcinoma in patients with hepatitis C. A systematic review and critical analysis. Ann Intern Med 139(1):46–50

Hsu SH, Wang B, Kota J, Yu J, Costinean S, Kutay H et al (2012) Essential metabolic, anti-inflammatory, and anti-tumorigenic functions of miR-122 in liver. J Clin Invest 122(8):2871–2883

Huggins C, Steven RE, Hodges CV (1941) Studies on prostatic cancer. II. The effects of castration on advanced carcinoma of the prostate gland. Arch Surg 43(2):209–223

Iorio MV, Ferracin M, Liu CG, Veronese A, Spizzo R, Sabbioni S et al (2005) MicroRNA gene expression deregulation in human breast cancer. Cancer Res 65(16):7065–7070

Iorio MV, Visone R, Di Leva G, Donati V, Petrocca F, Casalini P et al (2007) MicroRNA signatures in human ovarian cancer. Cancer Res 67(18):8699–8707

Janssen EA, Slewa A, Gudlaugsson E, Jonsdottir K, Skaland I, Søiland H et al (2010) Biologic profiling of lymph node negative breast cancers by means of microRNA expression. Mod Pathol 23(12):1567–1576

Janssen HL, Reesink HW, Lawitz EJ, Zeuzem S, Rodriguez-Torres M, Patel K et al (2013) Treatment of HCV infection by targeting microRNA. N Engl J Med 368(18):1685–1694

Jiang X, Huang H, Li Z, He C, Li Y, Chen P et al (2012) MiR-495 is a tumor-suppressor microRNA down-regulated in MLL-rearranged leukemia. Proc Natl Acad Sci U S A 109(47):19397–19402

Jopling CL, Yi M, Lancaster AM, Lemon SM, Sarnow P (2005) Modulation of hepatitis C virus RNA abundance by a liver-specific MicroRNA. Science 309(5740):1577–1581

Kaduthanam S, Gade S, Meister M, Brase JC, Johannes M, Dienemann H et al (2013) Serum miR-142-3p is associated with early relapse in operable lung adenocarcinoma patients. Lung Cancer 80(29):223–227

Klein U, Lia M, Crespo M, Siegel R, Shen Q, Mo T et al (2010) The DLEU2/miR-15a/16-1 cluster controls B cell proliferation and its deletion leads to chronic lymphocytic leukemia. Cancer Cell 17(1):28–40

Knudson AG Jr (1971) Mutation and cancer: statistical study of retinoblastoma. Proc Natl Acad Sci U S A 68(4):820–823

Köberle V, Kronenberger B, Pleli T, Trojan J, Imelmann E, Peveling-Oberhag J et al (2013) Serum microRNA-1 and microRNA-122 are prognostic markers in patients with hepatocellular carcinoma. Eur J Cancer. pii:S0959-8049(13)00462-0

Lanford RE, Hildebrandt-Eriksen ES, Petri A, Persson R, Lindow M, Munk ME et al (2010) Therapeutic silencing of microRNA-122 in primates with chronic hepatitis C virus infection. Science 327(5962):198–201

Lech A, Daneva T, Pashova S, Gagov H, Crayton R, Kukwa W et al (2013) Ovarian cancer as a genetic disease. Front Biosci (Landmark Ed) 18:543–563

Lee EJ, Gusev Y, Jiang J, Nuovo GJ, Lerner MR, Frankel WL et al (2007) Expression profiling identifies microRNA signature in pancreatic cancer. Int J Cancer 120(5):1046–1054

Lee CH, Subramanian S, Beck AH, Espinosa I, Senz J, Zhu SX et al (2009) MicroRNA profiling of BRCA1/2 mutation-carrying and non-mutation-carrying high-grade serous carcinomas of ovary. PLoS One 4(10):e7314

Leidinger P, Keller A, Borries A, Reichrath J, Rass K, Jager SU et al (2010) High-throughput miRNA profiling of human melanoma blood samples. BMC Cancer 10:262

Li LM, Hu ZB, Zhou ZX, Chen X, Liu FY, Zhang JF et al (2010) Serum microRNA profiles serve as novel biomarkers for HBV infection and diagnosis of HBV-positive hepatocarcinoma. Cancer Res 70(23):9798–9807

Li J, Wang Y, Yu W, Chen J, Luo J (2011) Expression of serum miR-221 in human hepatocellular carcinoma and its prognostic significance. Biochem Biophys Res Commun 406(1):70–73

Li L, Guo Z, Wang J, Mao Y, Gao Q (2012) Serum miR-18a: a potential marker for hepatitis B virus-related hepatocellular carcinoma screening. Dig Dis Sci 57(11):2910–2916

Lia M, Carette A, Tang H, Shen Q, Mo T, Bhagat G, Dalla-Favera R et al (2012) Functional dissection of the chromosome 13q14 tumor-suppressor locus using transgenic mouse lines. Blood 119(13):2981–2990

Lichtenstein P, Holm NV, Verkasalo PK, Iliadou A, Kaprio J, Koskenvuo M et al (2000) Environmental and heritable factors in the causation of cancer–analyses of cohorts of twins from Sweden, Denmark, and Finland. N Engl J Med 343(2):78–85

Lin Q, Chen T, Lin Q, Lin G, Lin J, Chen G et al (2013) Serum miR-19a expression correlates with worse prognosis of patients with non-small cell lung cancer. J Surg Oncol 107(7):767–771

Liu Q, Fu H, Sun F, Zhang H, Tie Y, Zhu J et al (2008) miR-16 family induces cell cycle arrest by regulating multiple cell cycle genes. Nucleic Acids Res 36(16):5391–5404

Liu AM, Yao TJ, Wang W, Wong KF, Lee NP, Fan ST et al (2012) Circulating miR-15b and miR-130b in serum as potential markers for detecting hepatocellular carcinoma: a retrospective cohort study. BMJ Open 2(2):e000825

Lowery AJ, Miller N, Devaney A, McNeill RE, Davoren PA, Lemetre C et al (2009) MicroRNA signatures predict oestrogen receptor, progesterone receptor and HER2/neu receptor status in breast cancer. Breast Cancer Res 11(3):R27

Lu J, Getz G, Miska EA, Alvarez-Saavedra E, Lamb J, Peck D et al (2005) MicroRNA expression profiles classify human cancers. Nature 435(7043):834–838

Luo X, Burwinkel B, Tao S, Brenner H (2011) MicroRNA signatures: novel biomarker for colorectal cancer? Cancer Epidemiol Biomarkers Prev 20(7):1272–1286

Meng F, Henson R, Lang M, Wehbe H, Maheshwari S, Mendell JT et al (2006) Involvement of human micro-RNA in growth and response to chemotherapy in human cholangiocarcinoma cell lines. Gastroenterology 130(7):2113–2129

Meng F, Henson R, Wehbe-Janek H, Ghoshal K, Jacob ST, Patel T (2007) MicroRNA-21 regulates expression of the PTEN tumor suppressor gene in human hepatocellular cancer. Gastroenterology 133(2):647–658

Mezzanzanica D, Canevari S, Cecco LD, Bagnoli M (2011) miRNA control of apoptotic programs: focus on ovarian cancer. Expert Rev Mol Diagn 11(3):277–286

Michael MZ, O' Connor SM, van Holst Pellekaan NG, Young GP, James RJ (2003) Reduced accumulation of specific microRNAs in colorectal neoplasia. Mol Cancer Res 1(12):882–891

Migliore C, Petrelli A, Ghiso E, Corso S, Capparuccia L, Eramo A et al (2008) MicroRNAs impair MET-mediated invasive growth. Cancer Res 68(24):10128–10136

Mitchell PS, Parkin RK, Kroh EM, Fritz BR, Wyman SK, Pogosova-Agadjanyan EL et al (2008) Circulating microRNAs as stable blood-based markers for cancer detection. Proc Natl Acad Sci U S A 105(30):10513–10518

Munker R, Calin GA (2011) MicroRNA profiling in cancer. Clin Sci (Lond) 121(4):141–158
Nakano H, Yamada Y, Miyazawa T, Yoshida T (2013) Gain-of-function microRNA screens identify miR-193a regulating proliferation and apoptosis in epithelial ovarian cancer cells. Int J Oncol 42(6):1875–1882
Ohyashiki JH, Ohyashiki K, Sandberg AA, Minowada J, Kinniburgh AJ (1987) Human-fms gene is retained in acute lymphoblastic leukemia cells with del(5)(q32). Cancer Genet Cytogenet 25(2):341–350
Pagliuca A, Valvo C, Fabrizi E, di Martino S, Biffoni M, Runci D et al (2012) Analysis of the combined action of miR-143 and miR-145 on oncogenic pathways in colorectal cancer cells reveals a coordinate program of gene repression. Oncogene 32(40):4806–4813
Penna E, Orso F, Cimino D, Tenaglia E, Lembo A, Quaglino E et al (2011) microRNA-214 contributes to melanoma tumour progression through suppression of TFAP2C. EMBO J 30(10):1990–2007
Qi P, Cheng SQ, Wang H, Li N, Chen YF, Gao CF (2011) Serum microRNAs as biomarkers for hepatocellular carcinoma in Chinese patients with chronic hepatitis B virus infection. PLoS One 6(12):e28486
Qian X, Yu J, Yin Y, He J, Wang L, Li Q et al (2013) MicroRNA-143 inhibits tumor growth and angiogenesis and sensitizes chemosensitivity to oxaliplatin in colorectal cancers. Cell Cycle 12(9):1385–1394
Qu KZ, Zhang K, Li H, Afdhal NH, Albitar M (2011) Circulating microRNAs as biomarkers for hepatocellular carcinoma. J Clin Gastroenterol 45(4):355–360
Radojicic J, Zaravinos A, Vrekoussis T, Kafousi M, Spandidos DA, Stathopoulos EN (2011) MicroRNA expression analysis in triple-negative (ER, PR and Her2/neu) breast cancer. Cell Cycle 10(3):507–517
Rothé F, Ignatiadis M, Chaboteaux C, Haibe-Kains B, Kheddoumi N, Majjaj S et al (2011) Global microRNA expression profiling identifies MiR-210 associated with tumor proliferation, invasion and poor clinical outcome in breast cancer. PLoS One 6(6):e20980
Sachdeva M, Zhu S, Wu F, Wu H, Walia V, Kumar S et al (2009) p53 represses c-Myc through induction of the tumor suppressor miR-145. Proc Natl Acad Sci U S A 106(9):3207–3212
Schetter AJ, Leung SY, Sohn JJ, Zanetti KA, Bowman ED, Yanaihara N et al (2008) MicroRNA expression profiles associated with prognosis and therapeutic outcome in colon adenocarcinoma. JAMA 299(4):425–436
Schotte D, Chau JC, Sylvester G, Liu G, Chen C, van der Velden VH et al (2009) Identification of new microRNA genes and aberrant microRNA profiles in childhood acute lymphoblastic leukemia. Leukemia 23(2):313–322
Schrauder MG, Strick R, Schulz-Wendtland R, Strissel PL, Kahmann L, Loehberg CR et al (2012) Circulating micro-RNAs as potential blood-based markers for early stage breast cancer detection. PLoS One 7(1):e29770
Schultz J, Lorenz P, Gross G, Ibrahim S, Kunz M (2008) MicroRNA let-7b targets important cell cycle molecules in malignant melanoma cells and interferes with anchorage-independent growth. Cell Res 18(5):549–557
Segura MF, Belitskaya-Lévy I, Rose AE, Zakrzewski J, Gaziel A, Hanniford D et al (2010) Melanoma MicroRNA signature predicts post-recurrence survival. Clin Cancer Res 16(5):1577–1586
Selth LA, Townley S, Gillis JL, Ochnik AM, Murti K, Macfarlane RJ et al (2012) Discovery of circulating microRNAs associated with human prostate cancer using a mouse model of disease. Int J Cancer 131(3):652–661
Sempere LF, Christensen M, Silahtaroglu A, Bak M, Heath CV, Schwartz G et al (2007) Altered MicroRNA expression confined to specific epithelial cell subpopulations in breast cancer. Cancer Res 67(24):11612–11620
Shi XB, Xue L, Yang J, Ma AH, Zhao J, Xu M et al (2007) An androgen-regulated miRNA suppresses Bak1 expression and induces androgen-independent growth of prostate cancer cells. Proc Natl Acad Sci U S A 104(50):19983–19988
Shi XB, Xue L, Ma AH, Tepper CG, Kung HJ, White RW (2011) miR-125b promotes growth of prostate cancer xenograft tumor through targeting pro-apoptotic genes. Prostate 71(5): 538–549

Shigoka M, Tsuchida A, Matsudo T, Nagakawa Y, Saito H, Suzuki Y et al (2010) Deregulation of miR-92a expression is implicated in hepatocellular carcinoma development. Pathol Int 60(5):351–357

Sieuwerts AM, Mostert B, Bolt-de Vries J, Peeters D, de Jongh FE, Stouthard JM et al (2011) mRNA and microRNA expression profiles in circulating tumor cells and primary tumors of metastatic breast cancer patients. Clin Cancer Res 17(11):3600–3618

Sorrentino A, Liu CG, Addario A, Peschle C, Scambia G, Ferlini C (2008) Role of microRNAs in drug-resistant ovarian cancer cells. Gynecol Oncol 111(3):478–486

Taylor DD, Gercel-Taylor C (2008) MicroRNA signatures of tumor-derived exosomes as diagnostic biomarkers of ovarian cancer. Gynecol Oncol 110(1):13–21

Teng Y, Wang Y, Fu J, Cheng X, Miao S, Wang L (2011) Cyclin T2: a novel miR-15a target gene involved in early spermatogenesis. FEBS Lett 585(15):2493–2500

Tjensvoll K, Svendsen KN, Reuben JM, Oltedal S, Gilje B, Smaaland R et al (2012) miRNA expression profiling for identification of potential breast cancer biomarkers. Biomarkers 17(5):463–470

Toiyama Y, Takahashi M, Hur K, Nagasaka T, Tanaka K, Inoue Y et al (2013) Serum miR-21 as a diagnostic and prognostic biomarker in colorectal cancer. J Natl Cancer Inst 105(12):849–859

Tomimaru Y, Eguchi H, Nagano H, Wada H, Tomokuni A, Kobayashi S et al (2010) MicroRNA-21 induces resistance to the anti-tumour effect of interferon-α/5-fluorouracil in hepatocellular carcinoma cells. Br J Cancer 103(10):1617–1626

Tomimaru Y, Eguchi H, Nagano H, Wada H, Kobayashi S, Marubashi S et al (2012) Circulating microRNA-21 as a novel biomarker for hepatocellular carcinoma. J Hepatol 56(1):167–175

Tong MJ, Blatt LM, Kao VW (2001) Surveillance for hepatocellular carcinoma in patients with chronic viral hepatitis in the United States of America. J Gastroenterol Hepatol 16(5):553–559

Tsai WC, Hsu SD, Hsu CS, Lai TC, Chen SJ, Shen R et al (2012) MicroRNA-122 plays a critical role in liver homeostasis and hepatocarcinogenesis. J Clin Invest 122(8):2884–2897

Tsimberidou AM, Iskander NG, Hong DS, Wheler JJ, Falchook GS, Fu S et al (2012) Personalized medicine in a phase I clinical trials program: the MD Anderson Cancer Center initiative. Clin Cancer Res 18(22):6373–6383

Van der Auwera I, Limame R, van Dam P, Vermeulen PB, Dirix LY, Van Laere SJ (2010) Integrated miRNA and mRNA expression profiling of the inflammatory breast cancer subtype. Br J Cancer 103(4):532–541

Volinia S, Calin GA, Liu CG, Ambs S, Cimmino A, Petrocca F et al (2006) A microRNA expression signature of human solid tumors defines cancer gene targets. Proc Natl Acad Sci U S A 103(7):2257–2261

Wang P, Chen L, Zhang J, Chen H, Fan J, Wang K et al (2013) Methylation-mediated silencing of the miR-124 genes facilitates pancreatic cancer progression and metastasis by targeting Rac1. Oncogene. doi:10.1038/onc.2012.598. [Epub ahead of print]

Wiggins JF, Ruffino L, Kelnar K, Omotola M, Patrawala L, Brown D et al (2010) Development of a lung cancer therapeutic based on the tumor suppressor microRNA-34. Cancer Res 70(14):5923–5930

Wyman SK, Parkin RK, Mitchell PS, Fritz BR, O'Briant K, Godwin AK et al (2009) Repertoire of microRNAs in epithelial ovarian cancer as determined by next generation sequencing of small RNA cDNA libraries. PLoS One 4(4):e5311

Xu J, Wu C, Che X, Wang L, Yu D, Zhang T et al (2011) Circulating microRNAs, miR-21, miR-122, and miR-223, in patients with hepatocellular carcinoma or chronic hepatitis. Mol Carcinog 50(2):136–142

Yamamichi N, Shimomura R, Inada K, Sakurai K, Haraguchi T, Ozaki Y et al (2009) Locked nucleic acid in situ hybridization analysis of miR-21 expression during colorectal cancer development. Clin Cancer Res 15(12):4009–4016

Yamamoto Y, Kosaka N, Tanaka M, Koizumi F, Kanai Y, Mizutani T et al (2009) MicroRNA-500 as a potential diagnostic marker for hepatocellular carcinoma. Biomarkers 14(7):529–538

Yaman Agaoglu F, Kovancilar M, Dizdar Y, Darendeliler E, Holdenrieder S, Dalay N et al (2011) Investigation of miR-21, miR-141, and miR-221 in blood circulation of patients with prostate cancer. Tumour Biol 32(3):583–588

Yan J, Jiang N, Huang G, Tay JL, Lin B, Bi C et al (2013) Deregulated MIR335 that targets MAPK1 is implicated in poor outcome of paediatric acute lymphoblastic leukaemia. Br J Haematol 163(1):93–103

Yanaihara N, Caplen N, Bowman E, Seike M, Kumamoto K, Yi M et al (2006) Unique microRNA molecular profiles in lung cancer diagnosis and prognosis. Cancer Cell 9(3):189–198

Yu S, Lu Z, Liu C, Meng Y, Ma Y, Zhao W et al (2010) miRNA-96 suppresses KRAS and functions as a tumor suppressor gene in pancreatic cancer. Cancer Res 70(14):6015–6025

Zanette DL, Rivadavia F, Molfetta GA, Barbuzano FG, Proto-Siqueira R, Silva-Jr WA et al (2007) miRNA expression profiles in chronic lymphocytic and acute lymphocytic leukemia. Braz J Med Biol Res 40(11):1435–1440

Zhang L, Volinia S, Bonome T, Calin GA, Greshock J, Yang N et al (2008) Genomic and epigenetic alterations deregulate microRNA expression in human epithelial ovarian cancer. Proc Natl Acad Sci U S A 105(19):7004–7009

Zhang H, Cai X, Wang Y, Tang H, Tong D, Ji F (2010) microRNA-143, down-regulated in osteosarcoma, promotes apoptosis and suppresses tumorigenicity by targeting Bcl-2. Oncol Rep 24(5):1363–1369

Zhao WG, Yu SN, Lu ZH, Ma YH, Gu YM, Chen J (2010) The miR-217 microRNA functions as a potential tumor suppressor in pancreatic ductal adenocarcinoma by targeting KRAS. Carcinogenesis 31(10):1726–1733

Zheng J, Dong P, Gao S, Wang N, Yu F (2012) High expression of serum miR-17-5p associated with poor prognosis in patients with hepatocellular carcinoma. Hepatogastroenterology 60(123):549–552

[illegible] Dizdar Y, Darendeliler E, Holdenrieder S, Dalay N et al (2011) [illegible] miR-21, miR-141, and miR-221 in blood circulation of patients with prostate cancer [illegible]

[illegible]

Chapter 14
MicroRNA Expression Profiling and Its Clinical Impact in Breast Cancer

Manuela Ferracin and Laura Lupini

Contents

Abstract Breast cancer is the leading cause of cancer death in women worldwide. Gene expression studies have been used over the last decades to define the signature of different breast cancer subtypes and to predict outcome and response to therapies. Recently, microRNAs (miRNAs) have been linked to several human diseases, including cancer. An aberrant miRNA expression in breast cancer was first reported in 2005. Now, an increasing body of experimental evidences supports the role of these small molecules in the tumorigenic process and their potential use as cancer specific biomarkers. Indeed, miRNAs are detectable as circulating molecules in the blood. In this chapter, we summarize our knowledge about the involvement of miRNAs in breast cancer and their potential as diagnostic, prognostic and therapeutic tools.

Keywords Breast cancer • microRNA expression profiling • Circulating microRNAs • Diagnosis • Prognosis • Therapy

M. Ferracin (✉) • L. Lupini
Department of Morphology, Surgery and Experimental Medicine, Laboratory for Technologies of Advanced Therapies, University of Ferrara, Ferrara, Italy
e-mail: manuela.ferracin@unife.it

S. Babashah (ed.), *MicroRNAs: Key Regulators of Oncogenesis*,
DOI 10.1007/978-3-319-03725-7_14,

1 Introduction

Breast cancer is one of the most frequent cancers worldwide and the most frequent affecting women. In 2012, 227,000 new cases of breast cancer were counted and 39,500 people died from this neoplasm in United States (Siegel et al. 2012). Breast cancer is a heterogeneous disease, that comprehends several histotypes, characterized by different biological and phenotypic features and that presents different prognostic and therapeutic procedures. The currently used factors for breast cancer diagnosis, classification and treatment include patient age, gender, menopausal status, lymph node status, tumor size, histological features (grade and type, peritumoral vascular invasion), hormone receptor status [Estrogen receptor (ER) and progesterone receptor (PR) expression], proliferation index (Mib1) and HER2 over-expression/amplification, as provided by St. Gallen criteria (Goldhirsch et al. 2009), Nottingham Prognostic Index (Galea et al. 1992) and Adjuvant Online (www.adjuvantonline.com). The definition of these guidelines for patient risk stratification, together with the improvement of chemotherapeutic regimens, enhanced breast cancer survival rate. However, these rules do not consider the individual molecular complexity of each neoplasm and for this reason some tumors belonging to a risk group do not behave in the expected way or do not respond to the chosen therapeutic strategy. The right prediction of aggressiveness and metastatic potential of a lesion are key factors in breast cancer patient management. The recent development of technologies (i.e. microarray, deep sequencing) helped in overcoming this problem. These methods are able to evaluate the global genomic and transcriptomic profile of tissues and tumors. Indeed, specific genome-wide profiles created for each kind of breast tumor subtype, improved histotype classification and prognosis definition (Perou et al. 2000; Koboldt et al. 2012; Curtis et al. 2012).

MicroRNAs (miRNAs) are a class of regulatory, non-coding small RNA that mainly post-transcriptionally regulate gene expression. The evaluation of cancer-specific miRNA profiles revealed to be useful in stratifying breast tumors. Breast cancer was among the first tumor types for which the evaluation of miRNA profile was performed. The miRNA microarray analysis has initially showed a panel of 29 miRNAs that were deregulated in breast cancer, if compared to healthy breast tissue (Iorio et al. 2005). Interestingly, further studies on other human cancers found that several of the 29 miRNAs were deregulated in other neoplasms, indicating that this first set of miRNAs could affect pathogenetic mechanisms potentially shared among tumors and suggesting the important role of miRNA in tumor development. Nowadays, the involvement of miRNAs in each step of cancerogenesis, from transformation to metastatic spreading, is well known and universally recognized.

MiRNA profiling has allowed for the identification of signatures associated with the diagnosis, staging, progression, prognosis, and response to treatment of human tumors (Dvinge et al. 2013). The miRNA-based classifier is much better than the mRNA classifier at establishing the correct diagnosis for metastatic cancer of unknown primary site and for metastases coming from or retrieved inside breast tissue (Ferracin et al. 2011). The maintenance of a strong nucleus of tissue-specific miRNAs after the spreading of cancer cells to distant sites is probably the reason why miRNA profiling is more effective in the diagnosis of cancers of unknown primary site. In this chapter we review the role of miRNAs in breast cancer transformation, progression and their possible application in breast cancer diagnosis, prognosis and therapy.

2 Deregulation of MicroRNAs in Breast Cancer

Several miRNAs were demonstrated to play important roles in breast cancer development (Le Quesne and Caldas 2010; Yu et al. 2010a). Like protein-coding genes, miRNAs could be classified as oncogenes (oncomiRs) or tumor suppressor genes, according to their expression levels in cancer and the cellular functions of miRNAs and their targets. In breast cancer, miR-145, miR-125, let-7 family and miR-200 family are the most known tumor suppressor miRNAs, while among the oncomiRs we can remind miR-21 and miR-155 (Fig. 14.1).

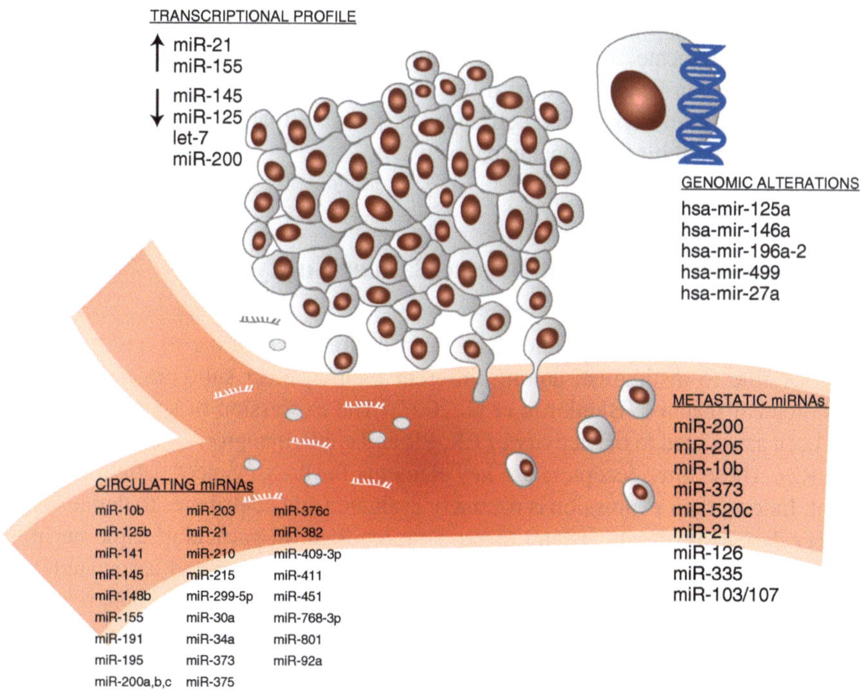

Fig. 14.1 MicroRNAs involved in specific processes linked to breast cancer, such as cancer predisposition (genomic alterations), cancer phenotype (transcriptional profile), cancer progression (metastatic miRNAs) and their release as circulating miRNAs in the blood of cancer patients (circulating miRNAs) (Figure modified from Elsevier (Ferracin and Calin 2011))

MiR-125 is considered as a tumor suppressor in breast cancer, since it was found down-regulated in cancer, compared to normal mammary tissue (Iorio et al. 2005). Its tumor suppressor function was confirmed by the finding that many miR-125 targets, among which ERBB2, ERBB3 and MUC1, are frequently over-expressed in breast cancers, contributing to aggressiveness of the pathology (Scott et al. 2007).

Let-7 is a tumor suppressor miRNA and it was found to be down-regulated in several human neoplasms including breast cancer. It was demonstrated that let-7 is a negative regulator of several stemness properties and is reduced in breast tumor-initiating cells (BT-ICs). Indeed, its forced up-regulation decreased proliferation, self-renewal and metastasizing capability of cells (Yu et al. 2007). Known targets of let-7 are ESR1 (Bhat-Nakshatri et al. 2009) and RAS (Johnson et al. 2005), two known oncogenes, supporting its tumor suppressive function.

The well-known oncogenic miR-17-92 cluster plays a controversial role in breast cancer. Li and colleagues demonstrated that the over-expression of miR-17-5p promotes invasion of breast cancer cell lines (Li et al. 2011). In contrast, Yu and colleagues found a reduced expression of this cluster in breast tumors, if compared to normal tissue, and they proved that the cluster was able to inhibit cell proliferation and metastasizing capability by targeting cyclin D1 and IL-8 (Yu et al. 2008, 2010b), suggesting a potential tumor suppressive role for these miRNAs. Further studies will be necessary to better elucidate the role of this cluster in breast cancer. However, these results indicate that the classification of a miRNA, as tumor suppressor miRNA or oncomiR, is sometime difficult, since miRNAs are complex players in the molecular pathways of the cell and it is possible that they cooperate both in inducing and inhibiting oncogenic pathways.

3 MicroRNA Profile of Breast Cancer Subtypes

Some miRNA expression studies have been performed in order to find molecular profiles able to distinguish among different breast cancer subtypes and clinical-pathological features (Blenkiron et al. 2007). The expression of specific miRNAs has been associated to the activity of ER, PR and epidermal growth factor 2 receptor (HER2). These three receptors are key elements for breast cancer patient management. Indeed, their expression is routinely evaluated and contributes to define breast cancer diagnosis, prognosis and therapeutic regimen. In particular, triple-negative breast cancers (ER-/PR-/HER2-) are usually more aggressive and associated with poor prognosis, while neoplasms over-expressing at least one of the three receptors are characterized by better prognosis (because drugs targeting these molecules are available and effective in treating positive cancers).

It has been demonstrated that specific miRNAs can regulate the expression of estrogen, progesterone and epidermal growth factor 2 receptors. Several groups performed microarray miRNA profiling of breast cancers characterized by different ER, PR and HER2 status, in order to identify miRNAs associated with these features (Iorio et al. 2005; Mattie et al. 2006; Lowery et al. 2009). We discuss here some results that emerged from these and other studies, in order to highlight the miRNAs that influence receptors expression in breast cancer.

Among the miRNAs targeting ER, we remind miR-22 (Pandey and Picard 2009; Xiong et al. 2010) and miR-145 (Spizzo et al. 2010). On the other hand, some miRNAs are regulated by ERα itself (miR-21, miR-181, miR-26) (Bhat-Nakshatri et al. 2009; Maillot et al. 2009; Wickramasinghe et al. 2009), establishing an interesting regulatory loop. Finally, there are some miRNAs including let-7 (Zhao et al. 2011; Bhat-Nakshatri et al. 2009), miR-17-92 cluster, miR-106a/363 (Castellano et al. 2009), miR-221/222 (Zhao et al. 2008; Di Leva et al. 2010) and miR-206 that are both regulated and regulating, suggesting the presence of feedback loops acting on ER pathway. Just to give an example, miR-206 (a miRNA found up-regulated in ER- tumors) targets ERα mRNA and its expression is impaired by ER agonists (Adams et al. 2007; Kondo et al. 2008). Moreover, this miRNA seems to play a role in the repression of estrogenic response mediated by epidermal growth factor (EGF) in MCF-7 that is responsible for the switch from luminal-A to basal-like phenotype (Adams et al. 2009).

To what concern the other two receptors, miR-26 and miR-181, two miRNAs down-regulated by estrogen treatment in MCF-7 cells, are able to modulate PR (Maillot et al. 2009). HER2 is a direct target of miR-125 (Scott et al. 2007) and its expression is indirectly reduced by miR-205, a miRNA that targets HER3, belonging to the same family of HER2 (Iorio et al. 2009). All these findings highlight the important contribution of miRNAs in defining the molecular subtypes and the clinical-pathological features of breast neoplasms and could represent important molecular targets in breast cancer management.

4 Role of MicroRNAs in Epithelial-Mesenchymal Transition and Metastasis

Important features of tumor aggressiveness are epithelial-mesenchymal transition (EMT) and metastatic capability. EMT is a process where epithelial cells lose their epithelial features, cell polarity and cell adhesion, and become mesenchymal cells, gaining migratory capability, invasive potential and, finally, metastasize. Breast cancers presenting an invasive phenotype or metastasis at diagnosis are characterized by poor prognosis. Several miRNAs play important role in regulating EMT and metastasis in breast cancer (Fig. 14.1).

MiR-10b was found highly expressed in breast cancer metastases and it was demonstrated that its expression is under the control of Twist, a transcription factor implicated in EMT process. MiR-10b up-regulation inhibits HOXD10 protein translation and induces RHOC expression, leading to an invasive and metastatic behavior of tumor cells (Ma et al. 2007). Furthermore, antagomiR-10b administration in a mouse-model of breast cancer induced a reduction in tumor proliferation and metastasis formation, introducing the promising role of miRNAs as therapeutic targets (Ma et al. 2010).

Other metastasis-promoting miRNAs are miR-373 and miR-520c (Huang et al. 2008). It was demonstrated that they are able to promote migration and invasion capability of MCF-7 breast cancer cell lines, by directly suppressing of CD44, a cell-surface glycoprotein involved in cell-cell interactions. Supporting this thesis, increased levels of miR-373 and CD44 reduction were found together in human metastatic breast cancers.

The well-known oncomiR-21 was extensively studied for its involvement in breast cancer development, because it is a master regulator of cell proliferation and survival. In addition, it is mightily implicated in metastasis, since it targets TPM1, a component of cytoskeleton (Zhu et al. 2008), maspin and PDCD4 (Lu et al. 2008; Qi et al. 2009) that are inhibitors of the pro-metastatic factor UPAR, and PTEN and TIMP3 (Qi et al. 2009), inhibitors of matrix metalloproteinases.

Other important pro-metastatic miRNAs belong to miR-103/107 family, whose high levels are associated with metastasis and poor outcome in breast cancer. Martello and colleagues proved that injection of miR-103/107 antagomiR in a breast cancer xenograft in mouse induced a reduction of number of metastatic foci in the animal. They also demonstrated that down-regulation of Dicer, a molecular target of miR-103/107, and the consequent global down-regulation of miRNAs biosynthesis, played an important role in metastatic effects induced by these miRNAs. Further, miR-103/107 promoted epithelial-to-mesenchymal transition by down-regulating miR-200 levels (Martello et al. 2010).

MiR-200 and miR-205 are frequently down-regulated in breast cancer, compared to healthy mammary tissue. It was demonstrated that they are able to protect cells from EMT by targeting ZEB1 and ZEB2, two strong activators of epithelial to mesenchymal transition (Gregory et al. 2008; Korpal et al. 2008; Park et al. 2008). Indeed, down-regulation of miR-200 family was associated with metastasis development in breast cancer, while miR-205 induction was responsible for a reduced invasion capability of breast cancer cell lines and *in vivo* metastasis formation (Gregory et al. 2008) and it has an important role in triple negative breast cancers (Piovan et al. 2012).

MiR-335 and miR-126 were found to be anti-metastatic miRNAs through the comparison of metastatic nodules versus primary tumors. Indeed, the two miRNAs showed reduced levels in metastasis and their normal expression in primary tumors correlated with increasing metastasis-free time. They were able to counteract metastasis development by blocking expression of SOX4 and TNC (Tavazoie et al. 2008). Similarly, miR-17-92 clusters seemed to play an anti-metastatic role in breast cancer, as previously discussed.

Altogether, these findings suggest that miRNAs play very crucial roles in invasion capability and metastatic development of cancer and for this reason they could be considered interesting targets for anti-cancer-tailored therapies.

5 Circulating MicroRNAs as Novel Diagnostic Markers

One of the major challenges in molecular oncology is the employment of miRNAs as biomarkers for early diagnosis of cancer. Until now, only few proteins have been used in the clinic as blood cancer biomarkers. MiRNAs are released in the blood from cancer cells, and they have a good potential of being employed for this purpose, since it was demonstrated that they are stable in human serum and plasma (Mitchell et al. 2008). Furthermore, it has been found that they are differentially represented in cancer patients compared to healthy controls. Concerning breast cancer, blood levels of several miRNAs were found differentially released in cancer patients compared to healthy controls (Chan et al. 2013; Cuk et al. 2013). Figure 14.1 summarizes the miRNAs that are deregulated in breast cancer and released into the circulation. As can be noticed, the circulating miRNA profile does not match the expression profile of solid tumors (Cookson et al. 2012).

Detection of miRNAs in serum or plasma for the early diagnosis of cancer is a very promising tool. However many issues should be solved before this tool could enter into the clinic. First of all, different technologies (RT-qPCR, digital PCR, microarray, next-generation sequencing) are now available for the evaluation of miRNA expression in the blood, but it was recently demonstrated that results of the same pathological condition achieved with different technologies and different preparation methods are very different. This is due to the fact that technical issues can dramatically influence the results of analysis. For this reason, the optimization of many preanalytical and analytical variables is necessary. As a consequence of the lack of optimized protocols, the majority of published studies do not reciprocally confirm their findings (Chan et al. 2013; Madhavan et al. 2013).

6 MicroRNAs as Prognostic Tools in Cancer

Several studies have demonstrated the potential of using miRNA to predict prognosis of human cancers. Some miRNAs have been negatively or positively associated with prognostic endpoints. MiR-21 is one of the most studied oncomiRs and its expression was linked to poor prognosis in several human cancers (Krichevsky and Gabriely 2009). In breast cancer, miR-21 over-expression correlates with negative hormone receptor status, advanced tumor stage, high grade, lymph node metastasis and poor survival (Yan et al. 2008; Qian et al. 2009). These findings clearly suggest that miR-21 over-expression is a negative prognostic factor. A similar function is suspected for miR-210, a miRNA whose transcription is induced by HIF-1 in hypoxic conditions (Kulshreshtha et al. 2007). A higher expression of this miRNA was associated with a shorter disease-free time, overall-survival (Camps et al. 2008) and time-to-metastasis in lymph node negative patients (Foekens et al. 2008). An opposite role has been described for miR-30c, whose expression was found to be an independent predictor of progression-free survival in a large cohort of advanced ER+ breast cancer patients treated with tamoxifen (Rodriguez-Gonzalez et al. 2011). Despite these promising findings, the use of miRNA expression as prognostic tool is not yet in clinical practice.

Being better than individual miRNAs, miRNA profiles (i.e. a panel of miRNAs) could constitute more effective prognostic factors. Gene expression profiling has been used since 2000 to study breast cancer biology and to find gene signatures able to distinguish between different subtypes. Perou et al. (2000) performed the first microarray analysis of 65 breast cancer samples allowing the identification of five different "intrinsic" subtypes, characterized by different gene expression profiles: luminal A, luminal B, basal-like, HER2-enriched and normal-like. These profiles, confirmed also by subsequent studies, were able to depict different kinds of breast cancer, associated with different prognosis and clinical course. Later, also Blenkiron and colleagues performed gene-expression and miRNA analysis using microarray technology on 51 breast cancer samples. Starting from gene-expression results, he classified the samples according to Perou's profile and compared miRNA expression analysis in samples belonging to different subtypes, allowing the identification of several subtypes-related miRNAs (Blenkiron et al. 2007). This discovery suggested that Perou's classification could be improved adding miRNA expression results to gene expression profiles.

Besides predicting prognosis of breast cancer, miRNA profiling was demonstrated to be a good tool also to predict therapeutic response. The identification of neoplasms that will respond to a given chemotherapeutic drug is extremely important to reduce breast cancer morbidity and mortality. Using the gene-expression data from NCI-60 panel of cell lines tested for sensitivity to several drugs, Salter and colleagues developed mRNA and miRNA profiles associated with resistance to chemotherapeutic agents and validated their results in a cohort of 133 breast cancer patients treated with TFAC (paclitaxel, 5-fluorouracil, adriamycin, and cyclophosphamide) regimen. Profiles-derived predictions were compared with predictions based on traditional criteria (ER, PR, HER2 and Topoisomerase IIA expression levels), suggesting the importance of integrating classical prediction methods with new molecular tests and concluding that molecular profiling could represent a rational strategy to identify alternative therapeutic opportunities (Salter et al. 2008). In another study, miRNA expression profile of doxorubicine-resistant versus sensitive MCF-7 breast cancer cell line was performed and it was found that resistant cells over-express miR-106, miR-21, miR-206 and miR-28, while they down-regulate miR-127, miR-200c, miR-34a, miR-27b and let-7. Moreover, they demonstrated that miR-451, whose expression was lost in resistant cells, targets multidrug resistance 1 gene (mdr1/ABCB1) (Kovalchuk et al. 2008).

With similar approaches, two studies found miRNAs involved in tamoxifen and cisplatin resistance: Miller and colleagues studied miRNAs related to tamoxifen sensitivity, finding miR-221, miR-222 and miR-181 among the up-regulated miRNAs in resistant cells and miR-21, miR-342 and miR-489 among the down-regulated ones (Miller et al. 2008). Pogribny et al. (2010) performed miRNA profiling of cisplatin-resistant and sensitive MCF-7 cell lines and identified a large panel of miRNAs between the two phenotypes. In another study, aimed at identifying miRNAs associated with chemotherapy response in ovarian cancers, let-7i down-regulation emerged to be associated to cisplatin resistance both in ovarian and breast tumors (Yang et al. 2008). Other miRNAs related to chemoresistance are miR-125b, that was able to reduce taxol sensitivity (Zhou et al. 2010), and miR-155 that targets the proapoptotic gene FOXO3 and was able to induce doxorubicine resistance (Kong et al. 2010).

Many of the above mentioned studies were performed only in cellular models of breast cancer. A further validation in clinical samples is therefore necessary to identify miRNAs that could really influence therapeutic decisions. Indeed, these findings clearly highlight the complexity of the scenario where several miRNA concur in determining sensitivity or resistance to chemotherapeutics. However, the potential of miRNAs expression profile as a tool to predict prognosis seems to be achievable very soon and constitutes a real tool to improve clinical patient management.

7 MicroRNAs as Therapeutic Targets or Therapeutic Agents

Since their first discovery, miRNAs were considered of special interest as cancer therapeutics. Indeed, the modulation (induction or inhibition) of a single miRNA is able to change the expression of hundreds of target genes in the cell. Moreover, their small size, high stability and endogenous nature, indicate they have suitable features to be used as therapeutic molecules. MiRNAs represent interesting targets in cancer treatment with two possible approaches: silencing of an over-expressed miRNA or introduction of a down-regulated miRNA.

To inhibit miRNA activity, different molecules have been tested and demonstrated their efficacy in *in vitro* and/or in *in vivo* models: anti-miRs (AMO), antagomiRs, locked-nucleic acid (LNA), sponge vectors; they have different molecular structures, but they are all able to block their target miRNA by directly linking mature miRNA sequences via complementarity. For example, the administration of antagomiR-10b in breast tumor-bearing mice was demonstrated to reduce metastasis formation (Ma et al. 2010).

The restoration of a down-regulated miRNA in tumor cells could be achieved in different ways. In cellular models, the simple transfection of pre-miR precursors is effective, while for *in vivo* delivery different kind of miRNA-expressing vectors have been studied. It was demonstrated that miR-145 transfection in different kind of breast cancer cell lines induced a significant block of cancer cell proliferation and enhanced apoptosis (Spizzo et al. 2010). In another study adeno-associated virus (AAV) expressing miR-26a was proved to be effective in reducing liver-tumor growth in mice, without relevant toxicity (Kota et al. 2009). Another interesting possibility for cancer therapy is the use of miRNA-dependent oncolytic vectors, that take advantage of our knowledge on cancer-specific miRNAs to control the viral infection (Callegari et al. 2013).

Even if some issues have yet to be extensively considered, like target delivery and systemic toxicity, several studies involving miRNAs as cancer therapeutics have been performed (Tong and Nemunaitis 2008) and some clinical trials have been approved (www.clinicaltrials.gov), highlighting the promising role of miRNAs in tumor treatment in the future.

8 Conclusions

MiRNA global expression profiling could be used for the classification of breast cancers, establishing specific diagnoses and offering prognostic values. Recent studies have demonstrated that circulating miRNAs could serve as biomarkers for the early diagnosis of tumors. Moreover, they represent interesting targets for a new-generation of targeted therapies. For all these reasons, miRNAs appear to be a very promising tool in each step of breast cancer management.

Acknowledgements Our work was supported by the Italian Association for Cancer Research (AIRC MFAG11676). We thank Miriam Ferracin for her assistance in generating the graphical illustration.

References

Adams BD, Furneaux H, White BA (2007) The micro-ribonucleic acid (miRNA) miR-206 targets the human estrogen receptor-alpha (ERalpha) and represses ERalpha messenger RNA and protein expression in breast cancer cell lines. Mol Endocrinol 21(5):1132–1147

Adams BD, Cowee DM, White BA (2009) The role of miR-206 in the epidermal growth factor (EGF) induced repression of estrogen receptor-alpha (ERalpha) signaling and a luminal phenotype in MCF-7 breast cancer cells. Mol Endocrinol 23(8):1215–1230

Bhat-Nakshatri P, Wang G, Collins NR, Thomson MJ, Geistlinger TR, Carroll JS et al (2009) Estradiol-regulated microRNAs control estradiol response in breast cancer cells. Nucleic Acids Res 37(14):4850–4861

Blenkiron C, Goldstein LD, Thorne NP, Spiteri I, Chin SF, Dunning MJ et al (2007) MicroRNA expression profiling of human breast cancer identifies new markers of tumor subtype. Genome Biol 8(10):R214

Callegari E, Elamin BK, D'Abundo L, Falzoni S, Donvito G, Moshiri F et al (2013) Anti-tumor activity of a miR-199-dependent oncolytic adenovirus. PloS One 8(9):e73964

Camps C, Buffa FM, Colella S, Moore J, Sotiriou C, Sheldon H et al (2008) hsa-miR-210 is induced by hypoxia and is an independent prognostic factor in breast cancer. Clin Cancer Res 14(5):1340–1348

Castellano L, Giamas G, Jacob J, Coombes RC, Lucchesi W, Thiruchelvam P et al (2009) The estrogen receptor-alpha-induced microRNA signature regulates itself and its transcriptional response. Proc Natl Acad Sci U S A 106(37):15732–15737

Chan M, Liaw CS, Ji SM, Tan HH, Wong CY, Thike AA et al (2013) Identification of circulating MicroRNA signatures for breast cancer detection. Clin Cancer Res 19(16):4477–4487

Cookson VJ, Bentley MA, Hogan BV, Horgan K, Hayward BE, Hazelwood LD et al (2012) Circulating microRNA profiles reflect the presence of breast tumours but not the profiles of microRNAs within the tumours. Cell Oncol 35(4):301–308

Cuk K, Zucknick M, Heil J, Madhavan D, Schott S, Turchinovich A et al (2013) Circulating microRNAs in plasma as early detection markers for breast cancer. Int J Cancer 132(7):1602–1612

Curtis C, Shah SP, Chin SF, Turashvili G, Rueda OM, Dunning MJ et al (2012) The genomic and transcriptomic architecture of 2,000 breast tumours reveals novel subgroups. Nature 486(7403):346–352

Di Leva G, Gasparini P, Piovan C, Ngankeu A, Garofalo M, Taccioli C et al (2010) MicroRNA cluster 221–222 and estrogen receptor alpha interactions in breast cancer. J Natl Cancer Inst 102(10):706–721

Dvinge H, Git A, Graf S, Salmon-Divon M, Curtis C, Sottoriva A et al (2013) The shaping and functional consequences of the microRNA landscape in breast cancer. Nature 497(7449):378–382

Ferracin M, Calin GA (2011) Principles of MicroRNA involvement in breast cancer. Breast Dis Year Book Q 22(3):238–243

Ferracin M, Pedriali M, Veronese A, Zagatti B, Gafa R, Magri E et al (2011) MicroRNA profiling for the identification of cancers with unknown primary tissue-of-origin. J Pathol 225(1):43–53

Foekens JA, Sieuwerts AM, Smid M, Look MP, de Weerd V, Boersma AW et al (2008) Four miRNAs associated with aggressiveness of lymph node-negative, estrogen receptor-positive human breast cancer. Proc Natl Acad Sci U S A 105(35):13021–13026

Galea MH, Blamey RW, Elston CE, Ellis IO (1992) The Nottingham Prognostic Index in primary breast cancer. Breast Cancer Res Treat 22(3):207–219

Goldhirsch A, Ingle JN, Gelber RD, Coates AS, Thurlimann B, Senn HJ (2009) Thresholds for therapies: highlights of the St Gallen International Expert Consensus on the primary therapy of early breast cancer 2009. Ann Oncol 20(8):1319–1329

Gregory PA, Bert AG, Paterson EL, Barry SC, Tsykin A, Farshid G et al (2008) The miR-200 family and miR-205 regulate epithelial to mesenchymal transition by targeting ZEB1 and SIP1. Nat Cell Biol 10(5):593–601

Huang Q, Gumireddy K, Schrier M, le Sage C, Nagel R, Nair S et al (2008) The microRNAs miR-373 and miR-520c promote tumour invasion and metastasis. Nat Cell Biol 10(2):202–210

Iorio MV, Ferracin M, Liu CG, Veronese A, Spizzo R, Sabbioni S et al (2005) MicroRNA gene expression deregulation in human breast cancer. Cancer Res 65(16):7065–7070

Iorio MV, Casalini P, Piovan C, Di Leva G, Merlo A, Triulzi T et al (2009) microRNA-205 regulates HER3 in human breast cancer. Cancer Res 69(6):2195–2200

Johnson SM, Grosshans H, Shingara J, Byrom M, Jarvis R, Cheng A et al (2005) RAS is regulated by the let-7 microRNA family. Cell 120(5):635–647

Koboldt DC, Fulton RF, McLellan MD, Schmidt M, Kalicki-Veizer J, TCGA N (2012) Comprehensive molecular portraits of human breast tumours. Nature 490(7418):61–70

Kondo N, Toyama T, Sugiura H, Fujii Y, Yamashita H (2008) miR-206 expression is down-regulated in estrogen receptor alpha-positive human breast cancer. Cancer Res 68(13):5004–5008

Kong W, He L, Coppola M, Guo J, Esposito NN, Coppola D et al (2010) MicroRNA-155 regulates cell survival, growth, and chemosensitivity by targeting FOXO3a in breast cancer. J Biol Chem 285(23):17869–17879

Korpal M, Lee ES, Hu G, Kang Y (2008) The miR-200 family inhibits epithelial-mesenchymal transition and cancer cell migration by direct targeting of E-cadherin transcriptional repressors ZEB1 and ZEB2. J Biol Chem 283(22):14910–14914

Kota J, Chivukula RR, O'Donnell KA, Wentzel EA, Montgomery CL, Hwang HW et al (2009) Therapeutic microRNA delivery suppresses tumorigenesis in a murine liver cancer model. Cell 137(6):1005–1017

Kovalchuk O, Filkowski J, Meservy J, Ilnytskyy Y, Tryndyak VP, Chekhun VF et al (2008) Involvement of microRNA-451 in resistance of the MCF-7 breast cancer cells to chemotherapeutic drug doxorubicin. Mol Cancer Ther 7(7):2152–2159

Krichevsky AM, Gabriely G (2009) miR-21: a small multi-faceted RNA. J Cell Mol Med 13(1):39–53

Kulshreshtha R, Ferracin M, Wojcik SE, Garzon R, Alder H, Agosto-Perez FJ et al (2007) A microRNA signature of hypoxia. Mol Cell Biol 27(5):1859–1867

Le Quesne J, Caldas C (2010) Micro-RNAs and breast cancer. Mol Oncol 4(3):230–241

Li H, Bian C, Liao L, Li J, Zhao RC (2011) miR-17-5p promotes human breast cancer cell migration and invasion through suppression of HBP1. Breast Cancer Res Treat 126(3):565–575

Lowery AJ, Miller N, Devaney A, McNeill RE, Davoren PA, Lemetre C et al (2009) MicroRNA signatures predict oestrogen receptor, progesterone receptor and HER2/neu receptor status in breast cancer. Breast Cancer Res 11(3):R27

Lu Z, Liu M, Stribinskis V, Klinge CM, Ramos KS, Colburn NH et al (2008) MicroRNA-21 promotes cell transformation by targeting the programmed cell death 4 gene. Oncogene 27(31): 4373–4379

Ma L, Teruya-Feldstein J, Weinberg RA (2007) Tumour invasion and metastasis initiated by microRNA-10b in breast cancer. Nature 449(7163):682–688

Ma L, Reinhardt F, Pan E, Soutschek J, Bhat B, Marcusson EG et al (2010) Therapeutic silencing of miR-10b inhibits metastasis in a mouse mammary tumor model. Nat Biotechnol 28(4):341–347

Madhavan D, Cuk K, Burwinkel B, Yang R (2013) Cancer diagnosis and prognosis decoded by blood-based circulating microRNA signatures. Front Genet 4:116

Maillot G, Lacroix-Triki M, Pierredon S, Gratadou L, Schmidt S, Benes V et al (2009) Widespread estrogen-dependent repression of micrornas involved in breast tumor cell growth. Cancer Res 69(21):8332–8340

Martello G, Rosato A, Ferrari F, Manfrin A, Cordenonsi M, Dupont S et al (2010) A MicroRNA targeting dicer for metastasis control. Cell 141(7):1195–1207

Mattie MD, Benz CC, Bowers J, Sensinger K, Wong L, Scott GK et al (2006) Optimized high-throughput microRNA expression profiling provides novel biomarker assessment of clinical prostate and breast cancer biopsies. Mol Cancer 5:24

Miller TE, Ghoshal K, Ramaswamy B, Roy S, Datta J, Shapiro CL et al (2008) MicroRNA-221/222 confers tamoxifen resistance in breast cancer by targeting p27Kip1. J Biol Chem 283(44):29897–29903

Mitchell PS, Parkin RK, Kroh EM, Fritz BR, Wyman SK, Pogosova-Agadjanyan EL et al (2008) Circulating microRNAs as stable blood-based markers for cancer detection. Proc Natl Acad Sci U S A 105(30):10513–10518

Pandey DP, Picard D (2009) miR-22 inhibits estrogen signaling by directly targeting the estrogen receptor alpha mRNA. Mol Cell Biol 29(13):3783–3790

Park SM, Gaur AB, Lengyel E, Peter ME (2008) The miR-200 family determines the epithelial phenotype of cancer cells by targeting the E-cadherin repressors ZEB1 and ZEB2. Genes Dev 22(7):894–907

Perou CM, Sorlie T, Eisen MB, van de Rijn M, Jeffrey SS, Rees CA et al (2000) Molecular portraits of human breast tumours. Nature 406(6797):747–752

Piovan C, Palmieri D, Di Leva G, Braccioli L, Casalini P, Nuovo G et al (2012) Oncosuppressive role of p53-induced miR-205 in triple negative breast cancer. Mol Oncol 6(4):458–472

Pogribny IP, Filkowski JN, Tryndyak VP, Golubov A, Shpyleva SI, Kovalchuk O (2010) Alterations of microRNAs and their targets are associated with acquired resistance of MCF-7 breast cancer cells to cisplatin. Int J Cancer 127(8):1785–1794

Qi L, Bart J, Tan LP, Platteel I, Sluis T, Huitema S et al (2009) Expression of miR-21 and its targets (PTEN, PDCD4, TM1) in flat epithelial atypia of the breast in relation to ductal carcinoma in situ and invasive carcinoma. BMC Cancer 9:163

Qian B, Katsaros D, Lu L, Preti M, Durando A, Arisio R et al (2009) High miR-21 expression in breast cancer associated with poor disease-free survival in early stage disease and high TGF-beta1. Breast Cancer Res Treat 117(1):131–140

Rodriguez-Gonzalez FG, Sieuwerts AM, Smid M, Look MP, Meijer-van Gelder ME, de Weerd V et al (2011) MicroRNA-30c expression level is an independent predictor of clinical benefit of endocrine therapy in advanced estrogen receptor positive breast cancer. Breast Cancer Res Treat 127(1):43–51

Salter KH, Acharya CR, Walters KS, Redman R, Anguiano A, Garman KS et al (2008) An integrated approach to the prediction of chemotherapeutic response in patients with breast cancer. PloS One 3(4):e1908

Scott GK, Goga A, Bhaumik D, Berger CE, Sullivan CS, Benz CC (2007) Coordinate suppression of ERBB2 and ERBB3 by enforced expression of micro-RNA miR-125a or miR-125b. J Biol Chem 282(2):1479–1486

Siegel R, Naishadham D, Jemal A (2012) Cancer statistics, 2012. CA Cancer J Clin 62(1): 10–29

Spizzo R, Nicoloso MS, Lupini L, Lu Y, Fogarty J, Rossi S et al (2010) miR-145 participates with TP53 in a death-promoting regulatory loop and targets estrogen receptor-alpha in human breast cancer cells. Cell Death Differ 17(2):246–254

Tavazoie SF, Alarcon C, Oskarsson T, Padua D, Wang Q, Bos PD et al (2008) Endogenous human microRNAs that suppress breast cancer metastasis. Nature 451(7175):147–152

Tong AW, Nemunaitis J (2008) Modulation of miRNA activity in human cancer: a new paradigm for cancer gene therapy? Cancer Gene Ther 15(6):341–355

Wickramasinghe NS, Manavalan TT, Dougherty SM, Riggs KA, Li Y, Klinge CM (2009) Estradiol down regulates miR-21 expression and increases miR-21 target gene expression in MCF-7 breast cancer cells. Nucleic Acids Res 37(8):2584–2595

Xiong J, Yu D, Wei N, Fu H, Cai T, Huang Y et al (2010) An estrogen receptor alpha suppressor, microRNA-22, is down regulated in estrogen receptor alpha-positive human breast cancer cell lines and clinical samples. FEBS J 277(7):1684–1694

Yan LX, Huang XF, Shao Q, Huang MY, Deng L, Wu QL et al (2008) MicroRNA miR-21 over expression in human breast cancer is associated with advanced clinical stage, lymph node metastasis and patient poor prognosis. RNA 14(11):2348–2360

Yang N, Kaur S, Volinia S, Greshock J, Lassus H, Hasegawa K et al (2008) MicroRNA microarray identifies Let-7i as a novel biomarker and therapeutic target in human epithelial ovarian cancer. Cancer Res 68(24):10307–10314

Yu F, Yao H, Zhu P, Zhang X, Pan Q, Gong C et al (2007) let-7 regulates self renewal and tumorigenicity of breast cancer cells. Cell 131(6):1109–1123

Yu Z, Wang C, Wang M, Li Z, Casimiro MC, Liu M et al (2008) A cyclin D1/microRNA 17/20 regulatory feedback loop in control of breast cancer cell proliferation. J Cell Biol 182(3):509–517

Yu Z, Baserga R, Chen L, Wang C, Lisanti MP, Pestell RG (2010a) microRNA, cell cycle, and human breast cancer. Am J Pathol 176(3):1058–1064

Yu Z, Willmarth NE, Zhou J, Katiyar S, Wang M, Liu Y et al (2010b) microRNA 17/20 inhibits cellular invasion and tumor metastasis in breast cancer by heterotypic signaling. Proc Natl Acad Sci U S A 107(18):8231–8236

Zhao JJ, Lin J, Yang H, Kong W, He L, Ma X et al (2008) MicroRNA-221/222 negatively regulates estrogen receptor alpha and is associated with tamoxifen resistance in breast cancer. J Biol Chem 283(45):31079–31086

Zhao Y, Deng C, Wang J, Xiao J, Gatalica Z, Recker RR et al (2011) Let-7 family miRNAs regulate estrogen receptor alpha signaling in estrogen receptor positive breast cancer. Breast Cancer Res Treat 127(1):69–80

Zhou M, Liu Z, Zhao Y, Ding Y, Liu H, Xi Y et al (2010) MicroRNA-125b confers the resistance of breast cancer cells to paclitaxel through suppression of pro-apoptotic Bcl-2 antagonist killer 1 (Bak1) expression. J Biol Chem 285(28):21496–21507

Zhu S, Wu H, Wu F, Nie D, Sheng S, Mo YY (2008) MicroRNA-21 targets tumor suppressor genes in invasion and metastasis. Cell Res 18(3):350–359

Chapter 15
MicroRNAs and Their Clinical Impact on Resistance to Anticancer Treatment

Kirsten Lindner, Joerg Haier, and Richard Hummel

Contents

Abstract The clinical use of microRNAs (miRNAs) as diagnostic tools for example for tumor classification or as prognostic markers is becoming increasingly established. In addition, recent studies demonstrated that miRNAs could be used as new therapeutic approach in anticancer treatment including the highly interesting aspect that is regulated by miRNAs: resistance to chemo- and radiotherapy. This chapter aims to elucidate the impact of miRNAs on drug resistance from a clinical point of view,

K. Lindner • R. Hummel (✉)
Department of General and Visceral Surgery, University Hospital Muenster, Muenster, Germany
e-mail: richard.hummel@ukmuenster.de

J. Haier
Comprehensive Cancer Center Muenster (CCCM), University Hospital Muenster, Muenster, Germany

S. Babashah (ed.), *MicroRNAs: Key Regulators of Oncogenesis*,
DOI 10.1007/978-3-319-03725-7_15,

and to highlight their potential role as predictors or modifiers of resistance towards chemotherapeutics and radiotherapy. Therefore, we selected exemplary two different tumor types that present either high or low resistance to chemotherapeutic treatment: esophageal cancer (highly therapy resistant tumor) and ovarian cancer (quite therapy sensitive tumor).

Keywords MicroRNA • Drug resistance • Esophageal cancer • Ovarian cancer • Predictor • Modifier

1 Introduction

MicroRNAs (miRNAs) are a novel class of regulatory molecules that control translation and stability of mRNAs on a post-transcriptional level. MiRNAs are involved in almost all physiological processes such as cell development or differentiation. So far, more over 1,000 human miRNAs have been identified (Kozomara and Griffith-Jones 2011), and each single miRNA can target hundreds of mRNAs (Li and Yang 2013). However, most importantly from a clinical point of view, miRNAs are highly involved in the initiation and progression of cancer by regulating for example metastasis and angiogenesis amongst others. Interesting in this context is that hundreds of miRNAs map to regions of the human genome that are known to be altered in cancer (Calin et al. 2004). Esquela-Kerschner et al. even established the term "oncomiRs" for miRNAs with oncogenic function, implying that abnormalities in miRNA expression might directly result in the de-differentiation of cells, allowing tumor formation to occur (Esquela-Kerschner and Slack 2006).

1.1 General Considerations: miRNAs and Their Clinical Use in Cancer Diagnostics and Treatment

Most oncological studies in the recent past investigated miRNA expression pattern in fresh frozen samples such as tumor biopsies or resection specimens, but miRNAs can also be detected and extracted from other sample types such as paraffin-embedded tissues (Iorio and Croce 2012). Most interestingly for clinicians, miRNAs can further be found as so called "Circulating miRNA" in a variety of human body fluids in healthy volunteers and cancer patients (Fang et al. 2012; Weber et al. 2010; Xiao et al. 2013; Allegra et al. 2012; Iorio and Croce 2012). These circulating miRNAs, which are surprisingly stable (Mo et al. 2012; Kim and Reitmair 2013) and can be detected even in 10-year-old human serum samples or in un-refrigerated dried serum blots (Cortez et al. 2011), are tissue-specific, stable, reproducible and consistent among individuals in the same species (Fang et al. 2012). In cancer patients, circulating miRNAs are thought to be mainly related to apoptosis and necrosis of cancer cells in the tumor microenvironment (Kim and Reitmaier 2013), and miRNA expression profiles of primary tumors and metastases seem very similar (Rosenwald et al. 2010).

With this in mind, the clinical use of miRNAs as diagnostic tools for example for tumor classification or as prognostic markers seems logical and feasible. In addition, latest research provides first very promising data that miRNAs could be used as new therapeutic approach in the fight against cancer as for example knockdown of oncomirs can affect tumor growth. In this context, only recently the focus has been turned on another highly interesting aspect that is regulated by miRNAs: resistance to chemo- and radiotherapy.

This chapter aims to elucidate the impact of miRNAs on drug resistance from a clinical point of view, and to highlight their potential role as predictors or modifiers of resistance towards chemotherapeutics and radiotherapy. Therefore, we selected exemplary two different tumor types that present either high or low resistance to chemotherapeutic treatment: (a) esophageal cancer, as a highly therapy resistant tumor with a complete response to neoadjuvant therapy in only about 13–25 % of all patients (Courrech Staal et al. 2010) and (b) ovarian cancer, as a quite therapy sensitive tumor with complete response to adjuvant therapy in about 80–90 % of the patients (Ozols 2005).

1.2 High Versus Low Resistant Tumors: On Overview About the Clinical Background

1.2.1 Esophageal Cancer

Esophageal cancer (EC) is characterized by a poor overall prognosis. Because of the high incidence of advanced disease at the time of diagnosis, the 5-year survival rate remains below 15 % and only about 15–20 % of patients finally qualify for curative surgical resection. In an attempt to improve outcome of patients after surgery and to potentially increase the number of patients who qualify for surgery by downstaging of the tumor, neoadjuvant therapy including chemotherapy and radiotherapy has been demonstrated to potentially advance overall survival for both histological subtypes adenocarcinoma (AdenoCA) and squamous cell carcinoma (SCC) (Urschel and Vasan 2003; Fiorica et al. 2004; Sjoquist et al. 2011). However, a complete pathologic response as determined by the "tumor regression grade TRG" can only be achieved in about 13–25 % of all patients (Courrech Staal et al. 2010).

There are several miRNAs that have been reported to be differentially expressed in esophageal cancer, with (prognostic and diagnostic) associations to tumor stage (Lin et al. 2012), histological differentiation (Hummel et al. 2011a; Lin et al. 2012), distant lymph node metastasis (Hummel et al. 2011a; Liu et al. 2012a), vascular invasion (Komatsu et al. 2011), overall and disease-free survival (Komatsu et al. 2012; Takeshita et al. 2013; Zhang et al. 2011) and tumor recurrence (Komatsu et al. 2011; Hummel et al. 2011a). Additionally, a number of serum miRNAs were described as potential diagnostic biomarkers (Komatsu et al. 2011; Takeshita et al. 2013; Zhang et al. 2011; Liu et al. 2012a; Cai et al. 2012; Wang and Zhang 2012; Zhang et al. 2010, 2012; Revilla-Nuin et al. 2013).

1.2.2 Ovarian Cancer

Ovarian cancer is one of the most aggressive female reproductive tract tumors. The prognosis depends on the stage of the disease and on histological and molecular characteristics. Platinum based chemotherapy agents, namely cisplatin and carboplatin, are widely used for the treatment of ovarian cancer. For advanced-stage disease, taxanes (e.g. paclitaxel, docetaxel) are often supplemented (van Jaarsveld et al. 2010). Despite the fact that a complete clinical response can be achieved in 80–90 and 50 % of patients with early-stage or advanced-stage disease, respectively, (Ozols 2005), ovarian cancer patients frequently develop resistance to chemotherapy, often resulting in a poor overall 5-year survival of only 30 % (Moss and Kaye 2002).

Several studies showed aberrantly expressed miRNAs in ovarian cancer and established a connection to histological subtypes (Lee et al. 2012; Zaman et al. 2012; Iorio et al. 2007), tumor stage or grade (Yang et al. 2012), primary or recurrent tumors (Hu et al. 2009), and survival (Hong et al. 2013; Peng et al. 2012; Lee et al. 2012; Marchini et al. 2011; Nam et al. 2008). Again, miRNAs could also be identified as diagnostic and prognostic biomarkers in serum (Xu et al. 2013; Resnick et al. 2009; Chang et al. 2012; Peng et al. 2012).

2 Clinical Application of miRNAs as Diagnostic Tools: MicroRNAs as Predictors of Response to Conventional Treatments

As outlined above, one highly promising clinical application of miRNAs as diagnostic tools involves their potential to predict response to conventional treatment such as chemotherapy and radiotherapy. If it might be possible to identify responders and non-responders before the start of neoadjuvant or adjuvant treatment, cancer therapy could be tailored more individually. Patients who do not benefit from chemotherapy or irradiation would not have to undergo this toxic treatment, and could be referred immediately to curative surgical resection. The use of miRNAs as predictors of therapy response implicates however that chemotherapy or radiotherapy resistant tumors exhibit distinct miRNA expression pattern that distinguishes them from sensitive tumors. These differences in miRNA expression would be necessary to allocate patients into the responder and non-responder groups. And indeed, there is growing evidence that chemo- and radiotherapy resistant tumors show specific pattern of miRNA deregulation, both *in vitro* and *in vivo*.

2.1 Experimental In Vitro Data

We found only one study that investigated the direct effect of chemotherapy treatment on miRNA expression in one adenocarcinoma and one squamous cell carcinoma cell lines after treatment with either cisplatin or 5-fluorouracil for 24/72 h. The authors could show that 13 miRNAs (miR-199a-5p, miR-302f, miR-320a, miR-342-3p, miR-425, miR-455-3p, miR-486-3p, miR-519c-5p, miR-548d-5p, miR-617, miR-758, miR-766, miR-1286) were deregulated after short-term or long-term treatment in either of the cell lines (Hummel et al. 2011b). However, a number of studies were published that report different miRNA expression pattern between drug resistant and sensitive cells. For example, comparing two cisplatin resistant human esophageal squamous cell lines (that were generated via exposure of sensitive cells to the chemotherapeutic drug) with controls, Sugimura et al. identified a total of 365 miRNAs to be differentially expressed between resistant and sensitive cells, with more than 1.7-fold changes in expression of 128 respectively 177 miRNAs. Most interestingly, 15 miRNAs showed an overlap between the two resistant cell lines with regards to their deregulation: miR-135a, miR-96, miR-141, miR-101, miR-146a, miR-489 and miR-545 were up-regulated, whereas miR-99a, let-7b, miR-204, let-7c, miR-202, miR-10a, miR-136 and miR-145 were down-regulated in both cisplatin-resistant cell limes (Sugimura et al. 2012). Another study reported that miR-141, miR-21, miR-19b, miR-200a, miR-19a, miR-27a, miR-20a and miR-20b were expressed at significantly higher levels, and miR-205 and miR-224 at significantly lower levels in cells with increasing resistance towards cisplatin. In this context, most profound deregulation was found for miR-141 (Imanaka et al. 2011). Concerning altered miRNA profiles in radioresistant cells, Zheng et al. compared a radioresistant squamous cell carcinoma cell line with controls and found 35 miRNAs to be deregulated: 10 miRNAs (miR-1539, miR-1237, miR-92b, etc.) were up-regulated, and 25 miRNAs (miR-185, miR-18b, miR-17, etc.) were down-regulated (Zheng et al. 2011).

For ovarian cancer, there are a number of reports published on deregulated miRNAs in chemotherapy resistant cell lines. For example, Kumar et al. compared sensitive human ovarian cancer cells and their cisplatin-resistant counterparts and found changes in the expression of 11 miRNAs out of 1,500 miRNAs, with miRplus-F1064, miR-300, miR-193b, miR-642 and miR-1299 being up-regulated and miR-625, miR-20b, miRPlus-F1147, let-7c, miRPlus-F1231 and miR-542-3p being down-regulated (Kumar et al. 2011). Van Jaarsveld reported 27 miRNAs to be differentially expressed in cells with increasing resistance towards cisplatin: miR-214, miR-412, miR-645, miR-17, miR-106a, miR-199a-5p, miR-215, miR-199a/b-3p, miR-335, miR-338-5p, miR-493, miR-135b, miR-130a, miR-186, miR-942, miR-18b, miR-20b, miR-196a, miR-10b, miR-19a, miR-421, miR-19b, miR-518e, miR-631, miR-222, miR-141 and miR-200c (van Jaarsveld et al. 2012). Another study identified diversely expressed miRNAs in resistant cell lines using one cisplatin-resistant and three

Table 15.1 Summary of microRNA profiles in drug resistance cells for two types of tumor (esophageal cancer vs. ovarian cancer)[a]

	Imanaka et al. (2011)	Sugimura et al. (2012)	Kumar et al. (2011)	Sorrentino et al. (2008)	Yang et al. (2012)	van Jaarsveld et al. (2012)	van Jaarsveld et al. (2012)
	SCC	SCC	OvarianCA	OvarianCA	OvarianCA	OvarianCA	OvarianCA
	Cisplatin	Cisplatin	Cisplatin	Cisplatin	Cisplatin	Cisplatin	Paclitaxel
miR-19a	red				red	green	
miR-20a	red				red		
miR-20b	red		green			green	
miR-21	red				green		
miR-27a	red				red		
miR-30c				green			green
miR-99a		green			red		
miR-101		red					
miR-106a					red	green	
miR-125b				red			green
miR-130a				green	red	green	green
miR-141	red	red				red	
miR-193b			red		green		
miR-205	green				green		
miR-222					red	green	
miR-335				green		green	green
Let-7c		green	green		red		
Let-7e				green	red	red	red

[a] To provide a better overview only 18 out of 177 miRNAs, which were found more than once were cited. Red cell: up-regulation of miRNA; green cell: down-regulation of miRNA. SCC: esophageal squamous cell carcinoma

paclitaxel-resistant cell lines: let-7e, miR-30c, miR-125b, miR-130a and miR-335. Interestingly, let-7e was up-regulated in one of the paclitaxel resistant cell line, while it was down-regulated in the other resistant cell lines. The opposite phenomenon was described for miR-125b, which was down-regulated in one paclitaxel resistant cell line and up-regulated in the other ones. MiR-30c, miR-130a and miR-335 were down-regulated in all the resistant cell lines (Sorrentino et al. 2008). Yang et al. detected 79 differently expressed miRNAs in a cisplatin-resistant cell line, including miR-130a associated with MDR1/P-gp-mediated drug resistance (Yang et al. 2012).

These experimental *in vitro* data clearly support the hypothesis that chemotherapy or radiotherapy resistant tumors present unique miRNA expression pattern that might allow identification of therapy responders based on profiling information in tumor cells. With regard to either high or low resistant tumors, 53 miRNAs were described in esophageal cancer to correlate with resistance, and 124 miRNAs were associated with response to therapy in ovarian cancer cells. Most interestingly, 18 miRNAs were identified in more than one study to present altered expression between resistant and sensitive tumors *in vitro*, with a few of them even presenting similar deregulation pattern when comparing esophageal (high resistant) and ovarian (low resistant) cancer. Table 15.1 provides an overview about these 18 miRNAs that seem to highly impact on drug resistance.

2.2 First Clinical Data

Odenthal et al. examined 80 patients with esophageal cancer (AdenoCA and SCC), who underwent multimodal therapy. Comprehensive miRNA profiling identified a number of miRNAs in pretherapeutic biopsies that were significantly differently expressed between major and minor responders. The pretherapeutic intratumoral expression of miR-192 and miR-194 was significantly associated with the histopathologic response of esophageal squamous cell carcinoma to neoadjuvant treatment (Odenthal et al. 2013). Also using pretreatment biopsy specimen of 25 patients who underwent irinotecan/cisplatin based chemotherapy and radiotherapy followed by surgical treatment, 71 miRNAs were found to be significantly differently expressed between pathologic complete responders and non-responders. Five of these miRNAs had a greater than two-fold difference in expression: HAS-240, miR-296, miR-141, miR-31 and miR-217. Comparison of post-treatment biopsies of responders versus non-responder patients further revealed that 52 miRNAs were significantly up-regulated or down-regulated after induction therapy, and nine of these had a greater than two-fold change in expression: miR-1238, miR-938, HS_228.1, HS_282, miR-200a, miR-200b, miR-429 and miR-141 amongst others. Patients with high levels of miR-135b or miR-145 in the posttreatment biopsy specimens had a significantly shorter median disease-free survival compared to patients with low levels (11.5 versus 5.1 months; $p=0.04$; 11.5 versus 2.8 months; $p=0.03$) (Ko et al. 2012). Furthermore, miR-31 expression was found to be significantly reduced in patients presenting a poor histomorphologic response to neoadjuvant therapy. In addition, Lynam-Lennon et al. could demonstrate an influence of miR-31 on the modulation of radioresistance (Lynam-Lennon et al. 2012). Other studies evaluated the impact of miRNA expression on therapy response by defining therapy response as longer survival. Hamano et al. investigated the expression of 9 miRNAs (let-7a, let-7 g, miR-21, miR-134, miR-145, miR-155, miR-200c, miR-203 and miR-296) in esophageal cancer patients who had received preoperative chemotherapy with cisplatin and 5-FU followed by surgery. The expression of miR-200c correlated inversely and significantly with the response to chemotherapy. Furthermore, the overexpression of miR-200c and miR-21 respectively the underexpression of miR-145 correlated significantly with shorter overall survival (Hamano et al. 2011). Additionally and highly interesting regarding a potential clinical use of miRNAs as response predictors, miRNA-21 levels in serum were shown to be significantly reduced in esophageal squamous cell carcinoma patients who responded to chemotherapy (Kurashige et al. 2012). Another study demonstrated an inverse correlation between expression levels of miR-483 and miR-214 and overall survival (Zhou et al. 2013). Finally, low expression of let-7c correlated with poor prognosis and was able to predict response to cisplatin-based chemotherapy (Sugimura et al. 2012).

A similar situation regarding miRNA expression and its correlation to response to treatment can be observed in ovarian cancer patients. Eitan et al. compared the miRNA profile of surgically treated ovarian cancer patients that received either solely platinum based chemotherapy ($n=21$), or paclitaxel with carboplatin ($n=34$)

and cyclophosphamid with cisplatin as first line treatment. Based on outcome, the authors divided the patients into platinum-sensitive vs. platinum-resistant patients. Seven miRNAs were identified to be significantly differently expressed between the two groups: miR-27a, miR-23a, miR-30c miR-7 g and miR-199-3p were increased in platinum-resistant patients, and miR-378 and miR-625 were increased in platinum-sensitive patients (Eitan et al. 2009). Lu et al. investigated let-7a expression in ovarian cancer patients receiving an adjuvant platinum based chemotherapy with or without paclitaxel in addition to surgical debulking. The let-7a expression did not correlate with disease stage, tumor grade, histology or debulking results. However, the authors could demonstrate that patients who responded well to platinum based chemotherapy combined with paclitaxel presented significantly lower let-7a levels. Conversely, survival analyses showed that patients with high let-7a levels presented a better survival compared to those with low levels (Lu et al. 2011). In another study, let-7i expression was found to be significantly reduced in chemotherapy resistant patients treated with paclitaxel and platinum (Yang et al. 2008). Most interestingly, patients who experienced a relapse of the disease showed a down-regulation of let-7 in samples collected after chemotherapy compared to prethera-peutic samples. Furthermore, a decrease in the expression of let-7 after chemotherapy negatively correlated with disease-free time before recurrence (Boyerinas et al. 2012). In this context, low levels of miR-199a may be another predictor for chemoresistance in recurrent tumors (Nam et al. 2008). Several studies further looked at different histological subtypes (serous versus non-serous) of ovarian cancer. For example, a signature of 23 miRNAs was associated with chemoresistance in patients with serous ovarian cancer treated with carboplatin and taxol, and PCR-based validation confirmed that three miRNAs were able to predict chemoresistance of these tumors: miR-484, miR-642 and miR-217 (Vecchione et al. 2013). Van Jaarsveld found higher miR-141 expression in patients with serous ovarian tumors who did not response to platinum-based chemotherapy (van Jaarsveld et al. 2012). Furthermore, low miR-376c expression was detected in patients with serous ovarian tumors who responded well to cisplatin based chemotherapy (Ye et al. 2011). Finally, another clinical study compared the expression of miR-21 and miR-214 in ascites and omental metastasis of patients with ovarian cancer treated with carboplatin. Malignant cells in ascites showed greater cell viability when treated with carboplatin compared to omental metastasis. Additionally, there was a significant up-regulation of miRNA-21 and miRNA-214 in tumor cells from ascites (Frederick et al. 2013).

Table 15.2 presents an overview about the data available so far on miRNAs as potential clinical predictors of chemotherapy resistance in esophageal and ovarian cancers. These first *in vivo* data clearly supports the results from the *in vitro* experi-ments and prove that, at least in the isolated patient populations of the respective publications, miRNAs can help separating responders from non-responders based on profiling information obtained from tumor specimen. However, the final clinical impact of miRNAs as potential response predictors remains to be determined, as the heterogeneity of the different treatment protocols in the studies, the respective experimental setups including the clinical response evaluation, and finally the obtained results in the different studies does not allow a definitive statement yet. For

Table 15.2 MicroRNA expression profiles in responder and non-responder to conventional chemotherapy in esophageal and ovarian cancers[a]

	Therapy responder		Therapy non-responder		
	Up-regulated	Down-regulated	Up-regulated	Down-regulated	References
SCC, AdenoCA, Ciplatin, 5-FU, RTX	miR-192, miR-194				Odenthal et al. (2013)
SCC, AdenoCA, Irinotecan, Ciplatin, RTX	HS_240, miR-296, miR-1238, miR-938, HS_228.1, HS_282	miR-141, **miR-31**, HS_217, miR-200a, miR-429, miR-200b, **miR-141**			Ko et al. (2012)
SCC, AdenoCA, Cisplatin, 5-FU, RTX	**miR-31**				Lynam-Lennon et al. (2012)
SCC, Cisplatin, 5-FU			miR-200c, **miR-21**	**miR-145**	Hamano et al. (2011)
SCC, Docetaxel, cisplatin, 5-FU		**miR-21**			Kurashige et al. (2012)
SCC			miR-483, miR-214		Zhou et al. (2013)
SCC, Cisplatin, 5-FU, Adriamycin	miR-135a, miR-96, miR-141, miR-146a, miR-489, miR-545, miR-10a, let-7b, **let-7c**, **miR-145**	miR-99a, miR-204		**let7-c**	Sugimura et al. (2012)
OvarianCA, Platinum, paclitaxel, carboplatin, cyclophosphamid	miR-378, miR-625		miR-27a, miR-23a, miR-30c, miR-79, miR-199-3p		Eitan et al. (2009)
OvarianCA, Platinum, paclitaxel		let-7a			Lu et al. (2011)
OvarianCA, platinum, paclitaxel				let-7i	Yang et al. (2008)
OvarianCA				miR-484, miR-642, miR-217	Vecchione et al. (2013)
OvarianCA, platinum			**miR-141**		Van Jaarsveld (2012)
OvarianCA		miR-376c			Ye et al. (2011)

Abbreviations: *AdenoCA* esophageal adenocarcinoma, *SCC* esophageal squamous cell carcinoma, *OvarianCA* Ovarian cancer, *5-FU* 5-fluorouracil, *RTX* radiotherapy

[a]Printed in bold type were miRNAs that were identified in more than one *in vivo* study to present altered expression between resistant and sensitive tumors

example, miR-31 was reported on the one hand to be up-regulated (Lynam-Lennon et al. 2012), and on the other hand to be down-regulated in esophageal cancer patients that respond to therapy (Ko et al. 2012). However, in summary these data are highly promising regarding a potential clinical benefit of miRNAs as response predictors in the future.

3 Clinical Application of miRNAs as Therapeutic Tools: MicroRNAs as Modifiers of Response to Conventional Treatments

Another possibly even more interesting approach for the clinical use of miRNAs might be their application as potential modifiers of chemotherapy. As the resistance of tumors to conventional treatment such as chemotherapy or radiotherapy represents a major obstacle in the fight against cancer, identification of a way to reverse drug resistance is one of the most important challenges for researchers all over the world. A solution to this problem might be a key breakthrough in the treatment of malignant diseases. If resistance to chemo- and radiotherapy can be overcome, toxicity of these treatments could be minimized by for example lowering the doses of chemotherapeutics, while achieving the same antitumor effects. And in fact, there is growing evidence that modulation of miRNA expression can affect resistance of various tumors to treatment. However, as this field of research is still very young, results are somewhat limited and refer so far only to *in vitro* experiments. But this does not affect its promising clinical significance.

A number of authors addressed the question whether miRNA modulation affects chemotherapy resistance in esophageal cancer. Zhang et al. for example demonstrated that down-regulation of miR-27a in esophageal cancer cells could significantly decrease the expression of P-glycoprotein [a well known drug efflux pump that influences on drug resistance (Wen et al. 2009)], Bcl-2 (an anti-apoptotic protein which is involved in tumor cell apoptosis and response to chemotherapy) (Kang and Reynolds 2009; Ballesta et al. 2013; Asakura and Ohkawa 2004; Ohkawa et al. 2004), and the transcription of the multidrug resistance gene. This conferred sensitivity towards P-glycoprotein related chemotherapeutics such as etoposide, doxorubicin and vinblastine (Zhang et al. 2010). Hong et al. discovered that decreased miR-296 expression improved the response of tumor cells to 5-FU and Ciplatin, probably due to changes in Bcl-2 and Bax levels, finally leading to an increase in apoptosis and decreased MDR-1 expression (Hong et al. 2010). Furthermore, miR-141 was shown to play an important role in the development of cisplatin resistance in esophageal squamous cell carcinoma by down-regulation of YAP1, which is known to have a crucial role in apoptosis induced by DNA-damaging agents (Imanaka et al. 2011). Another group found miR-148a to sensitize esophageal cancer cell lines to cisplatin and, to a lesser extent, to 5-flurouracil, and to attenuate resistance in chemotherapy-resistant variants (Hummel et al. 2011b). In a variety of tumors other than esophageal cancer, expression of miR-148a has been shown to negatively affect tumor growth, cell motility, invasion, migration and metastasis. One of the targets that can explain the effects of miR-148a modulation might be mitogen- and stress-activated kinase 1 (MSK1), which was identified in prostate cancer cells (Fujita et al. 2010). Other potential resistance-relevant targets of miR-148a include DNA methyltransferase-3B (DNMT3B) and DNA methyltransferase-1 (DNMT-1) (Merkerova et al. 2010; Duursma et al. 2009). Sugimura and colleagues demonstrated that let-7 modulated the chemosensitivity to cisplatin in esophageal cancer through the regulation of IL-6/STAT3 pathway (Sugimura et al. 2012). IL-6 is an inflammatory cytokine, which is released from macrophages and T-lymphocytes as well as from cancer cells. It modulates various cell functions (e.g., inflammatory reactions), and is a major activator of the JAK/STAT3 and PI3K/AKT signaling pathways. Various studies already demonstrated an association between IL-6 and resistance to chemotherapy for example in ovarian cancer (Wang et al. 2010), breast cancer (Iliopoulos et al. 2009) or gastrointestinal cancer (Chen et al. 2013). Additionally, with regards to esophageal cancer, one recent study showed that increasing intracellular IL-6 expression after cisplatin exposure is associated with reduced sensitivity to cisplatin treatment, and that knockdown of IL-6 expression restored sensitivity to cisplatin treatment (Chen et al. 2013). Finally, Wu et al. could show that overexpression of miRNA-200b, miR-200c and miR-429 correlated with resistance to cisplatin treatment. Chemotherapeutic drugs such as cisplatin induce expression of endogenous AP-2α, which contributes to chemosensitivity by enhancing therapy-induced apoptosis (Wu et al. 2011).

Again, a similar picture is found for ovarian cancer. For example, overexpression of miR-200c, miR-200a and miR-141 (the miR-200 family is known as the main suppressor of the epithelial-to-mesenchymal transition, EMT) was reported to enhance sensitivity to microtubule-targeting drugs (e.g. paclitaxel, vincristine, epothilone B) in ovarian cancer lines by repressing the class III β-tubulin TUBB3 (Cochrane et al. 2009; Prislei et al. 2013; Leskelä et al. 2010; Mateescu et al. 2011). TUBB3 is well known as a prominent mechanism of drug resistance found in a variety of solid tumors, but particularly in lung and ovarian cancer where it is associated with a perturbation in microtubule dynamics (Mariani et al. 2011). Accordingly, up-regulation of miR-200c levels in an ovarian cancer cell line increased the sensitivity towards micro-targeting drugs up to 85 % (Cochrane et al. 2010). Also, the family members miR-141/200c showed a correlation with cisplatin sensitivity in the NCI-60 panel. The NCI-60 cancer cell panel consists of 60 cancer cell lines of various histological origins, of which miRNA expression and drug sensitivity data can be obtained from a public database (Blower et al. 2007). Overexpression of miR-141 resulted in enhanced resistance to cisplatin in ovarian cancer cell lines as it directly targeted KEAP1, and induced cisplatin resistance via affection of the NF-κB pathway (van Jaarsveld et al. 2012). Another miRNA, miR-199a, was demonstrated to significantly increase the chemosensitivity of ovarian cancer-initiating cells to cisplatin, paclitaxel and adriamycin, and to reduce mRNA expression of the multidrug resistance gene ABCG2 (Chen et al. 2012). Further *in vitro* assays with knockdown of let-7i led to a decreased cisplatin-induced cell death in ovarian cancer cell lines (Yang et al. 2008). Let-7 g selectively affected the sensitivity of a drug resistant ovarian cancer cell line towards taxanes by targeting IMP-1, which in turn caused destabilization of MDR1 at the mRNA and protein level. This finally increased sensitivity of the multidrug resistant ovarian cancer cells to taxanes (Boyerinas et al. 2012). Furthermore, miR-125b targeted Bak1, a gene of the Bcl-2 protein family. Down-regulation of Bak1 resulted in an increased resistance to cisplatin by suppressing cisplatin-induced apoptosis (Kong et al. 2011). Down-regulation of miR-130b promoted the development of multidrug resistant ovarian cancer cells partly by targeting CSF-1, and silencing of miR-130b was demonstrated to be potentially mediated by DNA methylation. At the same time, low levels of miR-130b were associated with FIGO III-IV clinical stages in ovarian cancer patients and poor histological differentiation (Yang et al. 2012). MiR-152 and miR-185, which targets DNMT1 (Xiang et al. 2013), were also reported to be involved in chemotherapy resistance. DNMT1, the principal DNA methyltransferase, controls DNA methylation. In addition, miR-214 was shown to target PTEN, a regulator of cell proliferation, via the PI3K-Akt pathway. Knockdown of miR-214 reduced cell survival at around 20 % in cisplatin resistant ovarian cancer cells (Yang et al. 2008). MiR-376c was described to target al.K7 (a member of the TGF family that inhibits proliferation and induces apoptosis of epithelial ovarian cancer cells), and overexpression of miR-376c was found to block cisplatin-induced cell death, whereas anti-miR-376c treatment enhanced the effect of cisplatin (Ye et al. 2011). Another highly promising miRNA that is potentially involved in chemosensitivity is miR-21. As Liu et al. (2012b), found that berberine could inhibit miR-21 expression in several cancer cell lines. Subsequently, these authors investigated the influence of berberine on chemosensitivity of ovarian cancer cells to cisplatin. Interestingly, berberine could inhibit miR-21 expression and thereby modulate the sensitivity of cisplatin via regulating of the miR-21/PDCD4 axis (Liu et al. 2013). Other miRNAs involved in therapy resistance were miR-484 (target: vascular endothelial growth factor VEGFB and VEGFR2 pathways) (Vecchione et al. 2013), miR-23b, miR-27b, miR-424, and miR-503 (target: ALDH1) (Park et al. 2013), miR-106a (target: BCL10 and caspase-7) (Huh et al. 2013), miR-591 (target: ZEB1) (Huh et al. 2013), miR-31 (target: MET) (Mitamura et al. 2013), miR-130a (target: pro-metastatic and chemoresistance associated M-CSF) (Sorrentino et al. 2008), miR-484, miR-642 and miR-217 were described to be able to predict chemoresistance (Vecchione et al. 2013). Figure 15.1 shows a graphic presentation of a selection of resistance-relevant miRNAs and their downstream targets.

Despite the limitation of the *in vitro* character of these data, these experiments clearly demonstrate that miRNAs affect chemotherapy resistance on a cellular level. If these data can be reproduced in *in vivo* animal experiments, this would mean a major step towards a miRNA based new therapeutic approach for cancer patients. This new treatment option could be used either as additive treatment in conjunction with conventional therapies such as chemo- or radiotherapy, or maybe even as first line treatment if proven toxic to tumors *in vivo*. However, further experiments are highly warranted in order to unravel potential systemic side effects of miRNA based therapies, and to prove their success in complex organisms.

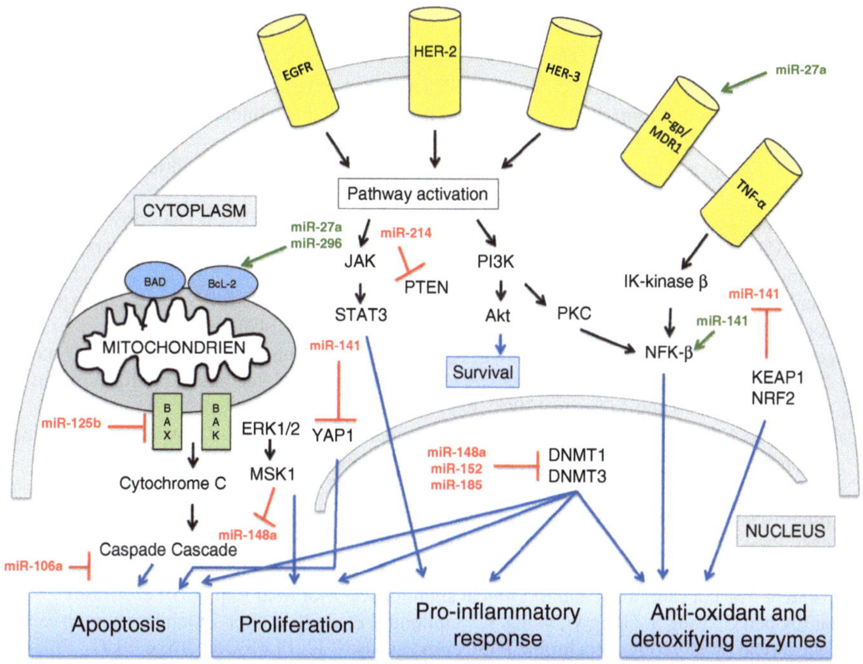

Fig. 15.1 MicroRNAs and intracellular signaling pathways with modulation of therapy response

4 Conclusion and Perspectives

The presented *in vitro* data and (so far limited) *in vivo* data draw a most promising picture of miRNAs as potential clinical predictors and modifiers of response to chemo- and (to a lesser extent) radiotherapy. Overall, these results are highly encouraging and outline the enormous clinical impact that might arise from the use of miRNAs in the near future. However, as hopeful as these data are, their limitations have definitively to be considered: as outlined in this chapter, data are still very limited and sometimes even somewhat contradictory. Furthermore, conclusive *in vivo* data on miRNA based therapies are missing to this very date in general, so that information about possible complex interactions of a systemic therapy and potential toxic side effects for the patients have to be investigated before bringing these molecules into the clinic. However, the data presented in this chapter highlight the enormous potential of miRNAs for clinical application, and we are very confident that soon first reports on the clinical (diagnostic or therapeutic) use of miRNAs in the context of chemotherapy resistance will be available.

References

Allegra A, Alonci A, Campo S, Penna G, Petrunago A, Gerace D et al (2012) Circulating microRNAs: new biomarkers in diagnosis, prognosis and treatment of cancer (review). Int J Oncol 41(6):1897–1912

Asakura T, Ohkawa K (2004) Chemotherapeutic agents that induce mitochondrial apoptosis. Curr Cancer Drug Targets 4(7):577–590

Ballesta A, Lopez J, Popgeorgiev N, Gonzalo P, Doumic M, Gillet G (2013) Data-driven modeling of SRC control on the mitochondrial pathway of apoptosis: implication for anticancer therapy optimization. PLoS Comput Biol 9(4):e1003011

Blower PE, Verducci JS, Lin S, Park JK, Dai Z, Liu CG et al (2007) MicroRNA expression profiles for the NCI-60 cancer cell panel. Mol Cancer Ther 6(5):1483–1491

Boyerinas B, Park SM, Murmann AE, Gwin K, Montag AG, Zillhardt M et al (2012) Let-7 modulates acquired resistance of ovarian cancer to Taxanes via IMP-1-mediated stabilization of multidrug resistance 1. Int J Cancer 130(8):1787–1797

Cai EH, Gao YX, Wei ZZ, Chen WY, Yu P, Li K (2012) Serum miR-21 expression in human esophageal squamous cell carcinomas. Asian Pac J Cancer Prev 13(4):1563–1567

Calin GA, Sevignani C, Dumitru CD, Hyslop T, Noch E, Yendamuri S et al (2004) Human microRNA genes are frequently located at fragile sites and genomic regions involved in cancers. Proc Natl Acad Sci U S A 101(9):2999–3004

Chang H, Zhou X, Wang ZN, Song YX, Zhao F, Gao P et al (2012) Increased expression of miR-148b in ovarian carcinoma and its clinical significance. Mol Med Rep 5(5):1277–1280

Chen F, Chen C, Yang S, Gong W, Wang Y, Cianflone K et al (2012) Let-7b inhibits human cancer phenotype by targeting cytochrome P450 epoxygenase 2 J2. PLoS One 7(6):e39197

Chen MF, Chen PT, Lu MS, Lin PY, Chen WC, Lee KD (2013) IL-6 expression predicts treatment response and outcome in squamous cell carcinoma of the esophagus. Mol Cancer 12:26

Cochrane DR, Howe EN, Spoelstra NS, Richer JK (2010) Loss of miR-200c: a marker of aggressiveness and chemoresistance in female reproductive cancers. J Oncol 2010:821717

Cochrane DR, Spoelstra NS, Howe EN, Nordeen SK, Richer JK (2009) MicroRNA-200c mitigates invasiveness and restores sensitivity to microtubule-targeting chemotherapeutic agents. Mol Cancer Ther 8(5):1055–1066

Cortez MA, Bueso-Ramos C, Ferdin J, Lopez-Berestein G, Sood AK, Calin GA (2011) MicroRNAs in body fluids–the mix of hormones and biomarkers. Nat Rev Clin Oncol 8(8):467–477

Courrech Staal EF, Aleman BM, Boot H, van Velthuysen ML, van Tinteren H, van Sandick JW (2010) Systematic review of the benefits and risks of neoadjuvant chemoradiation for oesophageal cancer. Br J Surg 97(10):1482–1496

Duursma AM, Kedde M, Schrier M, le Sage C, Agami R (2009) miR-148 targets human DNMT3b protein coding region. RNA 14(5):872–877

Eitan R, Kushnir M, Lithwick-Yanai G, David MB, Hoshen M, Glezerman M et al (2009) Tumor microRNA expression patterns associated with resistance to platinum based chemotherapy and survival in ovarian cancer patients. Gynecol Oncol 114(2):253–259

Esquela-Kerscher A, Slack FJ (2006) Oncomirs – microRNAs with a role in cancer. Nat Rev Cancer 6(4):259–269

Fang Y, Fang D, Hu J (2012) MicroRNA and its roles in esophageal cancer. Med Sci Monit 18(3):RA22–RA30

Fiorica F, Di Bona D, Schepis F, Licata A, Shahied L, Venturi A et al (2004) Preoperative chemoradiotherapy for oesophageal cancer: a systematic review and meta-analysis. Gut 53(7):925–930

Frederick P, Green H, Huang J, Egger M, Frieboes H, Grizzle W et al (2013) Chemoresistance in ovarian cancer linked to expression of microRNAs. Biotech Histochem 88(7):403–409

Fujita Y, Kojima K, Ohhashi R, Hamada N, Nozawa Y, Kitamoto A et al (2010) MiR-148a attenuates paclitaxel resistance of hormone-refractory, drug-resistant prostate cancer PC3 cells by regulating MSK1 expression. J Biol Chem 285(25):19076–19084

Hamano R, Miyata H, Yamasaki M, Kurokawa Y, Hara J, Moon JH et al (2011) Overexpression of miR-200c induces chemoresistance in esophageal cancers mediated through activation of the Akt signaling pathway. Clin Cancer Res 17(9):3029–3038
Hong L, Han Y, Zhang H, Li M, Gong T, Sun L et al (2010) The prognostic and chemotherapeutic value of miR-296 in esophageal squamous cell carcinoma. Ann Surg 251(6):1056–1063
Hong F, Li Y, Xu Y, Zhu L (2013) Prognostic significance of serum microRNA-221 expression in human epithelial ovarian cancer. J Int Med Res 41(1):64–71
Hu X, Macdonald DM, Huettner PC, Feng Z, El Naqa IM, Schwarz JK et al (2009) A miR-200 microRNA cluster as prognostic marker in advanced ovarian cancer. Gynecol Oncol 114(3):457–464
Huh JH, Kim TH, Kim K, Song JA, Jung YJ, Jeong JY et al (2013) Dysregulation of miR-106a and miR-591 confers paclitaxel resistance to ovarian cancer. Br J Cancer 109(2):452–461
Hummel R, Hussey DJ, Michael MZ, Haier J, Bruewer M, Senninger N et al (2011a) MiRNAs and their association with locoregional staging and survival following surgery for esophageal carcinoma. Ann Surg Oncol 18(1):253–260
Hummel R, Wang T, Watson DI, Michael MZ, Van der Hoek M, Haier J et al (2011b) Chemotherapy-induced modification of microRNA expression in esophageal cancer. Oncol Rep 26(4):1011–1017
Iliopoulos D, Hirsch HA, Struhl K (2009) An epigenetic switch involving NF-kappaB, Lin28, Let-7 MicroRNA, and IL6 links inflammation to cell transformation. Cell 139(4):693–706
Imanaka Y, Tsuchiya S, Sato F, Shimada Y, Shimizu K, Tsujimoto G (2011) MicroRNA-141 confers resistance to cisplatin-induced apoptosis by targeting YAP1 in human esophageal squamous cell carcinoma. J Hum Genet 56(4):270–276
Iorio MV, Croce CM (2012) MicroRNA dysregulation in cancer: diagnostics, monitoring and therapeutics. A comprehensive review. EMBO Mol Med 4(3):143–159
Iorio MV, Visone R, Di Leva G, Donati V, Petrocca F, Casalini P et al (2007) MicroRNA signatures in human ovarian cancer. Cancer Res 67(18):8699–8707
Kang MH, Reynolds CP (2009) Bcl-2 inhibitors: targeting mitochondrial apoptotic pathways in cancer therapy. Clin Cancer Res 15(4):1126–1132
Kim T, Reitmair A (2013) Non-coding RNAs: functional aspects and diagnostic utility in oncology. Int J Mol Sci 14(3):4934–4968
Ko MA, Zehong G, Virtanen C, Guindi M, Waddell TK, Keshavjee S et al (2012) MicroRNA expression profiling of esophageal cancer before and after induction chemoradiotherapy. Ann Thorac Surg 94(4):1094–1102
Komatsu S, Ichikawa D, Takeshita H, Konishi H, Nagata H, Hirajima S et al (2012) Prognostic impact of circulating miR-21 and miR-375 in plasma of patients with esophageal squamous cell carcinoma. Expert Opin Biol Ther 12(Suppl 1):53–59
Komatsu S, Ichikawa D, Takeshita H, Tsujiura M, Morimura R, Nagata H et al (2011) Circulating microRNAs in plasma of patients with oesophageal squamous cell carcinoma. Br J Cancer 105(1):104–111
Kong F, Sun C, Wang Z, Han L, Weng D, Lu Y, Chen G (2011) miR-125b confers resistance of ovarian cancer cells to cisplatin by targeting pro-apoptotic Bcl-2 antagonist killer 1. J Huazhong Univ Sci Technolog Med Sci 31(4):543–549
Kozomara A, Griffith-Jones S (2011) miRBase: integrating microRNA annotation and deep-sequencing data. Nucleic Acids Res 39(Database issue):D152–D157
Kumar S, Kumar A, Shah PP, Rai SN, Panguluri SK, Kakar SS (2011) MicroRNA signature of cis-platin resistant vs. cis-platin sensitive ovarian cancer cell lines. J Ovarian Res 4(1):17
Kurashige J, Kamohara H, Watanabe M, Tanaka Y, Kinoshita K, Saito S et al (2012) Serum microRNA-21 is a novel biomarker in patients with esophageal squamous cell carcinoma. J Surg Oncol 106(2):188–192
Lee H, Park CS, Deftereos G, Morihara J, Stern JE, Hawes SE et al (2012) MicroRNA expression in ovarian carcinoma and its correlation with clinicopathological features. World J Surg Oncol 10:174

Leskelä S, Leandro-García LJ, Mendiola M, Barriuso J, Inglada-Pérez L, Muñoz I et al (2010) The miR-200 family controls beta-tubulin III expression and is associated with paclitaxel-based treatment response and progression-free survival in ovarian cancer patients. Endocr Relat Cancer 18(1):85–95

Li H, Yang BB (2013) Friend or foe: the role of microRNA in chemotherapy resistance. Acta Pharmacol Sin 34(7):870–879

Lin RJ, Xiao DW, Liao LD, Chen T, Xie ZF, Huang WZ et al (2012) MiR-142-3p as a potential prognostic biomarker for esophageal squamous cell carcinoma. J Surg Oncol 105(2):175–182

Liu R, Liao J, Yang M, Shi Y, Peng Y, Wang Y, Pan E et al (2012a) Circulating miR-155 expression in plasma: a potential biomarker for early diagnosis of esophageal cancer in humans. J Toxicol Environ Health A 75(18):1154–1162

Liu S, Fang Y, Shen H, Xu W, Li H (2012b) Berberine sensitizes ovarian cancer cells to cisplatin through miR-21/PDCD4 axis. Acta Biochim Biophys Sin (shanghai) 45(9):756–762

Liu S, Fang Y, Shen H, Xu W, Li H (2013) Berberine sensitizes ovarian cancer cells to cisplatin through miR-21/PDCD4 axis. Acta Biochim Biophys Sin (Shanghai) 45(9):756–762

Lu L, Schwartz P, Scarampi L, Rutherford T, Canuto EM, Yu H et al (2011) MicroRNA let-7a: a potential marker for selection of paclitaxel in ovarian cancer management. Gynecol Oncol 122(2):366–371

Lynam-Lennon N, Reynolds JV, Marignol L, Sheils OM, Pidgeon GP, Maher SG (2012) MicroRNA-31 modulates tumour sensitivity to radiation in oesophageal adenocarcinoma. J Mol Med 90(12):1449–1458

Marchini S, Cavalieri D, Fruscio R, Calura E, Garavaglia D, Nerini IF et al (2011) Association between miR-200c and the survival of patients with stage I epithelial ovarian cancer: a retrospective study of two independent tumour tissue collections. Lancet Oncol 12(3):273–285

Mariani M, Shahabi S, Sieber S, Scambia G, Ferlini C (2011) Class III β-tubulin (TUBB3): more than a biomarker in solid tumors? Curr Mol Med 11(9):726–731

Mateescu B, Batista L, Cardon M, Gruosso T, de Feraudy Y, Mariani O et al (2011) MiR-141 and miR-200a act on ovarian tumorigenesis by controlling oxidative stress response. Nat Med 17(12):1627–1635

Merkerova M, Vasikova A, Belickova M, Bruchova H (2010) Micro-RNA expression profiles in umbilical cord blood cell lineages. Stem Cells Dev 19(1):17–26

Mitamura T, Watari H, Wang L, Kanno H, Hassan MK, Miyazaki M et al (2013) Downregulation of miRNA-31 induces taxane resistance in ovarian cancer cells through increase of receptor tyrosine kinase MET. Oncogenesis 2(3):e40

Mo MH, Chen L, Fu Y, Wang W, Fu SW (2012) Cell-free circulating miRNA biomarkers in cancer. J Cancer 3:432–448

Moss C, Kaye SB (2002) Ovarian cancer: progress and continuing controversies in management. Eur J Cancer 38(13):1701–1707

Nam EJ, Yoon H, Kim SW, Kim H, Kim YT, Kim JH et al (2008) MicroRNA expression profiles in serous ovarian carcinoma. Clin Cancer Res 14(9):2690–2695

Odenthal M, Bollschweiler E, Grimminger PP, Schröder W, Brabender J, Drebber U et al (2013) MicroRNA profiling in locally advanced esophageal cancer indicates a high potential of miR-192 in prediction of multimodality therapy response. Int J Cancer 133(10):2454–2463

Ohkawa K, Asakura T, Aoki K, Shibata S, Minami J, Fujiwara C et al (2004) Establishment and some characteristics of epoxomicin (a proteasome inhibitor) resistant variants of the human squamous cell carcinoma cell line, A431. Int J Oncol 24(2):425–433

Ozols RF (2005) Treatment goals in ovarian cancer. Int J Gynecol Cancer 15(Suppl 1):3–11

Park YT, Jeong JY, Lee MJ, Kim KI, Kim TH, Kwon YD, Lee C et al (2013) MicroRNAs overexpressed in ovarian ALDH1-positive cells are associated with chemoresistance. J Ovarian Res 6(1):18

Peng DX, Luo M, Qiu LW, He YL, Wang XF (2012) Prognostic implications of microRNA-100 and its functional roles in human epithelial ovarian cancer. Oncol Rep 27(4):1238–1244

Prislei S, Martinelli E, Mariani M, Raspaglio G, Sieber S, Ferrandina G et al (2013) MiR-200c and HuR in ovarian cancer. BMC Cancer 13:72

Resnick KE, Alder H, Hagan JP, Richardson DL, Croce CM, Cohn DE (2009) The detection of differentially expressed microRNAs from the serum of ovarian cancer patients using a novel real-time PCR platform. Gynecol Oncol 112(1):55–59

Revilla-Nuin B, Parrilla P, Lozano JJ, de Haro LF, Ortiz A, Martínez C, Munitiz V et al (2013) Predictive value of MicroRNAs in the progression of Barrett esophagus to adenocarcinoma in a long-term follow-up study. Ann Surg 257(5):886–893

Rosenwald S, Gilad S, Benjamin S (2010) Validation of a microRNA-based qRT-PCR test for accurate identification of tumor tissue origin. Mod Pathol 23(6):814–823

Sjoquist KM, Burmeister BH, Smithers BM, Zalcberg JR, Simes RJ, Barbour A et al (2011) Survival after neoadjuvant chemotherapy or chemoradiotherapy for resectable oesophageal carcinoma: an updated meta-analysis. Lancet Oncol 12(7):681–692

Sorrentino A, Liu CG, Addario A, Peschle C, Scambia G, Ferlini C (2008) Role of microRNAs in drug-resistant ovarian cells. Gynecol Oncol 111(3):478–486

Sugimura K, Miyata H, Tanaka K, Hamano R, Takahashi T, Kurokawa Y et al (2012) Let-7 expression is a significant determinant of response to chemotherapy through the regulation of IL-6/STAT3 pathway in esophageal squamous cell carcinoma. Clin Cancer Res 18(18):5144–5153

Takeshita N, Hoshino I, Mori M, Akutsu Y, Hanari N, Yoneyama Y et al (2013) Serum microRNA expression profile: miR-1246 as a novel diagnostic and prognostic biomarker for oesophageal squamous cell carcinoma. Br J Cancer 108(3):644–652

Urschel JD, Vasan H (2003) A meta-analysis of randomized controlled trials that compared neoadjuvant chemoradiation and surgery to surgery alone for resectable esophageal cancer. Am J Surg 185(6):538–543

van Jaarsveld MT, Helleman J, Berns EM, Wiemer EA (2010) MicroRNAs in ovarian cancer biology and therapy resistance. Int J Biochem Cell Biol 42(8):1282–1290

van Jaarsveld MT, Helleman J, Boersma AW, van Kuijk PF, van Ijcken WF, Despierre E (2012) miR-141 regulates KEAP1 and modulates cisplatin sensitivity in ovarian cancer cells. Oncogene 32(36):4284–4293

Vecchione A, Belletti B, Lovat F, Volinia S, Chiappetta G, Giglio S et al (2013) A microRNA signature defines chemoresistance in ovarian cancer through modulation of angiogenesis. Proc Natl Acad Sci 110(24):9845–9850

Wang B, Zhang Q (2012) The expression and clinical significance of circulating microRNA-21 in serum of five solid tumors. J Cancer Res Clin Oncol 138(10):1659–1666

Wang Y, Niu XL, Qu Y, Wu J, Zhu YQ, Sun WJ et al (2010) Autocrine production of interleukin-6 confers cisplatin and paclitaxel resistance in ovarian cancer cells. Cancer Lett 295(1):110–123

Weber JA, Baxter DH, Zhang S (2010) The microRNA spectrum in 12 body fluids. Clin Chem 56(11):1733–1741

Wen J, Zheng B, Hu Y, Zhang X, Yang H, Luo KJ et al (2009) Establishment and biological analysis of the EC109/CDDP multidrug-resistant esophageal squamous cell carcinoma cell line. Oncol Rep 22(1):65–71

Wu Y, Xiao Y, Ding X, Zhuo Y, Ren P, Zhou C et al (2011) A miR-200b/200c/429-binding site polymorphism in the 3′ untranslated region of the AP-2 α gene is associated with cisplatin resistance. PLoS One 6(12):e29043

Xiang Y, Ma N, Wang D, Zhang Y, Zhou J, Wu G et al (2013) MiR-152 and miR-185 co-contribute to ovarian cancer cells cisplatin sensitivity by targeting DNMT1 directly: a novel epigenetic therapy independent of decitabine. Oncogene. doi: 10.1038/onc.2012.575 [Epub ahead of print]

Xiao YF, Yong X, Fan YH, Lü MH, Yang SM, Hu CJ (2013) MicroRNA detection in feces, sputum, pleural effusion and urine: novel tools for cancer screening (Review). Oncol Rep 30(2):535–544

Xu YZ, Xi QH, Ge WL, Zhang XQ (2013) Identification of serum microRNA-21 as a biomarker for early detection and prognosis in human epithelial ovarian cancer. Asian Pac J Cancer Prev 14(2):1057–1060

Yang N, Kaur S, Volinia S, Greshock J, Lassus H, Hasegawa K et al (2008) MicroRNA microarray identifies Let-7i as a novel biomarker and therapeutic target in human epithelial ovarian cancer. Cancer Res 68(24):10307–10314

Yang C, Cai J, Wang Q, Tang H, Cao J, Wu L et al (2012) Epigenetic silencing of miR-130b in ovarian cancer promotes the development of multidrug resistance by targeting colony-stimulating factor 1. Gynecol Oncol 124(2):325–334

Ye G, Fu G, Cui S, Zhao S, Bernaudo S, Bai Y et al (2011) MicroRNA 376c enhances ovarian cancer cell survival by targeting activin receptor-like kinase 7: implications for chemoresistance. J Cell Sci 124(Pt 3):359–368

Zaman MS, Maher DM, Khan S, Jaggi M, Chauhan SC (2012) Current status and implications of microRNAs in ovarian cancer diagnosis and therapy. J Ovarian Res 5(1):44

Zhang C, Wang C, Chen X, Yang C, Li K, Wang J et al (2010) Expression profile of microRNAs in serum: a fingerprint for esophageal squamous cell carcinoma. Clin Chem 56(12):1871–1879

Zhang T, Wang Q, Zhao D, Cui Y, Cao B, Guo L et al (2011) The oncogenetic role of microRNA-31 as a potential biomarker in oesophageal squamous cell carcinoma. Clin Sci 121(10):437–447

Zhang T, Zhao D, Wang Q, Yu X, Cui Y, Guo L et al (2012) MicroRNA-1322 regulates ECRG2 allele specifically and acts as a potential biomarker in patients with esophageal squamous cell carcinoma. Mol Carcinog 52(8):581–590

Zheng ZF, Su HF, Zou Y, Peng Z, Wu SX (2011) Expression profiles of microRNAs in radioresistant esophageal cell line. Zhonghua Yi Xue Za Zhi 91(9):639–642

Zhou L, Yang ZX, Song WJ, Li QJ, Yang F, Wang DS et al (2013) MicroRNA-21 regulates the migration and invasion of a stem-like population in hepatocellular carcinoma. Int J Oncol 43(2):661–669

Chapter 16
Nanocarriers and MicroRNA-Based Scenarios for Cancer Therapy

Voyage of miRNA Delivery to the Land of Lilliputian Carriers

Babak Bakhshinejad, Mohammad Amin Javidi, Sahar Babashah, and Sadegh Babashah

Contents

B. Bakhshinejad • M.A. Javidi • S. Babashah (✉)
Department of Molecular Genetics, Faculty of Biological Sciences,
Tarbiat Modares University, Tehran, Iran
e-mail: sadegh.babashah@gmail.com

S. Babashah
Department of Pharmacology and Toxicology, Pharmaceutical Sciences Branch
and Pharmaceutical Sciences Research Center, Islamic Azad University, Tehran, Iran

S. Babashah (ed.), *MicroRNAs: Key Regulators of Oncogenesis*,
DOI 10.1007/978-3-319-03725-7_16, © Springer International Publishing Switzerland 2014

Abstract MicroRNAs (miRNAs) constitute a crucial category of endogenous, non-coding and regulatory RNA molecules which underlie a wide spectrum of cellular processes comprising differentiation, proliferation and apoptosis. Aberrant miRNA expression correlates well with a variety of pathological conditions including cancer thereby providing an illuminating signature for malignancies. Consistent with this notion, miRNAs have kindled interest to be harnessed as invaluable diagnostic and therapeutic tools. The development of efficacious delivery platforms exerts paradigm-changing effects on attaining miRNA-mediated therapeutic objectives. Many cancer-associated processes take place in nanoscale. The advent of cancer nanotechnology – delving into structures with a size ranging from 1 to 100 nm in at least one dimension – has proven to be a launching pad for tailoring efficient miRNA delivery strategies. These nanotechnology-based delivery vehicles add a new armamentarium to the arsenal of miRNA- associated modalities of cancer treatment. Herein, we address in detail nanocarriers exploited for miRNA delivery. The nanocarriers based on eukaryotic viruses, lipids, polymers, bacteriophage MS2, gold and silica nanoparticles are the focus of our attention. Undoubtedly, gaining insight into the opportunities and challenges offered by these vehicles for miRNA delivery is determining to clear the path for their future applications in the clinical framework.

Keywords Nanotechnology • Nanocarriers • MicroRNA delivery • Cancer therapy

1 Introduction

Famous writer Jonathan Swift in his worldly-known and widely-published novel "Gulliver's Travels" recounts the story of his protagonist's journey to the land of Lilliput, an island which was home to tiny inhabitants called Lilliputians. It seems that we can use this allegory to portray the relationship between microRNA (miRNA) delivery and Lilliputian nanotechnology-based vehicles known as nanocarriers. Overview of nanocarriers used for miRNA delivery is an odyssey through which the dramatic potential of a new generation of vehicles is unearthed. This journey takes us to the heart of a miniaturized world with miniscule entities that provide highly-valued tools to combat many challenges of gene and drug delivery. Currently, the ship of miRNA delivery has reached the shores of nanotechnology and following sterling efforts to go in search of useful vectors, the emergence of nanotechnology-based delivery vehicles promises the dawning of a new era. In this chapter, after a brief overview of miRNAs and their therapeutic implications and a glimpse of the advantages of nanosystems for delivery purposes, we go into detail about various nanocarriers used for miRNA delivery. The nanocarriers based on eukaryotic viruses, lipids, polymers, MS2 bacteriophage, gold and silica nanoparticles are the center of our attention.

Providing information regarding the general structure and characteristics of these nanocarriers, promises and problems they represent for miRNA delivery is addressed.

2 MicroRNAs and Their Therapeutic Implications

MiRNAs are a category of endogenous and small non-coding RNA molecules with an approximate length of 18–25 nucleotides. These regulatory molecules play a leading role in the post-transcriptional control of gene expression. They fulfill this function by binding to the 3′-untraslated regions (3′-UTRs) of their target mRNAs which leads to mRNA cleavage and degradation or translational repression. The idiosyncratic pattern of gene regulation generated by miRNAs provides support for a specific biology. In silico analyses have revealed that up to one-third of human proteins can be regulated by miRNAs (Filipowicz et al. 2008). Since the first discovery of miRNAs Lin-4 and Let-7 in *Caenorhabditis elegans*, a myriad of miRNAs have been recognized in a variety of species ranging from worms and flies to plants and human. In fact, studies conducted over the recent years have uncovered the evolutionary conservation of miRNAs across all species (Wang and Wu 2009).

MiRNAs are implicated in the regulation of a wide spectrum of cellular processes including development, differentiation, cell cycle, proliferation, and apoptosis. A wealth of information has indicated, like protein-coding genes, aberrant expression of miRNAs correlates with a variety of pathological conditions encompassing cardiovascular diseases (van Rooij et al. 2007), inflammation (O'Connell et al. 2007) and viral infection (Umbach and Cullen 2009). Cancer is also among the diseases with deregulated miRNA expression and abnormality in the expression of these functional RNAs has been detected in numerous cancers. Aberrantly expressed miRNAs can be employed as a signature for tumor development, progression, invasion and metastasis. Expression of miRNAs associated with a certain cancer contributes to forming a distinctive tumor-specific miRNA expression profile with functional relevance. This has laid the groundwork for miRNAs to emerge as advantageous prognostic and diagnostic markers (Calin and Croce 2006; Babashah et al. 2012). These disease-associated miRNAs have been demonstrated to represent potential to be exploited as useful targets for therapeutic intervention.

The development of RNA-based therapeutic approaches has received enormous attention in recent years. These agents as a novel class of molecular therapies have enriched the therapeutic armaments against diverse diseases particularly cancer. RNA molecules such as miRNAs provide highly effective and potent tools to remedy cancer. These compounds offer promising avenues to specifically inhibit the expression of genes connected to carcinogenesis or selectively regulate the pathways underlying the emergence and progression of malignant status (Chen et al. 2010). Taking advantage of miRNA molecules as a novel class of pharmaceutical compounds is still in its infancy. However, new therapeutic perspectives presented by these molecules promise to transform the portrait of biomedical science in the future. The fascinating characteristic of miRNA machinery to extensively modulate proliferation and survival of cancer cells has rendered them ideally suited to be exploited as anticancer agents and new drug development.

MiRNAs act as both oncogene and tumor suppressor. From the therapeutic point of view, two scenarios can be used in miRNA-based treatments. Over-expression of oncogenic miRNAs (oncomiRs) contributes to carcinogenesis, while expression decline of tumor suppressor miRNAs accounts for tumorigenesis (Babashah and Soleimani 2011). Consistent with this notion, two medicinal modalities can serve to rectify gene networks in cancer cells. For oncomiRs, a rational means is down-regulation through antagomirs (Krutzfeldt et al. 2005). Moreover, observation of pathological reduction of suppressive miRNAs has inspired the concept of miRNA replacement therapy to enhance the amount of these miRNAs thereby restoring them to normal levels (Kim et al. 2011).

One of the hallmark characteristics of miRNAs to be considered as emerging medicines is that these molecules have the capacity to concurrently interact with numerous target genes and target diverse genetic pathways which control cellular processes such as proliferation, differentiation and apoptosis. Furthermore, this mode of action dramatically limits the development of resistance mechanisms because specific blocking of multiple pathways contributing to oncogenesis poses greater challenges for tumor cells to set escape mechanisms in motion (Ibrahim et al. 2011). In this context, cancer cells need the occurrence of several simultaneous mutations to counteract the effects of expression of incoming miRNA. Estimations have reflected the fact that some miRNAs may affect over 100 target mRNAs. Fabricating a complex pattern of gene regulation, this captivating attribute potentially promotes potency of miRNA-based treatments via exerting influence on different genes and pathways. On the other hand, the "one hit, multiple targets" concept presents new perspectives to gene therapy and expands the repertoire of therapeutic regimens for cancer (Wang and Wu 2009).

Devising efficacious and highly efficient strategies for delivery purposes is of great relevance in the development and evolvement of treatments associated with miRNAs. These strategies will establish a robust foundation for harnessing miRNAs in cancer therapy and drastically expedite their applications in the clinical context. However, it is necessary to preclinically validate miRNAs-based treatments for tailoring suitable approaches to minimize their toxic effects.

3 Advantages of Nanocarriers for Therapeutic Delivery

Lipid-based nanovehicles which their discovery dates back to 1960s were the first nanocarriers used for delivery (Bangham et al. 1965). The years of 1970s witnessed exploitation of the first controlled-release polymer systems for the delivery of macromolecules (Langer and Folkman 1976). In 1980, the first cases of liposome targeting were demonstrated. Moreover, pH-sensitive liposomes with their possible clinical implications were introduced (Leserman et al. 1980; Yatvin et al. 1980; Heath et al. 1980). In 1990s, polyethylene glycol (PEG) was applied to generate long-circulating liposomes and polymeric nanoparticles which fathered the concept of stealth nanoparticles (Klibanov et al. 1990; Gref et al. 1994). Nanocarriers confer advantages for the delivery of various therapeutics which has triggered the surge of interest in these agents as miRNA delivery platforms. Herein, we explain the delivery-oriented opportunities offered by nanosystem-based vehicles.

3.1 Size

Having small physical dimensions is a very attractive feature for nanocarriers that potentiates them to bypass many physiological barriers. Tight epithelial junctions in skin and intestinal tract, blood brain barrier, branching pathways of the pulmonary system and endothelium of blood vessels are some of these biological hurdles which pose challenges to many therapeutic agents to reach targeted site (e.g. tumor) (Hughes 2005; Panyam and Labhasetwar 2003; Lockman et al. 2002). Nanostructured carriers present a solution to surmount these obstacles allowing for efficient uptake of therapeutic cargoes such as miRNAs by various cells and their selective localized accumulation at sites of interest.

3.2 Surface Modification

Easily modified surface chemistry is one characteristic of nanocarriers which can be used to make alteration and adjustment in their pharmacokinetic and pharmacodynamic features. Surface coupling of PEG known as pegylation is one example of such modification. Pegylation leads to escape from reticuloendothelial system (RES) and longer half-life in blood circulation (Gref et al. 1994).

3.3 Targeting

Tractability to manipulation of surface chemistry can be employed to achieve targeting of nanocarriers through both active and passive targeting mechanisms. Surface attachment of various targeting ligands including antibodies, peptides and oligonucleotide sequences capacitates nanocarriers to target their therapeutic cargoes to specific cells thereby fostering enhanced payload, elevated intracellular concentration of cargo and reduced risk of adverse systemic side effects (Ferrari 2005). Tight junctions of endothelial cells in the wall of normal blood vessels act as a mechanism to block extravasation of large particles from blood circulation. Tumor vessels become leaky arising from weakening of these tight junctions. This characteristic of tumor vasculature is known as the enhanced permeability and retention (EPR) effect. In addition to the capacity of nanocarriers to undergo targeting-oriented modifications, EPR phenomenon also contributes to enhanced intratumoral concentrations of cargo. While drugs used in conventional chemotherapy have impact on both normal and cancer cells, nanocarriers through EPR effect of the vascular system become preferentially accumulated in tumors, inflammatory sites and antigen sampling sites (Singh and Lillard 2009; Wang et al. 2012). This eventually boosts anticancer effects and therapeutic efficacy of the delivered therapeutic miRNA.

3.4 Increased Water Solubility

Limited solubility in water is a major factor in meager pharmacokinetic profile of many drugs. Encapsulation of cargo molecules within a nanocarrier dramatically raises solubility and stability of the drugs. On the other hand, another attractive advantage nanotechnology provides for the development of delivery platforms is the capability to formulate existing drugs at nonosized scale. This achievement unifies cargo and carrier and nanosized drug acts as its own carrier. Nanoscaled drugs have higher water solubility and dissolution rate which is very important for *in vivo* applications (Merisko-Liversidge et al. 2003).

3.5 Controlled Release

Nanocarriers can support the controlled release of the encapsulated drug over time. This property sustains cargo concentration within a therapeutic window and positively affects biodistribution of therapeutic agent. In some cases, delivery nanosystem can be devised in a manner that a stimulus specific to the site of cargo delivery activates cargo release. Exerting more rigorous control over the process of drug release presents overwhelming efficacy in the treatment procedure. Moreover, the utilization of biodegradable nanomaterials provides the basis for sustained drug release at the target tumor site over a longer time period (days or weeks). Poly lactic acid (PLA) and poly lactic-co-glycolic acid (PLGA) have been employed to formulate biodegradable nanoparticles for controlled drug delivery (Panyam et al. 2003; Moghimi 2006). These nanomaterials are of particular interest for greater antiproliferative activity of drugs with an intracellular target.

3.6 Multi-functionality

Nanocarriers can be manipulated to carry multiple drugs. This leads to simultaneous transportation of several drugs with one carrier. Multiple-drug cancer therapy necessitates sophisticated dosing regimens. However, nanocarriers due to offering the benefit of releasing each of drugs in a controlled manner abrogate the need for convoluted multidrug dosing regimens. This multimodal capacity of nanocarriers lays the groundwork for materialization of the important concept of combination therapy in cancer treatment. The combinatorial capability of nanoparticle-based carriers allows to deliver multiple compounds with various characteristics (such as imaging and therapeutic properties) into tumors and acts as a formidable obstacle against the emergence of drug resistance mechanisms in cancer cells (Alexis et al. 2010). As well, the synergistic effects of drugs in combination therapy target cell heterogeneity present in tumors thus substantially bolstering the efficiency of the therapeutic regimen.

4 Nanocarriers for MicroRNA Delivery

Various nanocarriers have found application in miRNA delivery. These nanoscale carrier systems come in two main types that embrace organic and inorganic. In the former nanovehicles based on eukaryotic viruses, lipids, polymers and bacteriophages and in the latter nanosystems derived from gold and silica nanoparticles are covered and their application for miRNA delivery is discussed. Figure 16.1 provides a schematic presentation of these nanocarrier systems.

4.1 Eukaryotic Virus-Based Nanocarriers

Eukaryotic virus-based vectors are the most frequently used nanocarriers for therapeutic delivery purposes. Thus far, many viral vectors have been elaborated for the delivery of miRNA pharmaceutics. Different viruses used as delivery platforms

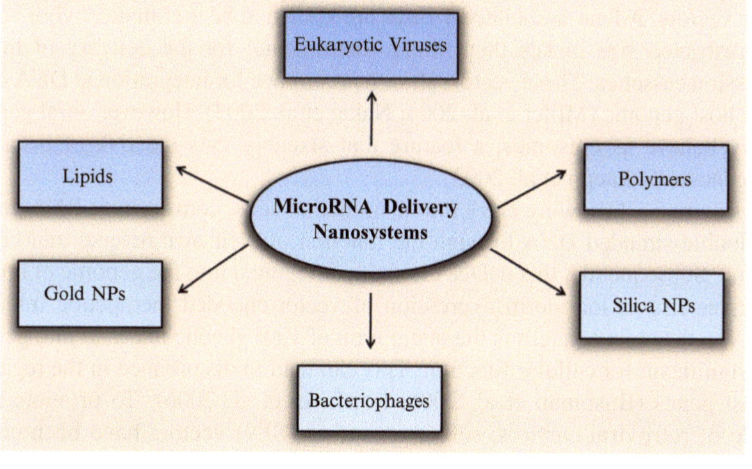

Fig. 16.1 Various nanocarriers used for microRNA delivery

harbor their own distinctive properties which render them well-suited for certain applications. The category of the used viral vector is largely dependent upon the therapeutic objective we pursue and the type of targeted cell. Viral vector-associated toxicity observed in gene therapy-oriented clinical trials sparked much controversy with regard to the safety of these carriers for medical goals. These bottlenecks have been a stimulus to manipulate viral vectors to maximize their safety profile. Nonetheless, there have been several recent publications reporting the involvement of vector sequences and vector-derived immunogenic epitopes in the stimulation of immune responses (Onion et al. 2009; Lamers et al. 2011). Four types of viruses are the most broadly exploited viral nanocarriers for delivery purposes. These include vectors based on adenovirus, adeno-associated virus, retrovirus, and lentivirus (Liu and Berkhout 2011). Providing an encyclopedic and in-depth description of viral vectors and their pros and cons is beyond the confines of this chapter. Herein, we offer a brief overview of viral-based vectors. To gain thorough and detailed information concerning the problems and prospects of viral nanocarriers and their diverse applications for the delivery of therapeutic cargoes, readers are referred to these references (Giacca and Zacchigna 2012; Young et al. 2006; Walther and Stein 2000).

Adenoviruses are capable of infecting both dividing and nondividing cells, replicate in the nucleus of mammalian cells and commonly do not integrate their genetic material into the genome of host cell. This lack of genomic integration spawns short-term expression of a transgene such as miRNA. The most important challenge posed by adenoviruses is their immunostimulatory effects that raise safety issues. To mitigate stimulation of immune responses and as well as enhance packaging capacity of adenoviruses, second-generation and third-generation adenovirus vectors have been developed (Cao et al. 2004; Liu and Berkhout 2011).

Adeno-associated viral nanocarriers provide the prime benefit of lack of immunogenicity. This stems from their minor ability to transduce antigen presenting cells. This non-pathogenicity renders them very captivating to be utilized as therapeutic vectors. Adeno-associated viruses are known to be the smallest viral vectors. This restricted size makes them much instrumental for the delivery of miRNA expression cassettes. These vectors show a preference for integration at DNA breaks of the host genome (Miller et al. 2004; Nakai et al. 2003). However, viral genomes mainly behave as episomes, a feature that sizeably cuts the risk of insertional mutagenesis (Schnepp et al. 2003).

Retroviruses, following entry into mammalian cells, convert their RNA genome into double-stranded DNA through the function of their own reverse transcriptase enzyme. Subsequently, this dsDNA is stably integrated into the genome of host cell that gives rise to long-term expression of vector-encoded therapeutic transgene. However, the main concern is the integration of viral genetic material into genomic sites significant for cellular function. This can lead to disturbance in the regulation of host genes (Bushman et al. 2005; Lewinski et al. 2006). To promote safety profile of retroviral carriers, self-inactivating (SIN) vectors have been crafted.

In SIN vectors, transcriptional enhancer elements affecting gene expression of the host cell have been removed (Yu et al. 1986).

Lentiviruses own characteristics such as stable transgene expression due to integration into the host genome, the capacity for harboring large inserts, and the capability of transducing non-dividing cells. Removal of U3 region in the 3′ long terminal repeat region and supporting integration into introns of active transcriptional units are contributing factors to the diminution of insertional oncogenesis (Zufferey et al. 1997; Laufs et al. 2006).

Various viral nanocarriers have been used for the delivery of miRNA therapeutics. Administration of miR-26a in a mouse model of hepatocellular carcinoma (HCC) through adeno-associated virus was carried out to explore the efficacy of a miRNA replacement therapy for liver cancer. The results showed that this strategy is able to impede the proliferation of cancer cells, induction of tumor specific apoptosis and substantial protection from disease progression without toxicity (Kota et al. 2009). Adenovirus-mediated delivery of a plasmid harboring miRNA let-7 was demonstrated to repress cancer growth in lung. In this study, let-7 delivered via adenoviral vector hampered the growth of multiple human lung cancer cell lines *in vitro*. Also, administration of this miRNA trough intranasal route in an established orthotopic mouse model of lung cancer diminished tumor formation (Esquela-Kerscher et al. 2008). Adeno-associated virus-mediated transfer of miRNA-based hairpin to silence peripherin-2, a photo-receptor specific gene in retina that its gain-of-function mutations contribute to both autosomal dominant retinitis pigmentosa and dominant maculopathies, in the mouse retina efficiently and specifically inhibited expression of this gene (Georgiadis et al. 2010). The utilization of lentiviruses for miRNA delivery has been also reported in a number of studies. For instance, a miRNA-based lentiviral vector was used to express miR-30. Also, a conditional lentiviral miRNA expression with a doxycycline-inducible vector was developed that exhibited the capacity of tight control on gene expression *in vitro* and *in vivo* (Stegmeier et al. 2005). In another study a lentiviral vector encoding three artificial miRNAs targeted against the Abl fraction of the chimeric Bcr-Abl oncogene mRNA was generated. This system showed success in reducing the expression of Bcr-Abl over 200-fold higher than that of control. This substantial decline in the oncogene expression through inhibitory miRNAs portrays the efficacy of this strategy to control population of malignant cells and hinder growth of leukemia cells (McLaughlin et al. 2007). In a study conducted by Babashah et al. (2013), lentivirus-mediated over-expression of miR-326 led effectively to down-regulation of Smo (the signal transducer of the oncogenic Hedgehog pathway) in chronic myeloid leukemia (CML) $CD34^+$ cells. This strategy resulted in decreased cell proliferation and elevated rate of apoptosis in CML $CD34^+$ stem/progenitor cells that represent a potential source of relapse in patients suffering from CML. Another study revealed that lentiviral vectors encoding miRNAs against osteopontin (OPN) gene, as one of the key genes implicated in promoting the metastasis of HCC, potently down-regulate the OPN expression level that triggers *in vitro* proliferation and *in vivo* tumor growth of HCC (Sun et al. 2008).

4.2 Lipid-Based Nanocarriers

Lipid-based materials embrace one of the most important categories of nanocarriers for the delivery of therapeutics. Among various lipid-based materials with the potential to be employed for delivery purposes, liposomes are regarded as the most widely-used carriers and have been the first nanoparticle-based drug delivery platforms enjoying application in the clinical setting. Since their initial description, liposome-based delivery systems have met with high success in the delivery of anticancer agents. Huge diversity in structure and composition has rendered them very tunable and versatile for biological and medical objectives. These carriers harbor appealing attributes such as general biocompatibility, biodegradability, and the merit of encapsulating both hydrophilic (within their aqueous core) and hydrophobic drugs (inside their lamellae) (Torchilin 2005; Peer et al. 2007). Liposomes are applied to the delivery of various macromolecular structures including gene therapy vehicles, oligonucleotides, proteins and other drugs into malignant cells and tissues. Their amphiphilic nature allows the accommodation of both hydrophilic and lipophilic anticancer therapeutics. Hydrophilic cargoes can be incorporated into the internal aqueous part and hydrophobic drugs can be transported in the hydrophobic portion of the bilayer.

Liposomes are self-assembled, spherical and closed-membrane structures. These vesicles have a central aqueous core surrounded by an outer lipid bilayer and are shaped by dispersion of phospholoipids in aqueous media. In the amphiphilic architecture of liposomes, hydrophobic domains of lipid molecules are brought together by hydrophobic interactions and hydrophilic head groups of the molecules are positioned towards the inner aqueous milieu. This arrangement of lipid compounds contributes to forming bilayer anatomy of lipid molecules. The size of liposomes, dependent on the design, can be from 50 nm to several micrometers (Torchilin 2005; Cho et al. 2008). Lipid constituent plays a major role in determining different structural and functional traits of liposomes. For instance, higher cholesterol content yields increased retention rate leading to subsequent mitigated drug loss in blood circulation (Gregoriadis 1995).

In general terms, two significant physical properties, – size and the degree of lamellarity (number of bilayers) – serve as a basis for classification of liposomes. According to this notion, liposomes come in four types: small unilamellar vesicles (SUV), large unilamellar vesicles (LUV), multilamellar vesicles (MLV) and multivesicular liposomes (MVL) (Szoka and Papahadjopoulos 1980). Unique structural properties of liposomes assist in augmenting delivery efficiency through the mechanism of "contact-facilitated delivery", a phenomenon which entails interaction of lipid elements of liposome with the membrane of the targeted cell. The increase of these lipid-lipid interchanges markedly facilitates the convective flux of hydrophobic agents to travel across lipid membranes of both liposome and targeted cell. Physicochemical characteristics of liposomes such as size, composition, surface charge and functionality can be conveniently altered through adding agents to the lipid membrane or modifying the surface chemistry. Modified liposomes have been demonstrated to offer favorable pharmacokinetic properties for the delivery of antisense oligonucleotides, siRNA, proteins and chemotherapeutic drugs (Papahadjopoulos et al. 1991). PEG conjugation is one of the most significant modifications made on the surface of liposomes. PEG molecules create a protective cover on the surface of liposomes. These PEGylated carriers (also known as stealth liposomes) can escape from mononuclear phagocytes of the RES and avoid interacting with plasma proteins which ultimately spawns prolonged circulation period (Kontermann 2006). Furthermore, active targeting can be used to provoke surface modifications on liposomes. Targeted delivery can be achieved through conjugation of targeting ligands such as antibodies or folate. Attaching antibodies to liposomes generates immunoliposomes with the capability of selective drug delivery to the action site of interest (Cryan 2005; Park et al. 2002; Gabizon et al. 2003; Wu et al. 2006). These liposome-based carriers have recently received a remarkable interest and are considered as the next generation of liposomal drugs. Also, Leamon et al. (2003) have demonstrated the efficacy of folate-decorated liposomes for the delivery of antisense oligodeoxynucleotide to cancer cells.

Solid lipid nanoparticles (SLNs) are another category of lipid-based nanocarriers which comprise lipid compounds being solid at room temperature. The solid lipid matrix is structurally stabilized by incorporating surfactant agents (tween 20, sodium dodecyl sulphate, etc.). A great number of solid lipids are naturally occurring which are implicated in a variety of physiological processes. This is of great relevance in the decline of their cellular toxicity. The challenge of toxicity enhancement resulting from the addition of surfactants can be circumvented by making use of nontoxic surfactants such as lecithin. SLNs, due to the role of their solid core in blocking the merge of particles, present higher stability compared to liquid droplets (Almeida and Souto 2007; Bondi and Craparo 2010; Qi et al. 2012).

Lipid-based nanocarriers have been used for the delivery of different miRNAs. A lipid-based nanoparticle was utilized for systemic delivery of plasmids harboring miRNA to cancer cells. Two miRNAs, miR-34a and the miR-143/145 cluster, were delivered through the liposomal nanovector. These two miRNAs are down-regulated in most types of pancreatic cancer. Systemic administration of both nanovectors through intravenous route led to growth inhibition of subcutaneous and orthotopic xenografts of pancreatic cancer. This inhibition of tumor growth was also accompanied by augmentation of apoptosis and decline of proliferation offering potential of this lipid-based platform for systemic delivery of therapeutic miRNAs to cancer cells (Pramanik et al. 2011). In another report, miR-7 -expressing plasmid was transfected into several lung cancer cell lines. This miRNA targets several sites in the 3′ UTR region of EGFR (epidermal growth factor receptor) mRNA. The results demonstrated anti-proliferative effects of the liposome-delivered miRNA. Furthermore, antitumor properties of miR-7 was analyzed via *in vivo* liposomal delivery against EGFR-TKI– resistant xenograft models of lung cancer in mouse which exhibited significant tumor regression (Rai et al. 2011). In another study miR-133b, which directly targets MCL-1 protein, was delivered via cationic lipoplexes into A549 non-small cell lung cancer (NSCLC) cells. Evaluation of delivery efficiency showed higher capacity of lipoplex platform with enhancement of miR-133b expression and decrease of MCL-1 expression *in vitro. In vivo* biodistribution study also revealed that mice treated with lipoplexes containing pre-miR-133b had a marked elevation of miR-133b expression in lung in comparison with untreated control mice (Wu et al. 2011). Shi et al. (2013) developed a SLN-based system to deliver miR-34a, a miRNA significant for cancer stem cell (CSC) therapy, into B16F10-CD44^{+}-bearing tumors in mouse lungs. They showed that miSLN dramatically inhibits tumor growth offering potential to be an efficient strategy to compromise growth and migration of CD44-positive CSCs.

4.3 Polymer-Based Nanocarriers

Polymers are one of the most frequently addressed nanocarriers for delivery purposes. In 1979, Couvreur et al. described one of the first examples of exploiting polymeric nanoparticles to develop a drug carrier system for cancer therapy (Couvreur et al. 1979, 1980). They attached antineoplastric drug doxorubicin to polyalkylcyanoacrylate nanoparticles and demonstrated the mechanism of drug release from the polymeric agent in calf serum. This watershed report was a starting point for a series of studies employing polymer-based materials for therapeutic delivery. In 1990s, Gref et al. (1994) elaborated polymeric nanocarriers comprising PLA/PLGA and PEG as long-circulating nanoparticles with stealth characteristics. This study created a flurry of interest in the utilization of polymeric nanocarriers to achieve therapeutic goals.

From the structural standpoint, polymeric nanocarriers can have spherical, branched or core-shell architecture. Polymeric nanocarriers are primarily formed through a spontaneous self-assembly process of block-copolymers in an aqueous surrounding spawning a core-shell anatomy. In this structure, the hydrophobic blocks constitute the core and the hydrophilic blocks establish the shell. Each of these sections plays its own role in the general structure and function of polymeric nanocarriers. The core considerably reduces the exposure of cargo to the surrounding environment while the shell makes an essential contribution to the core stabilization. The hydrophobic core with its high loading capability is responsible for carrying therapeutic cargoes and the hydrophilic shell creates a steric protection for nanocarrier platform. Polymeric nanocarriers are capable of entrapping hydrophilic and hydrophobic small molecules and as well as protein- and nucleic acid-based macromolecular structures (Chan et al. 2010). Polymeric vehicles are considered as one of the most significant nanoparticle-based therapeutics in preclinical and clinical surveys of delivery and their peculiar architecture assists in shaping a platform well-suited to the delivery of therapeutic cargoes.

Within this context, targeting of polymer-based carriers is worthy of consideration. A variety of molecules including peptides, antibodies and oligosaccharides can be used for active targeting. The strategy of active targeting promotes the localized and sustained delivery of therapeutic molecules to the biological site of interest and minimizes the side effects. Structural modification and functionalization of polymers enable us to exert greater control on the pharmacokinetic profile of therapeutic cargoes. Furthermore, one of the most significant achievements of recent years in the territory of polymer science has been the design of smart polymer-based delivery systems. These signal-sensitive platforms are capable of altering their physicochemical characteristics in response to environmental stimuli. A diverse array of physical (temperature, light, and electricity), chemical (pH, ionic concentration) and biological (enzymes, biomolecules) signals can be applied to achieve this end. The procedure of stimuli responsiveness of polymer-based delivery platforms can be more meticulously and precisely controlled within a narrow range through versatility of polymer sources and their convenient combination (Bamrungsap et al. 2012; Parveen and Sahoo 2008).

Polymeric nanocarriers can be constructed from natural polymers like chitosan, gelatin, heparin, collagen, dextran, cellulose, cyclodextrin or synthetic polymers such as polyethyleneimine (PEI), poly-L-lysine (PLL), poly(amidoamine) (PAA), poly(amino-co-ester) (PAE) and poly(2-N,N-dimethylaminoethylmethacrylate) (PDMAEMA) (Panyam and Labhasetwar 2003). Also, polymeric nanocarriers, depending on their electrical charge, are divided into main categories of anionic (negatively charged) and cationic (positively charged) polymers. Both of these carrier types, with their unique physicochemical properties, have been comprehensively explored for a variety of therapeutic applications. Anionic polymers interact with cationic molecules such as positively charged drugs and basic proteins. On the contrary, cationic polymers are able to form electrostatic complexes with anionic biomolecules such as negatively charged proteins and nucleic acids. Within the recent years, a great deal of attention has been directed toward cationic polymeric nanocarriers. This interest has roots in some hallmark characteristics of cationic polymers including flexibility, easy synthesis and efficiency in gene and drug delivery (Samal et al. 2012). Cationic polymers have the capacity to condense nucleic acids through forming polyelectrolyte complexes with them. This condensation enhances cellular entry and confers dramatic protection against enzymatic degradation and consequent endolysosomal escape of nucleic acid molecules. These attributes render cationic polymeric nanocarriers invaluable tools for the delivery of miRNA molecules. Among various polymeric nanocarriers, PEI and PLL have been exploited for miRNA delivery. In PEI, backbone bears positive charges while in PLL side groups harbor positive charges.

PEI has been previously applied in various studies to deliver DNA plasmids (Boussif et al. 1995), ribozymes (Aigner et al. 2002), and siRNA (Urban-Klein et al. 2005; Gunther et al. 2011) for therapeutic purposes. The condensed complexation between negatively charged miRNA and positively charged PEI polymer gives rise to the protection of miRNA against RNAse enzymes present in cells. PEI-based complexes enter the cells through caveolae- or clathrin-dependent pathways. A major complication of PEI polymer to be utilized as a nanocarrier is its cellular toxicity. This originates from nonbiodegradability of polymer molecules inside cells feeding into the formation of negatively charged protein aggregates. Son et al. (2011) introduced disulfide linkage in the branched PEI and proved that this modification enhances biocompatibility and biodegradability of the manipulated carrier

(SSPEI polymer). This is due to disulfide-mediated induction of carrier degradation by endogenous enzymes such as glutathione reductase. Another study carried out by this group indicated the capability of the rabies virus glycoprotein (RVG)-labeled SSPEI polymeric carrier for the neuron-specific delivery of miR-124a *in vivo* (Hwang do et al. 2011). Injection of RVG-SSPEI complexed with miR-124a into tail veins of mice triggered increased accumulation of this miRNA in isolated brain. The nanosized polyelectrolyte complexes between PEI and miRNA were internalized into cells through the mechanism of general endocytosis. Jin et al. (2012) crafted a RNA- PLL complex for the delivery of anti-miR-RNA-10b into breast cancer cell line and observed the effective delivery of RNA molecules into the cytoplasm of target cells in a concentration-dependent manner. The potent interaction between RNA backbone and cationic PLL has been suggested as the molecular mechanism underlying the delivery of anti-miR-RNA-10b into breast cancer cells. Ibrahim et al. (2011) introduced a low molecular weight branched PEI (PEI F25-LMW) as a strategy for the delivery of miR-145 and miR-33a in a miRNA replacement therapy. Following systemic or local administration of PEI/miRNA complex, intact miRNA molecules were delivered into a mouse xenograft model of colon carcinoma that led to decreased tumor proliferation and enhanced apoptosis. Yang et al. (2012) demonstrated that miR-145 delivery through a polyurethane-short branch polyethyleneimine (PU-PEI) to glioblastoma cells (brain malignant cells) dramatically inhibits tumorigenic and CSC-like capacities of these cells. This was accompanied by effective down-regulation of the genes implicated in resistance to drug and apoptosis. Furthermore, *in vivo* delivery of PU-PEI-miR145 remarkably suppressed tumorigenesis and promoted survival rate of mice.

4.4 Bacteriophage VLP-Based Nanocarriers

MS2 is an icosahedral single-stranded RNA bacteriophage with a 27–34 nm diameter that places it in the category of carriers with nanostructures. This bacteriophage has a 3560 nucleotide genome. A 19-nucleotide sequence residing at the 5′ end of the replicase cistron, called the *pac* site, plays a pivotal role in shaping the general architecture of this bacteriophage. The sequence of *pac* site is able to establish a stem-loop structure. Specific interaction between this stem-loop structure and the bacteriophage coat protein accounts for the precise assembly of bacteriophage particles in bacterial systems (Stockley et al. 1994). The self-assembly of MS2 has been applied to create virus-like particles (VLPs) that can be loaded with different pharmaceutical and imaging molecules thereby offering a new drug delivery platform. Up to now, there have been few efforts for the development of delivery vehicles based on VLPs. However, there has been an increasing attention towards repurposing these nanocarriers for the delivery of therapeutic agents. Information accumulated within the last recent years has unveiled numerous gripping characteristics of bacteriophage MS2 rendering it a well-suited carrier to be employed for delivery purposes. There is a large body of evidence regarding the packaging and self-assembly of MS2. This information demonstrates that bacteriophage capsids provide a convenient and efficacious modality to package RNAs, DNAs and other medicines. Bacteriophage MS2 confers a protective shield on miRNAs. Rapid degradation of naked miRNAs by nucleolytic enzymes makes them very unstable in biological fluids. To surmount this obstacle and promote the stability of miRNAs, these molecules have undergone diverse modifications (Orom et al. 2006; Bader et al. 2011). Although these manipulations ameliorate the half-life of miRNAs, they also have a negative impact on the hybridization kinetics of miRNA-based drugs. Packaging of RNA drugs into MS2-based VLPs safeguards them against enzymolyis and improves their pharmacologic activity.

Bacteriophage MS2 offers some other advantages for delivery-oriented objectives. Surface modification of bacteriophage capsid can be easily achieved through genetic insertion or chemical conjugation of outer capsid with different ligands. This enables bacteriophage particles for multivalent display of targeting ligands allowing selective delivery to a variety of cells (Wei et al. 2009). Also, the presence of a comparatively vast interior space renders it possible for MS2 VLPs to be loaded with large agents. Numerous pores existing in the capsid of bacteriophage facilitate access of therapeutic agents to the interior volume (Kovacs et al. 2007). On the other hand, the natural ability of MS2 coat protein to spontaneously self-assemble in the presence of nucleic acids provides a rational basis for their application in the encapsidation of RNAs and RNA-conjugated drugs with therapeutic potential. It is interesting to note that a variety of chemotherapeutic non-nucleic acid cargoes, when attached to the MS2 *pac* site, can also be encapsulated within MS2 VLPs. MS2 VLPs are biocompatible and biodegradable. They also bear dramatic stability in a wide spectrum of environmental conditions (temperature, pH, ionic strength and solvents) (Ashley et al. 2011).

Pan et al. (2012b) developed a system based on MS2-VLP containing pre-miR 146a. Subsequently, they conjugated these particles with a cell penetrating peptide called Tat, derived from HIV-1. The chemical conjugation of Tat was used to achieve the intracellular transduction of the delivered miRNA. MS2 VLP-Tat complex successfully transferred pre-miR 146a into HeLa, HepG2, Huh-7 and peripheral blood mononuclear cells (PBMCs). High levels of miR 146a were detected *in vitro* and *in vivo*. Furthermore, MS2-VLP nanocarrier was able to specifically silence target genes of the delivered miRNA. This group in another study used Tat-conjugated MS2 VLP-miR146a to reach therapeutic effects in lupus-prone mice. They witnessed high levels of miR-146a in PBMCs, lung, spleen, and kidney tissues of the mice following a 12-week therapeutic procedure and demonstrated the capability of this treatment system on lupus-prone mice through diminishing the production of pathogenic autoantibody (Pan et al. 2012a). Ashley et al. (2011) reported the delivery of several chemotherapeutic drugs (doxorubicin, cisplatin, and 5- fluorouracil), siRNA cocktails, and protein toxins to human HCC by using a nanocarrier based on bacteriophage MS2. They decorated MS2 particles with high densities of SP94, an HCC-specific peptide previously isolated by affinity selection of a phage display library against HCC targets. Their findings revealed that MS2 VLPs are able to selectively elicit toxicity in liver cancer.

4.5 Gold Nanoparticle-Based Nanocarriers

Inorganic nanoparticles have been the focus of vast investigations to be employed in drug delivery for cancer therapy. One of the foremost advantages offered by inorganic nanocarriers is their capacity to be synthesized with near monodispersity. Inorganic nanoparticles are mainly metal based. Metallic nanostructures have been applied to various purposes in biomedicine. Metal-based biosensors for the detection of antigenic agents and DNA and as well as metallic probes as imaging, diagnostic and therapeutic elements are several examples of metallic nanomaterials in the medical context (Elghanian et al. 1997; Hirsch et al. 2003). Metal particles are biocompatible. However, their application is not without difficulties. Accumulation of a large amount of metal particles in the body following administration may result in cytotoxicity thereby imposing restriction on their exploitation for clinical aims. Toxicity can be especially intensified in the case of repeated administration (Wang et al. 2012). For this reason, most of the investigations on the potential of metal nanoparticles for drug delivery purposes are in the preclinical stage.

Carriers based on gold nanoparticles (AuNPs) are one of the most important metal-based nanosystems for drug delivery. AuNPs are conveniently synthesized and manipulated. Surface modification and functionalization of AuNPs have served to the delivery of therapeutics. The attachment of numerous biomolecules onto the surface of AuNPs is a relatively facile procedure being feasible through a variety of affinity interactions. Also, physical characteristics of AuNPs render it possible to exert greater control on the rate of drug release. Due to this, AuNPs have attracted remarkable attention for the delivery of therapeutics. Rosi et al. (2006) employed functionalized AuNPs with thiolated oligonucleotide sequences for gene therapy through antisense mechanism. Intracellular delivery of these AuNP-oligonucleotide complexes strongly declined the expression of targeted mRNA encoding enhanced green fluorescent protein (EGFP). Tumor necrosis factor (TNF)-alpha, a robust cytokine with anticancer properties, has been conjugated to PEG-decorated AuNPs (Visaria et al. 2006). This antineoplastic agent in a nanocarrier formulation has shown reduced cytotoxicity and high efficacy in retarding the growth of SCK mammary carcinomas in a xenograft mouse model.

AuNPs have also found utility for miRNA delivery and several studies have highlighted their potential in this regard. In a study, Crew et al. (2012) demonstrated successful conjugation of miR-130b to AuNPs and exploited this miRNA-AuNP construct for cell transfection assays on multiple myeloma cells. Their findings exhibited the relatively high stability of miRNAs immobilized on AuNPs and effective transfection of miR-130b into myeloma cells. Functional luciferase assay was carried out for quantitative determination of miRNA delivery. The results indicated the knockdown of luciferase implying that miRNA delivery into target cells occurred by AuNPs. Ghosh et al. (2013) used cysteamine-functionalized AuNPs to deliver miRNA into two various tumor models, neuroblastoma and ovarian cancer cell lines. Moreover, they used PEG coating for miR-AuNP platform to inhibit nanoparticle aggregation and enhance half-life of miRNAs on AuNPs. This was due to the fact that in the absence of PEG, AuNPs indicated higher tendency to form aggregates. Their results suggested that AuNPs were successful in the delivery of miRNAs leading to a 10–20 fold over-expression of mature miRNAs in comparison with liposome-mediated transfection. Also, the miRNAs delivered via AuNP strategy were capable of potent down-regulation of target genes and modulation of cell proliferation rate.

4.6 Silica Nanoparticle-Based Nanocarriers

Silicon-based nanostructures are one of the other inorganic carrier systems that have received remarkable attention as a drug delivery platform. Biocompatibility, convenient synthetic procedures, stability, surface modification and the well-recognized silane chemistry as a means to simplify cross-linking of different drugs are benefits of these structures for drug delivery (Ohulchanskyy et al. 2007; Couleaud et al. 2010). The most frequently explored silicon-based structures for drug delivery are porous silicon and silica (silicon dioxide). Porous silicon embedded with platinum as a device for cancer therapy, silicon nanopores as antibody delivery platforms and porous silica nanoparticles harboring therapeutic cargoes such as antibiotics, enzymes and DNA are instances that show potential of silicon-based structures for treatment purposes (Chen et al. 2004; Giri et al. 2007). In porous silica nanoparticles, the density and size of the pores can be precisely tuned. This control over the structural properties of nanoparticles makes it possible to attain a constant and gradual drug release. Roy and coworkers successfully made use of silica nanoparticles surface functionalized with EGFP-encoding plasmid for gene therapy purpose (Roy et al. 2005). Lu et al. (2007) were successful in inducing apoptosis in various cancer cells though mesoporous silica nanoparticles inside which camptothecin (a hydrophobic anticancer agent) had been entrapped. This study demonstrated that mesoporous silica nanoparticles may offer potential to address the challenge of water insolubility of many chemotherapeutic drugs.

A number of reports have indicated the ability of silica nanoparticles for the delivery of miRNA therapeutics. In a study conducted by Tivnan et al. (2012), they reported the use of a silica nanoparticle-based platform for the delivery of a tumor suppressive and proapoptotic miRNA, miR-34a, to neuroblastoma tumors in an orthotopic xenograft mouse model. An antibody against GD2 was conjugated to these silica nanoparticles. GD2 is a glycolipid with considerable expression on the surface of neuroblastoma cells and offers a potential target for immunotherapy and targeting with therapeutic purposes. GD2 antibody attachment to the surface of porous silica nanocarriers triggered targeted delivery of miR-34a to xenograft tumors in mouse. Systemic administration of silica nanocarrier-miRNA platform into tumor-bearing mice elicited a substantial decrease in tumor growth and vascularization and a marked enhancement in inducing apoptosis in tumor cells implying the anti-tumorigenic effects of the delivered miRNA.

Table 16.1 lists various nanocarrier systems used for miRNA delivery together with the delivered miRNAs. Also, this table contains information regarding target cell or tissue and the related reference. It is important to note that this table involves

Table 16.1 A complete list of all nanocarrier systems discussed in the chapter

Nanocarrier system	Delivered miRNA/ miRNA- based therapeutic	Target cell/tissue/ cancer	Reference
Adeno-associated virus	miR-26a	Hepatocellular carcinoma	Kota et al. (2009)
Adenovirus	Let-7	Lung cancer	Esquela-Kerscher et al. (2008)
Adeno-associated virus	microRNA-based hairpin against peripherin-2	Retina	Georgiadis et al. (2010)
Lentivirus	miR-30	Ovarian cancer	Stegmeier et al. (2005)
Lentivirus	Artificial miRNAs against the Abl fraction of the chimeric Bcr-Abl oncogene mRNA	Leukemia	McLaughlin et al. (2007)
Lentivirus	miRNAs against osteopontin	Hepatocellular carcinoma	Sun et al. (2008)
Liposome	miR-34a, miR-143/145 cluster	Pancreatic cancer	Pramanik et al. (2011)
Liposome	miR-7	Lung cancer	Rai et al. (2011)
Cationic lipoplex	miR-133b	Non-Small Cell Lung Cancer (NSCLC)	Wu et al. (2011)
Solid lipid nanoparticle	miR-34a	Lung	Shi et al. (2013)
Polyethyleneimine	miR-124a	Neuron	Hwang do et al. (2011)
Poly-L-lysine	anti-miR-RNA-10b	Breast cancer	Jin et al. (2012)
Polyethyleneimine	miR-145 and miR-33a	Breast and colon cancer	Ibrahim et al. (2011)
Polyethyleneimine	miR145	Brain cancer	Yang et al. (2012)
MS2 bacteriophage-VLP	miR 146a	Ovary, liver, lung, spleen, kidney and Peripheral Blood Mononuclear Cells (PBMCs)	Pan et al. (2012b)
Gold nanoparticle	miR-130b	Multiple myeloma	Crew et al. (2012)
Gold nanoparticle	miR-AuNP-S-PEG	Brain and ovarian cancer	Ghosh et al. (2013)
Silica nanoparticle	miR-34a	Brain cancer	Tivnan et al. (2012)

This table also provides information concerning the delivered miRNAs by these nanocarriers, target cell/tissue/cancer into which delivery has been conducted together with the related reference

a complete list of all nanocarrier systems and their delivered miRNAs that have been discussed in this chapter. In fact, the table enables readers to gain a quick view of all the nanocarrier systems covered throughout the chapter.

5 Concluding Remarks

A wealth of information highlights the momentousness of alterations in miRNA expression in a huge variety of neoplastic disorders. This documented role has sparked interest in the utilization of dramatic capability of miRNAs as tools for the diagnosis and treatment of malignancies. To take advantage of this tremendous potential, there remains a great necessity to formulate efficient systems for miRNA delivery. Nanotechnology, with its multidisciplinary essence, has been a major driving force for the development and evolvement of vehicles ideally suited to the delivery of diagnostics and therapeutics and has ventured the field of delivery into a new era. Application of nanotechnology inspired carriers was primarily concentrated on the delivery of siRNA molecules and miRNA delivery remained unexplored. Within the past years, flourishing the scientific area of miRNA has provoked a flurry of interest in mining suitable and efficacious nanocarriers for the delivery of these regulatory molecules. It is interesting to note that a major portion of studies dedicated to miRNA delivery has occurred over the most recent years. Numerous nanocarriers have served to the delivery of miRNA pharmaceutics. Each of these nanosystems harbors its own benefits and pitfalls. Selection of each nanocarrier largely relies on the circumstances and the aims we pursue. In general terms, there is not any nanocarrier system that can be used as an omnipresent solution. For instance, carrier nanosystems based on eukaryotic viruses are highly efficient in the transduction of human cells. This extreme ability for cellular transduction originates from the deeply rooted evolutionary links between these viruses and human cells. This has been a driver for adaptation of viruses to infect human cells. The long-established connections of human viruses with their host cells have empowered these biological entities to obtain required tools for efficient binding, entry and subsequent intracellular delivery. However, the safety issues of eukaryotic viruses have been a great source of concern which has planted the thought of using nonviral carrier systems. Nonviral systems are less efficient in comparison with viral carriers. However, their application remarkably reduces toxicity concerns. Biological nonviral systems such as lipids, organic polymers and bacteriophages display high biocompatibility. Among nonviral systems, bacteriophages represent one of the most newly-arisen delivery systems. Bacteriophages, with their unique characteristics and excellent safety profile, hold enormous promise for the development of a novel generation of miRNA delivery platforms. The application of inorganic nonviral systems such as gold and silica nanoparticles has also met with success in a number of studies. These reports attest to the efficacy of metal-based nanocarriers for the development of suitable miRNA delivery platforms.

Taken together, different nanocarrier systems create different opportunities for the delivery of miRNA-based therapies. Each system provides a new perspective to the landscape of miRNA delivery and opens new routes to this evolving field. There has been an extreme tendency towards development of new delivery systems which combine the advantages of various platforms. On the other hand, the area of targeting of nanocarriers has flourished making us closer to the cherished goal of targeted delivery. These breakthroughs are great strides to translate the achievements of miRNA delivery to the clinic.

Acknowledgements This work was supported by a grant from Tarbiat Modares University.

References

Aigner A, Fischer D, Merdan T, Brus C, Kissel T, Czubayko F (2002) Delivery of unmodified bioactive ribozymes by an RNA-stabilizing polyethylenimine (LMW-PEI) efficiently down-regulates gene expression. Gene Ther 9(24):1700–1707

Alexis F, Pridgen EM, Langer R, Farokhzad OC (2010) Nanoparticle technologies for cancer therapy. Handb Exp Pharmacol 197:55–86

Almeida AJ, Souto E (2007) Solid lipid nanoparticles as a drug delivery system for peptides and proteins. Adv Drug Deliv Rev 59(6):478–490

Ashley CE, Carnes EC, Phillips GK, Durfee PN, Buley MD, Lino CA et al (2011) Cell-specific delivery of diverse cargos by bacteriophage MS2 virus-like particles. ACS Nano 5(7):5729–5745

Babashah S, Soleimani M (2011) The oncogenic and tumour suppressive roles of microRNAs in cancer and apoptosis. Eur J Cancer 47(8):1127–1137

Babashah S, Sadeghizadeh M, Tavirani MR, Farivar S, Soleimani M (2012) Aberrant microRNA expression and its implications in the pathogenesis of leukemias. Cell Oncol (Dordr) 35(5):317–334

Babashah S, Sadeghizadeh M, Hajifathali A, Tavirani MR, Zomorod MS, Ghadiani M et al (2013) Targeting of the signal transducer Smo links microRNA-326 to the oncogenic hedgehog pathway in CD34+ CML stem/progenitor cells. Int J Cancer 133(3):579–589

Bader AG, Brown D, Stoudemire J, Lammers P (2011) Developing therapeutic microRNAs for cancer. Gene Ther 18(12):1121–1126

Bamrungsap S, Zhao Z, Chen T, Wang L, Li C, Fu T et al (2012) Nanotechnology in therapeutics: a focus on nanoparticles as a drug delivery system. Nanomed (Lond) 7(8):1253–1271

Bangham AD, Standish MM, Watkins JC (1965) Diffusion of univalent ions across the lamellae of swollen phospholipids. J Mol Biol 13(1):238–252

Bondi ML, Craparo EF (2010) Solid lipid nanoparticles for applications in gene therapy: a review of the state of the art. Expert Opin Drug Deliv 7(1):7–18

Boussif O, Lezoualc'h F, Zanta MA, Mergny MD, Scherman D, Demeneix B et al (1995) A versatile vector for gene and oligonucleotide transfer into cells in culture and in vivo: polyethylenimine. Proc Natl Acad Sci U S A 92(16):7297–7301

Bushman F, Lewinski M, Ciuffi A, Barr S, Leipzig J, Hannenhalli S et al (2005) Genome-wide analysis of retroviral DNA integration. Nat Rev Microbiol 3(11):848–858

Calin GA, Croce CM (2006) MicroRNA signatures in human cancers. Nat Rev Cancer 6(11):857–866
Cao H, Koehler DR, Hu J (2004) Adenoviral vectors for gene replacement therapy. Viral Immunol 17(3):327–333
Chan JM, Valencia PM, Zhang L, Langer R, Farokhzad OC (2010) Polymeric nanoparticles for drug delivery. Methods Mol Biol 624:163–175
Chen JF, Ding HM, Wang JX, Shao L (2004) Preparation and characterization of porous hollow silica nanoparticles for drug delivery application. Biomaterials 25(4):723–727
Chen Y, Zhu X, Zhang X, Liu B, Huang L (2010) Nanoparticles modified with tumor-targeting scFv deliver siRNA and miRNA for cancer therapy. Mol Ther J Am Soc Gene Ther 18(9):1650–1656
Cho K, Wang X, Nie S, Chen ZG, Shin DM (2008) Therapeutic nanoparticles for drug delivery in cancer. Clin Cancer Res Off J Am Assoc Cancer Res 14(5):1310–1316
Couleaud P, Morosini V, Frochot C, Richeter S, Raehm L, Durand JO (2010) Silica-based nanoparticles for photodynamic therapy applications. Nanoscale 2(7):1083–1095
Couvreur P, Kante B, Roland M, Speiser P (1979) Adsorption of antineoplastic drugs to polyalkylcyanoacrylate nanoparticles and their release in calf serum. J Pharm Sci 68(12):1521–1524
Couvreur P, Kante B, Lenaerts V, Scailteur V, Roland M, Speiser P (1980) Tissue distribution of antitumor drugs associated with polyalkylcyanoacrylate nanoparticles. J Pharm Sci 69(2):199–202
Crew E, Tessel MA, Rahman S, Razzak-Jaffar A, Mott D, Kamundi M et al (2012) MicroRNA conjugated gold nanoparticles and cell transfection. Anal Chem 84(1):26–29
Cryan SA (2005) Carrier-based strategies for targeting protein and peptide drugs to the lungs. AAPS J 7(1):E20–E41
Elghanian R, Storhoff JJ, Mucic RC, Letsinger RL, Mirkin CA (1997) Selective colorimetric detection of polynucleotides based on the distance-dependent optical properties of gold nanoparticles. Science 277(5329):1078–1081
Esquela-Kerscher A, Trang P, Wiggins JF, Patrawala L, Cheng A, Ford L et al (2008) The let-7 microRNA reduces tumor growth in mouse models of lung cancer. Cell Cycle 7(6):759–764
Ferrari M (2005) Cancer nanotechnology: opportunities and challenges. Nat Rev Cancer 5(3):161–171
Filipowicz W, Bhattacharyya SN, Sonenberg N (2008) Mechanisms of post-transcriptional regulation by microRNAs: are the answers in sight? Nat Rev Genet 9(2):102–114
Gabizon A, Horowitz AT, Goren D, Tzemach D, Shmeeda H, Zalipsky S (2003) In vivo fate of folate-targeted polyethylene-glycol liposomes in tumor-bearing mice. Clin Cancer Res Off J Am Assoc Cancer Res 9(17):6551–6559
Georgiadis A, Tschernutter M, Bainbridge JW, Robbie SJ, McIntosh J, Nathwani AC et al (2010) AAV-mediated knockdown of peripherin-2 in vivo using miRNA-based hairpins. Gene Ther 17(4):486–493
Ghosh R, Singh LC, Shohet JM, Gunaratne PH (2013) A gold nanoparticle platform for the delivery of functional microRNAs into cancer cells. Biomaterials 34(3):807–816
Giacca M, Zacchigna S (2012) Virus-mediated gene delivery for human gene therapy. J Controll Release Off J Controll Release Soc 161(2):377–388
Giri S, Trewyn BG, Lin VS (2007) Mesoporous silica nanomaterial-based biotechnological and biomedical delivery systems. Nanomed (Lond) 2(1):99–111
Gref R, Minamitake Y, Peracchia MT, Trubetskoy V, Torchilin V, Langer R (1994) Biodegradable long-circulating polymeric nanospheres. Science 263(5153):1600–1603
Gregoriadis G (1995) Engineering liposomes for drug delivery: progress and problems. Trends Biotechnol 13(12):527–537
Gunther M, Lipka J, Malek A, Gutsch D, Kreyling W, Aigner A (2011) Polyethylenimines for RNAi-mediated gene targeting in vivo and siRNA delivery to the lung. Eur J Pharm Biopharm Off J Arbeitsgemeinschaft fur Pharmazeutische Verfahrenstechnik eV 77(3): 438–449

Heath TD, Fraley RT, Papahdjopoulos D (1980) Antibody targeting of liposomes: cell specificity obtained by conjugation of F(ab′)2 to vesicle surface. Science 210(4469):539–541

Hirsch LR, Stafford RJ, Bankson JA, Sershen SR, Rivera B, Price RE et al (2003) Nanoshell-mediated near-infrared thermal therapy of tumors under magnetic resonance guidance. Proc Natl Acad Sci U S A 100(23):13549–13554

Hughes GA (2005) Nanostructure-mediated drug delivery. Nanomed Nanotechnol Biol Med 1(1):22–30

Hwang do W, Son S, Jang J, Youn H, Lee S, Lee D et al (2011) A brain-targeted rabies virus glycoprotein-disulfide linked PEI nanocarrier for delivery of neurogenic microRNA. Biomaterials 32(21):4968–4975

Ibrahim AF, Weirauch U, Thomas M, Grunweller A, Hartmann RK, Aigner A (2011) MicroRNA replacement therapy for miR-145 and miR-33a is efficacious in a model of colon carcinoma. Cancer Res 71(15):5214–5224

Jin H, Yu Y, Chrisler WB, Xiong Y, Hu D, Lei C (2012) Delivery of MicroRNA-10b with polylysine nanoparticles for inhibition of breast cancer cell wound healing. Breast Cancer Basic Clin Res 6:9–19

Kim M, Kasinski AL, Slack FJ (2011) MicroRNA therapeutics in preclinical cancer models. Lancet Oncol 12(4):319–321

Klibanov AL, Maruyama K, Torchilin VP, Huang L (1990) Amphipathic polyethyleneglycols effectively prolong the circulation time of liposomes. FEBS Lett 268(1):235–237

Kontermann RE (2006) Immunoliposomes for cancer therapy. Curr Opin Mol Ther 8(1):39–45

Kota J, Chivukula RR, O'Donnell KA, Wentzel EA, Montgomery CL, Hwang HW et al (2009) Therapeutic microRNA delivery suppresses tumorigenesis in a murine liver cancer model. Cell 137(6):1005–1017

Kovacs EW, Hooker JM, Romanini DW, Holder PG, Berry KE, Francis MB (2007) Dual-surface-modified bacteriophage MS2 as an ideal scaffold for a viral capsid-based drug delivery system. Bioconjugate Chem 18(4):1140–1147

Krutzfeldt J, Rajewsky N, Braich R, Rajeev KG, Tuschl T, Manoharan M et al (2005) Silencing of microRNAs in vivo with 'antagomirs'. Nature 438(7068):685–689

Lamers CH, Willemsen R, van Elzakker P, van Steenbergen-Langeveld S, Broertjes M, Oosterwijk-Wakka J et al (2011) Immune responses to transgene and retroviral vector in patients treated with ex vivo-engineered T cells. Blood 117(1):72–82

Langer R, Folkman J (1976) Polymers for the sustained release of proteins and other macromolecules. Nature 263(5580):797–800

Laufs S, Guenechea G, Gonzalez-Murillo A, Zsuzsanna Nagy K, Luz Lozano M, del Val C et al (2006) Lentiviral vector integration sites in human NOD/SCID repopulating cells. J Gene Med 8(10):1197–1207

Leamon CP, Cooper SR, Hardee GE (2003) Folate-liposome-mediated antisense oligodeoxynucleotide targeting to cancer cells: evaluation in vitro and in vivo. Bioconjugate Chem 14(4):738–747

Leserman LD, Barbet J, Kourilsky F, Weinstein JN (1980) Targeting to cells of fluorescent liposomes covalently coupled with monoclonal antibody or protein A. Nature 288(5791):602–604

Lewinski MK, Yamashita M, Emerman M, Ciuffi A, Marshall H, Crawford G et al (2006) Retroviral DNA integration: viral and cellular determinants of target-site selection. PLoS Pathog 2(6):e60

Liu YP, Berkhout B (2011) miRNA cassettes in viral vectors: problems and solutions. Biochimica et biophysica acta 1809(11–12):732–745

Lockman PR, Mumper RJ, Khan MA, Allen DD (2002) Nanoparticle technology for drug delivery across the blood–brain barrier. Drug Dev Ind Pharm 28(1):1–13

Lu J, Liong M, Zink JI, Tamanoi F (2007) Mesoporous silica nanoparticles as a delivery system for hydrophobic anticancer drugs. Small 3(8):1341–1346

McLaughlin J, Cheng D, Singer O, Lukacs RU, Radu CG, Verma IM et al (2007) Sustained suppression of Bcr-Abl-driven lymphoid leukemia by microRNA mimics. Proc Natl Acad Sci U S A 104(51):20501–20506

Merisko-Liversidge E, Liversidge GG, Cooper ER (2003) Nanosizing: a formulation approach for poorly-water-soluble compounds. Eur J Pharm Sci Off J Eur Fed Pharm Sci 18(2):113–120
Miller DG, Petek LM, Russell DW (2004) Adeno-associated virus vectors integrate at chromosome breakage sites. Nat Genet 36(7):767–773
Moghimi SM (2006) Recent developments in polymeric nanoparticle engineering and their applications in experimental and clinical oncology. Anti-Cancer Agent Med Chem 6(6):553–561
Nakai H, Montini E, Fuess S, Storm TA, Grompe M, Kay MA (2003) AAV serotype 2 vectors preferentially integrate into active genes in mice. Nat Genet 34(3):297–302
O'Connell RM, Taganov KD, Boldin MP, Cheng G, Baltimore D (2007) MicroRNA-155 is induced during the macrophage inflammatory response. Proc Natl Acad Sci U S A 104(5):1604–1609
Ohulchanskyy TY, Roy I, Goswami LN, Chen Y, Bergey EJ, Pandey RK et al (2007) Organically modified silica nanoparticles with covalently incorporated photosensitizer for photodynamic therapy of cancer. Nano Lett 7(9):2835–2842
Onion D, Patel P, Pineda RG, James N, Mautner V (2009) Antivector and tumor immune responses following adenovirus-directed enzyme prodrug therapy for the treatment of prostate cancer. Hum Gene Ther 20(11):1249–1258
Orom UA, Kauppinen S, Lund AH (2006) LNA-modified oligonucleotides mediate specific inhibition of microRNA function. Gene 372:137–141
Pan Y, Jia T, Zhang Y, Zhang K, Zhang R, Li J et al (2012a) MS2 VLP-based delivery of microRNA-146a inhibits autoantibody production in lupus-prone mice. Int J Nanomed 7:5957–5967
Pan Y, Zhang Y, Jia T, Zhang K, Li J, Wang L (2012b) Development of a microRNA delivery system based on bacteriophage MS2 virus-like particles. FEBS J 279(7):1198–1208
Panyam J, Labhasetwar V (2003) Biodegradable nanoparticles for drug and gene delivery to cells and tissue. Adv Drug Deliv Rev 55(3):329–347
Panyam J, Sahoo SK, Prabha S, Bargar T, Labhasetwar V (2003) Fluorescence and electron microscopy probes for cellular and tissue uptake of poly(D, L-lactide-co-glycolide) nanoparticles. Int J Pharm 262(1–2):1–11
Papahadjopoulos D, Allen TM, Gabizon A, Mayhew E, Matthay K, Huang SK et al (1991) Sterically stabilized liposomes: improvements in pharmacokinetics and antitumor therapeutic efficacy. Proc Natl Acad Sci U S A 88(24):11460–11464
Park JW, Hong K, Kirpotin DB, Colbern G, Shalaby R, Baselga J et al (2002) Anti-HER2 immunoliposomes: enhanced efficacy attributable to targeted delivery. Clin Cancer Res Official J Am Assoc Cancer Res 8(4):1172–1181
Parveen S, Sahoo SK (2008) Polymeric nanoparticles for cancer therapy. J Drug Target 16(2): 108–123
Peer D, Karp JM, Hong S, Farokhzad OC, Margalit R, Langer R (2007) Nanocarriers as an emerging platform for cancer therapy. Nat Nanotechnol 2(12):751–760
Pramanik D, Campbell NR, Karikari C, Chivukula R, Kent OA, Mendell JT et al (2011) Restitution of tumor suppressor microRNAs using a systemic nanovector inhibits pancreatic cancer growth in mice. Mol Cancer Ther 10(8):1470–1480
Qi J, Lu Y, Wu W (2012) Absorption, disposition and pharmacokinetics of solid lipid nanoparticles. Curr Drug Metab 13(4):418–428
Rai K, Takigawa N, Ito S, Kashihara H, Ichihara E, Yasuda T et al (2011) Liposomal delivery of MicroRNA-7-expressing plasmid overcomes epidermal growth factor receptor tyrosine kinase inhibitor-resistance in lung cancer cells. Mol Cancer Ther 10(9):1720–1727
Rosi NL, Giljohann DA, Thaxton CS, Lytton-Jean AK, Han MS, Mirkin CA (2006) Oligonucleotide-modified gold nanoparticles for intracellular gene regulation. Science 312(5776):1027–1030
Roy I, Ohulchanskyy TY, Bharali DJ, Pudavar HE, Mistretta RA, Kaur N et al (2005) Optical tracking of organically modified silica nanoparticles as DNA carriers: a nonviral, nanomedicine approach for gene delivery. Proc Natl Acad Sci U S A 102(2):279–284
Samal SK, Dash M, Van Vlierberghe S, Kaplan DL, Chiellini E, van Blitterswijk C et al (2012) Cationic polymers and their therapeutic potential. Chem Soc Rev 41(21):7147–7194

Schnepp BC, Clark KR, Klemanski DL, Pacak CA, Johnson PR (2003) Genetic fate of recombinant adeno-associated virus vector genomes in muscle. J Virol 77(6):3495–3504

Shi S, Han L, Gong T, Zhang Z, Sun X (2013) Systemic delivery of microRNA-34a for cancer stem cell therapy. Angew Chem Int Ed Engl 52(14):3901–3905

Singh R, Lillard JW Jr (2009) Nanoparticle-based targeted drug delivery. Exp Mol Pathol 86(3):215–223

Son S, Hwang do W, Singha K, Jeong JH, Park TG, Lee DS et al (2011) RVG peptide tethered bioreducible polyethylenimine for gene delivery to brain. J Controll Release Off J Controll Release Soc 155(1):18–25

Stegmeier F, Hu G, Rickles RJ, Hannon GJ, Elledge SJ (2005) A lentiviral microRNA-based system for single-copy polymerase II-regulated RNA interference in mammalian cells. Proc Natl Acad Sci U S A 102(37):13212–13217

Stockley PG, Stonehouse NJ, Valegard K (1994) Molecular mechanism of RNA phage morphogenesis. Int J Biochem 26(10–11):1249–1260

Sun BS, Dong QZ, Ye QH, Sun HJ, Jia HL, Zhu XQ et al (2008) Lentiviral-mediated miRNA against osteopontin suppresses tumor growth and metastasis of human hepatocellular carcinoma. Hepatology 48(6):1834–1842

Szoka F Jr, Papahadjopoulos D (1980) Comparative properties and methods of preparation of lipid vesicles (liposomes). Ann Rev Biophys Bioeng 9:467–508

Tivnan A, Orr WS, Gubala V, Nooney R, Williams DE, McDonagh C et al (2012) Inhibition of neuroblastoma tumor growth by targeted delivery of microRNA-34a using anti-disialoganglioside GD2 coated nanoparticles. PloS one 7(5):e38129

Torchilin VP (2005) Recent advances with liposomes as pharmaceutical carriers. Nat Rev Drug Discov 4(2):145–160

Umbach JL, Cullen BR (2009) The role of RNAi and microRNAs in animal virus replication and antiviral immunity. Genes Dev 23(10):1151–1164

Urban-Klein B, Werth S, Abuharbeid S, Czubayko F, Aigner A (2005) RNAi-mediated gene-targeting through systemic application of polyethylenimine (PEI)-complexed siRNA in vivo. Gene Ther 12(5):461–466

van Rooij E, Sutherland LB, Qi X, Richardson JA, Hill J, Olson EN (2007) Control of stress-dependent cardiac growth and gene expression by a microRNA. Science 316(5824):575–579

Visaria RK, Griffin RJ, Williams BW, Ebbini ES, Paciotti GF, Song CW et al (2006) Enhancement of tumor thermal therapy using gold nanoparticle-assisted tumor necrosis factor-alpha delivery. Mol Cancer Ther 5(4):1014–1020

Walther W, Stein U (2000) Viral vectors for gene transfer: a review of their use in the treatment of human diseases. Drugs 60(2):249–271

Wang V, Wu W (2009) MicroRNA-based therapeutics for cancer. BioDrug Clin Immunother Biopharm Gene Ther 23(1):15–23

Wang AZ, Langer R, Farokhzad OC (2012) Nanoparticle delivery of cancer drugs. Ann Rev Med 63:185–198

Wei B, Wei Y, Zhang K, Wang J, Xu R, Zhan S et al (2009) Development of an antisense RNA delivery system using conjugates of the MS2 bacteriophage capsids and HIV-1 TAT cell-penetrating peptide. Biomed Pharmacothe 63(4):313–318

Wu J, Liu Q, Lee RJ (2006) A folate receptor-targeted liposomal formulation for paclitaxel. Int J Pharm 316(1–2):148–153

Wu Y, Crawford M, Yu B, Mao Y, Nana-Sinkam SP, Lee LJ (2011) MicroRNA delivery by cationic lipoplexes for lung cancer therapy. Mol Pharm 8(4):1381–1389

Yang YP, Chien Y, Chiou GY, Cherng JY, Wang ML, Lo WL et al (2012) Inhibition of cancer stem cell-like properties and reduced chemoradioresistance of glioblastoma using microRNA145 with cationic polyurethane-short branch PEI. Biomaterials 33(5):1462–1476

Yatvin MB, Kreutz W, Horwitz BA, Shinitzky M (1980) pH-sensitive liposomes: possible clinical implications. Science 210(4475):1253–1255

Young LS, Searle PF, Onion D, Mautner V (2006) Viral gene therapy strategies: from basic science to clinical application. J Pathol 208(2):299–318

Yu SF, von Ruden T, Kantoff PW, Garber C, Seiberg M, Ruther U et al (1986) Self-inactivating retroviral vectors designed for transfer of whole genes into mammalian cells. Proc Natl Acad Sci U S A 83(10):3194–3198

Zufferey R, Nagy D, Mandel RJ, Naldini L, Trono D (1997) Multiply attenuated lentiviral vector achieves efficient gene delivery in vivo. Nat Biotechnol 15(9):871–875

Chapter 17
Challenges and Strategies for Pulmonary Delivery of MicroRNA-Based Therapeutics

Yu Fujita, Kazuyoshi Kuwano, and Takahiro Ochiya

Contents

Abstract Therapeutic options for various lung diseases, especially lung cancer, continue to expand with the development of novel therapeutic strategies. RNA interference (RNAi)-based approaches provide a promising modality for the treatment of lung diseases. One of the greatest challenges in RNAi-based therapy continues to be the method for delivering the therapeutic small interfering RNAs (siRNAs) and

---

Y. Fujita
Division of Molecular and Cellular Medicine, National Cancer Center Research Institute, 5-1-1, Tsukiji, Chuo-ku, Tokyo 104-0045, Japan

Division of Respiratory Diseases, Department of Internal Medicine, Jikei University School of Medicine, 3-19-18, Nishi-shinbashi, Minato-ku, Tokyo 105-8471, Japan

K. Kuwano
Division of Respiratory Diseases, Department of Internal Medicine, Jikei University School of Medicine, 3-19-18, Nishi-shinbashi, Minato-ku, Tokyo 105-8471, Japan

T. Ochiya (✉)
Division of Molecular and Cellular Medicine, National Cancer Center Research Institute, 5-1-1, Tsukiji, Chuo-ku, Tokyo 104-0045, Japan
e-mail: tochiya@ncc.go.jp

S. Babashah (ed.), *MicroRNAs: Key Regulators of Oncogenesis*,
DOI 10.1007/978-3-319-03725-7_17,

microRNAs (miRNAs) to the target cells. The advance of pulmonary delivery systems into the clinic illustrates the notion that RNAi will be a valuable modality for the treatment of lung diseases. Currently, the development of miRNA-based therapies for lung cancer is rapidly advancing with the aid of new RNAi technologies. Given the important role of miRNAs in lung carcinogenesis, increasing effort is being dedicated to the research and development of miRNA-based therapies, including the restoration of tumor suppressive miRNA function and the inhibition of oncogenic miRNAs. In this chapter, we discuss the advantages of a pulmonary drug delivery system and the strategies for miRNA-based treatment of lung cancer.

Keywords RNA interference • Small interfering RNA • MicroRNA • Lung cancer
Pulmonary delivery

1 Introduction

Lung cancer is the leading cause of cancer mortality worldwide. Lung cancer can be classified into two main subtypes: non-small-cell lung cancer (NSCLC) and small-cell lung cancer (SCLC). Numerous differences are found between these two subtypes, including histological type, biological behavior, prevalence, prognosis and response to therapy. NSCLC accounts for more than 80 % of all lung cancer cases. Only a small percentage of patients with NSCLC present with early stage disease. In this circumstance, surgery remains the best therapeutic option for these patients. Approximately 70 % of all newly diagnosed patients present with locally advanced or metastatic disease and require systemic chemotherapy (Ramalingam et al. 2011). However, the commonly administered chemotherapeutics provide little benefit for patients with advanced stage disease and has reached a plateau in efficacy with a median survival of 8–10 months. The poor prognosis is due to late stage disease presentation, tumor heterogeneity within histological subtypes, and our relatively limited understanding of tumor biology. Furthermore, the high frequency of drug resistance is a key contributor to the poor survival rates of lung cancer patients; improvements in survival rely on continued elucidation of the molecular mechanisms underlying lung cancer tumorigenesis and drug response. Acquiring knowledge through genomic medicine raises the possibility of unraveling the remaining mysteries of lung cancer oncogenesis and opens the door to molecular classification and risk stratification based on gene expression profiles and microRNA (miRNA) signatures.

MiRNAs are short (19–23 nucleotides in length) non-coding RNAs found in multiple organisms that regulate gene expression primarily by decreasing the levels of their target mRNAs, through binding to specific target sites in the 3′ untranslated regions (3′UTRs) of these mRNAs (Winter et al. 2009). In the human genome, transcripts of approximately 60 % of all mRNAs are estimated to be targeted by miRNAs. Accumulating evidence shows that miRNAs are grossly dysregulated in human cancers, including NSCLC, and may serve as oncogenes or tumor suppressors (Croce 2009; Babashah and Soleimani 2011). Recent studies have not only shown that miRNAs are useful in lung cancer diagnosis but that specific miRNA profiles may also predict prognosis, drug response and disease recurrence (Yanaihara et al. 2006; Yu et al. 2008). These findings suggest that miRNAs are a promising technology for therapeutic development. In fact, given the significant role of miRNAs in multiple pathways governing lung carcinogenesis, increasing efforts are dedicated to the research and development of miRNA-based therapies, including the restoration of tumor suppressive miRNA function and the inhibition of oncogenic miRNAs (Bader et al. 2010).

The critical problems impeding the development of RNAi-based therapeutics are effective delivery to target sites, therapeutic potency, and elimination of off-target effects (Boudreau et al. 2009). The success of miRNA-based therapeutic delivery is also dependent upon uncovering a delivery route that yields efficient outcomes, is convenient, and promotes patient compliance. For this reason, direct administration of miRNA-based therapeutics to target organs is a promising approach to overcome the problems of systemic administration. Pulmonary delivery offers a new method for the treatment of various lung diseases (Fujita et al. 2013). We believe that delivery of miRNA-based therapeutics using this approach will potentially be useful in clinical practice. Here, we provide an overview of miRNAs as therapeutic targets in lung cancer and discuss the promise and limitations of pulmonary delivery strategies for miRNA-based therapeutics.

2 Role of MicroRNAs in Lung Cancer

Lung cancer biology has traditionally focused on genomic and epigenomic deregulation of protein-coding genes to identify oncogenes and tumor suppressors that are useful as diagnostic and therapeutic targets. Recently, miRNAs were also shown to up-regulate target gene expression by either directly binding to the target mRNA or indirectly repressing nonsense-mediated RNA decay (Vasudevan et al. 2007; Bruno et al. 2011). MiRNAs play an essential role in various cellular processes, such as development, proliferation and apoptosis, to ensure the cellular homeostasis of human cells. Alterations in miRNA expression are increasingly noted in relation to pathophysiological changes in cancer cells, thereby making miRNAs one of the most currently analyzed molecule types in cancer research. Numerous miRNAs are dysregulated in lung cancers, and a single miRNA can have multiple targets that are involved in different oncogenic pathways. A large body of evidence reveals that the aberrant expression of miRNAs in cancer patients can be taken advantage of innumerous ways, such as for potential use as diagnostic, clinicopathological, and/or prognostic markers and as promising therapeutic targets in lung cancer. Aberrant miRNA expression profiles provide additional insight into the clinical application of miRNA-directed therapies in lung cancer (Leidinger et al. 2011). Here, we focus on reviewing the known roles of miRNAs as regulators of cancer cell survival, drug sensitivity and tumorigenesis. These miRNAs hold great potential as targets in the treatment of lung cancer.

2.1 MiRNAs Function as Oncogenes in Lung Cancer

Many oncogenes important in controlling lung cancer tumorigenesis are targets of miRNAs. The miRNAs found in the miR-17-92 cluster (miR-17, miR-18a, miR-19a, miR-19b-1, miR-20a, miR-92-1) are oncogenic miRNAs (oncomiRs) that reside in the amplified chromosomal region 13q31.3 (He et al. 2005). These miRNAs cooperate with c-Myc to accelerate tumor development and promote tumor angiogenesis (Dews et al. 2006). It has been reported that the miR-17-92 cluster is over-expressed in SCLC (Hayashita et al. 2005). Moreover, Ebi et al. reported that miR-17-92 over-expression is associated with retinoblastoma (RB) inactivation (Ebi et al. 2009). Collectively, these results suggest that this miRNA cluster may be a potential therapeutic target in lung cancer.

The miR-21 gene is located on chromosome 17 and was one of the first miRNAs characterized as oncogenic, with its oncogenic function established in various types of cancers (Chan et al. 2005). MiR-21 has been suggested to be an independent negative prognostic factor for the overall survival of NSCLC patients (Markou et al. 2008). MiR-21 targets tumor suppressor genes such as programmed cell death 4 (PDCD4) and phosphatase and tensin homolog deleted from chromosome 10 (PTEN) (Lu et al. 2008; Zhang et al. 2010). Furthermore, miR-21 expression is up-regulated by epidermal growth factor receptor (EGFR) signaling in lung cancer. Antisense miR-21-enhanced EGFR tyrosine kinase inhibitors induce apoptosis of lung cancer cells (Seike et al. 2009). The critical function of miR-21 in regulating lung cancer tumorigenesis makes it a promising target for developing miRNA-based therapeutics and diagnostic tools. However, because miR-21 is also dysregulated in various type of cancer, it appears to be a general oncomiR without tissue specificity (Volinia et al. 2006).

MiR-31 is another miRNA with oncogenic properties in lung cancer. The host gene encoding miR-31 is located on chromosome 9. Liu et al. showed that miR-31 functions as an oncomiR by directly repressing large tumor suppressor 2 (LATS2) and Protein phosphatase 2, regulatory subunit B, Alpha isoform (PPP2R2A) and that knockdown of miR-31 represses lung cancer cell clonal growth and *in vivo* tumorigenicity (Liu et al. 2010).

2.2 MiRNAs Function as Tumor Suppressors in Lung Cancer

Among the numerous miRNAs that function as tumor suppressors, the let-7 family is one of the most studied. Let-7 was first identified in *C. elegans* as a regulator of the timing of cell fate determination (Reinhart et al. 2000). In humans, the let-7 family is a cluster of miRNAs whose encoding genes map to various chromosomal regions that are frequently deleted in lung cancer (Calin et al. 2004). Johnson et al. (2007) showed that let-7 over-expression in the A549 cell line inhibits cell growth and reduces cell-cycle progression. In mouse models of lung cancer, over-expression of let-7g reduces tumor growth (Kumar et al. 2008), and let-7a inhibits tumor growth via suppression of v-Ki-ras2 Kirsten rat sarcoma viral oncogene homolog (KRAS) and c-Myc (He et al. 2010). Furthermore, reduced let-7 gene expression in NSCLC patients correlates with poor prognosis (Yanaihara et al. 2006; Takamizawa et al. 2004). The 3′ UTR of members of the RAS GTPase family such as v-Ha-ras Harvey rat sarcoma viral oncogene homolog (HRAS), KRAS and neuroblastoma RAS viral oncogene homolog (NRAS) contains multiple putative let-7 binding sites. It has also been revealed that let-7 miRNAs negatively regulate multiple oncogenes, including MYC (Kumar et al. 2007), high mobility group AT-hook 2 (HMGA2) (Lee and Dutta 2007), B-cell leukemia/lymphoma 2 (BCL-2) (Xiong et al. 2011) and cell cycle proto-oncogenes such as cell division cycle 25A (CDC25A), cyclin-dependent kinase 6 (CDK6) and cyclin D2 (Johnson et al. 2007). These data show that let-7 miRNAs act as key tumor suppressors in regulating cell survival and proliferation in lung cancers.

The miR-34 family is another important group of miRNAs that function as tumor suppressors in many types of cancers (Hermeking 2010; Wong et al. 2011). The miR-34a gene is located on chromosome 1p36.22, and miR-34b/c are expressed from a polycistronic transcript encoded on chromosome 11q23.1. These genes are in chromosomal regions associated with fragile sites of the genome that are frequently altered in cancer (Calin et al. 2004). Structurally, miR-34 family members possess p53-binding sites, reflecting their function as tumor suppressors downstream of the p53 pathway. MiR-34a and miR-34b/c were found to be directly regulated by p53 to control apoptosis and cell cycle arrest in cancer cell lines, including lung cancer (Raver-Shapira et al. 2007; Wiggins et al. 2010). Subsequent studies demonstrated that the apoptotic function of miR-34a is mediated by the direct down-regulation of the expression of BCL-2 and sirtuin 1 (SIRT1) (Yamakuchi et al. 2008; Bommer et al. 2007). In addition, AXL (Mudduluru et al. 2011) and SNAIL1 (Kim et al. 2011) were identified as miR-34 direct targets in lung cancer cells; it is plausible that miR-34 expression inhibits lung cancer cell invasion and migration via repression of these genes. In various solid and hematological malignancies, including lung cancer, miR-34 antagonizes processes necessary for basic cancer cell viability as well as cancer stemness, metastasis and chemoresistance (Bader 2012). In the future, the utility of miR-34-directed therapeutics in the treatment of lung cancer will expand dramatically.

The anti-tumor activity of miR-143 and miR-145 in lung cancer is also well characterized. They are co-transcribed from a bicistronic gene cluster on chromosome 5 (Xin et al. 2009). MiR-143/145 have been identified as tumor suppressor in various types of cancer, including lung cancer. The restoration of miR-145 has been shown to inhibit cell growth in mouse and human lung cancer cells (Liu et al. 2009; Cho et al. 2009). It has also been reported that c-MYC, EGFR and nucleoside diphosphate linked moiety X-type motif 1 (NUDT1) are direct targets of miR-145 that regulate cell proliferation in lung cancer (Chen et al. 2010; Cho et al. 2011). Furthermore, miR-145 has also been shown to inhibit lung adenocarcinoma stem-like cell proliferation by targeting octamer-binding transcription factor 4 (OCT4) (Feng et al. 2011). Similarly, the expression of miR-143 was down-regulated in human lung tumor samples compared with normal tissues (Gao et al. 2010; Vosa et al. 2013).

Finally, miR-192 also might serve as a promising therapeutic target for lung cancer treatment. Retinoblastoma 1 (RB1) is a direct target of miR-192, and over-expression of miR-192 results in decreased expression of RB1 mRNA and protein. Caspase-7 and poly ADP-ribose polymerase (PARP) protein were activated by miR-192 over-expression, suggesting that miR-192 induces cell apoptosis through the caspase pathway. In addition, the analysis of miRNA expression in clinical samples has revealed that miR-192 is significantly down-regulated in lung cancer tissues compared with adjacent, normal lung tissues (Feng et al. 2011).

3 MicroRNA-Based Therapies for Lung Cancer

The development of miRNA-based therapeutics represents a new strategy in cancer treatment and is growing rapidly with the help of new RNAi technologies. Compared to siRNA-based therapies, which are already in clinical trials, miRNAs are less toxic and have the potential to target multiple genes. As presented above, miRNAs are generally classified as oncomiRs or tumor suppressors, with different therapeutic approaches developed for each class. Generally, the up-regulation of miRNA expression is achieved through administration of synthetic miRNA mimics or miRNA-expressing vectors. The down-regulation of miRNA expression is achieved through administration of antisense nucleotides, often chemically modified to ensure stability and specificity. Although each approach shares similarities with other therapies, each is sufficiently distinct such that miRNA-inhibitory and replacement approaches should be viewed as separate therapeutic modalities. In view of cancer as a heterogenic disease that cannot be successfully treated via single gene targeting, miRNA-based strategies may hold the key to therapeutic success. Table 17.1 shows a summary of miRNA-based therapeutic strategies for *in vivo* models of lung cancer.

3.1 MiRNA Inhibitor-Based Therapeutics

To reduce endogenous miRNA levels, anti-miRs are typically employed. Targeting miRNAs for suppression through the use of anti-miRs is possibly the best-studied modality to date. This approach is conceptually similar to other

Table 17.1 MicroRNA-based therapeutic strategies for *in vivo* models of lung cancer

MicroRNA	Administration	Modulation strategy	Delivery technology	Reference
let-7a	Intranasal	Replacement	Adenoviruses	Esquela-Kerscher et al. (2008)
let-7b	Systemic	Replacement	Neutral liposomes	Trang et al. (2011)
let-7g	Intratracheal	Replacement	Lentiviruses	Kumar et al. (2008)
miR-7	Intratumoral	Replacement	Cationic liposomes	Rai et al. (2011)
miR-29b	Systemic	Replacement	Cationic liposomes	Wu et al. (2013)
miR-34a	Intratumoral	Replacement	Neutral liposomes	Wiggins et al. (2010)
miR-145	Intratumoral	Replacement	Polyethyleneimines	Chiou et al. (2012)
miR-150	Intratumoral	Inhibition	Cationic liposomes	Li et al. (2012)

inhibitory therapeutics that target a single gene product, such as small molecule inhibitors and siRNAs. Various methods have been employed to render anti-miR constructs more stable *in vivo* and ensure adequate tissue availability and specificity (Krutzfeldt et al. 2005). Constructs can be modified with a cholesterol-conjugated 2′-*O*-methyl group to inhibit degradation and hence improve stability. Locked nucleic acid (LNA) is an additional method of antisense oligonucleotide modification whereby the 2′ oxygen and 4′ carbon of the nucleotide is bridged with methylene to form a cyclic structure. LNA is more resistant to endogenous nucleases, less toxic, and possess a stronger affinity for the target nucleotide (Elmen et al. 2008; Wahlestedt et al. 2000). Relative to studies on miRNA mimics, studies with antisense oligonucleotides have demonstrated greater efficacy using naked oligonucleotides. Furthermore, the LNA-anti-miR compound was well tolerated in both mice and primates, as no acute or subchronic toxicities in the treated animals were detected (Elmen et al. 2008). Recent data from the first Phase IIa study in patients with chronic HCV infection treated with the LNA-modified anti-miR-122 revealed that this compound was well tolerated and provided continuing viral suppression (Janssen et al. 2013). With regard to lung cancer, anti-miR-150 delivered to lung tumor xenografts in mice caused tumor growth inhibition (Li et al. 2012). Although there are few reports using LNA-anti-miR therapeutics in lung cancer mouse models, their inhibition of miRNA function is an important and widely used approach. Currently, miRNA sponges are a novel approach to miRNA inhibition, and this technology works with multiple complementary 3′-UTR mRNA sites of a specific miRNA (Ebert et al. 2007). MiRNA sponges specifically inhibit miRNAs with a complementary heptameric seed; thus, a single sponge can inhibit an entire miRNA seed family. In fact, the development of lung metastasis in a murine breast cancer model was significantly reduced via inhibition of the MYC driven miR-9 using a miRNA sponge (Ma et al. 2010). Furthermore, the use of miRNA sponges to inhibit miR-31 in a breast cancer model resulted in a significant induction of lung metastasis (Valastyan et al. 2009). Of potential concern is the possibility that the antagonist might also non-specifically bind to other RNAs, resulting in unwanted side effects. Therefore, adequate assessment of the functional effects of miRNA inhibition is of key importance for miRNA inhibitor-based loss-of-function studies and development of miRNA therapeutics. The high potency and metabolic stability of chemically modified anti-miRs highlights the utility of anti-miRs in the development of novel RNAi therapeutic modalities based on lung cancer associated miRNAs.

3.2 MiRNA Mimic-Based Therapeutics

Tumor suppressor miRNAs are responsible for down-regulating oncogenes and are primarily expressed in cancer (Croce 2009). In this context, miRNA replacement strategies have been developed to restore normal cellular expression levels via administration of tumor suppressor miRNA mimics (Bader et al. 2010). MiRNA mimics are synthetic RNA duplexes designed to imitate the endogenous functions of miRNAs. In addition, miRNAs may be unstable as a result of rapid degradation by endogenous nucleases or rapid elimination through renal and hepatic metabolism and extraction upon systemic administration (Bader et al. 2011). Local administration of RNAi-based therapeutics to the target cells is a promising approach to overcome the problems of systemic administration (see next section for details). Similarly, chemical modifications at specific positions or formulations with delivery vectors have been shown to improve stability. Lipid-based and polymer-based nanoparticles reduce the negative electrical charge of RNA nucleotides to promote cell uptake (Wu et al. 2011). Another strategy for efficient delivery of miRNA-based therapeutics is the use of viral vectors (Bonci et al. 2008). Indeed, adenoviral (Esquela-Kerscher et al. 2008) or lentiviral vectors (Kumar et al. 2008) can be used to transfer miRNAs to lung cancer cells. Successful delivery of miRNA-based therapeutics requires patient compliance with the intended delivery route and efficient delivery vectors. This approach has attracted much interest as it provides a novel opportunity to exploit tumor suppressors. The concept of miRNA replacement therapy is best exemplified by let-7 miRNA. Intranasal administration of a let-7 mimic into mouse models of lung cancer significantly reduced tumor growth, suggesting that miRNA replacement therapy is indeed promising (Trang et al. 2010). Based on these successful results, a clinical trial in non-small cell lung cancer using a let-7 based therapy will begin in the near future. As an additional example of the value of miRNA replacement strategies, miR-34a-based cancer therapies have powerful potential for clinical use. Both local and systemic delivery of a synthetic miR-34a mimic resulted in accumulation of miR-34a in the tumor tissue and inhibition of lung tumor growth. MiRNA therapeutics will initiate clinical trials of miR-34a mimics in 2013, making these mimics some of the first miRNA mimics to reach the clinic. Thus, the pharmacological delivery of miRNA mimics effectively inhibits tumor growth by targeting multiple genes. However, it is necessary to pay attention to any potential toxicities in normal tissues, given that therapeutic delivery of miRNA mimics can lead to an accumulation of exogenous miRNAs in normal cells. It will be important to investigate miRNA mimic-induced effects in normal cells and carefully assess the resultant toxicity before using such therapies in clinical practice.

4 Pulmonary Delivery of RNAi-Based Therapeutics

Despite the promise of miRNAs in cancer therapy, there are still hurdles to clear before clinical use, including safety, stability and successful delivery of therapeutic miRNAs to the appropriate tissue and into the appropriate cells. In general, the delivery of miRNAs can be achieved through systemic administration (via intravenous injection) or local administration (via a direct route). Conceptually, systemic delivery is an attractive option because it provides a simple route for miRNA administration to all tissues via the blood stream (Liu et al. 2007). Indeed, there have been some successful reports using systemic delivery of miRNAs in lung cancer models.

Nevertheless, this approach has more *in vivo* barriers to overcome, in addition to nuclease degradation. The delivery barriers are (i) renal clearance of molecules (<50 kDa), (ii) uptake by phagocytic immune cells, (iii) failure of molecules >5 nm in diameter to cross the capillary endothelium, (iv) limited passage through the extra-cellular matrix (polysaccharides and fibrous proteins), (v) inefficient endocytosis by target tumor cells, and (vi) inefficient endosomal release (Bader et al. 2011). Chemical modification and formulation with delivery vectors have been shown to improve stability and delivery to target tumor cells, but these alterations may attenuate the suppressive activity of oligonucleotides (Chernolovskaya and Zenkova 2010). In addition, systemic delivery of miRNAs may induce adverse events similar to those reported for other oligonucleotide-based therapies, such as aggregation and complement activation, liver toxicity and stimulation of the immune response (Kleinman et al. 2008). For these reasons, local administration of miRNAs to the target cancer cells is a promising approach to overcome the problems of systemic administration. Translation of locally administered modalities to the clinical setting is dependent upon the development of an efficient delivery system that is able to improve the pharmacokinetic and biodistribution properties of miRNAs. Thus far, locally administered modalities are available for ocular, transdermal, rectal and pulmonary delivery.

Dozens of RNAi-based therapeutics are being assessed in preclinical and clinical trials, and these studies provide further opportunities for successful results (Davidson and McCray 2011). Many of these studies are conducted using local administration to specific tissues. The lung is anatomically accessible to therapeutic drugs via the pulmonary route. Accessibility is a key requirement for successful RNAi-based *in vivo* and clinical studies, and this anatomical characteristic offers several important benefits over systemic delivery, including the use of lower doses of miRNAs, the reduction of undesirable systemic side effects, and improved miRNA stability due to reduced nuclease activity in the airways compared to serum. The local approach could potentially enhance the retention of RNAi-based therapeutics in the lungs. Because the delivery of siRNAs to the lungs is well studied using different routes and delivery strategies (Lam et al. 2012), many technologies developed for siRNAs may also be applicable to miRNAs. In most of the pulmonary RNAi-based therapy studies *in vivo*, agents were delivered intratracheally or intranasally. This approach has allowed remarkable progress in miRNA modulation in preclinical cancer models, bringing us closer to delivering on the promise of miRNAs as cancer therapeutics.

5 Strategies for Pulmonary Delivery of MicroRNA-Based Therapeutics

Pulmonary delivery approaches are very attractive because they tend to be non-invasive, locally restricted, and administered by the patient. With regard to siRNA-based therapeutics, Phase II clinical trials are underway for the treatment of respiratory syncytial virus (RSV) infection using an intranasal application of naked, chemically modified siRNA molecules that target viral gene products (DeVincenzo et al. 2008, 2010). To date, two successful studies of pulmonary delivery of miRNA-based therapeutics for lung cancer mouse models have been reported (Kumar et al. 2008; Esquela-Kerscher et al. 2008). These studies show that pulmonary delivery of miRNA from the let-7 family reduces lung tumor formation in an orthotopic lung cancer mouse model without systemic side effects (Table 17.1). These data suggest that intranasal or intratracheal administration of miRNAs may be a potent strategy for treating lung cancer. Although there are no reports of pulmonary delivery of miRNA-inhibitors in lung cancer at present, we predict that this delivery strategy will become a valuable resource for implementing miRNA-based therapies *in vivo* and in humans.

We believe that pulmonary delivery of miRNAs has two primary advantages over systemic delivery for clinical use. First, several sophisticated inhalation devices for lung diseases are already in clinical use. Inhaled therapeutics are used routinely to treat a variety of pulmonary conditions, including asthma, chronic obstructive pulmonary disease (COPD) and cystic fibrosis. Metered-dose inhalers (MDIs) and dry powder inhalers (DPIs) are the most common modes of inhaled delivery. The use of DPIs for the *in vivo* delivery of therapeutic macromolecules such as insulin (Mastrandrea and Quattrin 2006) and low-molecular-weight heparin (Bai et al. 2010) has yielded promising results. Currently, the use of spray-drying as a technique for engineering dry powder formulations of siRNA nanoparticles, which might allow the pulmonary delivery of biologically active siRNAs directly to the lung tissue, has been demonstrated (Jensen et al. 2010, 2012). Although a suitable carrier is also needed to protect miRNAs from degradation given the shear force and increased temperature of the drying process, these delivery technologies could open new avenues for pulmonary delivery of miRNAs and improve patient outcome. To make miRNA-based therapy practical in the treatment of lung cancer, we believe that the administration of inhaled miRNAs by DPIs is the best of choice of delivery strategy. Second, pulmonary delivery also offers the clinical benefit of a lower miRNA dose. The cost related to the development and application of a particular RNAi therapeutic delivery technology is undoubtedly an important factor (Dykxhoorn et al. 2006). Local administration is likely to be a more cost-efficient strategy for miRNA delivery *in vivo* and in the clinic than systemic administration. Furthermore, the advantage of pulmonary delivery is that it ensures high delivery efficiency with minimal drug loss. For this reason, pulmonary delivery of miRNAs has great potential for clinical use. However, the limitations of pulmonary delivery of miRNA-based therapeutics are important to consider. First, the pharmacokinetics of inhaled miRNAs in *in vivo* models and humans are estimated inaccurately. It is also unknown whether miRNA-based therapeutics delivered via the intrapulmonary route could also be delivered to other organs, such as the liver and kidneys. To prevent systemic side effects, the precise pharmacokinetics of miRNAs after intrapulmonary administration should be measured. Second, we also must pay attention to the pulmonary inflammatory and toxicological responses caused by the delivery vehicle. In fact, there are some reports that RNAi-based therapeutics with polyethyleneimine

(PEI) frequently cause inflammatory responses in the lungs (Beyerle et al. 2011). It has been reported that naked RNAi-therapeutic delivery possesses advantages over other delivery vectors, such as reduced toxicity and reduced inflammatory responses, as well as simple formulation (Heidel et al. 2004). However, the advantage of naked RNAi-therapeutics over delivery vectors in the treatment of lung diseases is controversial (Nielsen et al. 2010; Akinc et al. 2008). Therefore, we need to develop safer delivery technology for practical use in *in vivo* mouse models and humans.

6 Conclusions

During the past decade, miRNAs have quickly advanced from discovery to therapeutic development programs. This rapid progress reflects the importance of miRNA biology in cancer, leaving little doubt about the therapeutic potential of miRNAs in cancer treatment. Given the encouraging results of the profiled studies and preclinical testing, miRNAs are being integrated into human clinical trials. The first miRNA-targeted drug LNA-anti-miR-122 is successfully undergoing Phase II trials (Janssen et al. 2013). Accordingly, several companies are currently developing miRNA mimics or inhibitors for the treatment of cancer. The main focus in bringing miRNAs to cancer cells is the capacity of pharmacological drug delivery. The success of miRNA-based therapeutic delivery requires efficiency, convenience, and patient compliance using the delivery route. In this chapter, we showed that pulmonary delivery of miRNA-based therapeutics holds powerful potential for lung cancer treatment (Fig. 17.1). A realistic therapeutic intervention, such as inhalation, would

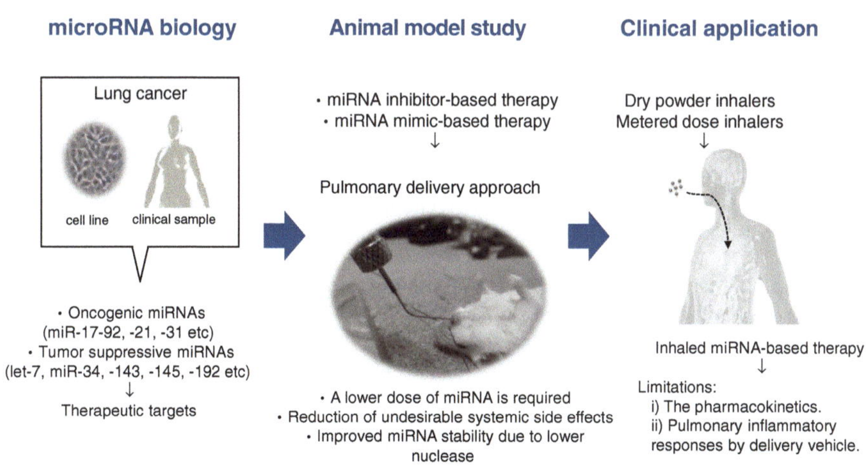

Fig. 17.1 Process for translating microRNA biology from bench to bedside in lung cancer

enhance drug delivery to the site of action and decrease systemic exposure, thereby reducing off-target effects. In the future, combining miRNA-based therapeutics with chemotherapy may potentiate the cancer treatment efficacy. Therefore, continued investigation on all fronts will be of equal importance to the eventual clinical application of miRNAs.

Acknowledgements This work was supported in part by a grant-in-aid for the Third-Term Comprehensive 10-Year Strategy for Cancer Control of Japan; Project for Development of Innovative Research on Cancer Therapeutics (P-Direct); Scientific Research on Priority Areas Cancer, Scientific Research on Innovative Areas ("functional machinery for non-coding RNAs") from the Japanese Ministry of Education, Culture, Sports, Science, and Technology; the National Cancer Center Research and Development Fund (23-A-2, 23-A-7, 23-C-6,); the Program for Promotion of Fundamental Studies in Health Sciences of the National Institute of Biomedical Innovation (NiBio), the Project for Development of Innovative Research on Cancer Therapeutics; and the Japan Society for the Promotion of Science (JSPS) through the "Funding Program for World-Leading Innovative R&D on Science and Technology (FIRST Program)" initiated by the Council for Science and Technology Policy (CSTP).

References

Akinc A, Zumbuehl A, Goldberg M, Leshchiner ES, Busini V, Hossain N et al (2008) A combinatorial library of lipid-like materials for delivery of RNAi therapeutics. Nat Biotechnol 26(5):561–569

Babashah S, Soleimani M (2011) The oncogenic and tumour suppressive roles of microRNAs in cancer and apoptosis. Eur J Cancer 47(8):1127–1137

Bader AG (2012) miR-34 – a microRNA replacement therapy is headed to the clinic. Front Genet 3:120

Bader AG, Brown D, Winkler M (2010) The promise of microRNA replacement therapy. Cancer Res 70(18):7027–7030

Bader AG, Brown D, Stoudemire J, Lammers P (2011) Developing therapeutic microRNAs for cancer. Gene Ther 18(12):1121–1126

Bai S, Gupta V, Ahsan F (2010) Inhalable lactose-based dry powder formulations of low molecular weight heparin. J Aerosol Med Pulm Drug Deliv 23(2):97–104

Beyerle A, Braun A, Merkel O, Koch F, Kissel T, Stoeger T (2011) Comparative in vivo study of poly(ethylene imine)/siRNA complexes for pulmonary delivery in mice. J Control Release 151(1):51–56

Bommer GT, Gerin I, Feng Y, Kaczorowski AJ, Kuick R, Love RE et al (2007) p53-mediated activation of miRNA34 candidate tumor-suppressor genes. Curr Biol 17(15):1298–1307

Bonci D, Coppola V, Musumeci M, Addario A, Giuffrida R, Memeo L et al (2008) The miR-15a-miR-16-1 cluster controls prostate cancer by targeting multiple oncogenic activities. Nat Med 14(11):1271–1277

Boudreau RL, Martins I, Davidson BL (2009) Artificial microRNAs as siRNA shuttles: improved safety as compared to shRNAs in vitro and in vivo. Mol Ther 17(1):169–175

Bruno IG, Karam R, Huang L, Bhardwaj A, Lou CH, Shum EY et al (2011) Identification of a microRNA that activates gene expression by repressing nonsense-mediated RNA decay. Mol Cell 42(4):500–510

Calin GA, Sevignani C, Dumitru CD, Hyslop T, Noch E, Yendamuri S et al (2004) Human microRNA genes are frequently located at fragile sites and genomic regions involved in cancers. Proc Natl Acad Sci U S A 101(9):2999–3004

Chan JA, Krichevsky AM, Kosik KS (2005) MicroRNA-21 is an antiapoptotic factor in human glioblastoma cells. Cancer Res 65(14):6029–6033

Chen Z, Zeng H, Guo Y, Liu P, Pan H, Deng A et al (2010) miRNA-145 inhibits non-small cell lung cancer cell proliferation by targeting c-Myc. J Exp Clin Cancer Res 29:151

Chernolovskaya EL, Zenkova MA (2010) Chemical modification of siRNA. Curr Opin Mol Ther 12(2):158–167

Chiou GY, Cherng JY, Hsu HS, Wang ML, Tsai CM, Lu KH et al (2012) Cationic polyurethanes-short branch PEI-mediated delivery of Mir145 inhibited epithelial-mesenchymal transdifferentiation and cancer stem-like properties and in lung adenocarcinoma. J Control Release 159(2):240–250

Cho WC, Chow AS, Au JS (2009) Restoration of tumour suppressor hsa-miR-145 inhibits cancer cell growth in lung adenocarcinoma patients with epidermal growth factor receptor mutation. Eur J Cancer 45(12):2197–2206

Cho WC, Chow AS, Au JS (2011) MiR-145 inhibits cell proliferation of human lung adenocarcinoma by targeting EGFR and NUDT1. RNA Biol 8(1):125–131

Croce CM (2009) Causes and consequences of microRNA dysregulation in cancer. Nat Rev Genet 10(10):704–714

Davidson BL, McCray PB Jr (2011) Current prospects for RNA interference-based therapies. Nat Rev Genet 12(5):329–340

DeVincenzo J, Cehelsky JE, Alvarez R, Elbashir S, Harborth J, Toudjarska I et al (2008) Evaluation of the safety, tolerability and pharmacokinetics of ALN-RSV01, a novel RNAi antiviral therapeutic directed against respiratory syncytial virus (RSV). Antiviral Res 77(3):225–231

DeVincenzo J, Lambkin-Williams R, Wilkinson T, Cehelsky J, Nochur S, Walsh E et al (2010) A randomized, double-blind, placebo-controlled study of an RNAi-based therapy directed against respiratory syncytial virus. Proc Natl Acad Sci U S A 107(19):8800–8805

Dews M, Homayouni A, Yu D, Murphy D, Sevignani C, Wentzel E et al (2006) Augmentation of tumor angiogenesis by a Myc-activated microRNA cluster. Nat Genet 38(9):1060–1065

Dykxhoorn DM, Palliser D, Lieberman J (2006) The silent treatment: siRNAs as small molecule drugs. Gene Ther 13(6):541–552

Ebert MS, Neilson JR, Sharp PA (2007) MicroRNA sponges: competitive inhibitors of small RNAs in mammalian cells. Nat Methods 4(9):721–726

Ebi H, Sato T, Sugito N, Hosono Y, Yatabe Y, Matsuyama Y et al (2009) Counterbalance between RB inactivation and miR-17-92 overexpression in reactive oxygen species and DNA damage induction in lung cancers. Oncogene 28(38):3371–3379

Elmen J, Lindow M, Schutz S, Lawrence M, Petri A, Obad S et al (2008) LNA-mediated microRNA silencing in non-human primates. Nature 452(7189):896–899

Esquela-Kerscher A, Trang P, Wiggins JF, Patrawala L, Cheng A, Ford L et al (2008) The let-7 microRNA reduces tumor growth in mouse models of lung cancer. Cell Cycle 7(6):759–764

Feng S, Cong S, Zhang X, Bao X, Wang W, Li H et al (2011) MicroRNA-192 targeting retinoblastoma 1 inhibits cell proliferation and induces cell apoptosis in lung cancer cells. Nucleic Acids Res 39(15):6669–6678

Fujita Y, Takeshita F, Kuwano K, Ochiya T (2013) RNAi therapeutic platforms for lung diseases. Pharmaceuticals 6(2):223–250

Gao W, Yu Y, Cao H, Shen H, Li X, Pan S et al (2010) Deregulated expression of miR-21, miR-143 and miR-181a in non small cell lung cancer is related to clinicopathologic characteristics or patient prognosis. Biomed Pharmacother 64(6):399–408

Hayashita Y, Osada H, Tatematsu Y, Yamada H, Yanagisawa K, Tomida S et al (2005) A polycistronic microRNA cluster, miR-17-92, is overexpressed in human lung cancers and enhances cell proliferation. Cancer Res 65(21):9628–9632

He L, Thomson JM, Hemann MT, Hernando-Monge E, Mu D, Goodson S et al (2005) A microRNA polycistron as a potential human oncogene. Nature 435(7043):828–833

He XY, Chen JX, Zhang Z, Li CL, Peng QL, Peng HM (2010) The let-7a microRNA protects from growth of lung carcinoma by suppression of k-Ras and c-Myc in nude mice. J Cancer Res Clin Oncol 136(7):1023–1028

Heidel JD, Hu S, Liu XF, Triche TJ, Davis ME (2004) Lack of interferon response in animals to naked siRNAs. Nat Biotechnol 22(12):1579–1582

Hermeking H (2010) The miR-34 family in cancer and apoptosis. Cell Death Differ 17(2):193–199

Janssen HL, Reesink HW, Lawitz EJ, Zeuzem S, Rodriguez-Torres M, Patel K et al (2013) Treatment of HCV infection by targeting microRNA. N Engl J Med 368(18):1685–1694

Jensen DM, Cun D, Maltesen MJ, Frokjaer S, Nielsen HM, Foged C (2010) Spray drying of siRNA-containing PLGA nanoparticles intended for inhalation. J Control Release 142(1):138–145

Jensen DK, Jensen LB, Koocheki S, Bengtson L, Cun D, Nielsen HM et al (2012) Design of an inhalable dry powder formulation of DOTAP-modified PLGA nanoparticles loaded with siRNA. J Control Release 157(1):141–148

Johnson CD, Esquela-Kerscher A, Stefani G, Byrom M, Kelnar K, Ovcharenko D et al (2007) The let-7 microRNA represses cell proliferation pathways in human cells. Cancer Res 67(16):7713–7722

Kim NH, Kim HS, Li XY, Lee I, Choi HS, Kang SE et al (2011) A p53/miRNA-34 axis regulates Snail1-dependent cancer cell epithelial-mesenchymal transition. J Cell Biol 195(3):417–433

Kleinman ME, Yamada K, Takeda A, Chandrasekaran V, Nozaki M, Baffi JZ et al (2008) Sequence- and target-independent angiogenesis suppression by siRNA via TLR3. Nature 452(7187):591–597

Krutzfeldt J, Rajewsky N, Braich R, Rajeev KG, Tuschl T, Manoharan M et al (2005) Silencing of microRNAs in vivo with 'antagomirs'. Nature 438(7068):685–689

Kumar MS, Lu J, Mercer KL, Golub TR, Jacks T (2007) Impaired microRNA processing enhances cellular transformation and tumorigenesis. Nat Genet 39(5):673–677

Kumar MS, Erkeland SJ, Pester RE, Chen CY, Ebert MS, Sharp PA et al (2008) Suppression of non-small cell lung tumor development by the let-7 microRNA family. Proc Natl Acad Sci U S A 105(10):3903–3908

Lam JK, Liang W, Chan HK (2012) Pulmonary delivery of therapeutic siRNA. Adv Drug Deliv Rev 64(1):1–15

Lee YS, Dutta A (2007) The tumor suppressor microRNA let-7 represses the HMGA2 oncogene. Genes Dev 21(9):1025–1030

Leidinger P, Keller A, Meese E (2011) MicroRNAs – important molecules in lung cancer research. Front Genet 2:104

Li YJ, Zhang YX, Wang PY, Chi YL, Zhang C, Ma Y et al (2012) Regression of A549 lung cancer tumors by anti-miR-150 vector. Oncol Rep 27(1):129–134

Liu N, Ding H, Vanderheyden JL, Zhu Z, Zhang Y (2007) Radiolabeling small RNA with technetium-99m for visualizing cellular delivery and mouse biodistribution. Nucl Med Biol 34(4):399–404

Liu X, Sempere LF, Galimberti F, Freemantle SJ, Black C, Dragnev KH et al (2009) Uncovering growth-suppressive microRNAs in lung cancer. Clin Cancer Res 15(4):1177–1183

Liu X, Sempere LF, Ouyang H, Memoli VA, Andrew AS, Luo Y et al (2010) MicroRNA-31 functions as an oncogenic microRNA in mouse and human lung cancer cells by repressing specific tumor suppressors. J Clin Invest 120(4):1298–1309

Lu Z, Liu M, Stribinskis V, Klinge CM, Ramos KS, Colburn NH et al (2008) MicroRNA-21 promotes cell transformation by targeting the programmed cell death 4 gene. Oncogene 27(31):4373–4379

Ma L, Young J, Prabhala H, Pan E, Mestdagh P, Muth D et al (2010) miR-9, a MYC/MYCN-activated microRNA, regulates E-cadherin and cancer metastasis. Nat Cell Biol 12(3):247–256

Markou A, Tsaroucha EG, Kaklamanis L, Fotinou M, Georgoulias V, Lianidou ES (2008) Prognostic value of mature microRNA-21 and microRNA-205 overexpression in non-small cell lung cancer by quantitative real-time RT-PCR. Clin Chem 54(10):1696–1704

Mastrandrea LD, Quattrin T (2006) Clinical evaluation of inhaled insulin. Adv Drug Deliv Rev 58(9–10):1061–1075

Mudduluru G, Ceppi P, Kumarswamy R, Scagliotti GV, Papotti M, Allgayer H (2011) Regulation of Axl receptor tyrosine kinase expression by miR-34a and miR-199a/b in solid cancer. Oncogene 30(25):2888–2899

Nielsen EJ, Nielsen JM, Becker D, Karlas A, Prakash H, Glud SZ et al (2010) Pulmonary gene silencing in transgenic EGFP mice using aerosolised chitosan/siRNA nanoparticles. Pharm Res 27(12):2520–2527

Rai K, Takigawa N, Ito S, Kashihara H, Ichihara E, Yasuda T et al (2011) Liposomal delivery of MicroRNA-7-expressing plasmid overcomes epidermal growth factor receptor tyrosine kinase inhibitor-resistance in lung cancer cells. Mol Cancer Ther 10(9):1720–1727

Ramalingam SS, Owonikoko TK, Khuri FR (2011) Lung cancer: new biological insights and recent therapeutic advances. CA Cancer J Clin 61(2):91–112

Raver-Shapira N, Marciano E, Meiri E, Spector Y, Rosenfeld N, Moskovits N et al (2007) Transcriptional activation of miR-34a contributes to p53-mediated apoptosis. Mol Cell 26(5):731–743

Reinhart BJ, Slack FJ, Basson M, Pasquinelli AE, Bettinger JC, Rougvie AE et al (2000) The 21-nucleotide let-7 RNA regulates developmental timing in *Caenorhabditis elegans*. Nature 403(6772):901–906

Seike M, Goto A, Okano T, Bowman ED, Schetter AJ, Horikawa I et al (2009) MiR-21 is an EGFR-regulated anti-apoptotic factor in lung cancer in never-smokers. Proc Natl Acad Sci U S A 106(29):12085–12090

Takamizawa J, Konishi H, Yanagisawa K, Tomida S, Osada H, Endoh H et al (2004) Reduced expression of the let-7 microRNAs in human lung cancers in association with shortened postoperative survival. Cancer Res 64(11):3753–3756

Trang P, Medina PP, Wiggins JF, Ruffino L, Kelnar K, Omotola M et al (2010) Regression of murine lung tumors by the let-7 microRNA. Oncogene 29(11):1580–1587

Trang P, Wiggins JF, Daige CL, Cho C, Omotola M, Brown D et al (2011) Systemic delivery of tumor suppressor microRNA mimics using a neutral lipid emulsion inhibits lung tumors in mice. Mol Ther 19(6):1116–1122

Valastyan S, Reinhardt F, Benaich N, Calogrias D, Szasz AM, Wang ZC et al (2009) A pleiotropically acting microRNA, miR-31, inhibits breast cancer metastasis. Cell 137(6):1032–1046

Vasudevan S, Tong Y, Steitz JA (2007) Switching from repression to activation: microRNAs can up-regulate translation. Science 318(5858):1931–1934

Volinia S, Calin GA, Liu CG, Ambs S, Cimmino A, Petrocca F et al (2006) A microRNA expression signature of human solid tumors defines cancer gene targets. Proc Natl Acad Sci U S A 103(7):2257–2261

Vosa U, Vooder T, Kolde R, Vilo J, Metspalu A, Annilo T (2013) Meta-analysis of microRNA expression in lung cancer. Int J Cancer 132(12):2884–2893

Wahlestedt C, Salmi P, Good L, Kela J, Johnsson T, Hokfelt T et al (2000) Potent and nontoxic antisense oligonucleotides containing locked nucleic acids. Proc Natl Acad Sci U S A 97(10):5633–5638

Wiggins JF, Ruffino L, Kelnar K, Omotola M, Patrawala L, Brown D et al (2010) Development of a lung cancer therapeutic based on the tumor suppressor microRNA-34. Cancer Res 70(14):5923–5930

Winter J, Jung S, Keller S, Gregory RI, Diederichs S (2009) Many roads to maturity: microRNA biogenesis pathways and their regulation. Nat Cell Biol 11(3):228–234

Wong MY, Yu Y, Walsh WR, Yang JL (2011) microRNA-34 family and treatment of cancers with mutant or wild-type p53 (Review). Int J Oncol 38(5):1189–1195

Wu Y, Crawford M, Yu B, Mao Y, Nana-Sinkam SP, Lee LJ (2011) MicroRNA delivery by cationic lipoplexes for lung cancer therapy. Mol Pharm 8(4):1381–1389

Wu Y, Crawford M, Mao Y, Lee RJ, Davis IC, Elton TS et al (2013) Therapeutic delivery of microRNA-29b by cationic lipoplexes for lung cancer. Mol Ther Nucleic Acids 2:e84

Xin M, Small EM, Sutherland LB, Qi X, McAnally J, Plato CF et al (2009) MicroRNAs miR-143 and miR-145 modulate cytoskeletal dynamics and responsiveness of smooth muscle cells to injury. Genes Dev 23(18):2166–2178

Xiong S, Zheng Y, Jiang P, Liu R, Liu X, Chu Y (2011) MicroRNA-7 inhibits the growth of human non-small cell lung cancer A549 cells through targeting BCL-2. Int J Biol Sci 7(6):805–814

Yamakuchi M, Ferlito M, Lowenstein CJ (2008) miR-34a repression of SIRT1 regulates apoptosis. Proc Natl Acad Sci U S A 105(36):13421–13426

Yanaihara N, Caplen N, Bowman E, Seike M, Kumamoto K, Yi M et al (2006) Unique microRNA molecular profiles in lung cancer diagnosis and prognosis. Cancer Cell 9(3):189–198

Yu SL, Chen HY, Chang GC, Chen CY, Chen HW, Singh S et al (2008) MicroRNA signature predicts survival and relapse in lung cancer. Cancer Cell 13(1):48–57

Zhang JG, Wang JJ, Zhao F, Liu Q, Jiang K, Yang GH (2010) MicroRNA-21 (miR-21) represses tumor suppressor PTEN and promotes growth and invasion in non-small cell lung cancer (NSCLC). Clin Chim Acta 411(11–12):846–852

Index

S. Babashah (ed.), *MicroRNAs: Key Regulators of Oncogenesis*,
DOI 10.1007/978-3-319-03725-7, © Springer International Publishing Switzerland 2014

MIX
Papier aus verantwortungsvollen Quellen
Paper from responsible sources
FSC® C105338

If you have any concerns about our products,
you can contact us on
ProductSafety@springernature.com

In case Publisher is established outside the EU,
the EU authorized representative is:
Springer Nature Customer Service Center GmbH
Europaplatz 3, 69115 Heidelberg, Germany

Printed by Libri Plureos GmbH
in Hamburg, Germany